SOLUTIONS MANUAL

for

ORGANIC CHEMISTRY

second edition

T. W. GRAHAM SOLOMONS
JACK E. FERNANDEZ
University of South Florida

JOHN WILEY & SONS
NEW YORK CHICHESTER BRISBANE TORONTO

ISBN 0 471 05770 3

Printed in the United States of America

10 9 8 7 6 5 4 3 2 1

TO THE STUDENT

This manual presents solutions to all the problems in the text including the end-of-chapter problems. In many instances we have given not only the answer, but also an explanation of the reasoning that leads to the solution. Although many problems have more than one solution, we have generally given only one. Thus, you should not necessarily assume that your answer is incorrect if it differs from the one given here.

The heart of organic chemistry lies in problem-solving. This is as true for the practicing organic chemist as for the beginning student. But, a problem in organic chemistry is like a riddle; once you have seen the answer, it is impossible for you to go through the process of solving it. The essential value of a problem lies in the mental exercise of the problem-solving process, and you cannot get this exercise if you already know the answer. The best way to use this manual, therefore, is to check the problems you have worked, or to find explanations for unsolved problems *only* after you have made serious attempts at working them.

Problems involving syntheses are always intriguing to chemists and students alike. These problems are probably unlike any that you have seen in other courses. In them you are asked to put together a series of reactions that will convert one compound into another. To do this, it is not enough to start your reasoning with the starting materials because you must also keep in mind the desired product. In some instances you can work from both ends simultaneously, and this is best done by trying to find an intermediate that will link the starting materials and products in a sequence of reactions. In many cases, however, a synthesis problem is best solved by reasoning backward, by starting your thinking with the product and by trying to discover a series of reactions that will lead back to the starting compounds. Sherlock Holmes in *A Study in Scarlet* said:

> Most people if you describe a train of events to them, will tell you what the result would be. They can put these events together in their minds, and argue from them that something will come to pass. There are a few people, however, who, if you told them a result, would be able to evolve from their own inner conciousness what the steps were which led up to that result. This power is what I mean when I talk of reasoning backward, or analytically.

This power is what the organic chemistry student must develop.

CHAPTER ONE

Carbon Compounds and Chemical Bonds

1.1

Using the elemental symbol to denote the nucleus and inner shell electrons:

(a) 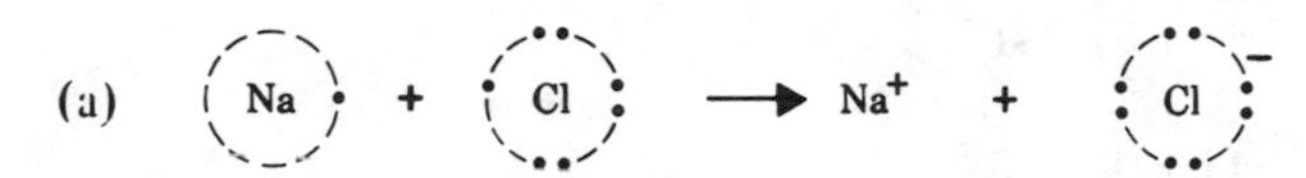

(b) $Mg\!:\; + 2\;\dot{\underset{\cdot\cdot}{\overset{\cdot\cdot}{F}}}\!: \longrightarrow Mg^{+2} + 2\;:\!\underset{\cdot\cdot}{\overset{\cdot\cdot}{F}}\!:^{-}$

(c) $K\cdot + \dot{\underset{\cdot\cdot}{\overset{\cdot\cdot}{Br}}}\!: \longrightarrow K^{+} + :\!\underset{\cdot\cdot}{\overset{\cdot\cdot}{Br}}\!:^{-}$

1.2

(a) $H\!:\!\underset{\cdot\cdot}{\overset{\cdot\cdot}{Br}}\!:$, $H{-}\underset{\cdot\cdot}{\overset{\cdot\cdot}{Br}}\!:$

(b) $:\!\underset{\cdot\cdot}{\overset{\cdot\cdot}{Br}}\!:\!\underset{\cdot\cdot}{\overset{\cdot\cdot}{Br}}\!:$, $:\!\underset{\cdot\cdot}{\overset{\cdot\cdot}{Br}}{-}\underset{\cdot\cdot}{\overset{\cdot\cdot}{Br}}\!:$

(c) $:\!\overset{\cdot\cdot}{O}\!::\!C\!::\!\overset{\cdot\cdot}{O}\!:$, $:\!\overset{\cdot\cdot}{O}{=}C{=}\overset{\cdot\cdot}{O}\!:$

(d) $H\!:\!\underset{H}{\overset{H}{C}}\!:\!H$, $H{-}\underset{H}{\overset{H}{C}}{-}H$

(e) $H\!:\!\underset{\cdot\cdot}{\overset{\cdot\cdot}{O}}\!:\!\underset{\cdot\cdot}{\overset{\cdot\cdot}{O}}\!:\!H$, $H{-}\underset{\cdot\cdot}{\overset{\cdot\cdot}{O}}{-}\underset{\cdot\cdot}{\overset{\cdot\cdot}{O}}{-}H$

(f) $H\!:\!\underset{H}{\overset{H}{Si}}\!:\!H$, $H{-}\underset{H}{\overset{H}{Si}}{-}H$

(g) $H\!:\!\underset{H}{\overset{\cdot\cdot}{N}}\!:\!H$, $H{-}\underset{H}{\overset{\cdot\cdot}{N}}{-}H$

(h) $:\!\underset{\cdot\cdot}{\overset{\cdot\cdot}{Cl}}\!:\!\underset{:\underset{\cdot\cdot}{Cl}:}{\overset{\cdot\cdot}{P}}\!:\!\underset{\cdot\cdot}{\overset{\cdot\cdot}{Cl}}\!:$, $:\!\underset{\cdot\cdot}{\overset{\cdot\cdot}{Cl}}{-}\underset{:\underset{\cdot\cdot}{Cl}:}{\overset{\cdot\cdot}{P}}{-}\underset{\cdot\cdot}{\overset{\cdot\cdot}{Cl}}\!:$

(i) $:\!\underset{\cdot\cdot}{\overset{\cdot\cdot}{F}}\!:\!\underset{:\underset{\cdot\cdot}{F}:}{\overset{\cdot\cdot}{N}}\!:\!\underset{\cdot\cdot}{\overset{\cdot\cdot}{F}}\!:$, $:\!\underset{\cdot\cdot}{\overset{\cdot\cdot}{F}}{-}\underset{:\underset{\cdot\cdot}{F}:}{\overset{\cdot\cdot}{N}}{-}\underset{\cdot\cdot}{\overset{\cdot\cdot}{F}}\!:$

(j) $H\!:\!\underset{H}{\overset{H}{C}}\!:\!\underset{\cdot\cdot}{\overset{\cdot\cdot}{Cl}}\!:$, $H{-}\underset{H}{\overset{H}{C}}{-}\underset{\cdot\cdot}{\overset{\cdot\cdot}{Cl}}\!:$

(k) $H\!:\!\underset{H}{\overset{\cdot\cdot}{O}}\!:$, $H{-}\underset{H}{\overset{\cdot\cdot}{O}}\!:$

(l) $:\!\underset{\cdot\cdot}{\overset{\cdot\cdot}{O}}\!:\!H$, $:\!\underset{\cdot\cdot}{\overset{\cdot\cdot}{O}}^{-}{-}H$

(m) $\left[\mathrm{H:\ddot{N}:H}\right]^+$ with H above and below N, $:\ddot{\mathrm{Cl}}:^-$, $\left[\mathrm{H{-}N{-}H}\right]^+$ with H above and below N, $:\ddot{\mathrm{Cl}}:^-$ (n) $\mathrm{Na^+\ :\ddot{O}:H^-}$, $\mathrm{Na^+\ :\ddot{O}{-}H^-}$

1.3

Formal charge = group number − [½ (number of shared electrons) + (number of unshared electrons)]

Charge on ion = sum of all formal charges;

		Formal Charge	Total Charge
(a) H–B–H (H above and below B)	H	1-(1 + 0) = 0	-1
	B	3-(4 + 0) = -1	
(b) $:\ddot{\mathrm{O}}{-}\mathrm{H}$	H	1-(1 + 0) = 0	-1
	O	6-(1 + 6) = -1	
(c) $:\ddot{\mathrm{F}}{-}\mathrm{B}{-}\ddot{\mathrm{F}}:$ ($:\ddot{\mathrm{F}}:$ above and below B)	F	7-(1 + 6) = 0	-1
	B	3-(4 + 0) = -1	
(d) $\mathrm{H{-}\ddot{O}{-}H}$ (H below O)	H	1-(1 + 0) = 0	+1
	O	6-(3 + 2) = +1	
(e) $:\ddot{\mathrm{O}}$ double-bonded to C, C bonded to two $:\ddot{\mathrm{O}}:$	top O	6-(2 + 4) = 0	-2
	C	4-(4 + 0) = 0	
	bottom O's	6-(1 + 6) = -1	
(f) $\mathrm{H{-}\ddot{C}{-}H}$ (H below C)	H	1-(1 + 0) = 0	-1
	C	4-(3 + 2) = -1	
(g) H–C–H (H below C)	H	1-(1 + 0) = 0	+1
	C	4-(3 + 0) = +1	
(h) $\mathrm{H{-}\dot{C}{-}H}$ (H below C)	H	1-(1 + 0) = 0	0
	C	4-(3 + 1) = 0	
(i) $\mathrm{H{-}\ddot{C}{-}H}$	H	1-(1 + 0) = 0	0
	C	4-(2 + 2) = 0	
(j) $:\ddot{\mathrm{N}}{-}\mathrm{H}$ (H below N)	H	1-(1 + 0) = 0	-1
	N	5-(2 + 4) = -1	

1.4

Zero formal charges are not shown.

(a) No formal charges (d) No formal charges (g) $CH_3-C(=\ddot{O}:)-\ddot{O}:^-$

(b) No formal charges (e) No formal charges (h) $CH_3CH_2-\overset{+}{\ddot{O}}(H)-H$

(c) $CH_3-\overset{+}{N}(CH_3)(:\ddot{O}:^-)-CH_3$ (f) $CH_3-\overset{+}{N}(:\ddot{O}:^-)=\ddot{O}:$ (i) $CH_3CH-CHCH_3$ bridged by $:\overset{+}{Br}:$

1.5

(a) $^-:\ddot{O}-\overset{+}{\ddot{S}}=\ddot{O}: \longleftrightarrow :\ddot{O}=\overset{+}{\ddot{S}}-\ddot{O}:^-$

(b) Yes. Both O–S bonds are hybrids of a single and a double bond. Therefore the two O–S bonds are equivalent and of equal length.

1.6

(a) $\overset{\longmapsto}{H-Br}$ (c) H–H (dipole moment = 0)

(b) $\overset{\longmapsto}{I-Cl}$ (d) Cl–Cl (dipole moment = 0)

1.7

(a) Yes. (b) No. Two square planar structures are possible:

X, X, H, H on C (X's adjacent) and X, H, H, X on C (X's opposite)

Therefore, if CH_2X_2 had a square planar structure we should observe two compounds (isomers) with the formula CH_2X_2

1.8

(a) Tetrahedral (c) Trigonal planar (e) Tetrahedral (g) Tetrahedral

(b) Tetrahedral (d) Tetrahedral (f) Linear (h) Linear

(i) Trigonal planar

1.9

(a) $:\ddot{O}=C=\ddot{O}:$ Linear

(b) $:\ddot{O}=\overset{+}{\ddot{S}}-\ddot{O}:^-$ Angular

(c) SO_3 with S^{+2} (one $S=\ddot{O}:$, two $S-\ddot{O}:^-$) Trigonal planar

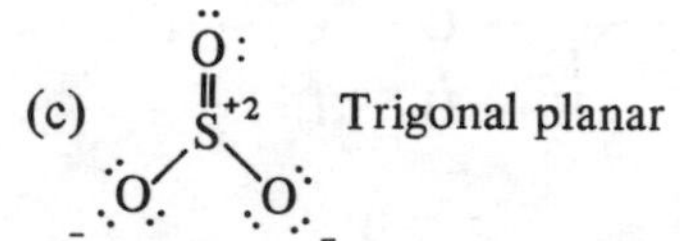

1.10

The two C=O bond moments are opposed and cancel each other: O = C = O

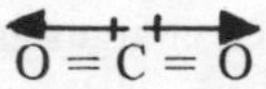

If the bond angle were other than 180°, then the individual bond moments would not cancel. There would be a resultant dipole moment.

1.11

That SO_2 is an angular molecule O S̈ O . Its S–O bond moments do not cancel each other.

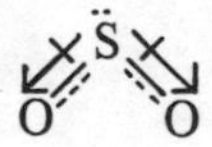

1.12

The direction of polarity of the N–H bond is opposite to that of the N–F bond.

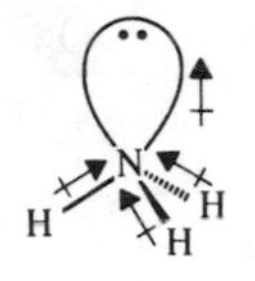

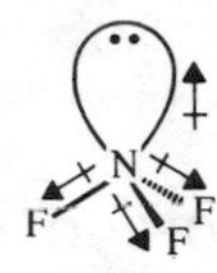

In NH_3, the resultant N–H bond polarities and the polarity of the unshared electron pair are in the same direction.

In NF_3 the resultant N–F bond polarities partially cancel the polarity of the unshared electron pair.

1.13

BF_3 is trigonal planar. Its B–F bonds are all necessarily equal in polarity, and the F–B–F bond angles are all equal (120°).

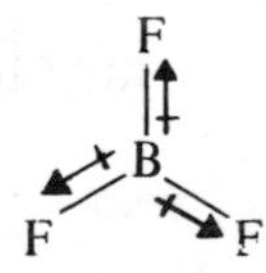

1.14

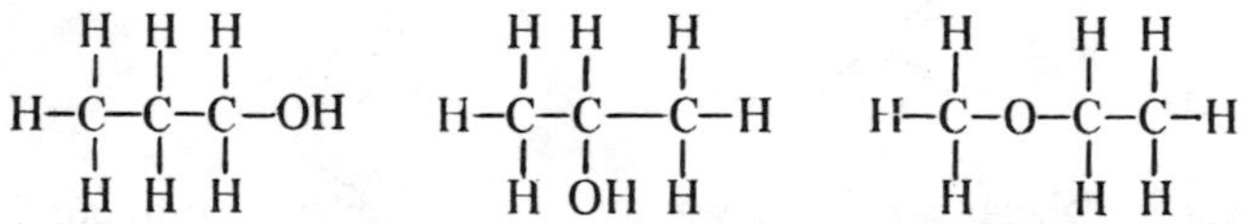

H H H   H H H   H H H
H–C–C–C–OH   H–C–C–C–H   H–C–O–C–C–H
H H H   H OH H   H H H

1.15

H :C̈l: H H
(a) H — C — C — C — C — H
H :C̤l: H H

Cl Cl

H
H H–C–H H H
(c) H — C — C — C — C — H
H H–C–H H H
H

H H H H
(b) 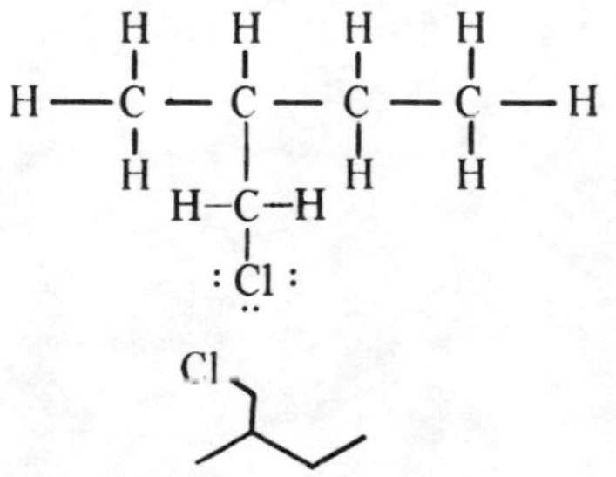 H — C — C — C — C — H
H H H
H–C–H
:C̤l:

Cl

:C̈l: :C̈l:
H :Cl: :Cl: H
(d) 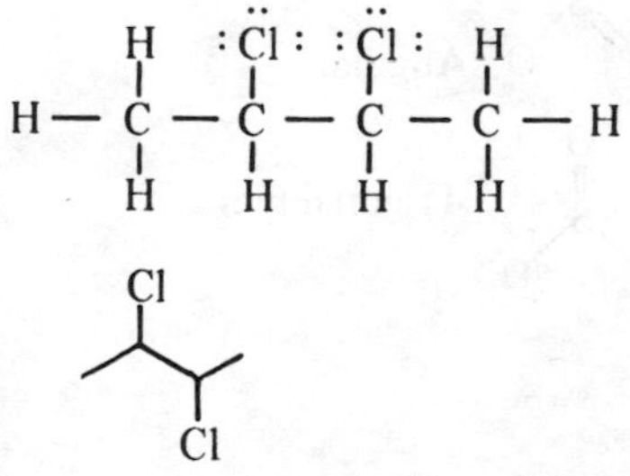 H — C — C — C — C — H
H H H H

Cl

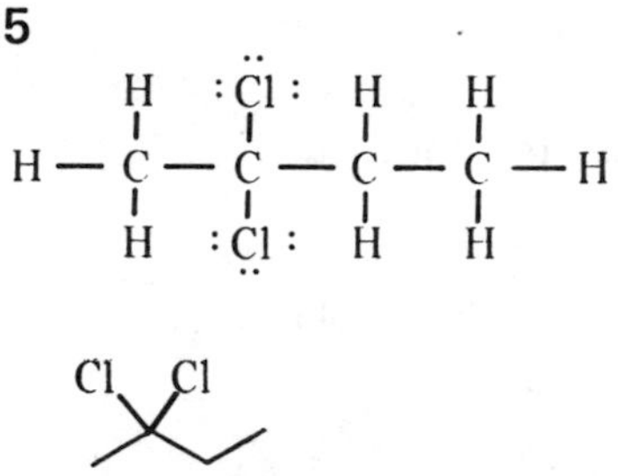

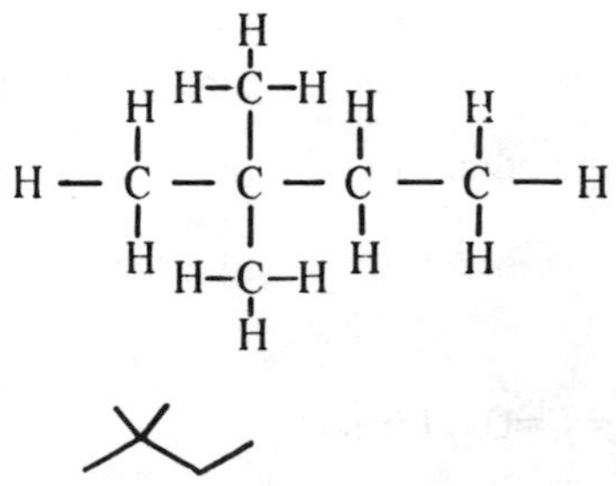

Cl

(e) 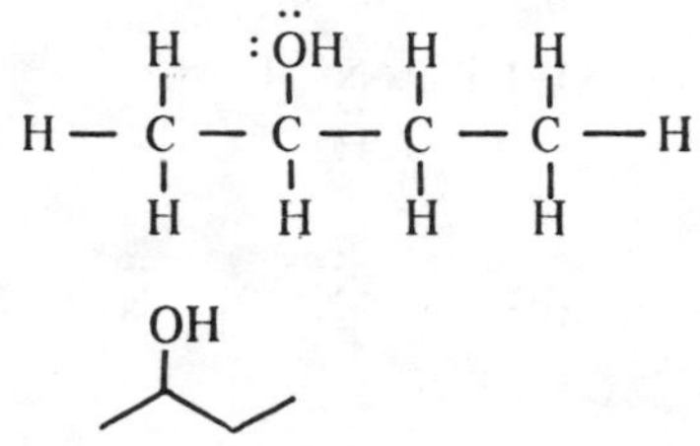 (g)

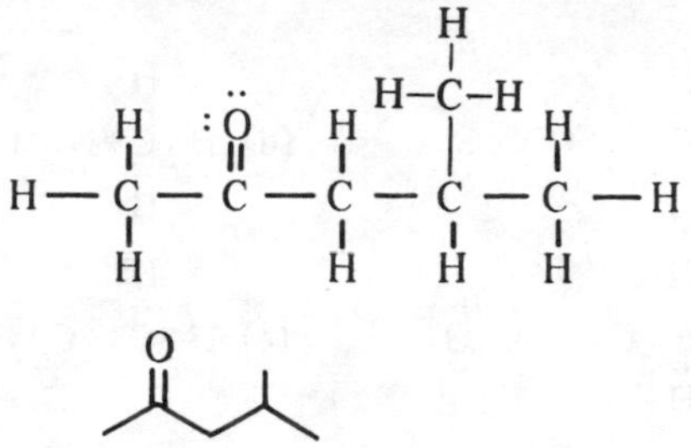

(f) 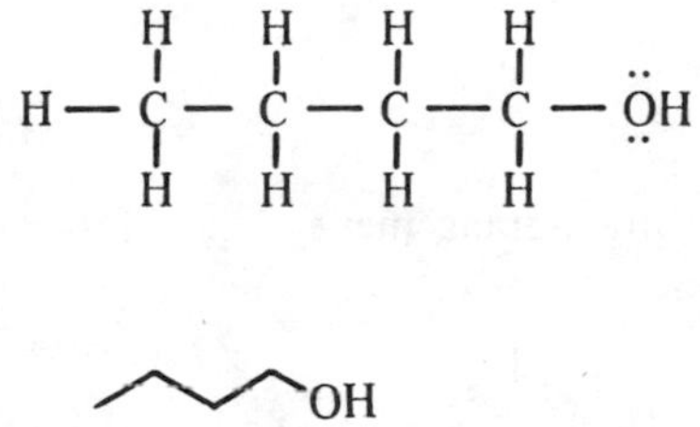 (h)

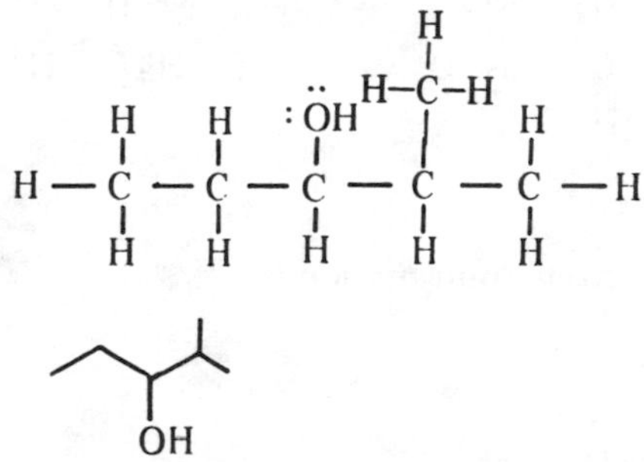

OH

O

OH

OH

1.16

(a) and (d) are structural isomers. (e) and (f) are structural isomers.

1.17

(a) 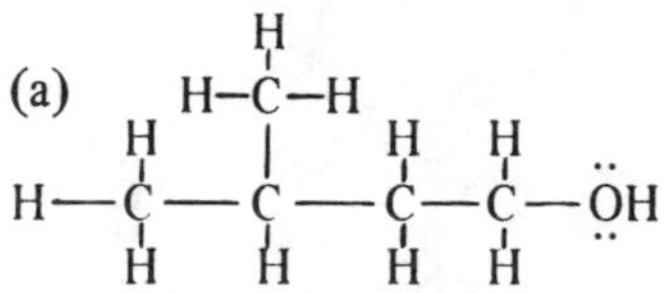 (b) 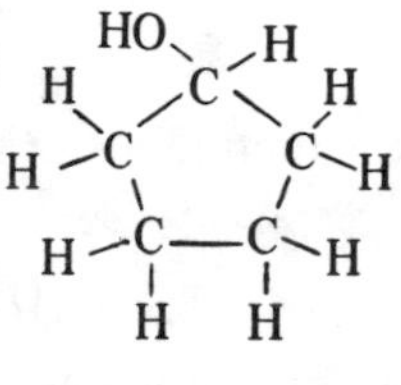 (c) 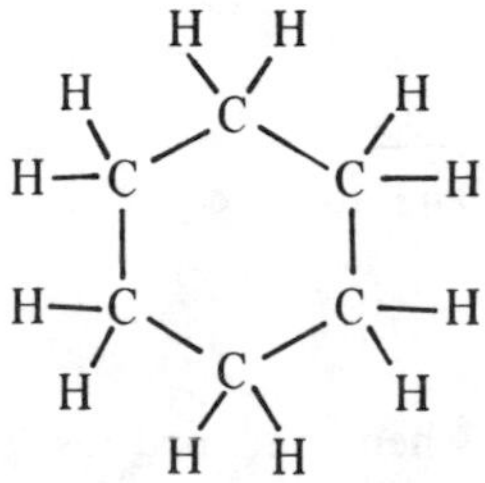 (d)

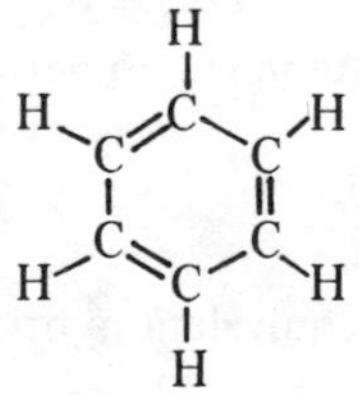

1.18

Each electron experiences less repulsion from other electrons if it is in an orbital by itself because the electrons can be further apart. Consider the three 2p orbitals as an example (see Fig. 1.21). With one electron in each 2p orbital each electron occupies a different region of space. This would not be true if two electrons were in the same 2p orbital.

1.19

(a) Monovalent because only one orbital (*2p*) contains a single electron. (b) The two *p* orbitals lie at 90° to one another; the resulting bonds would also lie at 90° to each other. Thus BF_3 based on an excited state of boron would have the following structure.

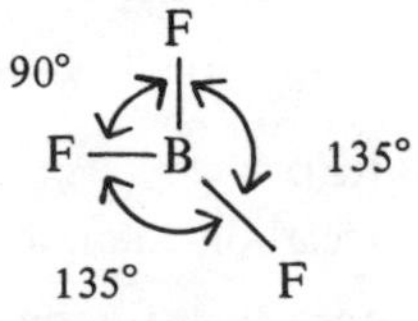

90° 135° 135°

1.20

```
     H                        H
     ..   ..        ..        ..  ..      ..            ..
(a) H:C:N::C::S:        (d) H:C:N::C::O:         (g) K+ :N:H
     ..        ..             ..                        ..
     H                        H                         H

     H                        H                             -     +   ..
     ..       +  ..           ..        ..                ..   ..
(b) H:C:C:::N:O:-       (e) H:C::C::O:          (h) Na+ :N::N::N:-
     ..          ..
     H

        ..
     H  :O
     .. .. :: ..              H
(c) H:C:O:N:O:-         (f) H:C::N::N:-
     .. ..  + ..              ..  +  ..
     H
```

1.21

(a) Electron Configuration

(1) Be $1s^2 2s^2$

(2) B $1s^2 2s^2 2p_x^1$

(3) C $1s^2 2s^2 2p_x^1 2p_y^1$

(4) N $1s^2 2s^2 2p_x^1 2p_y^1 2p_z^1$

(5) O $1s^2 2s^2 2p_x^2 2p_y^1 2p_z^1$

(b) Orbital Arrangement

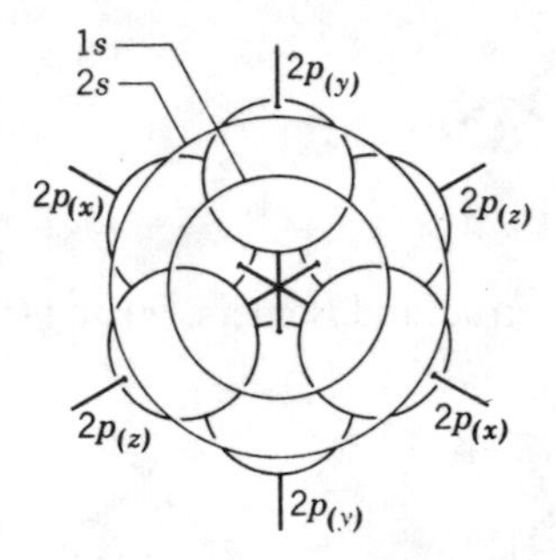

1.22

The three equivalent resonance structures,

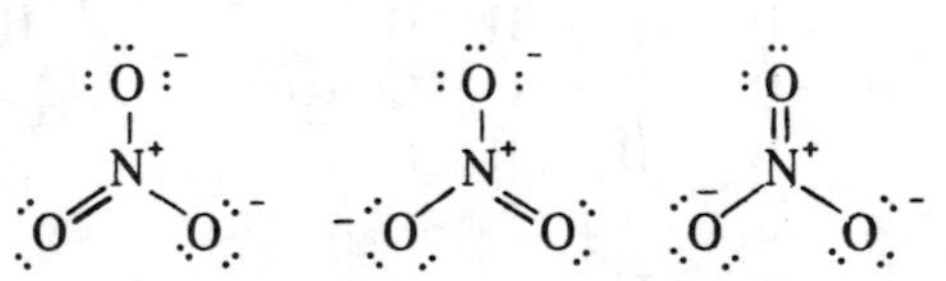

when combined result in equivalent N–O bonds that are intermediate in length between single and double bonds.

1.23

$^{-}$:Ö–N=Ö: :Ö=N–Ö:$^{-}$ Yes because the two O–N bonds are equivalent hybrids of a single and a double bond.

1.24

(a) An sp³ orbital. (b) sp³ Orbitals. In ammonia and in water the bond angles are close to the tetrahedral angle of 109½°; therefore the N and O atoms must be sp³ hybridized.

1.25

(a) $e = \frac{\mu}{d} = \frac{1.08 \times 10^{-18} \text{ esu-cm}}{1.27 \times 10^{-8} \text{ cm}}$

$e = 0.85 \times 10^{-10}$ esu

(b) $\frac{0.85 \times 10^{-10} \text{ esu}}{4.8 \times 10^{-10} \text{ esu/electron}} = 0.18$ electron

1.26

A carbon-chlorine bond is longer than a carbon fluorine bond because chlorine is a larger atom than fluorine. Thus in $\overset{\delta+}{CH_3}-\overset{\delta-}{Cl}$ the distance, d, that separates the charges is greater than in $\overset{\delta+}{CH_3}-\overset{\delta-}{F}$. The greater value of d for CH_3Cl more than compensates for the smaller value of e and thus the dipole moment ($e \times d$) is larger.

1.27

(a) While the structures differ in the position of their electrons they also differ in the positions of their nuclei and thus *they are not resonance structures.* (In cyanic acid the hydrogen nucleus is bonded to oxygen; in isocyanic acid it is bonded to nitrogen.)

(b) The anion obtained from either acid is a resonance hybrid of the following structures: $^{-}:\ddot{O}-C\equiv N: \longleftrightarrow :\ddot{O}=C=\ddot{N}:^{-}$

1.28

No. In each of the three examples both structures are the same if the four bonds of carbon are tetrahedrally arranged. [The problem here is with the way we write formulas. Sometimes we write them this way, as though the structure of carbon were square planar (which it is not). This representation suggests that isomers are possible when they are not.]

1.29

In He_2^+, two electrons are in a bonding molecular orbital, and only one electron is in an antibonding molecular orbital. The resultant He_2^+ ion has a lower energy than the non-bonded He : + $\cdot He^+$.

1.30

(a) BF_3 has an empty orbital which can accommodate the electron pair of $:NH_3$; also, formation of the new B—N bond stabilizes the system. (b) -1, (c) +1, (d) sp^3, (e) sp^3.

1.31

Acid strength increases with increasing (positive) formal charge on the central atom.

Acid Strength: $H_3O^+ > H_2O > OH^-$; $NH_4^+ > NH_3$; and $H_2S > HS^-$.

1.32

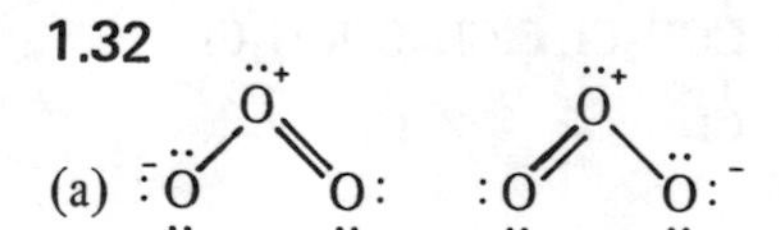

(b) Yes. (c) The ozone molecule is angular, thus the two O—O dipoles do not cancel. (d) Yes. The unshared electron pair on the central atom occupies space and repels the electrons of the oxygen-oxygen bonds.

1.33

The carbon atom in CH_3^+ utilizes only three of its four valence orbitals; therefore it is sp^2 hybridized. The vacant orbital is a p orbital.

CHAPTER TWO

Some Representative Carbon Compounds

2.1

(a) $CH_3-\underset{\displaystyle Cl}{\overset{\displaystyle H}{\underset{|}{\overset{|}{C}}}}-Cl$ and $Cl-CH_2CH_2-Cl$

(b) They differ only by rotation about single bonds. They are therefore interconvertible and are not different compounds.

2.2

(a) $CH_3CH_2CH_2Cl$ and $CH_3\underset{\displaystyle Cl}{\underset{|}{C}}HCH_3$

(b) $CH_3CH_2CHCl_2$, $CH_3\underset{\displaystyle Cl}{\underset{|}{C}}HCH_2Cl$, $ClCH_2CH_2CH_2Cl$, and $CH_3\underset{\displaystyle Cl}{\overset{\displaystyle Cl}{\underset{|}{\overset{|}{C}}}}CH_3$

(c) $CH_3CH_2CCl_3$, $CH_3\underset{\displaystyle Cl}{\underset{|}{C}}HCHCl_2$, $ClCH_2CH_2CHCl_2$, $CH_3\underset{\displaystyle Cl}{\overset{\displaystyle Cl}{\underset{|}{\overset{|}{C}}}}CH_2Cl$, $ClCH_2\underset{\displaystyle Cl}{\underset{|}{C}}HCH_2Cl$

2.3

Counting the double bond as two electron pairs located in the region of space between the two carbon atoms, each carbon has three atoms attached to it:

$$\begin{array}{ccccc} H & & & & H \\ & \diagdown & & \diagup & \\ & & C=C & & \\ & \diagup & & \diagdown & \\ H & & & & H \end{array}$$

The maximum separation of the electrons bonding the three atoms about each carbon atom occurs when they are equally spaced about that atom; i.e., when the bond angles are $\sim 120°$ and the molecule is planar.

2.4

(a) The C–X bond moments in the *trans*-isomers point in opposite directions and therefore cancel:

H X X X
C=C C=C X = Cl or Br
X H H H
trans - *cis*-

In the *cis*- isomers the bond moments are additive

(b) The C–Cl bond moment is larger than the C–Br bond moment because Cl is more electronegative than Br.

2.5

(a)

CH_3 Cl CH_3 H CH_3
C=C C=C $C{=}CH_2$ $Cl{-}CH_2CH{=}CH_2$
H H H Cl Cl

cis - *trans* isomers

(b)

CH_3 Cl CH_3 H $Cl{-}CH_2$ Cl $Cl{-}CH_2$ H
C=C C=C C=C C=C
Cl H Cl Cl H H H Cl

cis - *trans* isomers *cis* - *trans* isomers

$CH_3CH{=}CCl_2$ $Cl{-}\underset{Cl}{C}HCH{=}CH_2$ $Cl{-}CH_2\underset{Cl}{C}{=}CH_2$

(c)

$Cl_3CCH{=}CH_2$ $Cl{-}\underset{Cl}{C}H\underset{Cl}{C}{=}CH_2$ $Cl{-}CH_2CH{=}CCl_2$ $CH_3\underset{Cl}{C}{=}CCl_2$

Cl_2CH Cl Cl_2CH H $Cl{-}CH_2$ Cl $Cl{-}CH_2$ H
C=C C=C C=C C=C
H H H Cl Cl H Cl Cl

cis - *trans* isomers *cis* - *trans* isomers

(d)

$Cl_3C\underset{Cl}{C}{=}CH_2$

Cl_3C Cl Cl_3C H
C=C C=C $Cl_2CHCH{=}CCl_2$
H H H Cl

cis - *trans* isomers

Cl_2CH H Cl_2CH Cl
C=C C=C $ClCH_2\underset{Cl}{C}{=}CCl_2$
Cl Cl Cl H

cis - *trans* isomers

(e) 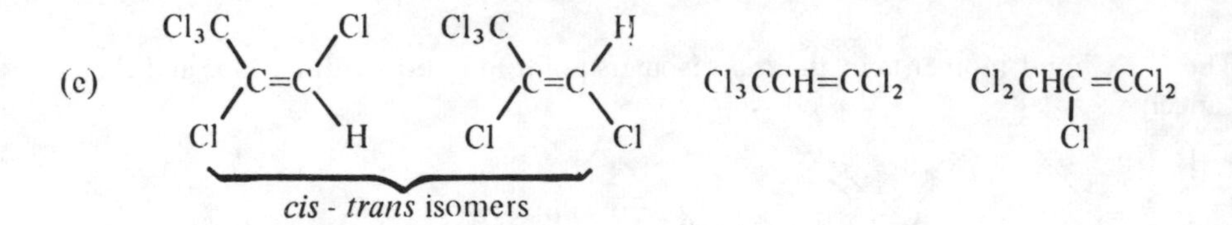

(f) See (a-e) above.

2.6

(a) Yes, counting the electrons in the triple bond together and as occupying the region between the two carbon atoms, the remaining electrons (in the C–H bonds) are at a maximum separation from the electrons in the triple bond when the bond angle is 180°.

(b) $H-C\equiv C-\overset{|}{\underset{|}{C}}-$

(c) $H-C\equiv C-CH_3$ (C bearing three H, drawn tetrahedrally) or H–◯≡◯–◯(H)(H)(H) (ball-and-stick model)

2.7

RH

2.8

(a) RCH_2OH, (b) $\underset{\underset{R}{|}}{RCHOH}$, (c) $\overset{\overset{R}{|}}{\underset{\underset{R}{|}}{RCOH}}$

2.9

(a) RCH_2X, (b) $\underset{\underset{R}{|}}{RCHX}$, (c) $\overset{\overset{R}{|}}{\underset{\underset{R}{|}}{RCX}}$, (d) RX

2.10

(a) $CH_3CH_2CH_2Cl$, (b) $\underset{\underset{Br}{|}}{CH_3CHCH_3}$, (c) ethyl fluoride, (d) isopropyl iodide,

(e) methyl iodide.

2.11

(a) $CH_3{-}O{-}CH_2CH_3$, (b) $CH_3CH_2CH_2{-}O{-}CH_2CH_2CH_3$, (c) $CH_3{-}O{-}\overset{\overset{CH_3}{|}}{C}HCH_3$,
(d) ethyl propyl ether, (e) isopropyl propyl ether

2.12

(a) $CH_3{-}\underset{\underset{CH_3}{|}}{N}{-}H$, (b) $CH_3CH_2{-}\underset{\underset{CH_2CH_3}{|}}{N}{-}CH_2CH_3$, (c) $CH_3CH_2{-}\underset{\underset{CH_3}{|}}{N}{-}CH_2CH_2CH_3$,

(d) isopropylmethylamine, (e) dipropylmethylamine, (f) isopropylamine

2.13

(a): (f) only, (b): (a), (d), (c): (b), (c), (e)

2.14

(a) sp^3, (b) sp^3

2.15

(a) $$K_a = \frac{[H_3O^+][CF_3COO^-]}{[CF_3COOH]} = 1$$

let $[H_3O^+] = [CF_3COO^-] = X$

then $[CF_3COOH] = 0.1 - X$

$$\therefore \quad \frac{(X)(X)}{0.1 - X} = 1 \quad \text{or} \quad X^2 = 0.1 - X$$

$$X^2 + X - 0.1 = 0$$

Using the quadratic formula, $X = \dfrac{-b \pm \sqrt{b^2 - 4ac}}{2a}$,

$$X = \frac{-1 \pm \sqrt{1 + 0.4}}{2} = \frac{-1 \pm \sqrt{1.4}}{2} = \frac{-1 \pm 1.183}{2} = \frac{+0.183}{2}$$

$X = 0.0915$ (We can exclude negative values of X.)

$$[H_3O^+] = [CF_3COO^-] = 0.0915\ \underline{M}$$

(b) Percentage ionized $= \dfrac{[H_3O^+]}{0.1} \times 100 = \dfrac{(0.0915)(100)}{0.1}$

Percentage ionized = 91.5%

2.16

Molecules of N-propylamine can form hydrogen bonds to each other,

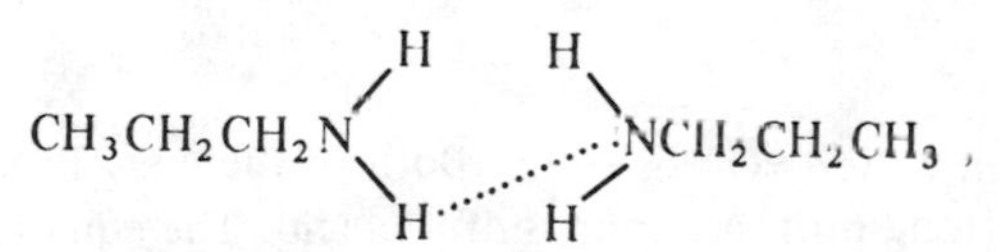

whereas molecules of trimethylamine, because they have no hydrogens attached to nitrogen, cannot form hydrogen bonds to each other.

2.17

(a) Alkyne (b) Carboxylic acid (c) Alcohol
(d) Aldehyde (e) Alkane (f) Ketone

2.18

(a) Carbon-carbon double bonds, hydroxyl group
(b) Ketone group, hydroxyl group, carbon-carbon double bond

(c) Carbon-carbon double bond, ester group

(d) Amide groups

(e) Aldehyde group, hydroxyl groups

(f) Carbon-carbon double bond, ether linkage

(g) Carbon-carbon double bond, hydroxyl group

2.19

(a) Primary (b) Secondary (c) Tertiary (d) Secondary

(e) Secondary (f) Tertiary

2.20

(a) Secondary (b) Primary (c) Tertiary (d) Primary

(e) Secondary

2.21

(a) $2H_3O^+ + CO_3^{-2} \longrightarrow [H_2CO_3] + 2H_2O \longrightarrow 3H_2O + CO_2$

(b) $H_3O^+ + CH_3\overset{\overset{O}{\|}}{C}O^- \longrightarrow H_2O + CH_3\overset{\overset{O}{\|}}{C}OH$

(c) $CO_3^{-2} + H_2O \longrightarrow HCO_3^- + OH^-$

(d) $:H^- + H_2O \longrightarrow H_2 + OH^-$

(e) $:CH_3^- + H_2O \longrightarrow CH_4 + OH^-$

(f) $:CH_3^- + HC{\equiv}CH \longrightarrow HC{\equiv}C:^- + CH_4$

(g) $H_3O^+ + NH_3 \longrightarrow NH_4^+ + H_2O$

(h) $NH_4^+ + NH_2^- \longrightarrow 2\,NH_3$

(i) $CH_3CH_2O^- + H_2O \longrightarrow CH_3CH_2OH + OH^-$

2.22

Oxygen-containing compounds contain either $=\ddot{O}:$ or $-\ddot{O}-$. Both of these are Brønsted-Lowry bases in the presence of the strong proton donor, sulfuric acid. The equation for the reaction using an ether as an example is

$$R-\ddot{O}-R + H_2SO_4 \rightleftharpoons \underbrace{R-\overset{\overset{H}{|}}{\ddot{O}^+}-R + HSO_4^-}_{\text{Salt}}$$

The salt is soluble in the highly polar H_2SO_4.

2.23

(a) Ethyl alcohol because its molecules can form hydrogen bonds to each other. Methyl ether molecules have no hydrogens attached to oxygen.

(b) Ethylene glycol because its molecules have more OH groups and will therefore participate in more extensive hydrogen bonding.

(c) Heptane because it has a higher molecular weight. (Neither compound can form hydrogen bonds.)

(d) 1-Propanol because its molecules can form hydrogen bonds to each other. Acetone molecules have no hydrogens attached to oxygen.

(e) *Cis*-1,2-dichloroethane because its molecules have a higher dipole moment.

(f) Propionic acid because its molecules can form hydrogen bonds to each other.

2.24

(a) $CH_3OCH_2CH_3$ (b) $CH_3CH_2CH_2OH$ (c) $CH_3\overset{OH}{\overset{|}{C}}HCH_3$

(d) $CH_3\overset{O}{\overset{\|}{C}}OCH_2CH_3$ $CH_3CH_2\overset{O}{\overset{\|}{C}}OCH_3$ (e) $CH_3CH_2CH_2CH_2X$

(f) $CH_3CH_2CHXCH_3$ (g) $CH_3\underset{X}{\underset{|}{\overset{CH_3}{\overset{|}{C}}}}CH_3$ (h) $CH_3\overset{CH_3}{\overset{|}{C}}H\underset{O}{\underset{\|}{C}}H$ or $CH_3CH_2CH_2\underset{O}{\underset{\|}{C}}H$

(i) $CH_3\overset{O}{\overset{\|}{C}}CH_2CH_3$ (j) $CH_3CH_2\overset{CH_3}{\overset{|}{C}}HNH_2$ (k) $CH_3CH_2CH_2NHCH_3$

(l) $CH_3CH_2N(CH_3)_2$ (m) $CH_3CH_2CH_2\overset{O}{\overset{\|}{C}}NH_2$ (n) $CH_3\overset{O}{\overset{\|}{C}}NHCH_2CH_3$

(o) 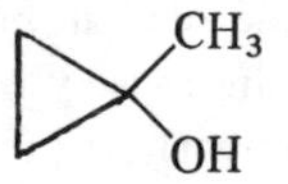 cyclopropane ring bearing CH_3 and OH on the same carbon

2.25

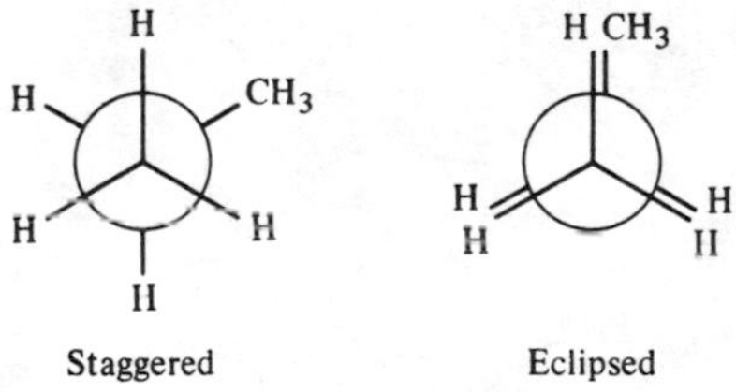

H, H, CH_3, H, H, H

Staggered

H CH_3, H, H, H, H

Eclipsed

2.26

Basic strength depends upon ability to accept a proton. In $(CF_3)_3N:$, the high electronegativity of fluorine reduces the availability of the lone electron pair on nitrogen. In $(CH_3)_3N:$, the lone pair is more available to bond with a proton. (An alternate view is that the conjugate acid is rendered less stable by the presence of the electronegative fluorines.)

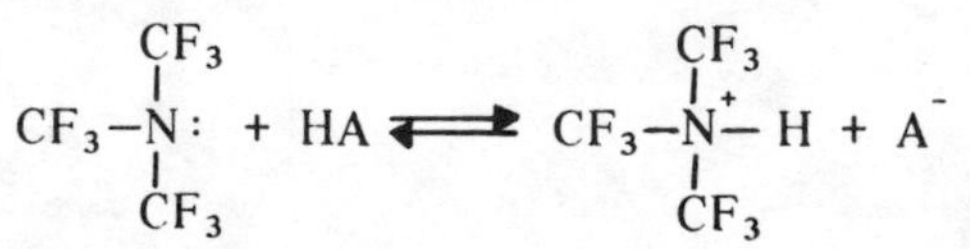

$$CF_3-\underset{CF_3}{\underset{|}{\overset{CF_3}{\overset{|}{N}}}}: + HA \rightleftharpoons CF_3-\underset{CF_3}{\underset{|}{\overset{CF_3}{\overset{|}{N^+}}}}-H + A^-$$

2.27

An ester group.

2.28

The attractive forces between hydrogen fluoride molecules are the very strong dipole-dipole attractions that we call *hydrogen bonds*. (The partial positive charge of a hydrogen fluoride molecule is relatively exposed because it resides on the hydrogen nucleus. By contrast, the positive charge of an ethyl fluoride molecule is buried in the ethyl group and is shielded by the surrounding electrons. Thus the positive end of one hydrogen fluoride molecule can approach the negative end of another hydrogen fluoride molecule much more closely with the result that the attractive force between them is much stronger.)

2.29

Since both molecules are nonpolar, the only intermolecular forces that we need to consider are van der Waals forces. Tetrafluoromethane is a generally spherical, compact molecule and its outer electrons (of the fluorine atoms) are very tightly held. Consequently, it will be difficult for a temporary dipole in one molecule of CF_4 to induce a very large temporary dipole in an adjacent molecule and, as a result, the van der Waals forces acting between them will be small. Hexane is a much larger molecule and the outer electrons are more loosely held. The outer electrons of hexane are more easily distorted—hexane is said to be more *polarizable*—thus the van der Waals forces acting between hexane molecules will be much larger.

2.30

In the molecules or atoms with greater molecular weight there are not only more electrons, but the outermost electrons are further from the nucleus where they are more loosely held. Thus the temporary dipoles that occur in a given molecule can be larger. Because the outermost electrons of adjacent molecules are also more loosely held, the induced dipoles in the adjacent molecules will also be larger. Consequently the attractive forces between the heavier molecules will be larger and the boiling points will be higher.

CHAPTER THREE

Alkanes and Cycloalkanes

3.1

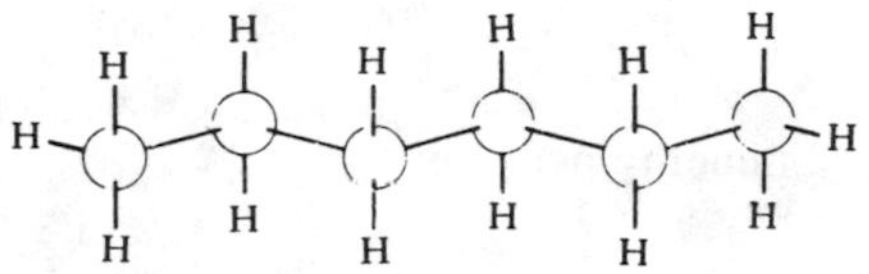

(1) $CH_3CH_2CH_2CH_2CH_2CH_3$

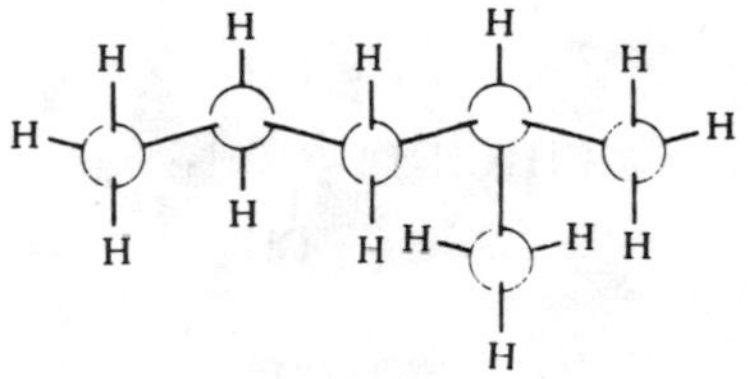

(2) $CH_3CH_2CH_2CHCH_3$
with CH_3 on the CH

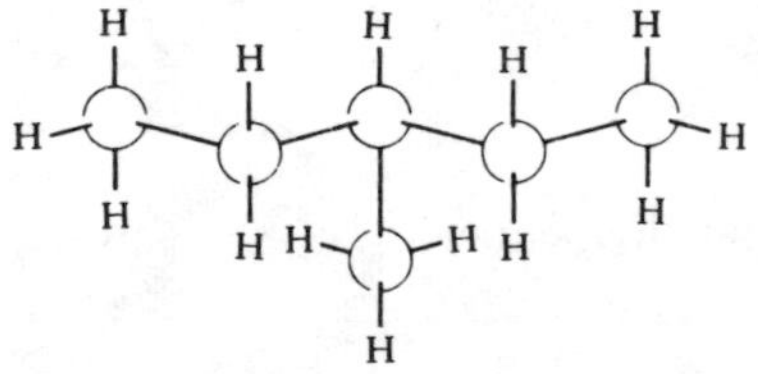

(3) $CH_3CH_2CHCH_2CH_3$
with CH_3 on the CH

(4) $CH_3CHCHCH_3$
with CH_3 on each CH

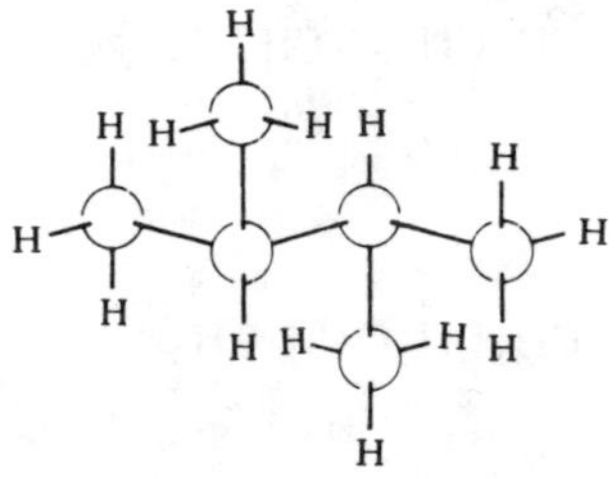

(5) $CH_3CH_2CCH_3$
with two CH_3 groups on the C

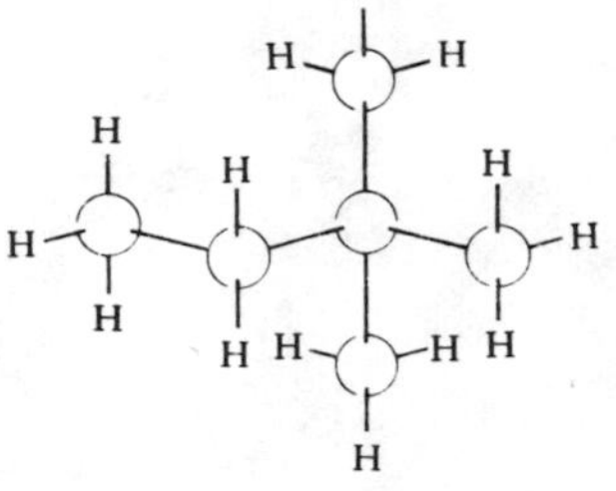

3.2

(a) Refer to problem 3.1 above:

(1) Hexane, (2) 2-methylpentane, (3) 3-methypentane, (4) 2,3-dimethylbutane, (5) 2,2-dimethylbutane.

(b)

$CH_3CH_2CH_2CH_2CH_2CH_2CH_3$ — heptane

$CH_3CH_2CH_2CH_2CHCH_3$ with CH_3 on C-2 — 2-methylhexane

$CH_3CH_2CH_2CHCH_2CH_3$ with CH_3 on C-3 — 3-methylhexane

$CH_3CH_2CH_2CCH_3$ with two CH_3 on C-2 — 2, 2-dimethylpentane

$CH_3CH_2CHCHCH_3$ with CH_3 on C-2 and C-3 — 2, 3-dimethylpentane

$CH_3CHCH_2CHCH_3$ with CH_3 on C-2 and C-4 — 2, 4-dimethylpentane

$CH_3CH_2CCH_2CH_3$ with two CH_3 on C-3 — 3, 3-dimethylpentane

$CH_3CH{-}CCH_3$ with CH_3 on CH and two CH_3 on C — 2, 2, 3-trimethylbutane

$CH_3CH_2CHCH_2CH_3$ with CH_2CH_3 on C-3 — 3-ethylpentane

3.3

(a) 1-*tert*-Butyl-3-methylcyclohexane
(b) 1, 3-dimethylcyclobutane
(c) 1-cyclohexylheptane

3.4

(a) Bicyclo[2.2.0]hexane
(b) bicyclo[4.4.0] decane
(c) bicyclo[2.2.2] octane

(d) Bicyclo[3.1.1]heptane; or

bicyclo[4.1.0]heptane

3.5

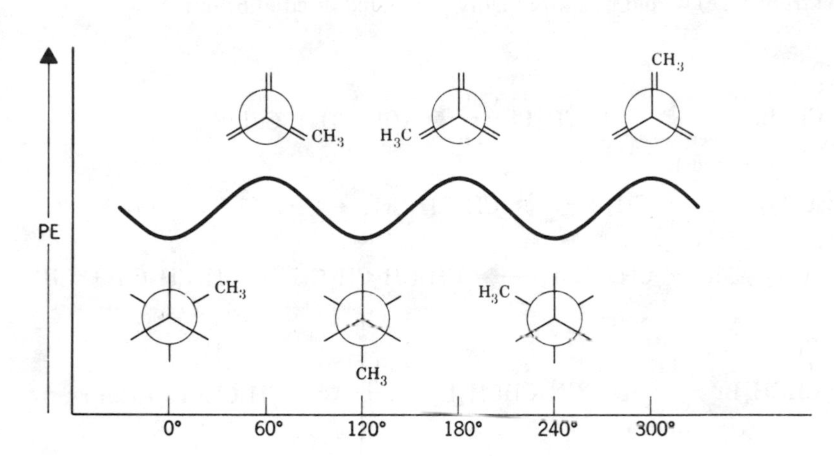

3.6

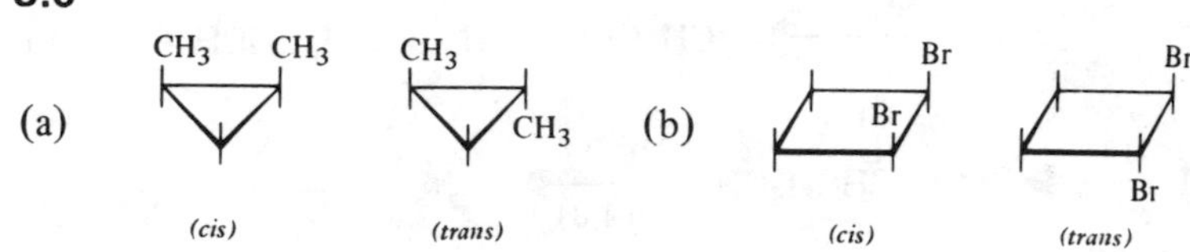

3.7

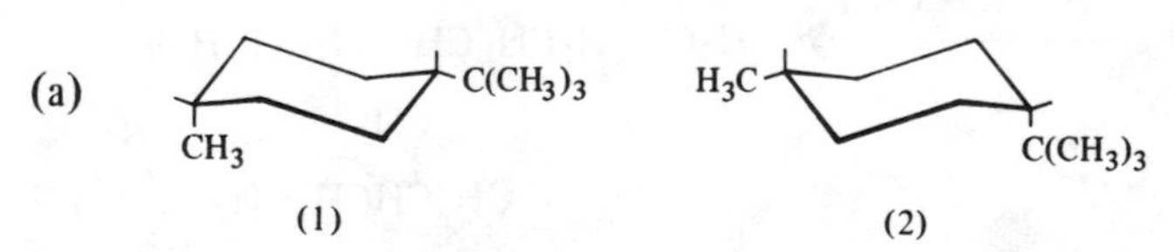

(b) No. In (1), the methyl group is axial and the *tert*-butyl group is equatorial; in (2) the situation is reversed.

(c) The *tert*-butyl group is larger than the methyl; conformation (1) is more stable because the *tert*-butyl group is equatorial.

(d) The preferred conformation at equilibrium is (1).

3.8

(a) Conformations of *cis*-isomer are equivalent, (e, a) and (a, e).

CH_3 CH_3 (a, e)

H_3C CH_3 (e, a)

(b) Conformations of *trans*-isomer are not equivalent, (e, e) and (a, a).

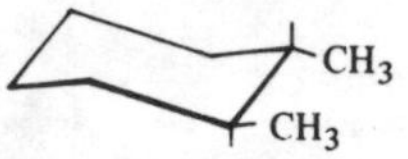

CH_3 CH_3 (e, e) CH_3 CH_3 (a, a)

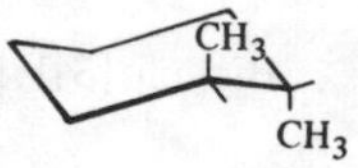

(c) The *trans*-(e, e) conformation is more stable than the *trans*-(a,a).

(d) The *trans*-(e, e) would be more highly populated at equilibrium.

3.9

(a) $$CH_3CH_2Br \xrightarrow[\text{ether (-LiBr)}]{2Li} 2CH_3CH_2Li \xrightarrow[\text{(-LiI)}]{CuI} (CH_3CH_2)_2CuLi$$

$$(CH_3CH_2)_2CuLi + CH_3I \longrightarrow CH_3CH_2CH_3 + CH_3CH_2Cu + LiI$$

(b) $$(CH_3CH_2)_2CuLi + CH_3CH_2I \longrightarrow CH_3CH_2CH_2CH_3 + CH_3CH_2Cu + LiI$$

(c) $$CH_3\overset{\overset{CH_3}{|}}{C}HCH_2Br \xrightarrow[\text{ether (-LiBr)}]{2Li} 2CH_3\overset{\overset{CH_3}{|}}{C}HCH_2Li \xrightarrow[\text{(-LiI)}]{CuI} (CH_3\overset{\overset{CH_3}{|}}{C}HCH_2)_2CuLi$$

$$\xrightarrow{CH_3I} CH_3\overset{\overset{CH_3}{|}}{C}HCH_2CH_3 + CH_3\overset{\overset{CH_3}{|}}{C}HCH_2Cu + LiI$$

(d) $$CH_3\overset{\overset{CH_3}{|}}{C}HCH_2CH_2I \xrightarrow[\text{ether (-LiI)}]{2Li} 2\,CH_3\overset{\overset{CH_3}{|}}{C}HCH_2CH_2Li \xrightarrow[\text{(-LiI)}]{CuI}$$

$$(CH_3\overset{\overset{CH_3}{|}}{C}HCH_2CH_2)_2CuLi \xrightarrow{CH_3\overset{\overset{CH_3}{|}}{C}HCH_2CH_2I} CH_3\overset{\overset{CH_3}{|}}{C}HCH_2CH_2CH_2CH_2\overset{\overset{CH_3}{|}}{C}HCH_3$$

$$+ CH_3\overset{\overset{CH_3}{|}}{C}HCH_2CH_2Cu + LiI$$

3.10

(a) $$CH_3CH_2CH_2Br \xrightarrow[\text{ether (-LiBr)}]{2Li} 2CH_3CH_2CH_2Li \xrightarrow[\text{(-LiI)}]{CuI}$$

$$(CH_3CH_2CH_2)_2CuLi \xrightarrow{CH_3CH_2CH_2Br} CH_3CH_2CH_2CH_2CH_2CH_3$$

$$+ CH_3CH_2CH_2Cu + LiBr$$

(b) $$CH_3CH_2CH_2CH_2Br \xrightarrow[\text{ether (-LiBr)}]{2Li} 2CH_3CH_2CH_2CH_2Li \xrightarrow[\text{(-LiI)}]{CuI}$$

$$(CH_3CH_2CH_2CH_2)_2CuLi \xrightarrow{CH_3CH_2Br} CH_3CH_2CH_2CH_2CH_2CH_3$$

$$+ CH_3CH_2CH_2CH_2Cu + LiBr$$

(c) $CH_3CH_2CH_2CH_2CH_2Br \xrightarrow[\text{ether (-LiBr)}]{2Li} CH_3CH_2CH_2CH_2CH_2Li \xrightarrow[\text{(-LiI)}]{CuI}$

$$(CH_3CH_2CH_2CH_2CH_2)_2CuLi \xrightarrow{CH_3Br} CH_3CH_2CH_2CH_2CH_2CH_3$$

$$+ CH_3CH_2CH_2CH_2CH_2Cu + LiBr$$

(d) $2CH_3CH_2CH_2CH_2CH_2CH_2Br \xrightarrow[Zn]{H^+} 2CH_3CH_2CH_2CH_2CH_2CH_3 + ZnBr_2$

(e) $CH_3CH_2CH{=}CHCH_2CH_3 \xrightarrow[\substack{C_2H_5OH \\ (25°,\ 50atm.)}]{Ni} CH_3CH_2CH_2CH_2CH_2CH_3$

3.11

(a) NH_2 (b) CH (c) C_2H_4O (d) C_5H_7N (e) CH_2 (f) CH

3.12

	Empirical Formula	Empirical Formula Weight	(Molecular Wt. / Emp. Form. Wt.)	Molecular Formula
(a)	CH_2O	30	$\frac{179}{30} \cong 6$	$C_6H_{12}O_6$
(b)	CHN	27	$\frac{80}{27} \cong 3$	$C_3H_3N_3$
(c)	CCl_2	83	$\frac{410}{83} \cong 5$	C_5Cl_{10}

3.13

Assuming a 100-g sample, the amounts of the elements are:

	Weight	Moles (A)	B		
C	57.45	$\frac{57.45}{12.01} = 4.78$	$\frac{4.78}{0.300}$	= 15.9	= 16
H	5.40	$\frac{5.40}{1.008} = 5.36$	$\frac{5.36}{0.300}$	= 17.9	= 18
N	8.45	$\frac{8.45}{14.01} = 0.603$	$\frac{0.603}{0.300}$	= 2.01	= 2
S	9.61	$\frac{9.61}{32.06} = 0.300$	$\frac{0.300}{0.300}$	= 1.00	= 1
O*	19.09	$\frac{19.09}{16.00} = 1.19$	$\frac{1.19}{0.300}$	= 3.97	= 4
	100.00				

(* by difference from 100)

The empirical formula is thus $C_{16}H_{18}N_2SO_4$. The empirical formula weight (334.4) is within the range given for the molecular weight (330 ± 10), thus the molecular formula for Penicillin G is the same as the empirical formula.

3.14

(a) $CH_3CHCHCH_2CH_3$ (with Cl on C-2 and Cl on C-3)

(b) CH_3CCH_3 (with CH_3 above and I below on the central C)

(c) $CH_3CH_2CHCH_2CH_3$ (with CH_2CH_3 on C-3)

(d) $CH_3CH{-}CH{-}CHCH_2CH_2CH_2CH_2CH_2CH_3$ (with CH_3, CH_3, CH_3 on C-2, C-3, C-4)

(e) $CH_3CH_2CH_2CHCH_2CH_2CH_2CH_2CH_3$ (with $CH(CH_3)_2$ on C-4)

(f) cyclopropane ring with CH_3 and CH_3 on the same carbon

(g) cyclohexane ring with CH_3 and CH_3

(h) H_3C … CH_3 (cyclohexane ring)

(i) cyclohexane ring with H and $CH(CH_3)_2$

(j) cyclohexane ring with CH_3, H, H and $CH(CH_3)_2$

(k) $CH_3CHCH_2CH_2CH_2{-}Cl$ (with CH_3 on C-2)

(l) $CH_3CCH_2CCH_2CH_2CH_2CH_3$ (with CH_3, CH_3 above and CH_3, CH_3 below on C-2 and C-4)

(m) $CH_3CCH_2{-}Cl$ (with CH_3 above and CH_3 below)

(n) $CH_3CHCH_2CH_3$ (with CH_3 on C-2)

3.15

(a) 3,4-dimethylhexane
(b) 2-methylbutane
(c) 2,4-dimethylpentane
(d) 3-methylpentane
(e) ethylcyclohexane
(f) cyclopentylcyclopentane
(g) 6-isobutyl-2-methyldecane

3.16

		Common	IUPAC
(a)	$CH_3CH_2CH_2{-}Cl$	*n*-propyl chloride	1-chloropropane
	CH_3CHCH_3 (Cl on C-2)	isopropyl chloride	2-chloropropane
(b)	$CH_3CH_2CH_2CH_2{-}Br$	*n*-butyl bromide	1-bromobutane
	$CH_3CH_2CHCH_3$ (Br on C-2)	sec-butyl bromide	2-bromobutane

$CH_3CHCH_2{-}Br$ (with CH_3 on C-2) isobutyl bromide 1-bromo-2-methylpropane

CH_3 above and CH_3 below: $CH_3C{-}Br$ tert-butyl bromide 2-bromo-2-methylpropane

3.17

(a) CH_3CCH_3 (with CH_3 above and CH_3 below the central C) 2,2-dimethylpropane (neopentane)

(b) $CH_3CHCH_2CH_3$ (with CH_3 above C-2) 2-methylbutane (isopentane)

(c) $CH_3CH_2CH_2CH_2CH_3$ pentane

(d) ring of five CH_2 groups: $CH_2{-}CH_2$, CH_2, CH_2, CH_2 cyclopentane

(e) $CH_3CH{-}CHCH_3$ (with CH_3 above C-2 and CH_3 above C-3) 2,3-dimethylbutane

3.18

(a) $CH_3CHCHCHCH_3$ (with CH_3 above C-2, CH_3 above C-4, and CH_3CHCH_3 below C-3) 2,4-dimethyl-3-isopropylpentane

(b) $CH_3CH_2CCH_2CH_3$ (with CH_2CH_3 above and CH_2CH_3 below C-3) 3,3-diethylpentane

(c) $CH_3CHCH_2CCH_2CHCH_3$ (with CH_3 above C-2, CH_3 above C-6, CH_2–CH_3CHCH_3 above C-4, and CH_2–CH_3CHCH_3 below C-4) 4,4-diisobutyl-2,6-dimethylheptane

(d) $CH_3CH_2CHCHCHCH_2CH_3$ (with CH_3 above C-3, CH_3 above C-5, and $CHCH_2CH_3$ with CH_3 below C-4) 4-*sec*-butyl-3,5-dimethylheptane

(e)
$$CH_3C(CH_3)_2CH_2C(CH_3)_2CH_3$$
2,2,4,4-tetramethylpentane

(f)
$$CH_3CH_2CH_2CH_2C(CH_2CH_2CH_2CH_3)_2CH_2CH_2CH_2CH_3$$
5,5-dibutylnonane

3.19

$$\left.\begin{array}{l} CH_2{=}C(CH_3)CH_2CH_3 \\ CH_3C(CH_3){=}CHCH_3 \\ CH_3CH(CH_3)CH{=}CH_2 \end{array}\right\} + H_2 \xrightarrow[C_2H_5OH]{Ni} CH_3CH(CH_3)CH_2CH_3$$

3.20

The alkane is 2-methylpentane, $CH_3CH(CH_3)CH_2CH_2CH_3$. The five alkyl chlorides are

$ClCH_2CH(CH_3)CH_2CH_2CH_3$, $CH_3CCl(CH_3)CH_2CH_2CH_3$, $CH_3CH(CH_3)CHClCH_2CH_3$,

$CH_3CH(CH_3)CH_2CHClCH_3$, and $CH_3CH(CH_3)CH_2CH_2CH_2Cl$

3.21

H_3C—$C(H)(CH_3)_2$ (tetrahedral carbon bearing three CH_3 groups and one H)

The methyl groups are larger than the hydrogen atom. The resulting mutual repulsions among the methyl groups cause a larger than tetrahedral bond angle.

3.22

(a)

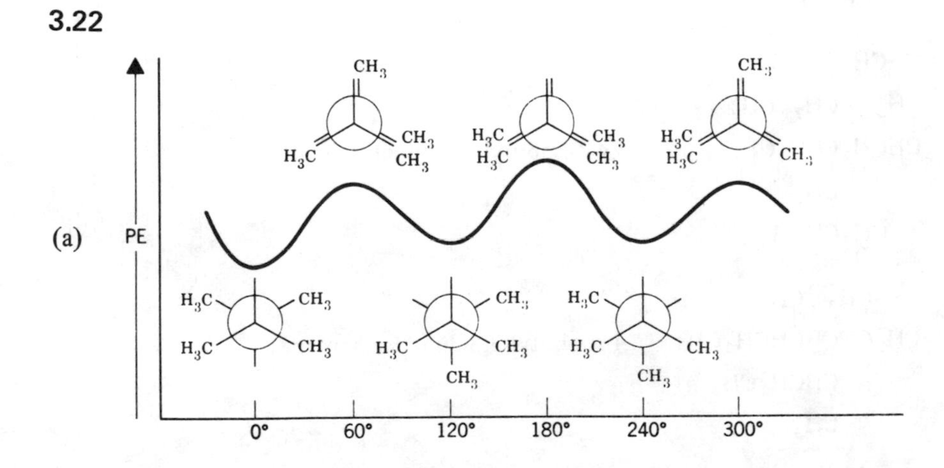

(b)

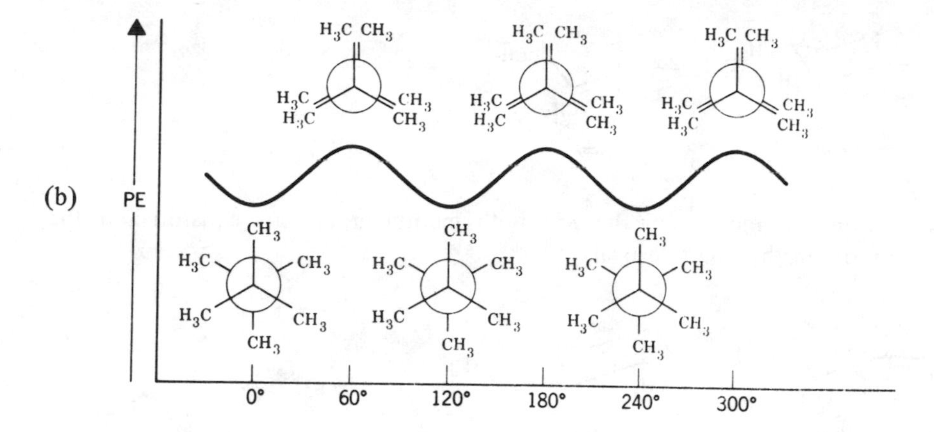

3.23

(a) Hexane. Branched chain hydrocarbons have lower boiling points than their normal isomers.

(b) Hexane. Boiling point increases with molecular weight.

(c) Pentane. (See (a) above).

(d) Chloroethane, because it has a higher molecular weight, and is more polar.

3.24

(a) The *trans*-isomer is more stable.

(b) Since they both yield the same combustion products and in the same molar amounts, the one that has the larger heat of combustion has the higher potential energy, and is therefore less stable (The *cis*-isomer is less stable because of the crowding that exists between the methyl groups on the same side of the ring.)

3.25

(a)

CH_3 $CH(CH_3)_2$

(1) (e, e)

CH_3 $CH(CH_3)_2$

(2) (a, a)

(b)

CH_3 $CH(CH_3)_2$

(3) (a, e)

H_3C $CH(CH_3)_2$

(4) (e, a)

(c) (1) is more stable than (2) because in (1), both substituents are equatorial. (3) is more stable than (4) because in (3), the larger group ($CH(CH_3)_2$) is equatorial.

3.26

(a) The *trans*-isomer is more stable because both methyl groups can be equatorial. In *cis*-1, 2-dimethylcyclohexane, one methyl must be axial.

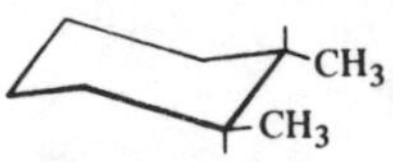

CH_3 CH_3 *(trans)* CH_3 CH_3 *(cis)*

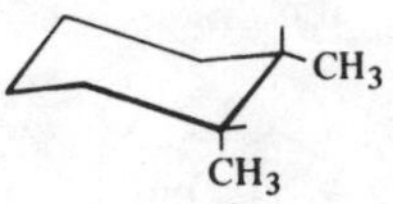

(b) The *cis*-isomer is more stable because both methyl groups are equatorial. In the *trans*-isomer, one methyl must be axial.

CH_3 CH_3 *(cis)* CH_3 CH_3 *(trans)*

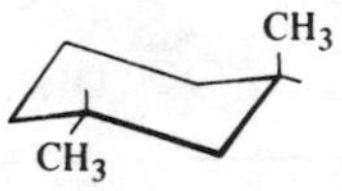

(c) The *trans*-isomer is more stable for the same reason as in (a).

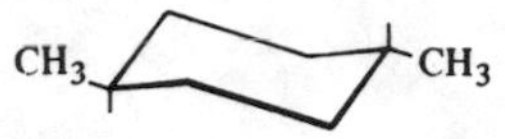

CH_3 CH_3 *(trans)* CH_3 CH_3 *(cis)*

3.27

In *cis*-1, 3-*di-tert*-butylcyclohexane, the two substituents are both equatorial (see problem 3.18 b above), whereas in the *trans*-isomer, one of the *tert*-butyl groups must be axial. The instability of a chair conformation with such a large group in an axial position forces the molecule into a less strained twist conformation:

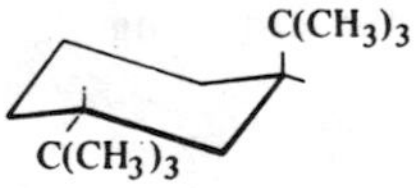

$C(CH_3)_3$ $C(CH_3)_3$ $C(CH_3)_3$ $C(CH_3)_3$

trans (chair conformation)

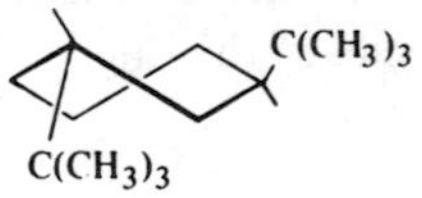

3.28

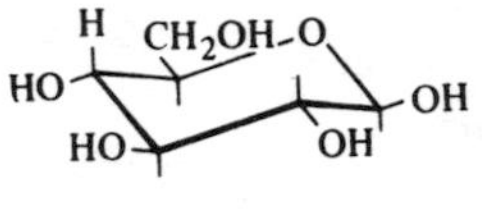

H CH_2OH–O HO OH HO OH

3.29

(a) To calculate the percentage composition from the molecular formula, first determine the weight of each element in one mole of the compound. For $C_6H_{12}O_6$,

$$C_6 = 6 \times 12.01 = 72.06 \qquad \frac{72.06}{180.2} = 0.400 = 40.0\%$$

$$H_{12} = 12 \times 1.008 = 12.10 \qquad \frac{12.10}{180.2} = 0.0671 = 6.7\%$$

$$O_6 = 6 \times 16.00 = 96.00 \qquad \frac{96.00}{180.2} = 0.533 = 53.3\%$$

$$\text{Molecular Wt.} \qquad 180.16$$

Then determine the percentage of each element using the formula.

$$\text{Percentage of A} = \frac{\text{Weight of A}}{\text{Molecular Weight}} \times 100$$

(b)

$$C_2 = 2 \times 12.01 = 24.02 \qquad \frac{24.02}{75.07} = 0.320 = 32.0\%$$

$$H_5 = 5 \times 1.008 = 5.04 \qquad \frac{5.04}{75.07} = 0.067 = 6.7\%$$

$$N = 1 \times 14.01 = 14.01 \qquad \frac{14.01}{75.07} = 0.187 = 18.7\%$$

$$O_2 = 2 \times 16.00 = 32.00 \qquad \frac{32.00}{75.07} = 0.426 = 42.6\%$$

(c)

$$C_3 = 3 \times 12.01 = 36.03 \qquad \frac{36.03}{280.77} = 0.128 = 12.8\%$$

$$H_5 = 5 \times 1.008 = 5.04 \qquad \frac{5.04}{280.77} = 0.018 = 1.8\%$$

$$Br_3 = 3 \times 79.90 = 239.70 \qquad \frac{239.70}{280.77} = 0.854 = 85.4\%$$

3.30

If the compound contains iron, each molecule must contain at least one atom of iron. This means that one mole of the compound must contain at least 55.85 grams of iron. Therefore,

$$\text{MW of ferrocene} = 55.85\ \frac{\text{grams Fe}}{\text{mole}} \times \frac{1.000\ \text{gram}}{0.3002\ \text{gram Fe}}$$

$$= 186.0\ \frac{\text{grams}}{\text{mole}}$$

3.31

First we must determine the empirical formula. Assuming that the difference between the percentages given and 100 percent is due to oxygen,

$$\text{C:} \quad 40.04 \qquad \frac{40.04}{12.01} = 3.33 \qquad \frac{3.33}{3.33} = 1$$

$$\text{H:} \quad 6.69 \qquad \frac{6.69}{1.008} = 6.64 \qquad \frac{6.64}{3.33} \cong 2$$

$$\text{O:} \quad \frac{53.27}{100.00} \qquad \frac{53.27}{16.00} = 3.33 \qquad \frac{3.33}{3.33} = 1$$

The empirical formula is thus CH_2O.

To determine the molecular formula we must first determine the molecular weight. At standard temperature and pressure, the volume of one mole of an ideal gas is 22.4 liters. Assuming ideal behavior,

$$\frac{1.00\text{ g}}{0.746\text{ liter}} = \frac{M}{22.4\text{ liters}} \quad \text{Where M = Molecular weight.}$$

$$M = \frac{(1.00)(22.4)}{0.746} = 30.0\text{ g}$$

The empirical formula weight (30.0) equals the molecular weight, thus the molecular formula is the same as the empirical formula.

3.32

As in problem 3.31, the molecular weight is found by the equation

$$\frac{1.251\text{ g}}{1.00\text{ liter}} = \frac{M}{22.4\text{ liter}}$$

$$M = (1.251)(22.4)$$

$$M = 28.02$$

To determine the empirical formula, we must determine the amount of carbon in 3.926 g of carbon dioxide, and the amount of hydrogen in 1.608 g of water.

$$\text{C:} \quad \left(3.926\,\cancel{\text{g }CO_2}\right)\left(\frac{12.01\text{ g C}}{44.01\,\cancel{\text{g }CO_2}}\right) = 1.071\text{ g carbon}$$

$$\text{H:} \quad \left(1.608\,\cancel{\text{g }H_2O}\right)\left(\frac{2.016\text{ g H}}{18.016\,\cancel{\text{g}H_2O}}\right) = \frac{0.179\text{ g hydrogen}}{1.250\text{ g sample}}$$

The weight of C and H in a 1.250 g sample is 1.250 g. Therefore there are no other elements present.

To determine the empirical formula we proceed as in problem 3.13 except that the sample size is 1.250 instead of 100 g.

$$\text{C:} \quad \frac{1.071}{12.01} = 0.0892 \qquad \frac{0.0892}{0.0892} = 1$$

$$\text{H:} \quad \frac{0.179}{1.008} = 0.178 \qquad \frac{0.178}{0.0892} = 2$$

The empirical formula is thus CH_2. The empirical formula weight (14) is one-half the molecular weight. Thus the molecular formula is C_2H_4.

3.33

Using the procedure of problem 3.13,

$$\text{C:} \quad 59.10 \qquad \frac{59.10}{12.01} = 4.92 \qquad \frac{4.92}{0.817} = 6.02 \cong 6$$

$$\text{H:} \quad 4.92 \qquad \frac{4.92}{1.008} = 4.88 \qquad \frac{4.88}{0.817} = 5.97 \cong 6$$

$$\text{N:} \quad 22.91 \qquad \frac{22.91}{14.01} = 1.64 \qquad \frac{1.64}{0.817} = 2$$

$$\text{O:} \quad \frac{13.07}{100.00} \qquad \frac{13.07}{16.00} = 0.817 \qquad \frac{0.817}{0.817} = 1$$

The empirical formula is thus $C_6H_6N_2O$. The empirical formula weight is 123.13 which is equal to the molecular weight within experimental error. The molecular formula is thus the same as the empirical formula.

3.34

C:	40.88	$\frac{40.88}{12.01} = 3.40$	$\frac{3.40}{0.619} = 5.5$	$5.5 \times 2 = 11$
H:	3.74	$\frac{3.74}{1.008} = 3.71$	$\frac{3.71}{0.619} = 6$	$6 \times 2 = 12$
Cl:	21.95	$\frac{21.95}{35.45} = 0.619$	$\frac{0.619}{0.619} = 1$	$1 \times 2 = 2$
N:	8.67	$\frac{8.67}{14.01} = 0.619$	$\frac{0.619}{0.619} = 1$	$1 \times 2 = 2$
O:	24.76	$\frac{24.76}{16.00} = 1.55$	$\frac{1.55}{0.619} = 2.5$	$2.5 \times 2 = 5$
	100.00			

The empirical formula is thus $C_{11}H_{12}Cl_2N_2O_5$. The empirical formula weight (323) is equal to the molecular weight, therefore the molecular formula is the same as the empirical formula.

*3.35

(a) CH_3 CH_3

(b) From Table 3.7 we find that this is *cis*-1,2-dimethylcyclohexane.

(c) Since catalytic hydrogenation produces the *cis* isomer both hydrogens must have added from the same side of the double bond. (As we will see in Sect. 7.8B, this type of addition is called a *syn* addition.)

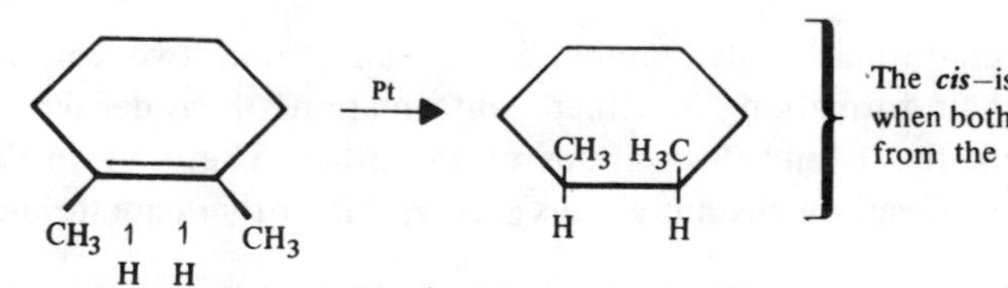

The *cis*–isomer is produced when both hydrogens add from the same side.

cis–1,2–dimethylcyclohexane

*3.36

(a) From Table 3.7 we find that this is *trans*-1,2-dichlorocyclohexane.

(b) Since the product is the *trans*-isomer we can conclude that the chlorine atoms have added from opposite sides of the double bond.

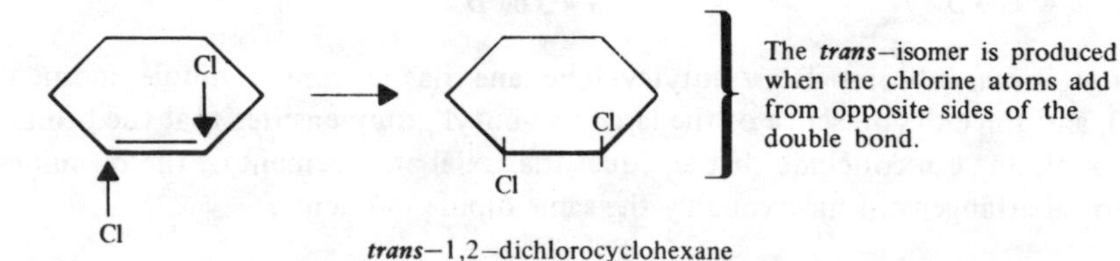

trans–1,2–dichlorocyclohexane

*3.37

(a)

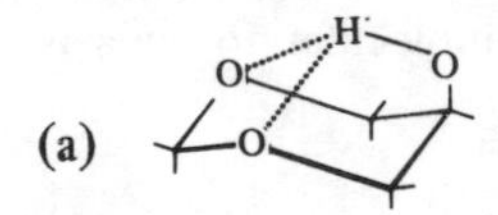

(b) Having the hydroxyl axial allows the formation of a hydrogen bond to either ring oxygen as shown above.

*3.38

(a) We can better understand the conformational rigidity of *trans*-decalin if we consider one ring (*A*, below) to be a *trans*-1,2-disubstituted cyclohexane where the 1,2-substituents are the two ends of a four-carbon chain, that is, $-CH_2CH_2CH_2CH_2-$.

Here we consider ring **B** *to be a four-carbon chain. It has no difficulty linking the 1- and 2-positions of ring* **A** *when its ends are diequatorial.*

However if ring **A** of *trans*-decalin were to be flipped into another chair conformation, the carbons of the other ring, **B**, would have to assume a 1,2-diaxial orientation. This is an impossible arrangement for the four carbons of the other ring to assume.

Here we have flipped ring **A** *into another chair conformation. This is an impossible arrangement for the four-carbon chain of ring* **B**, *however, because it cannot link the 1- and 2- positions when its ends are diaxial.*

(b) No, *cis*-decalin is conformationally "mobile" and can assume two conformations with each ring in a chair formation. In either conformation of *cis*-decalin the four carbons of one ring link the 1- and 2- positions of the other. They can do this easily because one end of the four-carbon chain is always axial and the other equatorial.

*3.39

cis-1,2-Dibromocyclohexane must exist in two equivalent conformations with one bromine equatorial and the other axial:

cis-3-Bromo-*trans*-4-bromo-1-*tert*-butylcyclohexane has a similar dipole moment (μ = 3.28 *D*), and since the presence of the large *tert*-butyl group ensures that the bromines are diequatorial, one can conclude that an equatorial-axial arrangement of the bromines and a diequatorial arrangement have roughly the same dipole moments.

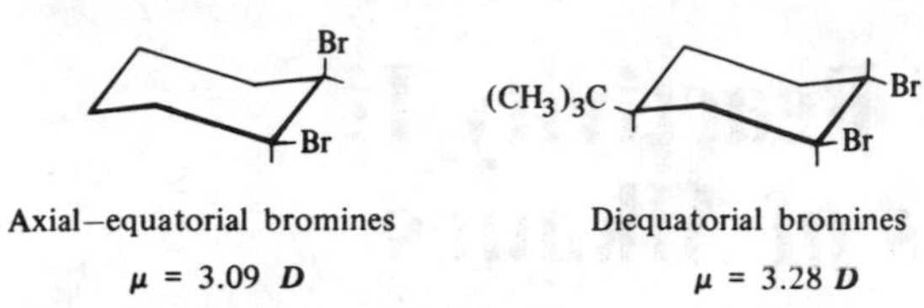

Axial–equatorial bromines $\mu = 3.09\ D$

Diequatorial bromines $\mu = 3.28\ D$

Thus were *trans*-1,2-dibromocyclohexane to exist primarily in a diequatorial conformation we would expect it to have a similar dipole moment (i.e., ~ 3.09 D). The fact that the dipole moment of *trans*-1,2-dibromo cyclohexane is much lower (2.11 D) suggests that the diaxial conformation (with $\mu \simeq 0$) is present in appreciable concentration.

$\mu \cong 3.09\ D$ $\rightleftarrows$ $\mu \cong 0$

(b) Because of bromine's electronegativity the bromine atoms will be partially negatively charged. In the diequatorial conformation, the bromines are closer together and therefore they repel each other. In the diaxial conformation the bromines are farther apart.

CHAPTER FOUR

Chemical Reactivity I: Reactions of Alkanes and Cycloalkanes

4.1

(a) $H{-}H + Br{-}Br \longrightarrow 2\,H{-}Br$

(DH°=104) (DH°=46) → 2(DH°=87.5)

+ 150 kcal/mole is required for bond cleavage

−175 kcal/mole is evolved in bond formation

$\Delta H° = +150 - 175 = -25$ kcal/mole (exothermic)

(b) $CH_3CH_2{-}H + F{-}F \longrightarrow CH_3CH_2{-}F + H{-}F$

(DH°=98) (DH°=38) → (DH°=106) (DH°=136)

+ 136 kcal/mole −242 kcal/mole = $\Delta H°$ = −106 kcal/mole (exothermic)

(c) $CH_3CH_2{-}H + I{-}I \longrightarrow CH_3CH_2{-}I + H{-}I$

(DH°=98) (DH°=36) → (DH°=53.5) (DH°=71)

+ 134 kcal/mole −124.5 kcal/mole = $\Delta H°$ = +9.5 kcal/mole (endothermic)

(d) $CH_3{-}H + Cl{-}Cl \longrightarrow CH_3{-}Cl + HCl$

(DH°=104) (DH°=58) → (DH°=83.5) (DH°=103)

+ 162 kcal/mole − 186.5 kcal/mole = $\Delta H°$ = −24.5 kcal/mole (exothermic)

(e) $(CH_3)_3C{-}H + Cl{-}Cl \longrightarrow (CH_3)_3C{-}Cl + H{-}Cl$

(DH°=91) (DH°=58) → (DH°=78.5) (DH°=103)

+ 149 kcal/mole − 181.5 kcal/mole = $\Delta H°$ = −32.5 kcal/mole (exothermic)

(f) $(CH_3)_3C{-}H + Br{-}Br \longrightarrow (CH_3)_3C{-}Br + H{-}Br$

(DH°=91) (DH°=46) → (DH°=63) (DH°=87.5)

+ 137 kcal/mole − 150.5 kcal/mole = $\Delta H°$ = −13.5 kcal/mole (exothermic)

(g) $CH_3CH_2{-}CH_3 \longrightarrow CH_3CH_2\cdot + CH_3\cdot$

(DH°=85) → (DH°=0) (DH°=0)

+ 85 kcal/mole − 0 − 0 = $\Delta H°$ = +85 kcal/mole (endothermic)

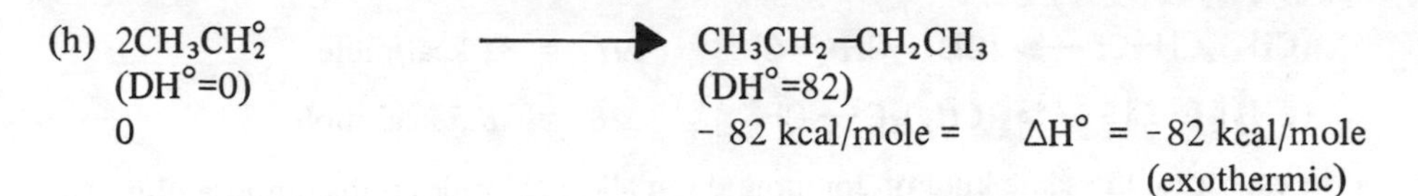

4.2

(a)

$\Delta H_2^\circ > \Delta H_1^\circ$; therefore isopropyl is more stable than ethyl.

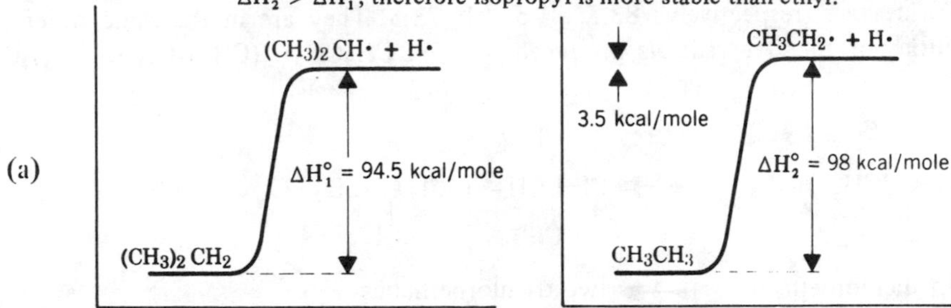

(b)

$\Delta H_3^\circ > \Delta H_2^\circ$; therefore ethyl is more stable than methyl.

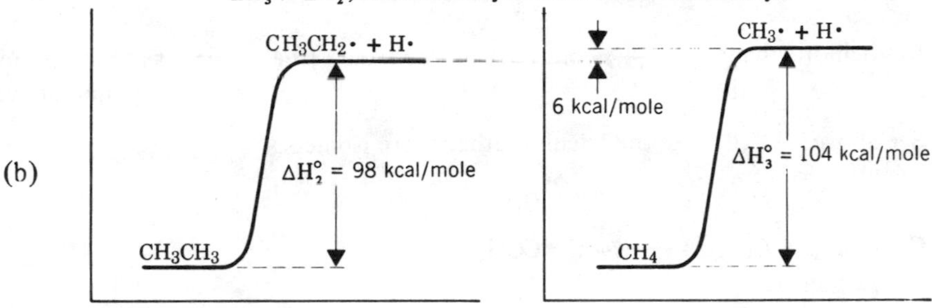

(c)

$\Delta H_2^\circ \simeq \Delta H_4^\circ$; therefore the two radicals have nearly equal stabilities.

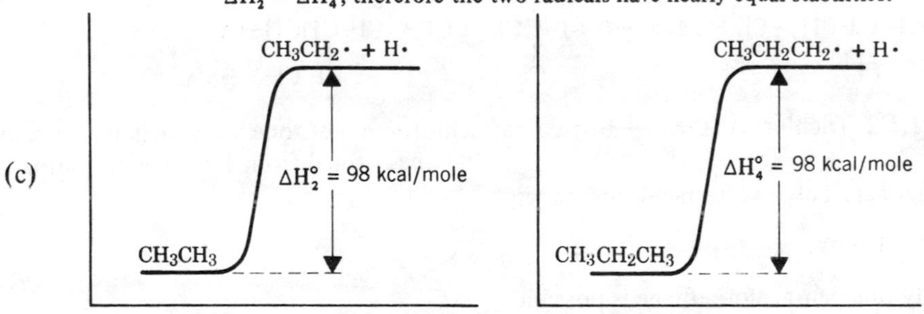

(d) The radicals produced are both primary radicals, and they are otherwise structurally similar, therefore they are of essentially equal stability.

4.3

Bond dissociation energies of the following C–Cl bonds are:

$CH_3{-}Cl \longrightarrow CH_3\cdot + Cl\cdot$ $\qquad \Delta H^\circ = 83.5$ kcal/mole

$CH_3CH_2{-}Cl \longrightarrow CH_3CH_2\cdot + Cl\cdot$ $\qquad \Delta H^\circ = 81.5$ kcal/mole

$$(CH_3)_2CH{-}Cl \longrightarrow (CH_3)_2CH\cdot + Cl\cdot \qquad \Delta H^\circ = 81 \text{ kcal/mole}$$

$$(CH_3)_3C{-}Cl \longrightarrow (CH_3)_3C\cdot + Cl\cdot \qquad \Delta H^\circ = 78.5 \text{ kcal/mole}$$

Since in each case the same kind of compound (an alkyl chloride) is decomposed into the same kinds of products (an alkyl free radical and a chlorine atom), it follows that the energy required (ΔH°) is a measure of the instability of the radical relative to the alkyl halide. In other words, the less stable the free radical, the more energy will be required to break the bond between it and the chlorine atom. Bond dissociation energies for these alkyl chlorides are, respectively, 83.5, 81.5, 81, 78.5. They are in the same order as the stabilities of the free radicals produced: $CH_3\cdot < CH_3CH_2\cdot < (CH_2)_2CH\cdot < (CH_3)_3C\cdot$

4.4

(a) 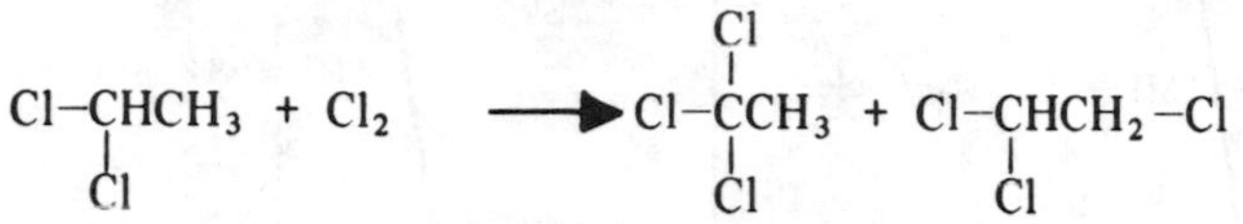 $Cl{-}\underset{\displaystyle Cl}{\underset{|}{C}}HCH_3 + Cl_2 \longrightarrow Cl{-}\overset{\displaystyle Cl}{\overset{|}{\underset{\displaystyle Cl}{\underset{|}{C}}}}CH_3 + Cl{-}\underset{\displaystyle Cl}{\underset{|}{C}}HCH_2{-}Cl$

1,1-dichloroethane → two trichloroethanes

(b) 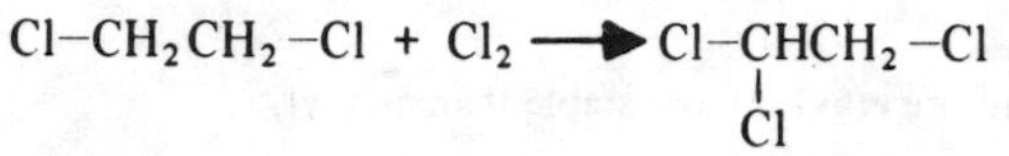 $Cl{-}CH_2CH_2{-}Cl + Cl_2 \longrightarrow Cl{-}\underset{\displaystyle Cl}{\underset{|}{C}}HCH_2{-}Cl$

1,2-dichloroethane → one trichloroethane (the same as the second compound above)

(c) See above (d) Yes, the trichloroethanes are isomers.

(e) 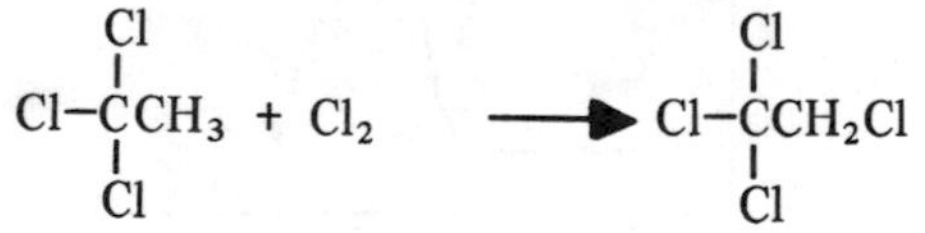 $Cl{-}\overset{\displaystyle Cl}{\overset{|}{\underset{\displaystyle Cl}{\underset{|}{C}}}}CH_3 + Cl_2 \longrightarrow Cl{-}\overset{\displaystyle Cl}{\overset{|}{\underset{\displaystyle Cl}{\underset{|}{C}}}}CH_2Cl$

1,1,1–trichloroethane → one tetrachloroethane

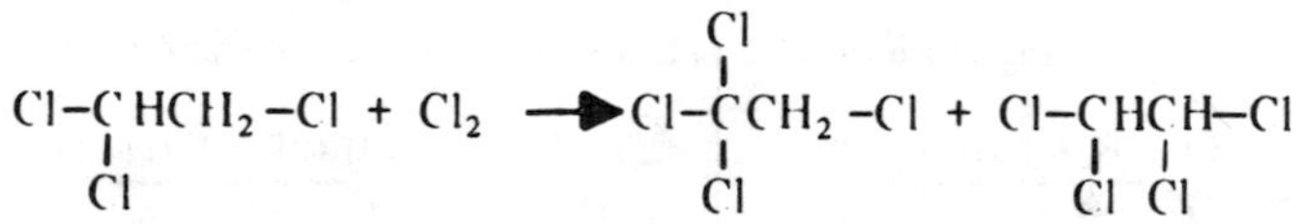

$Cl{-}\underset{\displaystyle Cl}{\underset{|}{C}}HCH_2{-}Cl + Cl_2 \longrightarrow Cl{-}\overset{\displaystyle Cl}{\overset{|}{\underset{\displaystyle Cl}{\underset{|}{C}}}}CH_2{-}Cl + Cl{-}\underset{\displaystyle Cl}{\underset{|}{C}}H\underset{\displaystyle Cl}{\underset{|}{C}}H{-}Cl$

1,1,2 trichloroethane → two tetrachloroethanes, one of which is the same as the one formed from 1,1,1-trichloroethane.

Thus there are only two tetrachloroethanes

(f) See (c) above.

(g) Only one pentachloroethane is possible:

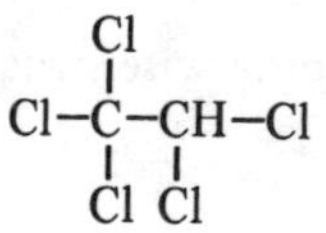

$$Cl{-}\overset{\displaystyle Cl}{\overset{|}{\underset{\displaystyle Cl}{\underset{|}{C}}}}{-}\underset{\displaystyle Cl}{\underset{|}{C}}H{-}Cl$$

4.5

(a) $CH_3CH_2CH_2Cl$ and $CH_3CHClCH_3$

(b) Not from boiling points alone.

(c) $CH_3CH_2CH_2Cl + Cl_2 \longrightarrow CH_3CH_2CHCl_2$
(b. p. 46.6°)

$+ CH_3CHClCH_2Cl + ClCH_2CH_2CH_2Cl$

three isomers with the formula $C_3H_6Cl_2$

$CH_3CHClCH_3 + Cl_2 \longrightarrow CH_3CHClCH_2Cl + CH_3CCl_2CH_3$ — two isomers with the formula $C_3H_6Cl_2$
(b. p. 36.5°)

The number of isomers produced in each reaction allows us to assign the structures without ambiguity.

(d) See (c) above

4.6

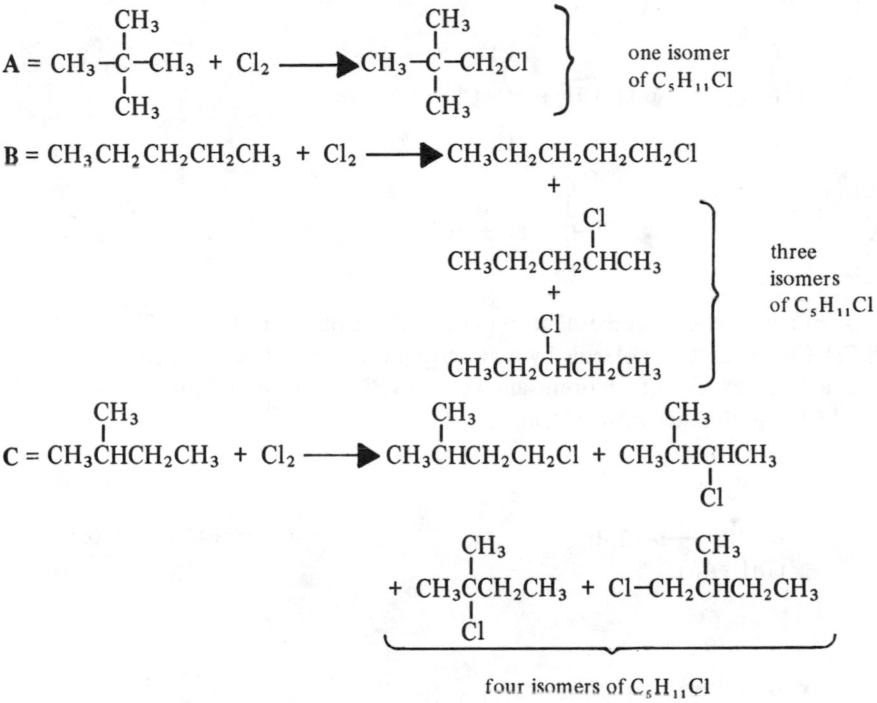

$A = CH_3C(CH_3)_2CH_3 + Cl_2 \longrightarrow CH_3C(CH_3)_2CH_2Cl$ — one isomer of $C_5H_{11}Cl$

$B = CH_3CH_2CH_2CH_2CH_3 + Cl_2 \longrightarrow CH_3CH_2CH_2CH_2CH_2Cl + CH_3CH_2CH_2CHClCH_3 + CH_3CH_2CHClCH_2CH_3$ — three isomers of $C_5H_{11}Cl$

$C = CH_3CH(CH_3)CH_2CH_3 + Cl_2 \longrightarrow CH_3CH(CH_3)CH_2CH_2Cl + CH_3CH(CH_3)CHClCH_3 + CH_3CCl(CH_3)CH_2CH_3 + Cl{-}CH_2CH(CH_3)CH_2CH_3$

four isomers of $C_5H_{11}Cl$

4.7

(a) $:\ddot{Cl}:\ddot{Cl}: \xrightarrow[\text{light}]{\text{heat or}} 2\ :\ddot{Cl}\cdot$

$:\ddot{Cl}\cdot + H:CH_3 \longrightarrow H:\ddot{Cl}: + \cdot CH_3$

$H_3C\cdot + :\ddot{Cl}:\ddot{Cl}: \longrightarrow H_3C:\ddot{Cl}: + \cdot\ddot{Cl}:$

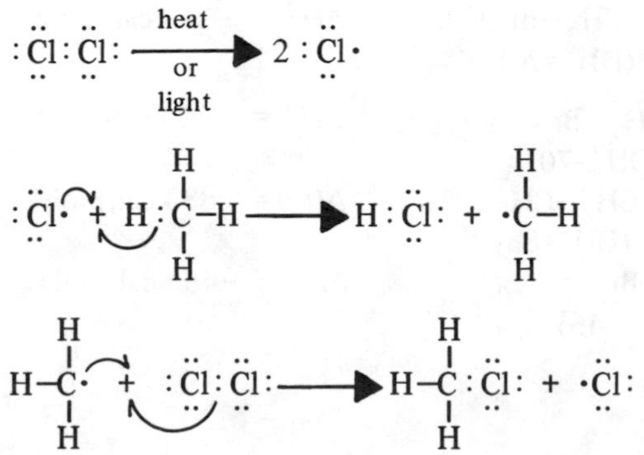

$$:\ddot{Cl}\cdot + H:\overset{H}{\underset{H}{C}}-Cl \longrightarrow H:\ddot{Cl}: + \cdot\overset{H}{\underset{H}{C}}-Cl$$

$$Cl-\overset{H}{\underset{H}{C}}\cdot + :\ddot{Cl}:\ddot{Cl}: \longrightarrow Cl-\overset{H}{\underset{H}{C}}:\ddot{Cl}: + \cdot\ddot{Cl}:$$

$$:\ddot{Cl}\cdot + H:\overset{H}{\underset{Cl}{C}}-Cl \longrightarrow H:\ddot{Cl}: + \cdot\overset{H}{\underset{Cl}{C}}-Cl$$

$$Cl-\overset{H}{\underset{Cl}{C}}\cdot + :\ddot{Cl}:\ddot{Cl}: \longrightarrow Cl-\overset{H}{\underset{Cl}{C}}:\ddot{Cl}: + \cdot\ddot{Cl}:$$

$$:\ddot{Cl}\cdot + H:\overset{Cl}{\underset{Cl}{C}}-Cl \longrightarrow H:\ddot{Cl}: + \cdot\overset{Cl}{\underset{Cl}{C}}-Cl$$

$$Cl-\overset{Cl}{\underset{Cl}{C}}\cdot + :\ddot{Cl}:\ddot{Cl}: \longrightarrow Cl-\overset{Cl}{\underset{Cl}{C}}:\ddot{Cl}: + \cdot\ddot{Cl}:$$

(b) The use of a large excess of methane minimizes the probability that a $:\ddot{Cl}\cdot$ will attack a CH_3Cl, CH_2Cl_2, or $CHCl_3$ molecule by maximizing the concentration of $\ddot{C}H_4$. Similarly the use of a large excess of chlorine allows all of the chlorinated methanes–CH_3Cl, CH_2Cl_2, and $CHCl_3$–to react with chlorine.

4.8

Chain-initiating step	$Br-Br \longrightarrow 2\,Br\cdot$ (DH°=46)	ΔH° = +46 kcal/mole
Chain-propagating steps	$Br\cdot + CH_3-H \longrightarrow CH_3\cdot + HBr$ (DH°=104) (DH°=87.5)	ΔH° = +16.5 kcal/mole
	$CH_3\cdot + Br-Br \longrightarrow CH_3-Br + Br\cdot$ (DH°=46) (DH°=70)	ΔH° = −24 kcal/mole
Chain-terminating steps	$CH_3\cdot + Br\cdot \longrightarrow CH_3-Br$ (DH°=70)	ΔH° = −70 kcal/mole
	$CH_3\cdot + CH_3\cdot \longrightarrow CH_3-CH_3$ (DH°=88)	ΔH° = −88 kcal/mole
	$Br\cdot + Br\cdot \longrightarrow Br-Br$ (DH°=46)	ΔH° = −46 kcal/mole

4.9

It would be incorrect to include chain-initiation and chain-termination in the calculation of the overall value of ΔH° because those steps occur only rarely (once for hundreds or thousands of propagation steps).

4.10

(a) E_{act} would equal zero for reactions (3) and (5) because radicals (in the gas phase) are combining to form molecules.

(b) E_{act} would be greater than zero for reactions (1), (2), and (4) because all of these involve bond breaking.

(c) E_{act} equals ΔH° for reaction (1) because this is a gas-phase reaction in which a bond is broken homolytically but no bonds are formed.

4.11

(a) $CH_3\cdot + H{-}Cl \longrightarrow CH_3{-}H + Cl\cdot$ (DH°=103) (DH°=104)

$\Delta H^\circ = -1$ kcal/mole
$E_{act} = +2.8$ kcal/mole
(See text, p 144; E_{act} for the reverse reaction is 3.8 kcal/mole)

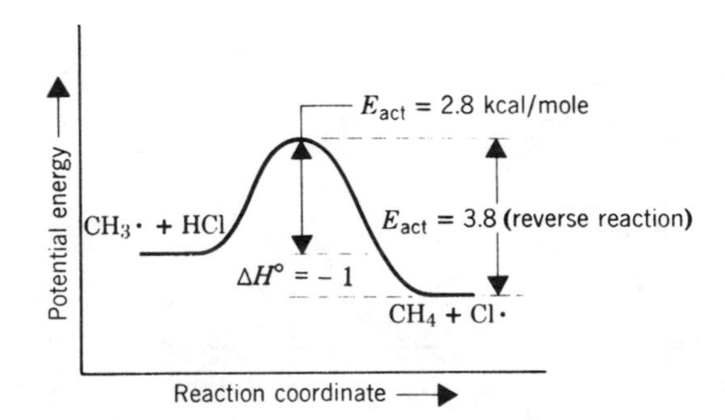

(b) $CH_3\cdot + H{-}Br \longrightarrow CH_3{-}H + Br\cdot$ (DH°=87.5) (DH°=104)

$\Delta H^\circ = -16.5$ kcal/mole
$E_{act} = +2.1$ kcal/mole

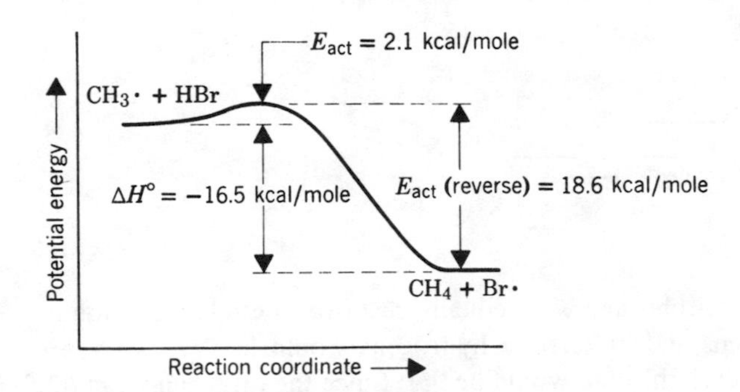

(c) $CH_3{-}CH_3 \longrightarrow 2\,CH_3\cdot$ (DH°=88)

$\Delta H° = +88$ kcal/mole
$E_{act} = +88$ kcal/mole

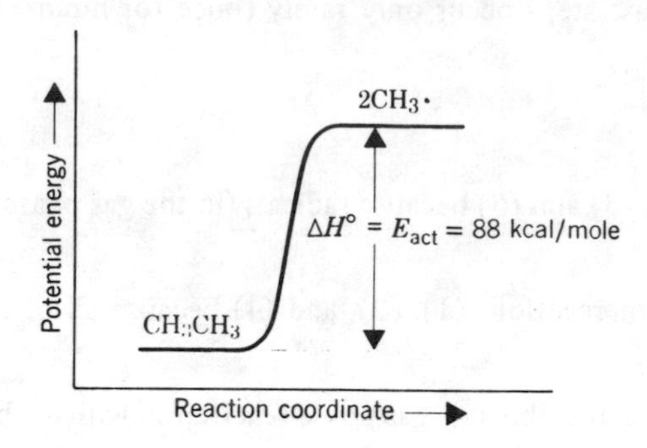

$\Delta H° = E_{act}$ for any reaction in which bonds are broken but no bonds are formed.

(d) $Br{-}Br \longrightarrow 2\,Br\cdot$ (DH°=46)

$\Delta H° = +46$ kcal/mole
$E_{act} = 46$ kcal/mole

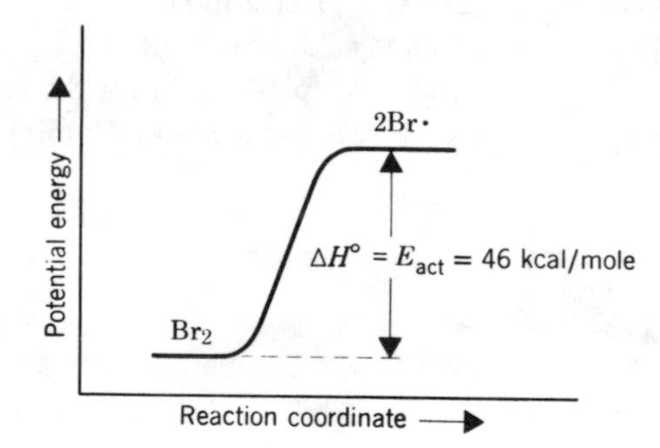

(e) $2\,Cl\cdot \longrightarrow Cl{-}Cl$ (DH°=58)

$\Delta H° = -58$ kcal/mole
$E_{act} = 0$ kcal/mole

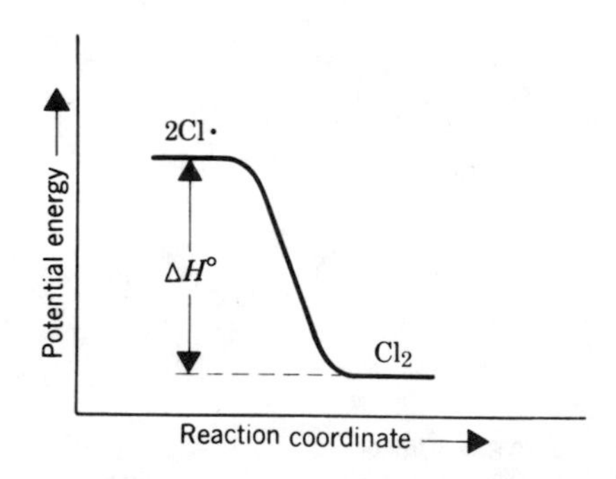

4.12

If all 10 hydrogen atoms of isobutane were equally reactive, the relative amounts of reaction at primary hydrogens and at tertiary hydrogens would be 9/1, i.e., the ratio of isobutyl chloride to *tert*-butyl chloride would be 9/1. Since the ratio is instead 62.5/37.5

(1.67), the tertiary hydrogen atom must be more reactive than the primary hydrogen atoms.

4.13

Laboratory preparation of alkyl halides by direct chlorination can be accomplished in good yield when all hydrogens in the alkane are equivalent. This is true of neopentane, and cyclopentane. (In these cases, the preparation would be practical only for monochlorination, where an excess of hydrocarbon would be employed, or for complete chlorination where an excess of chlorine would be used.)

4.14

(a) $Cl\cdot + CH_3CH_2{-}H \longrightarrow CH_3CH_2\cdot + H{-}Cl$
(DH°=98) (DH°=103)

$\Delta H^\circ = 98 - 103 = -5$ kcal/mole (exothermic)

(b) $Cl\cdot + (CH_3)_2CH{-}H \longrightarrow (CH_3)_2CH\cdot + H{-}Cl$
(DH°=94.5) (DH°=103)

$\Delta H^\circ = 94.5 - 103 = -8.5$ kcal/mole (exothermic)

(c) $Cl\cdot + CH_3CH_2CH_2{-}H \longrightarrow CH_3CH_2CH_2\cdot + H{-}Cl$
(DH°=98) (DH°=103)

$\Delta H^\circ = 98 - 103 = -5$ kcal/mole (exothermic)

4.15

The hydrogen abstraction steps in alkane fluorinations are always highly exothermic. Thus the transition states are even more reactant-like in structure and in energy than they are in alkane chlorinations. The type of C–H bond being broken (1°, 2°, or 3°) has practically no effect on the relative rates of the reactions.

4.16

(a) Homolysis is cleavage of a covalent bond in such a way that the electrons of the ruptured bond are divided equally between the atoms involved:

$$:\ddot{\underset{..}{Cl}}{-}\ddot{\underset{..}{Cl}}: \longrightarrow :\ddot{\underset{..}{Cl}}\cdot + \cdot\ddot{\underset{..}{Cl}}:$$

(b) Heterolysis is cleavage of a covalent bond in such a way that both electrons of the ruptured bond remain with one atom. Ions are formed:

$$H{-}\ddot{\underset{..}{Cl}}: \longrightarrow H^+ + :\ddot{\underset{..}{Cl}}:^-$$

(c) Bond dissociation energy (DH°) is the energy required to dissociate a covalent bond homolytically:

$$H{-}H \longrightarrow 2\,H\cdot \quad DH^\circ = 104 \text{ kcal/mole}$$

(d) A free radical is an atom or group that has an unpaired electron.

$:\ddot{\underset{..}{Br}}\cdot$ or $CH_3\cdot$

(e) A carbocation is an ion that has a trivalent carbon atom that bears a positive charge:

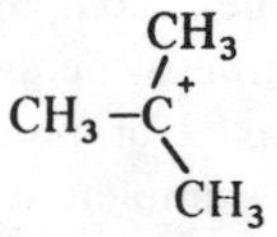

$$(CH_3)_3C^+$$

(f) A carbanion is an ion that has a trivalent carbon atom that bears an unshared electron pair and a negative charge.

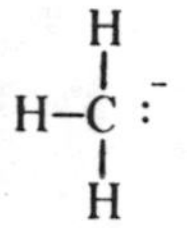

$$H_3C:^-$$

4.17

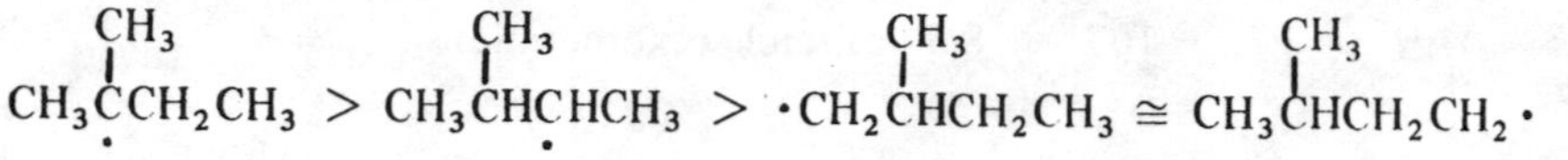

$$CH_3\dot{C}(CH_3)CH_2CH_3 > CH_3CH(CH_3)\dot{C}HCH_3 > \cdot CH_2CH(CH_3)CH_2CH_3 \cong CH_3CH(CH_3)CH_2CH_2\cdot$$

4.18

(a) 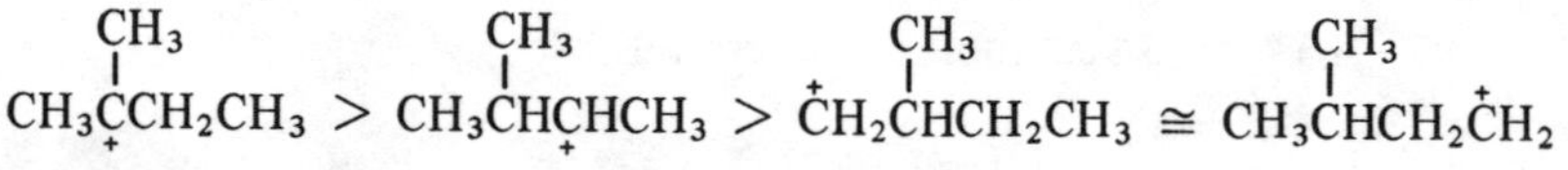 $$CH_3\overset{+}{C}(CH_3)CH_2CH_3 > CH_3CH(CH_3)\overset{+}{C}HCH_3 > \overset{+}{C}H_2CH(CH_3)CH_2CH_3 \cong CH_3CH(CH_3)CH_2\overset{+}{C}H_2$$

(b) 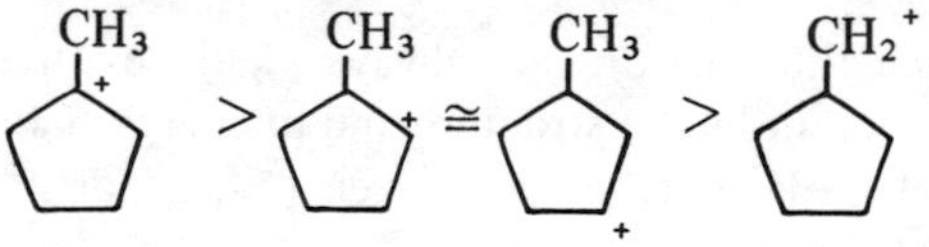 1-methylcyclopentyl cation (CH₃, + on C-1) > 2-methylcyclopentyl cation (+ on C-2) ≅ 3-methylcyclopentyl cation (+ on C-3) > cyclopentylmethyl cation (CH_2^+)

4.19

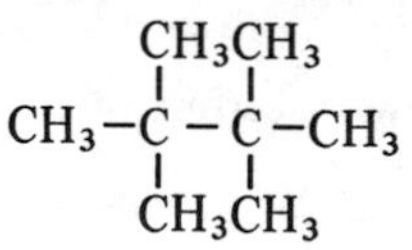

$$CH_3-C(CH_3)_2-C(CH_3)_2-CH_3$$

4.20

Six: CH_2Cl-cyclopentane, 1-chloro-1-methylcyclopentane, cis- and trans-1-chloro-2-methylcyclopentane (cis-trans isomers), cis- and trans-1-chloro-3-methylcyclopentane (cis-trans isomers)

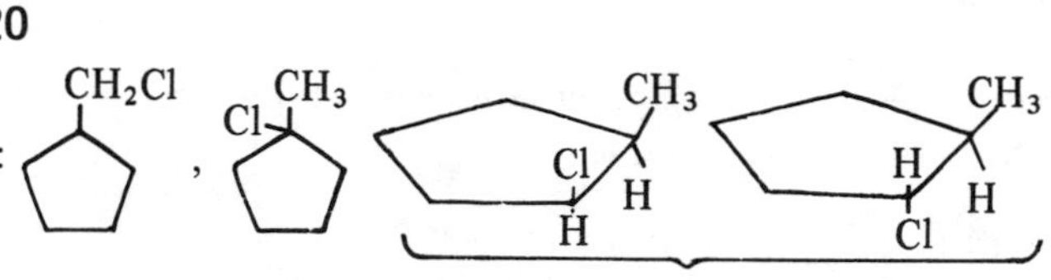

cis-trans isomers

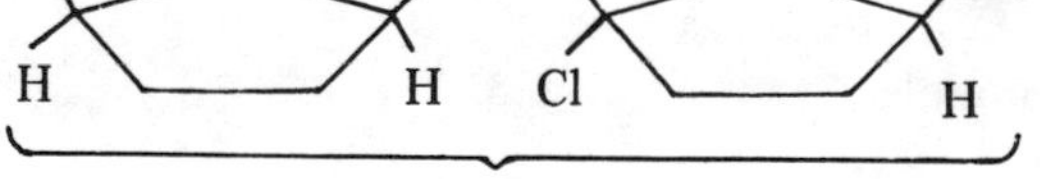

cis-trans isomers

4.21

Seven: 1,1-dichlorocyclohexane (Cl, Cl), 1,2-dichlorocyclohexane (Cl, Cl) (cis and trans),

1,3-dichlorocyclohexane (Cl, Cl) (cis and trans), and 1,4-dichlorocyclohexane (Cl, Cl) (cis and trans)

4.22

(a) $CH_3\overset{Br}{\underset{CH_3}{C}}CH_2CH_3$, because the tertiary hydrogen atom is much more reactive than either the primary or secondary hydrogen atoms.

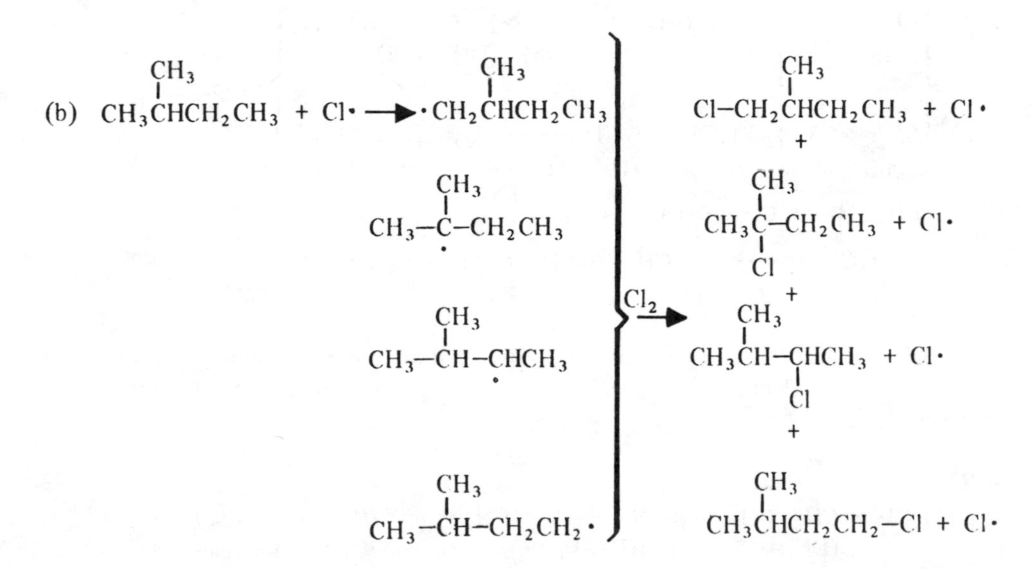

(c) Chlorine is more reactive than bromine, and is therefore less selective. See pages 153-156.

4.23

(a) $Cl_2 \longrightarrow 2\ Cl\cdot$ (b) $2\ Cl\cdot \longrightarrow Cl_2$

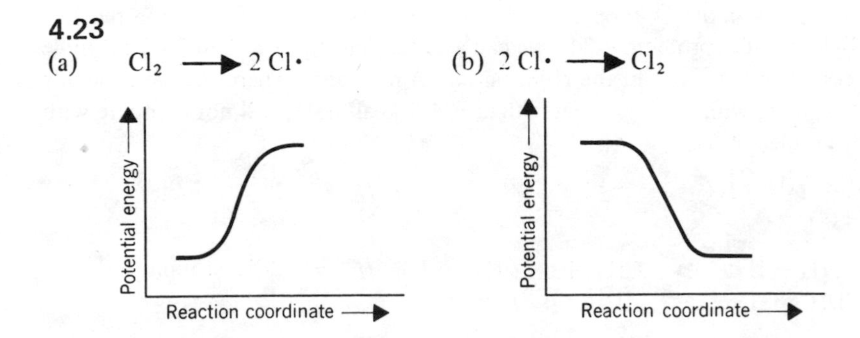

(c) $H\cdot + Cl_2 \longrightarrow HCl + Cl\cdot$ (d) $I\cdot + CH_4 \longrightarrow HI + CH_3\cdot$

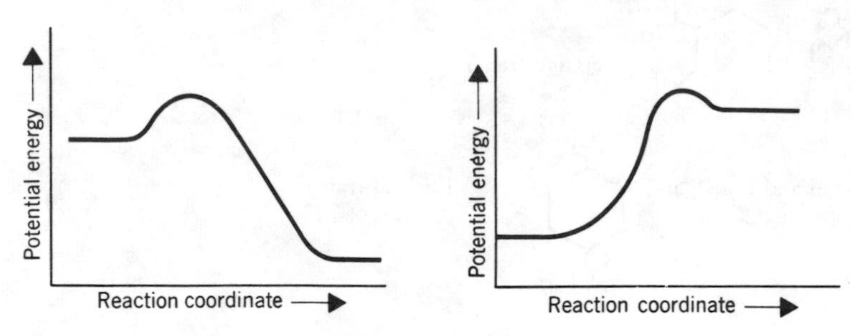

4.24

(a) $Cl_2 \xrightarrow[\text{light}]{\text{heat or}} 2\,Cl\cdot$ Chain-initiating step

(b) Chain-propagating steps:

$\Delta H^\circ = -5$ kcal/mole: $Cl\cdot + CH_3CH_2{-}H \longrightarrow H{-}Cl + CH_3CH_2\cdot$ (DH°=98) (DH°=103)

$\Delta H^\circ = -23.5$ kcal/mole: $CH_3CH_2\cdot + Cl_2 \longrightarrow CH_3CH_2{-}Cl + Cl\cdot$ (DH°=58) (DH°=81.5)

Chain-terminating steps:

$CH_3CH_2\cdot + Cl\cdot \longrightarrow CH_3CH_2Cl$

$2CH_3CH_2\cdot \longrightarrow CH_3CH_2CH_2CH_3$

$2Cl\cdot \longrightarrow Cl_2$

(c) The bond dissociation energy of the $CH_3CH_2{-}H$ bond (98 kcal/mole) is smaller than that of the $CH_3{-}H$ bond (104 kcal/mole), therefore ethane reacts with $Cl\cdot$ faster than methane does.

4.25

(a) (2) $Cl\cdot + CH_3{-}CH_3 \longrightarrow CH_3{-}Cl + CH_3\cdot$ (DH°=88) (DH°=83.5) $\Delta H^\circ = +4.5$ kcal/mole; $E_{act} > +4.5$ kcal/mole

(3) $CH_3\cdot + Cl_2 \longrightarrow CH_3{-}Cl + Cl\cdot$ (DH°=58) (DH°=83.5) $\Delta H^\circ = -25.5$ kcal/mole; E_{act} is small

In this reaction, step 2 is endothermic ($\Delta H = +4.5$ kcal/mole) and thus E_{act} must be greater than + 4.5 kcal/mole. Although we do not know the exact E_{act} of the reaction that yields ethyl chloride (problem 4.24), we can assume that it is less than 3.8 kcal/mole (E_{act} for the corresponding step in the chlorination of methane). Therefore we conclude that the reaction here, with an E_{act} greater than + 4.5 kcal/mole, will not compete with the reaction of problem 4.24.

(b) (1) $F{-}F \longrightarrow 2\,F\cdot$ (DH°=38) $\Delta H^\circ = +38$ kcal/mole

(2) $F\cdot + CH_3{-}CH_3 \longrightarrow CH_3{-}F + CH_3\cdot$ (DH°=88) (DH°=108) $\Delta H^\circ = -20$ kcal/mole; $E_{act} > 0$

(3) $CH_3\cdot + F{-}F \longrightarrow CH_3{-}F + F\cdot$ (DH°=38) (DH°=108) $\Delta H^\circ = -70$ kcal/mole; $E_{act} > 0$

Since the propagation steps are both highly exothermic it is possible for each E_{act} to be quite small, and therefore for the reaction to take place.

4.26

(a) $CH_3{-}H + F{-}F \longrightarrow CH_3\cdot + H{-}F + F\cdot$ $\Delta H° = +6$ kcal/mole
(DH°=104) (DH°=38) (DH°=136) $E_{act} > 6$ kcal/mole

$CH_3\cdot + F\cdot \longrightarrow CH_3{-}F$ $\Delta H° = -108$ kcal/mole
(DH°=108) $E_{act} = 0$

If E_{act} for the first step is not much greater than 6 kcal/mole, this mechanism is likely.

(b) $CH_3{-}H + Cl{-}Cl \longrightarrow CH_3\cdot + H{-}Cl + Cl\cdot$ $\Delta H° = +59$ kcal/mole
(DH°=104) (DH°=58) (DH°=103) $E_{act} \geq 59$ kcal/mole

$CH_3\cdot + Cl\cdot \longrightarrow CH_3{-}Cl$ $\Delta H° = -83.5$ kcal/mole
(DH°=83.5) $E_{act} = 0$

This mechanism is highly unlikely because the E_{act} for the first step must be ≥ 59 kcal/mole.

4.27

$$(CH_3)_3C{-}H + FSO_3H\cdot SbF_5 \longrightarrow \left[CH_3{-}\overset{CH_3\ \ CH_3}{C}\overset{\cdot\cdot}{\ \ }H\right]^+ \overset{-}{FSO_3\cdot SbF_5} \longrightarrow$$

$$CH_3\overset{+}{C}HCH_3 + CH_4 + \overset{-}{FSO_3\cdot SbF_5}$$

4.28

(a) $$\text{cyclo-}(CH_2CH_2CH_2) + HI \longrightarrow \left[CH_2{-}CH_2 \text{ bridged by } CH_2{-}{-}{-}H\right]^+ \ \overset{-}{I} \longrightarrow \overset{+}{C}H_2{-}CH_2{-}CH_2{-}H$$

(b) $CH_3CH_2\overset{+}{C}H_2 + \overset{-}{I} \longrightarrow CH_3CH_2CH_2{-}I$

4.29

(a) $CH_3{-}H$ DH° = 104, $CH_3CH_2{-}H$ DH° = 98 kcal/mole. (Recall that here, E_{act} = DH°.)

$CH_3CH_2{-}H$ bond rupture requires less energy, therefore spontaneous homolysis (cracking) occurs at a lower temperature.

(b) $CH_3{-}CH_3$ DH° = 88 kcal/mole = E_{act}

C–C bond rupture requires less energy than C–H bond rupture, therefore C–C bond rupture occurs more readily than $CH_3CH_2{-}H$ bond rupture.

(c) $CH_3CH_2{-}CH_2CH_3$ DH° = 82 kcal/mole = E_{act}

$CH_3CH_2CH_2{-}CH_3$ DH° = 85 kcal/mole = E_{act}

Here again the bond with the lower bond dissociation energy will undergo spontaneous homolysis (cracking) more readily.

4.30

(1) $CH_3CH_2CH_3 \rightarrow CH_3CH_2\cdot + CH_3\cdot$ DH° = 85 kcal/mole
(2) $CH_3CH_2CH_3 \rightarrow CH_3CH_2CH_2\cdot + H\cdot$ DH° = 98 kcal/mole
(3) $CH_3CH_2CH_3 \rightarrow CH_3\dot{C}HCH_3 + H\cdot$ DH° = 94.5 kcal/mole

(a) Since E_{act} is equal to DH°, we can assume that (1) is the most likely chain-initiating step.

(b) $CH_3\cdot + CH_3CH_2CH_3 \rightarrow CH_3{-}H + \cdot CH_2CH_2CH_3$ $\Delta H° = -6$ kcal/mole
(DH°=98) (DH°=104)

Since $\Delta H°$ is negative, E_{act} need not be large.

(c) $CH_3\cdot + CH_3CH_2CH_3 \rightarrow CH_4 + CH_3\dot{C}HCH_3$ $\Delta H° = -9.5$ kcal/mole
(DH°=94.5) (DH°=104)

On the basis of energy requirements, this is a likely alternative to step 1. On the basis of the probability factor, it is less likely because there are only two secondary hydrogen atoms compared with six primary hydrogen atoms.

4.31

(a) The valence shell of a carbocation contains only six electrons; the carbocation needs an additional electron pair to achieve a stable octet.

(b) 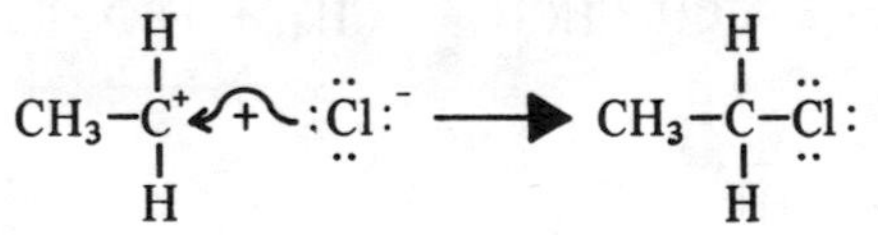 $CH_3{-}CH_2^{+} + :\ddot{Cl}:^{-} \rightarrow CH_3{-}CH_2{-}\ddot{Cl}:$

(c) 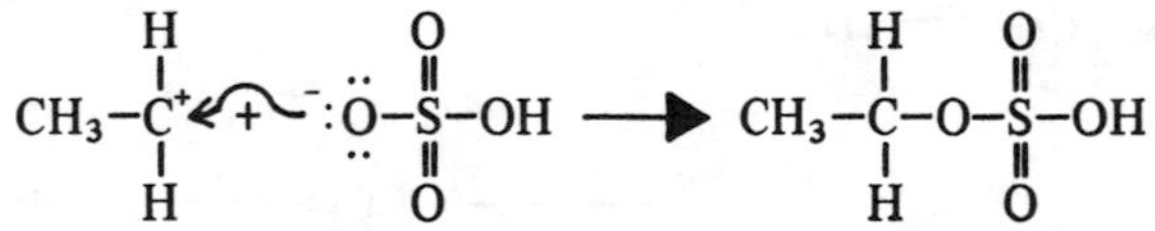 $CH_3{-}CH_2^{+} + {}^{-}:\ddot{O}{-}SO_2{-}OH \rightarrow CH_3{-}CH_2{-}O{-}SO_2{-}OH$

(d) 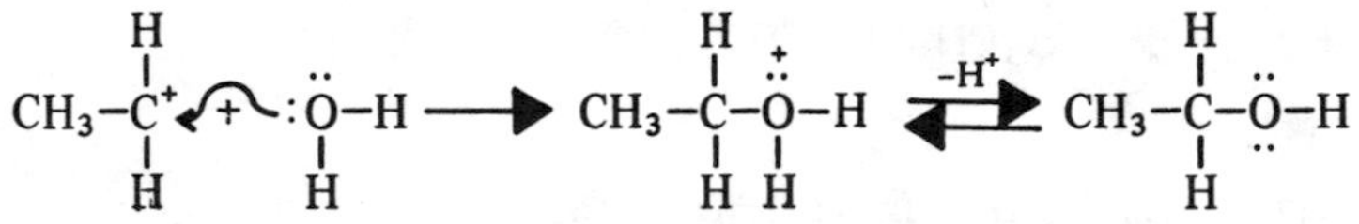 $CH_3{-}CH_2^{+} + :\ddot{O}H_2 \rightarrow CH_3{-}CH_2{-}\overset{+}{\ddot{O}}H_2 \underset{}{\overset{-H^{+}}{\rightleftharpoons}} CH_3{-}CH_2{-}\ddot{O}{-}H$

(e) Loss of a proton from a carbon adjacent to the carbon bearing the positive charge.

(f) 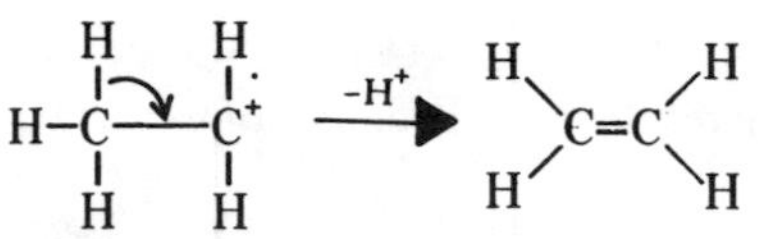 $H{-}CH_2{-}CH_2^{+} \xrightarrow{-H^{+}} H_2C{=}CH_2$

*4.32

(a) Oxygen-oxygen single bonds are especially weak, that is,

HO—OH DH° = 51 kcal/mole
$CH_3CH_2O{-}OCH_3$ DH° = 44 kcal/mole

This means that a peroxide will dissociate into free radicals at a relatively low temperature.

$$RO{-}OR \xrightarrow{100\text{-}200^\circ} 2RO\cdot$$

Oxygen-hydrogen single bonds, on the other hand, are very strong. (For HO–H, DH° = 119 kcal/mole.) This means that reactions like the following will be highly exothermic.

$$RO\cdot + R{-}H \longrightarrow RO{-}H + R\cdot$$

(b) (1) $(CH_3)_3CO{-}OC(CH_3)_3 \xrightarrow{heat} 2(CH_3)_3CO\cdot$
(2) $(CH_3)_3CO\cdot + R{-}H \longrightarrow (CH_3)_3COH + R\cdot$
} chain-initiating steps

(3) $R\cdot + Cl{-}Cl \longrightarrow R{-}Cl + Cl\cdot$
(4) $Cl\cdot + R{-}H \longrightarrow H{-}Cl + R\cdot$
} chain-propagating steps

***4.33**

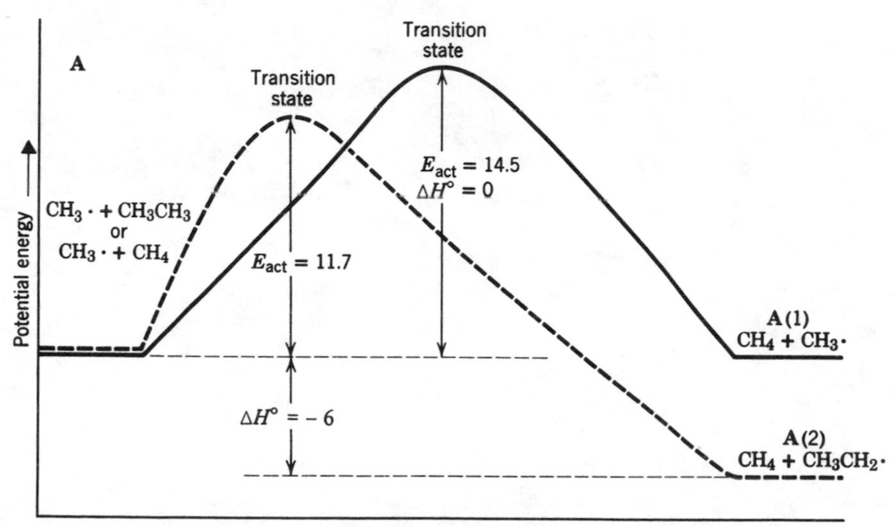

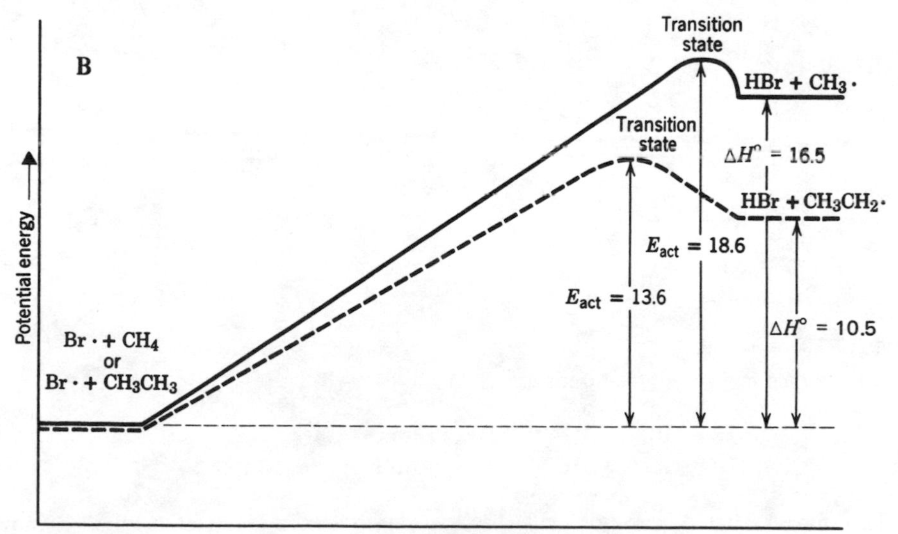

(b) Reaction **A** (2) since it is most exothermic.

(c) Reaction **B** (1) since it is most endothermic.

(d) Since $\Delta H^\circ = 0$, bond breaking should be approximately 50% complete.

(e) The reactions of set **B**.

(f) The difference in ΔH° simply reflects the difference in the C—H bond strengths of methane and ethane.

(g) Because the reactions of set **B** are highly endothermic the transition states show a strong resemblance to products in structure and *in energy*, and the products differ in energy by 6 kcal/mole. (In this instance, since the difference in E_{act} is five-sixths of the difference in ΔH°, we can estimate that bond breaking is about five-sixths complete when the transition states are reached.)

B

Special Topic

B.1

(a) $X \longrightarrow Y$, $K_{eq} = \frac{[Y]}{[X]} = 10$

Initial $[X] = 1.0$
Equilibrium $[Y] = a$
Equilibrium $[X] = 1.0 - a$

then $K_{eq} = 10 = \frac{a}{1.0 - a}$

$$10 - 10a = a$$
$$-11a = -10$$
$$a = \frac{10}{11} = 0.91 \text{ mole/liter}$$

At equilibrium, $[Y] = 0.91$ mole/liter,
$[X] = 0.09$ mole/liter,
and 91% of X is converted to product, Y.

(b) If $K_{eq} = 1$, $1 = \frac{a}{1.0 - a}$

$$1 - a = a$$
$$-2a = -1$$
$$a = 0.5$$

∴ At equilibrium, $[Y] = 0.5$ mole/liter,
$[X] = 0.5$ mole/liter,
and 50% of X is converted to product, Y.

(c) If $K_{eq} = 10^{-3}$, $10^{-3} = \frac{a}{1 - a}$

$$10^{-3} - 10^{-3}a = a$$
$$-1.001a = -10^{-3}$$
$$a = \frac{10^{-3}}{1.001} \cong 10^{-3}$$

At equilibrium, $[Y] = 10^{-3}$ mole/liter,
$[X] = 0.999$ mole/liter,
and 0.1% of X is converted to product, Y.

B.2

(a) The majority of the molecules (~99.99%) are in the chair form at equilibrium because the equilibrium, chair $\rightleftarrows$ boat, has a ΔG = -5 to -6 kcal/mole. See Table B.1, last column under -5.5 kcal/mole.

(b) For $\Delta G^\circ \sim -1.8$, Table B.1 tells us that 95% of ethylcyclohexane is in the equatorial form.

B.3

(a) $\Delta G^\circ = \Delta H^\circ - T\Delta S^\circ$

$\Delta G^\circ = -41{,}700 \text{ cal/mole} - 300\cancel{\text{deg}}(-26.6 \text{ cal/}\cancel{\text{deg}}\text{ mole})$

$\Delta G^\circ = -41{,}700 + 7980 = -33{,}720 \text{ cal/mole}$

or $\Delta G^\circ = -33.72 \text{ kcal/mole}$

(b) Yes, because a negative value of ΔG° tells us that the products are favored at equilibrium.

(c) No, a negative entropy tells us that the products are more ordered, and therefore less favored than the reactants.

(d) There are fewer degrees of freedom in the product molecule, ethene, than in the separate and independent molecules, ethyne and hydrogen.

CHAPTER FIVE

Chemical Reactivity II: An Introduction to Nucleophilic Substitution

5.1

(a) $CH_3CH_2-\ddot{\underset{..}{O}}-H$, (b) $CH_3CH_2-\ddot{\underset{..}{O}}:^-$, (c) $H-\ddot{N}(H)-H$,

(d) $CH_3-\ddot{N}(H)-H$, (e) $^-:C{\equiv}N:$, (f) $CH_3C(=\ddot{O}:)-\ddot{\underset{..}{O}}-H$, (g) $CH_3-C(=\ddot{O}:)-\ddot{\underset{..}{O}}:^-$,

(h) $H-C(=\ddot{O}:)-\ddot{\underset{..}{O}}-H$, (i) $H-C(=\ddot{O}:)-\ddot{\underset{..}{O}}:^-$, (j) $CH_3CH_2-\ddot{\underset{..}{S}}-H$,

(k) $CH_3CH_2-\ddot{\underset{..}{S}}:^-$, (l) $^-:\ddot{N}=\overset{+}{N}=\ddot{N}:^-$

5.2

[cyclopentane ring: C with CH_3 (up) and H; C with OH (up) and H]

, cis-3-methylcyclopentanol

5.3

(a) 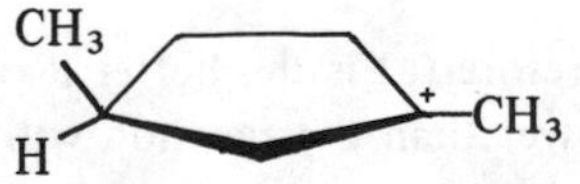 [cyclopentyl cation: C with CH_3 (up) and H; $\overset{+}{C}$ with $-CH_3$]

(b) 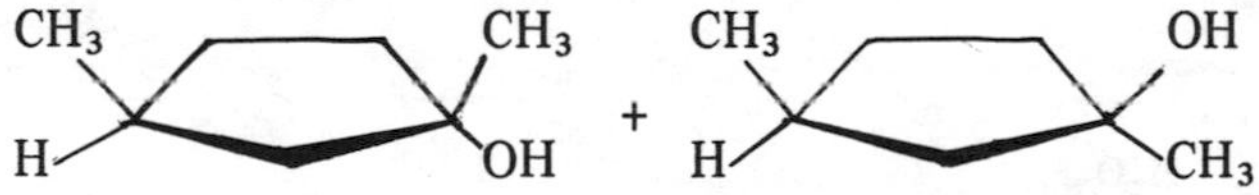 [ring: C with CH_3 and H; C with CH_3 (up) and OH] + [ring: C with CH_3 and H; C with OH (up) and CH_3]

5.4

(a) 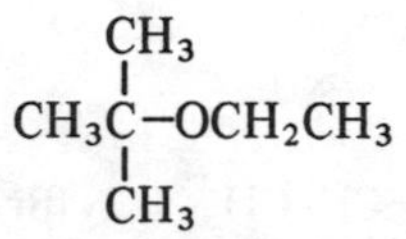 $CH_3C(CH_3)_2-OCH_2CH_3$

(b) 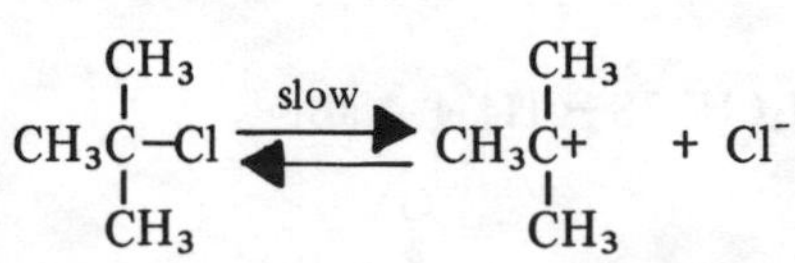 $CH_3C(CH_3)_2-Cl \underset{}{\overset{\text{slow}}{\rightleftharpoons}} CH_3\overset{+}{C}(CH_3)_2 + Cl^-$

$$(CH_3)_3C^+ + CH_3CH_2\ddot{O}H \underset{}{\overset{fast}{\rightleftharpoons}} (CH_3)_2C(CH_3)-\overset{+}{O}(H)CH_2CH_3$$

$$(CH_3)_3C-\overset{+}{O}(H)CH_2CH_3 + CH_3CH_2OH \overset{fast}{\rightleftharpoons} (CH_3)_3C-OCH_2CH_3 + CH_3CH_2\overset{+}{O}H_2$$

5.5

(a) NH_2^- , (b) RS^- , (c) PH_3

5.6

In the forward reaction, Cl^- is the leaving group; in the reverse reaction, OH^- would have

$$HO^- + CH_3-Cl \rightleftharpoons Cl^- + CH_3OH$$ (reverse arrow crossed out)

to be the leaving group. OH^- is very basic and therefore is such a poor leaving group that, for all practical purposes, the reverse reaction does not occur.

5.7

(a) Increase, (b) decrease, (c) increase
Note that (c) is an S_N1 reaction.

5.8

(a) $HOOCCH(NH_2)CH_2CH_2-\ddot{S}-CH_2$ — (ribose ring with O, Adenine, OH, OH)

(b) $HOOCCH(NH_2)CH_2CH_2-\ddot{S}:^-$

(c) The leaving group (a) is a weaker base than (b), therefore (a) is the better leaving group. The reaction with methionine would be much slower than the reaction with S-adenosylmethionine.

5.9

(a) $CH_3CH_2CH_2Br + NaOH \xrightarrow{H_2O} CH_3CH_2CH_2OH + NaBr$

(b) $CH_3CH_2CH_2Br + NaI \xrightarrow{H_2O} CH_3CH_2CH_2I + NaBr$

(c) $CH_3CH_2CH_2Br + CH_3CH_2ONa \xrightarrow{CH_3CH_2OH} CH_3CH_2CH_2-O-CH_2CH_3 + NaBr$

(d) $CH_3CH_2CH_2Br + CH_3SNa \xrightarrow{CH_3SH} CH_3CH_2CH_2-S-CH_3 + NaBr$

(e) $CH_3CH_2CH_2Br + CH_3\overset{O}{\overset{\|}{C}}-ONa \xrightarrow{CH_3COOH} CH_3CH_2CH_2-O-\overset{O}{\overset{\|}{C}}CH_3$

(f) $CH_3CH_2CH_2Br + NaN_3 \longrightarrow CH_3CH_2CH_2N_3 + NaBr$

(g) $CH_3CH_2CH_2Br + :N(CH_3)_3 \longrightarrow CH_3CH_2CH_2{-}\overset{+}{N}(CH_3)_3 \; Br^-$

(h) $CH_3CH_2CH_2Br + NaCN \longrightarrow CH_3CH_2CH_2CN + NaBr$

(i) $CH_3CH_2CH_2Br + NaSH \longrightarrow CH_3CH_2CH_2SH + NaBr$

5.10

(a) $CH_3CH_2CH_2CH_2Br$ because 1° halides are less hindered than 2° halides.

(b) $CH_3CH_2CH(Br)CH_3$ because 2° halides are less hindered than 3° halides.

(c) $CH_3CH_2CH_2Br$ because bromide ion is a better leaving group than chloride ion.

(d) $CH_3CH(CH_3)CH_2CH_2Br$ because it is less hindered than $CH_3CH_2CH(CH_3)CH_2Br$

(e) CH_3CH_2Cl because vinyl halides ($CH_2{=}CHCl$) are very unreactive.

5.11

(a) The second because CH_3O^- is a better nucleophile than CH_3OH.

(b) The second because SH^- is a better nucleophile than OH^-.

(c) The second because CH_3SH is a better nucleophile than CH_3OH.

(d) The second, CH_3S^- (2.0 molar) because the rate is proportional to $[CH_3S^-]$ as well as to $[CH_3CH_2I]$.

5.12

(a) The first because I^- is a better leaving group than Cl^-.

(b) The first because H_2O is a more polar solvent than CH_3OH.

(c) Both the same because $[CH_3O^-]$ does not affect the rate of an S_N1 reaction.

(d) The first because vinyl halides are unreactive.

5.13

(a) $H:^-$ is a very strong base and therefore is an extremely poor leaving group.

(b) $:CH_3^-$ is a very strong base and therefore is an extremely poor leaving group.

(c) $-\ddot{C}H_2^-$ is a very strong base and an extremely poor leaving group.

(d) With a relatively strong base like CN^-, elimination would predominate to yield $CH_2{=}C(CH_3)_2 + HCN + Br^-$. S_N2 attack cannot take place at the 3° carbon.

(e) Vinyl halides are unreactive in S_N1 and S_N2 reactions.

(f) CH_3O^- is a strong base and therefore a poor leaving group.

(g) $CH_3CH_2\overset{+}{O}H_2$ is a strong acid and would react with NH_3 to convert it to NH_4^+ which is not a nucleophile.

(h) $CH_3:^-$ would react with the acidic proton in CH_3CH_2OH to form $CH_4 + CH_3CH_2O^-$

5.14

$CH_3CHBrCH_3$ because a 2° halide is less likely to give an S_N2 reaction than a 1° halide and therefore an E2 reaction (dehydrohalogenation) would be more likely to predominate.

5.15

(a) $CH_3CH_2CH_2CH_2OCH_3$ (major) by S_N2; $CH_3CH_2CH{=}CH_2$ (minor) by E2.

(b) $CH_3CH_2CH_2CH_2{-}OC(CH_3)_3$ (minor) by S_N2; $CH_3CH_2CH{=}CH_2$ (major) by E2.

(c) $CH_3{-}O{-}C(CH_3)_3$ (only product) by S_N2.

(d) $CH_2{=}C(CH_3)_2$ (only product) by E2.

(e) CH_3, H (major) by E2; CH_3, H, H, OCH_3 (minor) by S_N2.

(f) CH_3, H, OCH_3, CH_3 + CH_3, H, CH_3, OCH_3 (major products) by S_N1;

CH_3, H and CH_3, H (minor products) by E1.

(g) $CH_3CH{=}CHCH_2CH_3$ (major) by E2; $CH_3CH_2\underset{|}{C}HCH_2CH_3$ (minor) by S_N2.
OC_2H_5

(h) $CH_2{=}CHCH_3$ (major) by E2; $CH_3\underset{|}{C}HCH_3$ (minor) by S_N2.
$OC(CH_3)_3$

5.16

Iodide ion is a good nucleophile and a good leaving group; it can rapidly convert an alkyl chloride or alkyl bromide into an alkyl iodide, and the alkyl iodide can then react rapidly with another nucleophile. With methyl bromide in water, for example, the following reaction can take place:

$$CH_3{-}Br \xrightarrow[\text{(slower)}]{H_2O \text{ alone}} CH_3OH_2^+ + Br^-$$

$$CH_3{-}Br \xrightarrow[\text{(faster)}]{H_2O \text{ containing } I^-} CH_3I \xrightarrow[\text{(faster)}]{H_2O} CH_3\overset{+}{O}H_2 + I^-$$

5.17

The rate of formation of *tert*-butyl alcohol does not increase with increasing $[OH^-]$ because the reaction is S_N1 and is therefore independent of $[OH^-]$. Increasing $[OH^-]$, however, increases the rate of the E2 reaction which consumes OH^- through the conversion of *tert*-butyl chloride into $CH_2{=}C(CH_3)_2$.

5.18

(a) Use a strong, hindered base such as $(CH_3)_3COK$ in a solvent of low polarity in order to bring about an E2 reaction.

(b) Here we want an S_N1 reaction. We use ethanol as the solvent *and as the nucleophile*, and we carry out the reaction at a low temperature so that elimination will be minimized.

5.19

(a) Backside attack by the nucleophile is prevented by the cyclic structure. (Notice, too, that the carbon bearing the leaving group is tertiary.)

(b) The bridged cyclic structure prevents the carbon bearing the leaving group from assuming the planar trigonal conformation required of a carbocation.

5.20

The nucleophile can be described by resonance structures that place a pair of electrons and a formal negative charge on either atom: $:\bar{C}:::N: \longleftrightarrow :C::\ddot{N}:^-$. Thus both atoms are nucleophilic.

5.21

(a) Since the halides are all primary, this is almost certainly an S_N2 reaction with ethanol acting as the nucleophile.

$$C_2H_5\ddot{O}H + \underset{H}{\overset{R}{H\cdots C}}{-}Br \xrightarrow[-HBr]{} C_2H_5O{-}\underset{H}{\overset{R}{C}}{\cdots}H$$

(b) Increasing the size of the R group increases steric hindrance to the approaching ethanol molecule and decreases the rate of reaction.

5.22

(a) This is another example of the relation between reactivity and selectivity that we first encountered in Chapter 4: generally speaking, highly reactive species are relatively unselective while less reactive species are more selective. In an S_N1 reaction the species that reacts with the nucleophile is a *carbocation*—a species that is electron deficient and thus is *highly reactive*. A carbocation, therefore, shows little tendency to discriminate between weak and strong nucleophiles—most often it simply reacts with the first nucleophile that it encounters. In S_N2 reactions, on the other hand, the species that reacts with the nucleophile is an alkyl halide or an alkyl tosylate. Such compounds are far less reactive toward nucleophiles than carbocations and they show much greater nucleophilic selectivities. An alkyl halide molecule, for example, might collide with a weak nucleophile thousands of times before a reaction takes place because few of the collisions will have sufficient energy to allow the weak nucleophile to displace the leaving group. On the other hand, an alkyl halide molecule might collide with a strong nucleophile only a few times before a collison leads to a reaction. This will be true because the strong nucleophile is better able to displace the leaving group and therefore a larger fraction of collisons will have sufficient energy to be fruitful.

(b) The reaction of $CH_3CH_2CH_2CH_2Cl$ is an S_N2 reaction and thus $CH_3CH_2CH_2CH_2Cl$ discriminates very effectively between the strongly nucleophilic CN^- ions and the weakly nucleophilic solvent molecules. By contrast, the reaction of $(CH_3)_3CCl$ is an S_N1 reaction and the carbocation that is formed shows little tendency to discriminate between solvent molecules and CN^- ions. Since solvent molecules are present in a much higher concentration the major product is $(CH_3)_3C{-}OCH_2CH_3$.

5.23

The rate-determining step in the S_N1 reaction of *tert*-butyl bromide is the following:

$$(CH_3)_3C{-}Br \underset{\times}{\overset{\text{slow}}{\rightleftharpoons}} (CH_3)_3C^+ + Br^-$$

$$(CH_3)_3C^+ \xrightarrow{H_2O} (CH_3)_3COH_2^+$$

$(CH_3)_3C^+$ is so unstable that it reacts almost immediately with one of the surrounding water molecules and, for all practical purposes, no reverse reaction with Br^- takes place. Adding a common ion (Br^- from NaBr) therefore, has no effect on the rate.

Because the $(C_6H_5)_2CH^+$ cation is more stable, a reversible first step occurs and

$$(C_6H_5)_2CHBr \rightleftharpoons (C_6H_5)_2CH^+ + Br^-$$

$$(C_6H_5)_2CH^+ \xrightarrow{H_2O} (C_6H_5)_2CHOH_2^+$$

adding a common ion (Br^-) slows the overall reaction by increasing the rate at which $(C_6H_5)_2CH^+$ is converted back to $(C_6H_5)_2CHBr$.

5.24

Two different mechanisms are involved. $(CH_3)_3CBr$ reacts by an S_N1 mechanism and apparently this reaction takes place fastest. The other three alkyl halides react by an S_N2 mechanism and their reactions are slower because the nucleophile (H_2O) is weak. The reaction rates of CH_3Br, CH_3CH_2Br, and $(CH_3)_2CHBr$ are affected by the steric hindrance and thus their order of reactivity is $CH_3Br > CH_3CH_2Br > (CH_3)_2CHBr$.

CHAPTER SIX

Alkenes: Structure and Synthesis

6.1

(a) 2-Methyl-2-butene

(b) *Cis* - 4-octene

(c) 1-Bromo-2-methylpropene

(d) 4-Methylcyclohexene

6.2

(a) CH_3CH_2 CH_2CH_3 C=C H H

(b)

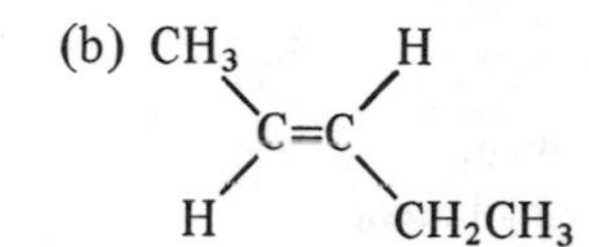

(c) CH_2CH_3

(d) $CH{=}CH_2$

(e) $CH_2{=}CHCH_2C(CH_3)_2CH_2CH_3$

6.3

Absorption of a photon of the correct frequency can excite a π electron into an anti-bonding orbital. Such an orbital has a nodal plane between the carbon atoms, thus rotation about the C–C bond can occur (see page 210).

6.4

(a) C_6H_{14} = formula of alkane
$\underline{C_6H_{12}}$ = formula of 2-hexene

H_2 = difference = 1 pair of hydrogens.
Index of hydrogen deficiency = 1

(b) C_6H_{14} = formula of alkane
$\underline{C_6H_{12}}$ = formula of methylcyclopentane

H_2 = difference = 1 pair of hydrogens.
Index of hydrogen deficiency = 1

(c) No, all isomers of C_6H_{12}, for example, have the same index of hydrogen deficiency.

(d) No.

(e) C_nH_{2n+2} = formula of alkane
C_nH_{2n-2} = formula of alkyne

H_4 = difference = 2 pairs of hydrogens.
Index of hydrogen deficiency = 2

(f) $C_{10}H_{22}$ (alkane)
$C_{10}H_{16}$ (compound)

H_6 = difference
Index of hydrogen deficiency = 3

The structural possibilities are thus

3 double bonds
1 double bond and one triple bond
2 double bonds and 1 ring
1 double bond and 2 rings
3 rings
1 triple bond and one ring

6.5

(a) $C_{15}H_{32}$ = formula of alkane
$C_{15}H_{24}$ = formula of zingiberene
H_8 = difference = 4 pairs of hydrogens

Index of hydrogen deficiency = 4

(b) Since one mole of zingiberene absorbs three moles of hydrogen, one molecule of zingiberene must contain three double bonds. (We are assuming here that molecules of zingiberene do not contain any triple bonds.)

(c) If a molecule of zingiberene has three double bonds and an index of hydrogen deficiency equal to 4, it must have one ring. (The structural formula for zingiberene can be found in Problem E.1.)

6.6

(a) $CH_2{=}C(CH_3)CH_3 + H_2 \longrightarrow CH_3CH(CH_3)CH_3$ (isobutane)

(b) $C_4H_8 + 6O_2 \longrightarrow 4CO_2 + 4H_2O$

(c) Yes, because the same molar amounts of the same combustion products are formed from all of the C_4H_8 isomers.

(d) $\begin{matrix} CH_2-CH_2 \\ | \quad\; | \\ CH_2-CH_2 \end{matrix}$ (cyclobutane) and CH_2—CH_2—$CHCH_3$ ring (methylcyclopropane

(e) Yes. (See answer to (c)).

6.7

(a) *cis*-2-Heptene, (b) *trans*-2-heptene, (c) 2-methyl-2-hexene, (d) 2,3-dimethyl-2-pentene

In (a), (c), and (d), the more highly substituted alkene is more stable. In (b), *trans* is more stable than *cis*.

6.8

You could use heats of hydrogenation to determine the relative stabilities of pairs (a) and (b). You would be required to use heats of combustion for pairs (c) and (d) because the members in pairs (c) and (d) give different alkanes on hydrogenation.

6.9

(a) 2-Butene, the more highly substituted alkene. (b) *trans*-2-Butene.

6.10

(a) OH^-, a strong base and an extremely poor leaving group. (b) The acid catalyst reacts with the alcohol to form the protonated alcohol $R\overset{+}{O}H_2$. When this ion undergoes dehydration, the leaving group is a weakly basic H_2O molecule—a much better leaving group.

6.11

(a) No.

(b) Protonation followed by loss of water gives a secondary carbocation which can lose a proton two different ways:

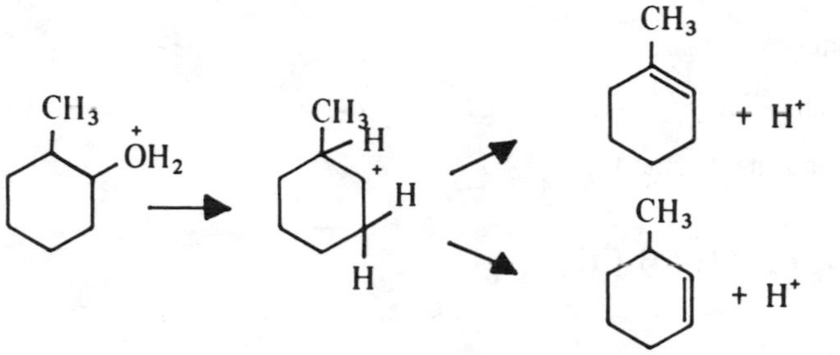

CH3 / $\overset{+}{O}H_2$ (2-methylcyclohexyloxonium) → CH3, H, +, H, H (secondary carbocation) → CH3 (1-methylcyclohexene) + H^+ ; → CH3 (3-methylcyclohexene) + H^+

(c) 1-Methylcyclohexene is the major product because it is the more stable (more highly substituted) alkene.

6.12

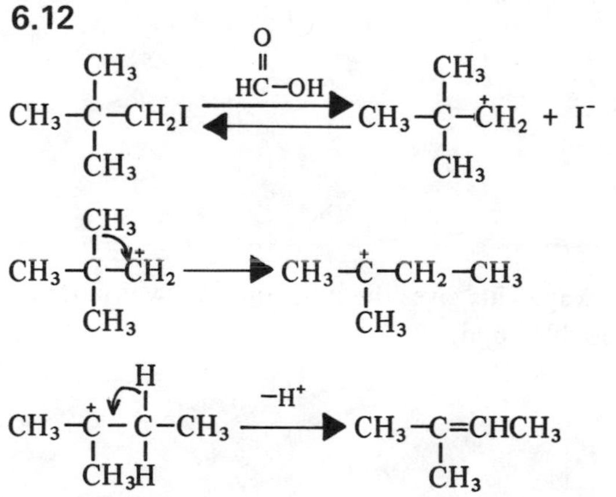

$$CH_3-\underset{CH_3}{\overset{CH_3}{C}}-CH_2I \underset{}{\overset{HCOOH}{\rightleftharpoons}} CH_3-\underset{CH_3}{\overset{CH_3}{C}}-\overset{+}{C}H_2 + I^-$$

$$CH_3-\underset{CH_3}{\overset{CH_3}{C}}-\overset{+}{C}H_2 \longrightarrow CH_3-\underset{CH_3}{\overset{+}{C}}-CH_2-CH_3$$

$$CH_3-\underset{CH_3}{\overset{+}{C}}-\underset{H}{\overset{H}{C}}-CH_3 \xrightarrow{-H^+} CH_3-\underset{CH_3}{C}=CHCH_3$$

6.13

(a) One must use the lower number to designate the location of the double bond.

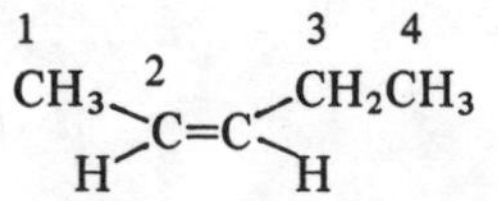

$\overset{1}{CH_3}\overset{2}{C}(H)=\overset{3}{C}(H)\overset{4}{CH_2}CH_3$ *cis*-2-pentene (not *cis*-3-pentene)

(b) One must select the longest chain as the base name.

1 4
$\overset{2}{} \overset{3}{}$
CH_3 CH_3
$\quad C{=}C$
CH_3 CH_3

2,3-dimethylbutene
(not 1,1,2,2-tetramethylethene)

(c) One must number the ring so as to give the carbons of the double bond numbers 1 and 2 *and to give the substituent the lower possible* number.

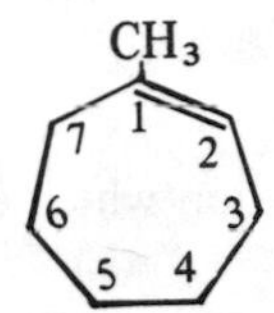

1-methylcycloheptene
(not 2-methylcycloheptene)

(d) One must select the longest chain.

1 2 3 4 5 6 7 8
$CH_3CH{=}CHCH_2CH_2CH_2CH_2CH_3$

2-octene
(not 1-methylheptene)

(e) One must number the chain from the other end. This gives the double bond the same number but it gives the methyl group a *lower* number.

$$\begin{array}{c} \qquad\quad CH_3 \\ \qquad\quad | \\ CH_3CH{=}CCH_3 \end{array}$$
4 3 2 1

2-methyl-2-butene
(not 3-methyl-2-butene)

(f) One must number the ring the other way. This gives the substituents lower numbers while retaining positions 1 and 2 for the double bond.

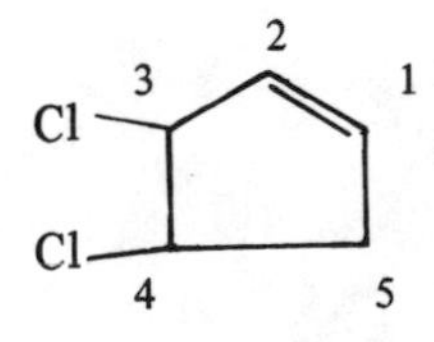

3,4-dichlorocyclopentene
(not 4,5-dichlorocyclopentene)

6.14

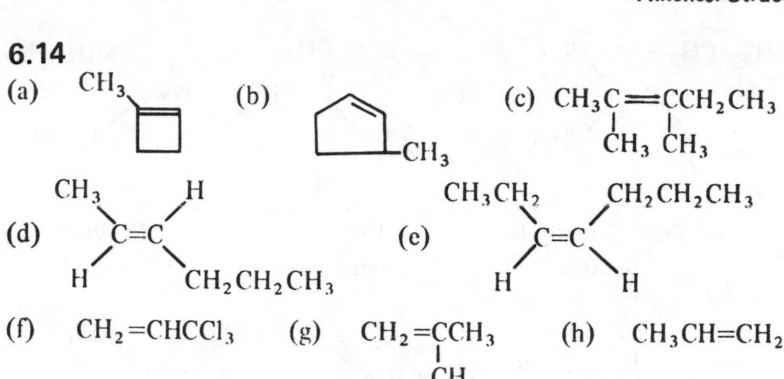

(i) $CH_2{=}CHCH_2CHCH_3$ (with a cyclopentyl group on C-4)

(j) $CH_2{=}CH$–(cyclopropyl)

6.15

(a) $CH_2{=}CHCH_2CH_2CH_3$ — 1-Pentene

$(CH_3)HC{=}CH(CH_2CH_3)$ — *Cis* -2-pentene

$(CH_3)HC{=}CH(CH_2CH_3)$ — *Trans* -2-pentene

$(CH_3)_2C{=}CH(CH_3)$ — 2-Methyl-2-butene

$CH_2{=}C(CH_3)CH_2CH_3$ — 2-Methyl-1-butene

$CH_2{=}CH{-}CH(CH_3)CH_3$ — 3-Methyl-1-butene

(b) $CH_2{=}CHCH_2CH_2CH_2CH_3$ — 1-Hexene

$(CH_3)HC{=}CH(CH_2CH_2CH_3)$ — *Cis* -2-hexene

$(CH_3)HC{=}CH(CH_2CH_2CH_3)$ — *Trans* -2-hexene

$(CH_3CH_2)HC{=}CH(CH_2CH_3)$ — *Cis* -3-hexene

$(CH_3CH_2)HC{=}CH(CH_2CH_3)$ — *Trans* -3-hexene

$CH_2{=}C(CH_3)CH_2CH_2CH_3$ — 2-Methyl-1-pentene

$CH_2{=}CHCH(CH_3)CH_2CH_3$ — 3-Methyl-1-pentene

$CH_2{=}CHCH_2CH(CH_3)CH_3$ — 4-Methyl-1-pentene

$CH_3C(CH_3){=}CHCH_2CH_3$ — 2-Methyl-2-pentene

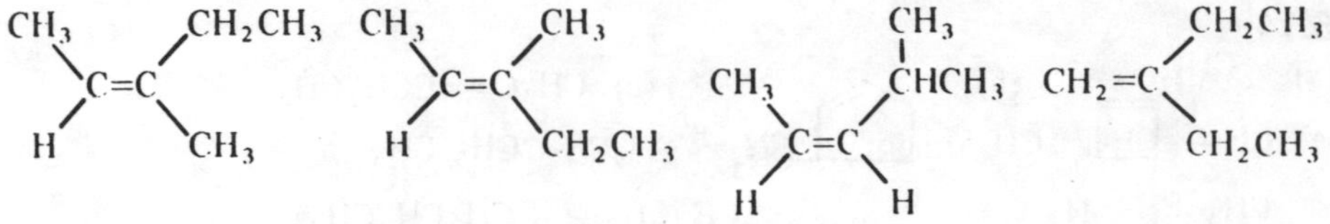

Cis-3-methyl-2-pentene*

Trans-3-methyl-2-pentene*

Cis-4-methyl-2-pentene

2-Ethyl-1-butene

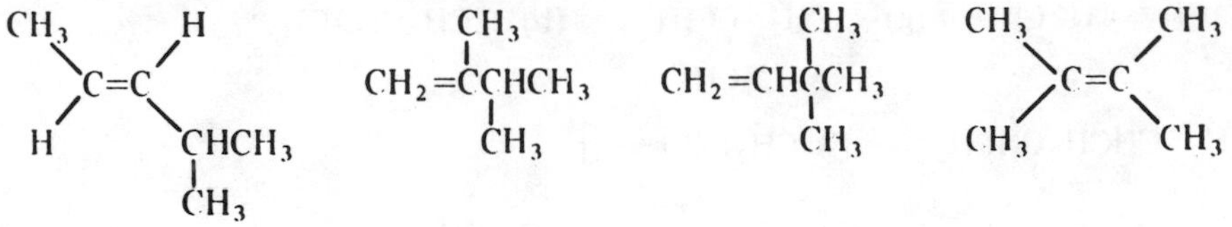

Trans-4-methyl-2-pentene

2, 3-dimethyl-1-Butene ($CH_2=C(CH_3)CH(CH_3)CH_3$)

3, 3-dimethyl-1-Butene ($CH_2=CHC(CH_3)_2CH_3$)

2, 3-Dimethyl-2-butene

(c)

C_5H_{10} :

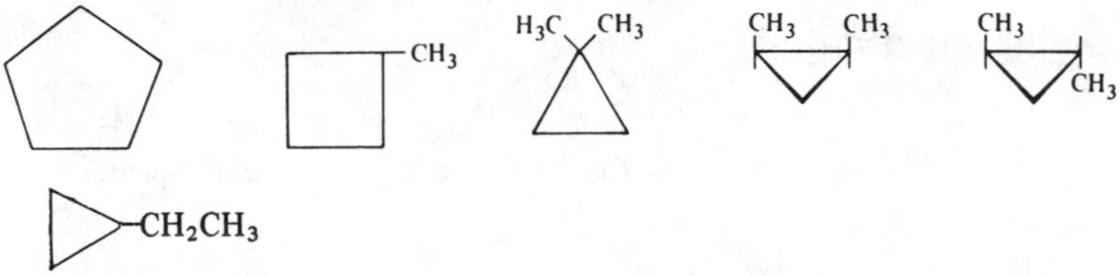

C_6H_{12} :

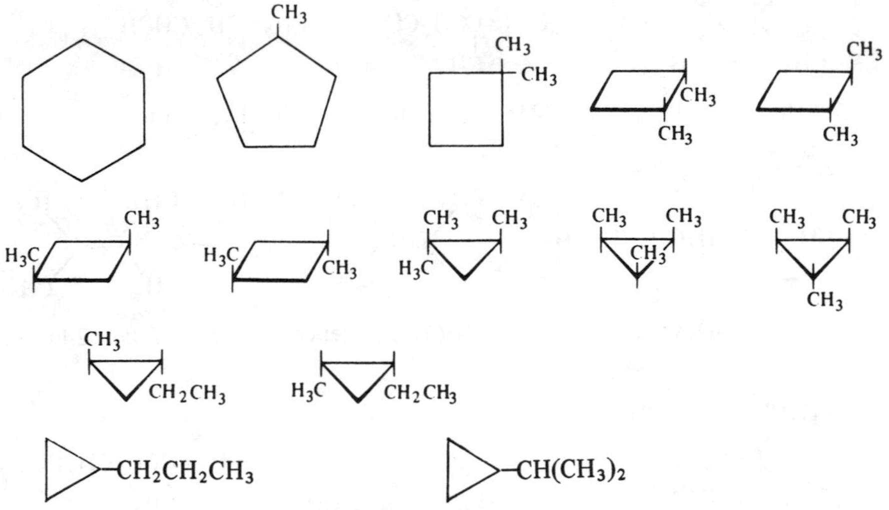

$\triangleright$—$CH_2CH_2CH_3$ $\qquad$ $\triangleright$—$CH(CH_3)_2$

6.16

(a) 1,3-Dimethylcyclohexene, (b) 2-ethyl-1-pentene, (c) 2-ethyl-1-pentene, (d) 1-ethyl-2-methylcyclopentene

* The *cis-trans* designation here is ambiguous. See pp. 325-327.

6.17

(a) $CH_3\overset{CH_3}{\overset{|}{C}}{=}CHCH_2CH_3$ (major) + $CH_2{=}\overset{CH_3}{\overset{|}{C}}CH_2CH_2CH_3$

(b) $CH_2{=}\overset{CH_3}{\overset{|}{C}}CH_2CH_2CH_3$ (major) + $CH_3\overset{CH_3}{\overset{|}{C}}{=}CHCH_2CH_3$

(c) $CH_3\overset{CH_3}{\overset{|}{C}}{=}CHCH_2CH_3$ (major) + $CH_2{=}\overset{CH_3}{\overset{|}{C}}CH_2CH_2CH_3$

(d) 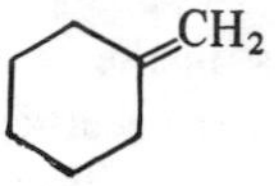 CH_2 , (e) 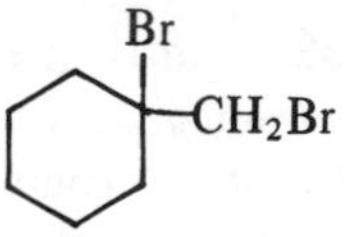 Br CH_2Br , (f) 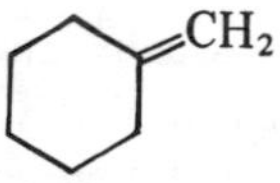 CH_2

(g) 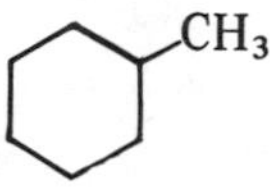 CH_3 , (h)

6.18

$$CH_2{=}CHCH_2CH_2CH_2CH_3 + Br_2 \xrightarrow[\text{(dark) R.T.}]{CCl_4} \underset{Br}{\underset{|}{C}}H_2\underset{Br}{\underset{|}{C}}HCH_2CH_2CH_2CH_3 \text{ (colorless)}$$

Cyclohexane does not react with Br_2 in the dark at room temperature, thus the red-brown color of the bromine will persist in the solution.

6.19

(a) No (b) No

(c) Yes $(CH_3)(H)C{=}C(CH_2CH_2CH_2CH_3)(H)$ and $(CH_3)(H)C{=}C(H)(CH_2CH_2CH_2CH_3)$

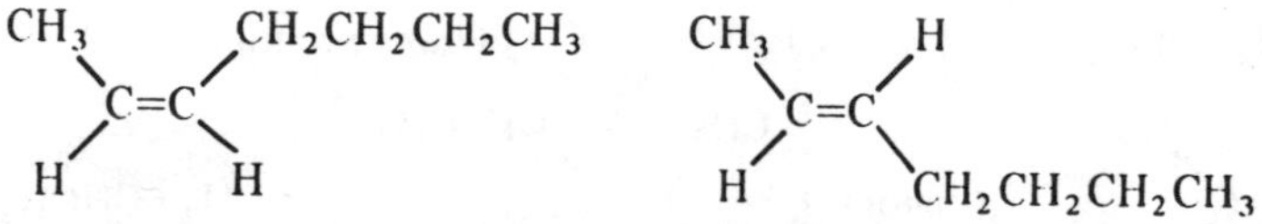

(d) No (e) No (f) No (g) Yes 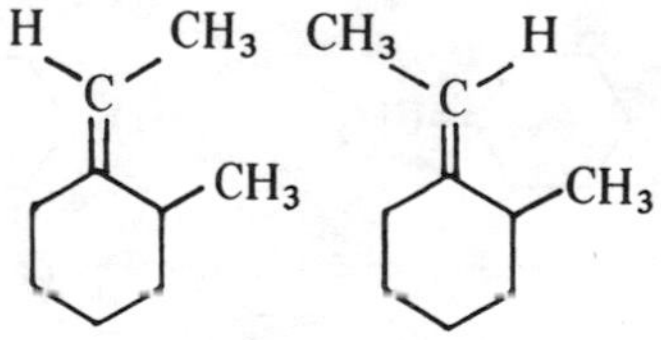 H CH_3 C CH_3 ; CH_3 H C CH_3

(h) Yes $(H)(Cl)C{=}C(H)(CH_2CH_3)$ and $(H)(Cl)C{=}C(CH_2CH_3)(H)$ (i) No

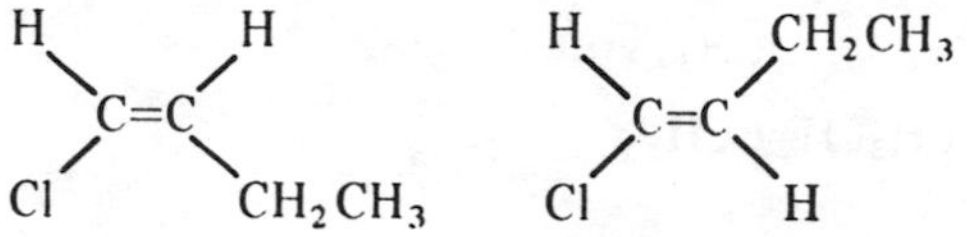

(j) No (k) No (l) Yes H CH_3 C CH_3 ; CH_3 H C CH_3

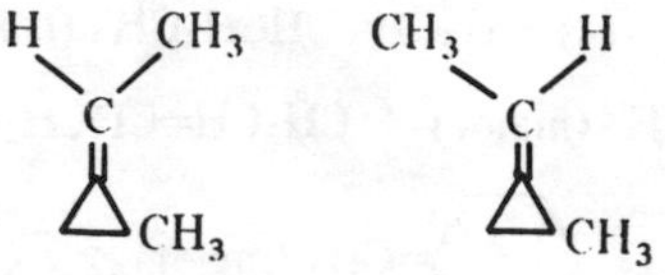

6.20

(a) 2,3-dimethyl-2-butene > 2-methyl-2-pentene > *trans*-3-hexene > *cis*-2-hexene > 1-hexene.

(b) The only alkenes whose relative stabilities could be measured by comparative heats of hydrogenation are those that yield the same hydrogenation product; i.e., *trans*-3-hexene, 1-hexene, *cis*-2-hexene all yield hexane on hydrogenation.

6.21

Although *trans* molecules are usually more stable than their *cis* isomers, in the case of cyclooctene, the *trans* isomer is probably more strained than the *cis* isomer because the ring is too small to allow a strain-free *trans* configuration. Therefore we would expect the *trans* isomer to have the higher heat of hydrogenation.

6.22

(a) *Cis-trans* isomerization caused by rupture of the π-bond.

(b) Equilibrium should favor the *trans* isomer because it is more stable than the *cis* isomer.

6.23

(a) $CH_3CH{=}\overset{\overset{CH_3}{|}}{C}CH_3$ (major) + $CH_2{=}CH\overset{\overset{CH_3}{|}}{C}HCH_3$

(b) $CH_3CH_2\overset{\overset{CH_3}{|}}{C}{=}CH_2$ (c) $CH_3CH{=}CHCH_2CH_3$ (*trans*-predominates)

(d) $\overset{CH_3}{\underset{H}{}}\!\!>C{=}C<\!\!\overset{H}{\underset{CH_2CH_3}{}}$ (major) + $\overset{CH_3}{\underset{H}{}}\!\!>C{=}C<\!\!\overset{CH_2CH_3}{\underset{H}{}}$ + $CH_2{=}CHCH_2CH_2CH_3$

(e) $=CH_2$ (f) $-CH_3$ (major product) + $=CH_2$

6.24

(a) $CH_2{=}CH\overset{\overset{CH_3}{|}}{C}HCH_3$ (major) + $CH_3CH{=}\overset{\overset{CH_3}{|}}{C}CH_3$

(b) $CH_3CH_2\overset{\overset{CH_3}{|}}{C}{=}CH_2$ (c) $CH_3CH{=}CHCH_2CH_3$ (*trans*-predominates)

(d) $CH_2{=}CHCH_2CH_2CH_3$ (major) + $CH_3CH{=}CHCH_2CH_3$ (*trans*-predominates)

(e) $=CH_2$ (f) $=CH_2$ (major) + $-CH_3$

6.25

(a) 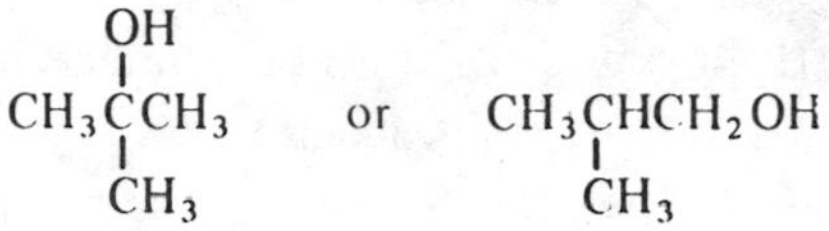 $CH_3C(OH)(CH_3)CH_3$ or $CH_3CH(CH_3)CH_2OH$

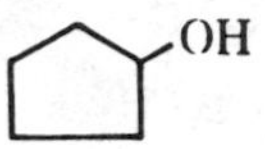

(b) 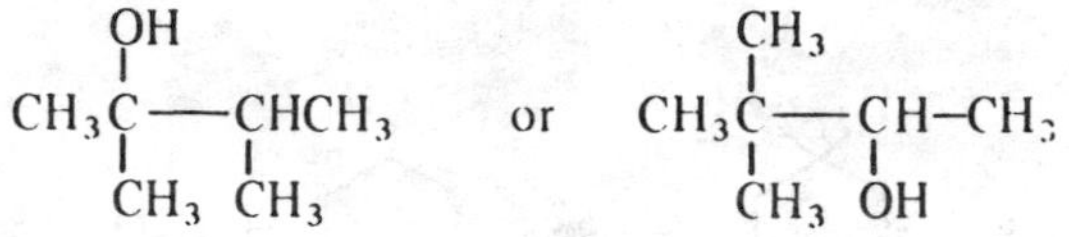 $CH_3C(OH)(CH_3)—CH(CH_3)CH_3$ or $CH_3C(CH_3)(CH_3)—CH(OH)—CH_3$

(c) cyclopentyl–OH

(d) 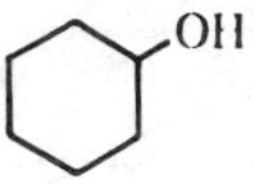 cyclohexyl–OH

6.26

$$CH_3C(OH)(CH_3)CH_2CH_3 > CH_3CH(CH_3)CH(OH)CH_3 > CH_3CH(CH_3)CH_2CH_2OH$$

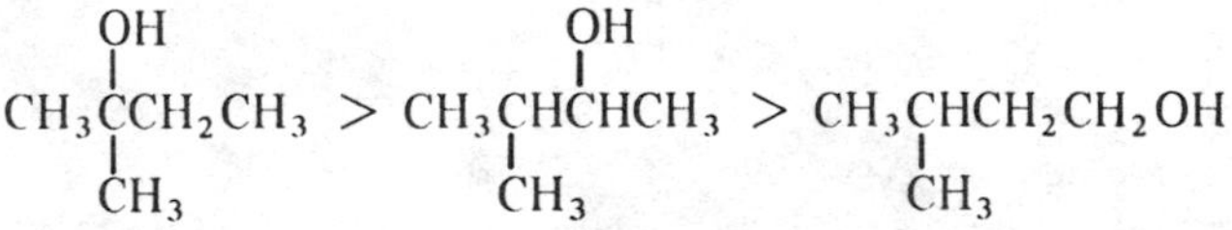

The order of reactivity is dictated by the order of stability of the intermediate carbocations: tertiary > secondary > primary

6.27

(a) *Cis*-1, 2-dimethylcyclopentane

(b) *Cis*-1, 2-dimethylcyclohexane: [chair with CH_3, CH_3]

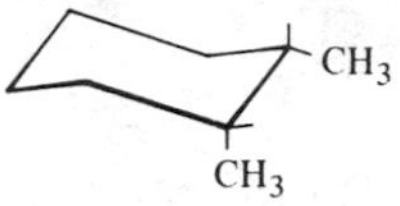

(c) *Cis*-1, 2-dideuteriocyclohexane: [chair with D, D]

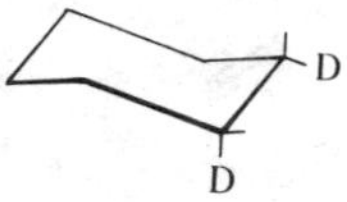

6.28

(a) 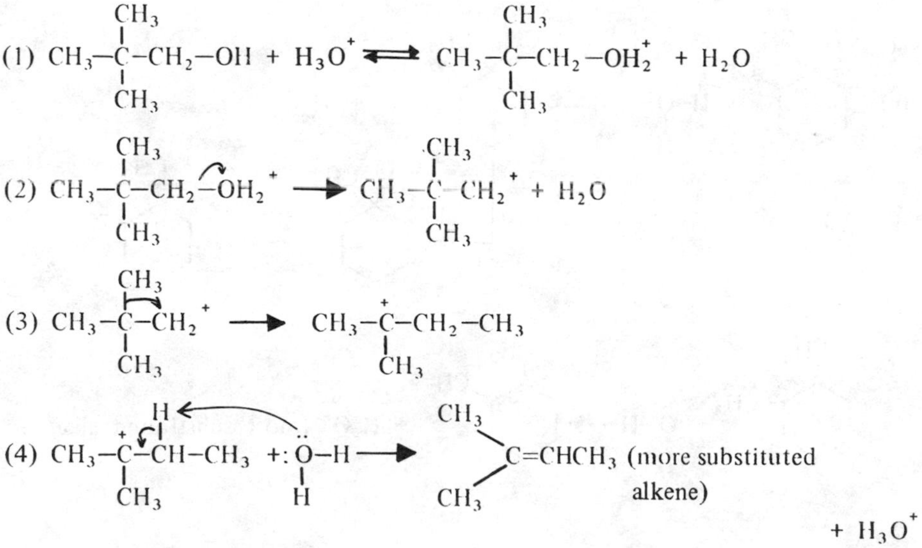 (1) $CH_3—C(CH_3)_2—CH_2—OH + H_3O^+ \rightleftharpoons CH_3—C(CH_3)_2—CH_2—OH_2^+ + H_2O$

(2) $CH_3—C(CH_3)_2—CH_2—OH_2^+ \longrightarrow CH_3—C(CH_3)_2—CH_2^+ + H_2O$

(3) $CH_3—C(CH_3)_2—CH_2^+ \longrightarrow CH_3—\overset{+}{C}(CH_3)—CH_2—CH_3$

(4) $CH_3—\overset{+}{C}(CH_3)—CH(H)—CH_3 + :\ddot{O}(H)—H \longrightarrow (CH_3)_2C{=}CHCH_3$ (more substituted alkene) $+ H_3O^+$

(4a) $CH_2\text{-}\overset{+}{C}(CH_3)\text{-}CH_2\text{-}CH_3 + :\ddot{O}H_2 \longrightarrow CH_2{=}C(CH_2CH_3)(CH_3)$ (less substituted alkene) $+ H_3O^+$

(Steps 2 and 3 may occur at the same time.)

(b) 1,4-dimethylcyclohexanol $+ H\text{-}O\text{-}P(=O)(OH)\text{-}OH \rightleftarrows$ protonated alcohol ($-\overset{+}{O}H_2$) $+ H_2PO_4^- \longrightarrow$ carbocation $+ H_2O$

carbocation $+ H_2\ddot{O}: \longrightarrow$ methylenecyclohexane (4-methyl) (less substituted alkene) $+ H_3O^+$

carbocation $+ H_2\ddot{O}: \longrightarrow$ 1,4-dimethylcyclohexene (more substituted alkene) $+ H_3O^+$

(c) 2,2-dimethylcyclohexanol $+ H\text{-}\overset{+}{O}(H)\text{-}H \rightleftarrows$ protonated alcohol ($-\overset{+}{O}H_2$) $+ H_2O$

$-H_2O$ → secondary carbocation → (methyl shift) → tertiary carbocation

tertiary carbocation $+ :\ddot{O}H_2 \longrightarrow$ 1,2-dimethylcyclohexene $+ H_3O^+$ (most substituted alkene)

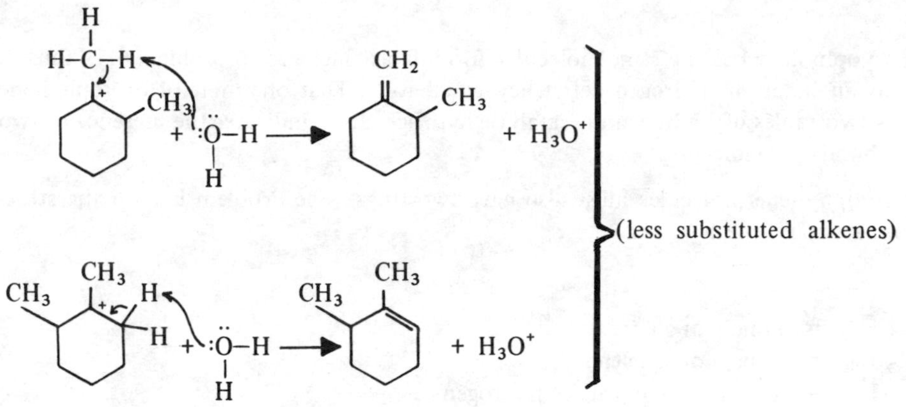

CH_3, CH_2, CH_3, H, $H-C-H$, $+ :\ddot{O}-H \longrightarrow$, $+ H_3O^+$

CH_3, CH_3, H, H, $+ :\ddot{O}-H \longrightarrow$, $+ H_3O^+$

(less substituted alkenes)

6.29

The alkyl halide has the structural feature $>CHCH_2Br$. Dehydrobromination gives only one alkene and no *cis-trans* isomers, therefore the bromine must be on the end of the chain. The remainder of the molecule may vary, so there are several answers possible. Three of them are

$$CH_3CH_2CH_2CH_2CH_2CH_2CH_2Br\ ,\ CH_3\overset{CH_3}{\overset{|}{C}}HCH_2CH_2CH_2CH_2Br$$

$$\text{and}\ CH_3CH_2CH_2CH_2\overset{CH_3}{\overset{|}{C}}HCH_2Br$$

6.30

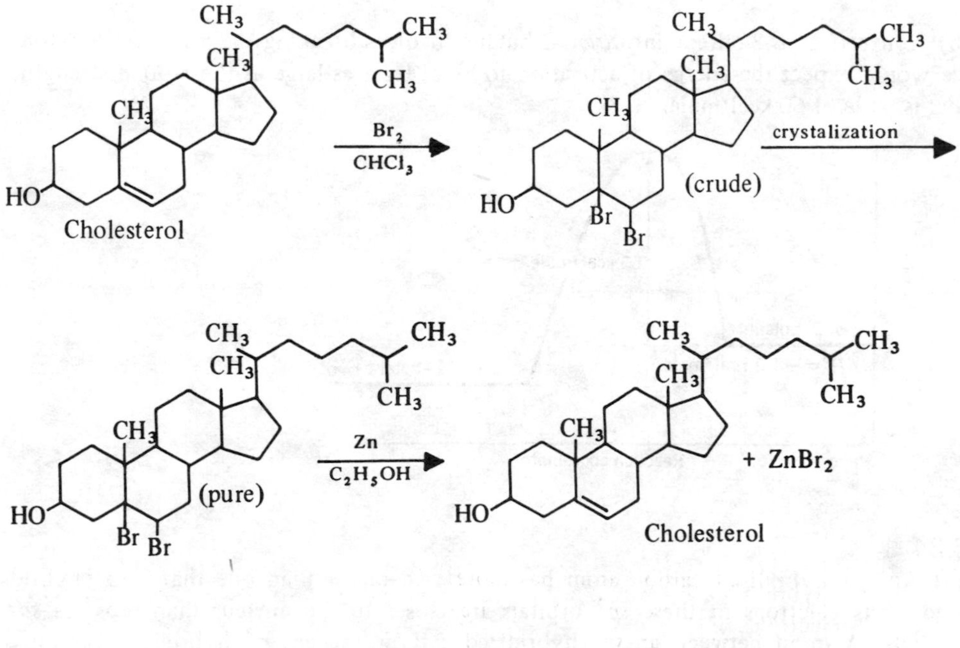

Cholesterol $\xrightarrow[CHCl_3]{Br_2}$ (crude) dibromide $\xrightarrow{\text{crystalization}}$ (pure) dibromide $\xrightarrow[C_2H_5OH]{Zn}$ Cholesterol $+ ZnBr_2$

(HO, CH_3, Br, Br labels on structures)

6.31

(a) Caryophyllene has the same molecular formula as zingiberene (problem 6.5), thus it, too, has an index of hydrogen deficiency equal to 4. That one mole of caryophyllene absorbs two moles of hydrogen on catalytic hydrogenation indicates the presence of two double bonds per molecule.

(b) Caryophyllene molecules must also have two rings. (See Problem E.1 for the structure of caryophyllene.)

6.32

(a) $C_{30}H_{62}$ = formula of alkane
$C_{30}H_{50}$ = formula of squalene
H_{12} = difference = 6 pairs of hydrogens

Index of hydrogen deficiency = 6

(b) Molecules of squalene contain six double bonds.

(c) Squalene molecules contain no rings. (See Problem E.1 for the structural formula of squalene.)

6.33

(a) We are given (on p. 214) the following heats of hydrogenation:

cis-2-Butene + H_2 ⟶ butane $\Delta H° = -28.6$ kcal/mole

trans-2-Butene + H_2 ⟶ butane $\Delta H° = -27.6$ kcal/mole

thus for

cis-2-Butene ⟶ *trans*-2-butene $\Delta H° = -1.0$ kcal/mole

(b) Converting *cis*-2-butene into *trans*-2-butene involves breaking the π bond. Therefore we would expect the energy of activation to be at least as large as the π-bond strength, that is, at least 60 kcal/mole.

(c)

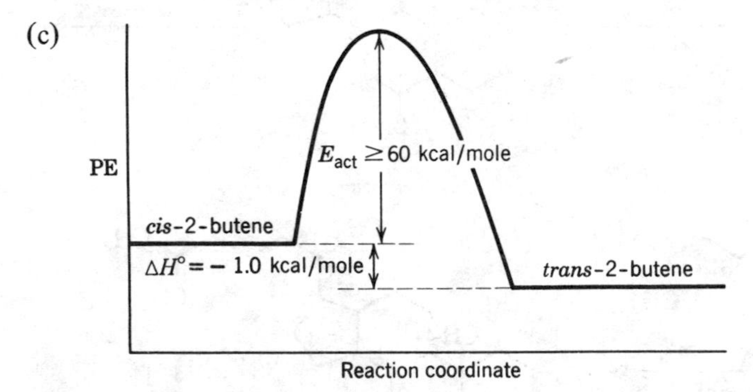

*6.34

(a) An sp^2 hybridized carbon atom has more s-character than one that is sp^3 hybridized, thus electrons in these sp^2 orbitals are closer to the nucleus than those in sp^3 orbitals. A bond between an sp^2 hybridized carbon and an sp^3 hybridized carbon is polarized toward the sp^2 hybridized carbon.

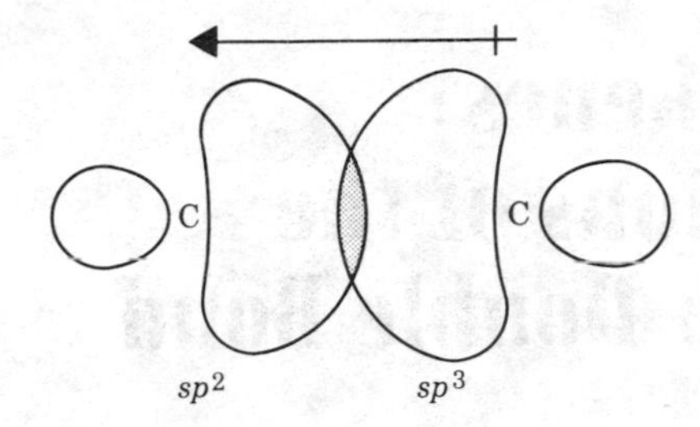

(b) The greater acidity of ethene can be explained by the greater stability of its conjugate base, $CH_2{=}\ddot{C}H^-$. The greater stability of $CH_2{=}\ddot{C}H^-$ compared with $CH_3\ddot{C}H_2^-$ can be explained by the greater *s*-character of the sp^2 orbital that holds the unbonded electron pair in $CH_2{=}\ddot{C}H^-$. In $CH_3\ddot{C}H_2^-$, the electrons are in an sp^3 orbital. Being farther from the nucleus, the sp^3 electrons have a higher potential energy.

***6.35**

3β-Friedelanol $\xrightarrow{H^+}$ $\xrightarrow{-H_2O}$ ⟶ (see below)

The migrations occur in the following sequence:

(1) $H\colon^-$ from C_4 to C_3 leaves (+) at C_4.
(2) $CH_3\colon^-$ from C_5 to C_4 leaves (+) at C_5.
(3) $H\colon^-$ from C_{10} to C_5 leaves (+) at C_{10}.
(4) $CH_3\colon^-$ from C_9 to C_{10} leaves (+) at C_9.
(5) $H\colon^-$ from C_8 to C_9 leaves (+) at C_8.
(6) $CH_3\colon^-$ from C_{14} to C_8 leaves (+) at C_{14}.
(7) $CH_3\colon^-$ from C_{13} to C_{14} leaves (+) at C_{13}.
(8) Loss of H^+ from C_{18} leaves double bond at $C_{13}{-}C_{18}$:

13(18)-Oleanene

The groups which migrate remain on the same face of the molecule after migration as before migration (see page 225 for transition state and see also Corey and Ursprung, *J. Am. Chem. Soc.*, 78, 5041 (1956)).

CHAPTER SEVEN

Reactions of Alkenes: Addition Reactions of the Carbon-Carbon Double Bond

7.1

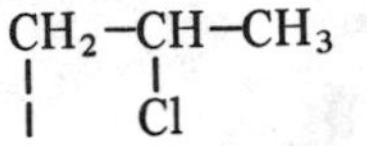

$$CH_2I{-}CH(Cl){-}CH_3 \quad \text{2-chloro-1-iodopropane}$$

7.2

(a) 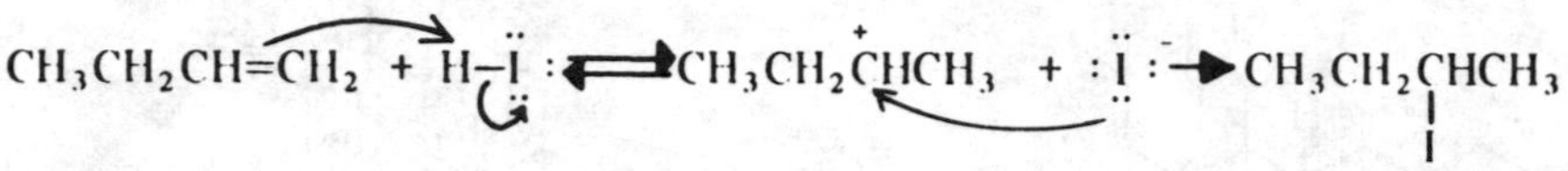 $$CH_3CH_2CH{=}CH_2 + H{-}\ddot{I}: \rightleftharpoons CH_3CH_2\overset{+}{C}HCH_3 + :\ddot{I}:^- \longrightarrow CH_3CH_2CH(I)CH_3$$

(b) $$(CH_3)_2C{=}CHCH_3 + :\overset{\delta+}{\ddot{I}}{-}\overset{\delta-}{\ddot{Br}}: \rightleftharpoons (CH_3)_2\overset{+}{C}{-}CH(I)CH_3 + :\ddot{Br}:^-$$

$$\longrightarrow CH_3C(CH_3)(Br){-}CH(I)CH_3$$

(c) $$\text{1-methylcyclohexene} + H\ddot{Cl}: \rightleftharpoons \text{1-methylcyclohexyl cation} + :\ddot{Cl}:^- \longrightarrow \text{1-chloro-1-methylcyclohexane}$$

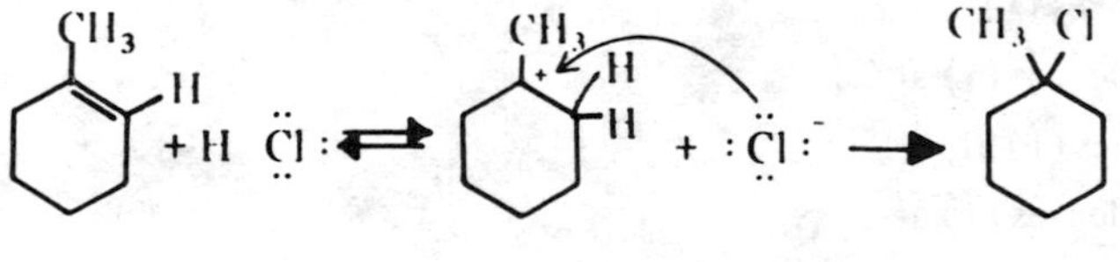

7.3

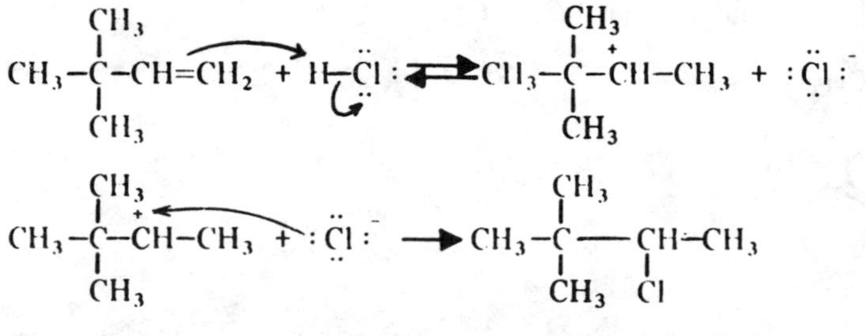

$$CH_3{-}C(CH_3)_2{-}CH{=}CH_2 + H{-}\ddot{Cl}: \rightleftharpoons CH_3{-}C(CH_3)_2{-}\overset{+}{C}H{-}CH_3 + :\ddot{Cl}:^-$$

$$CH_3{-}C(CH_3)_2{-}\overset{+}{C}H{-}CH_3 + :\ddot{Cl}:^- \longrightarrow CH_3{-}C(CH_3)_2{-}CH(Cl){-}CH_3$$

or

$$CH_3{-}C(CH_3)_2{-}\overset{+}{C}H{-}CH_3 \longrightarrow CH_3{-}\overset{+}{C}(CH_3){-}CH(CH_3){-}CH_3 \xrightarrow{:\ddot{Cl}:^-} CH_3{-}C(Cl)(CH_3){-}CH(CH_3){-}CH_3$$

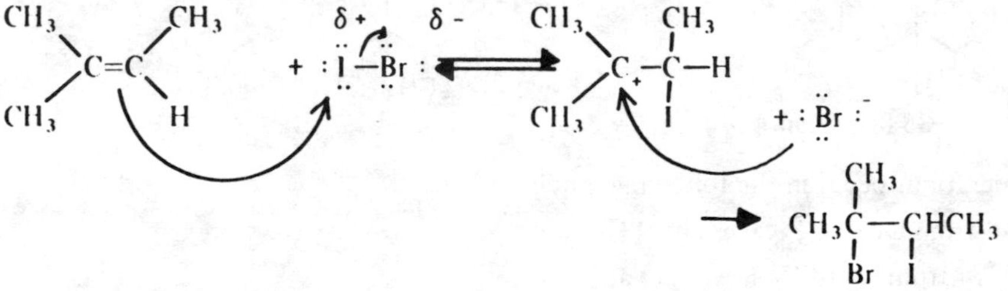

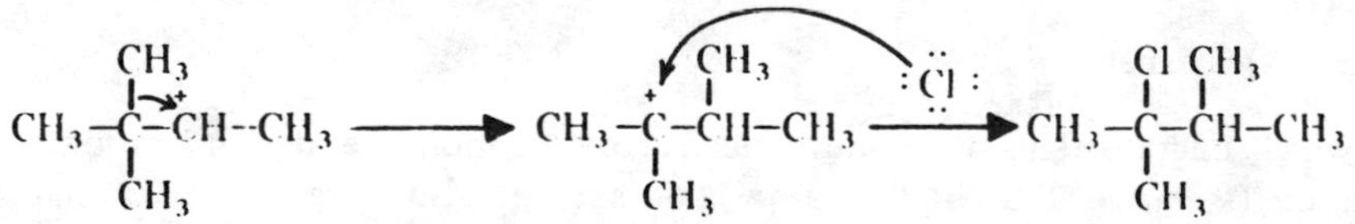

7.4

(a) $CH_3-CH=CH_2 + H-\ddot{O}^+-H \rightleftharpoons CH_3-\overset{+}{C}H-CH_3 + H_2O$ (the $\ddot{O}^+$ also bears an H)

$CH_3-\overset{+}{C}H-CH_3 + :\ddot{O}H_2 \rightleftharpoons CH_3-CH(\overset{+}{O}H_2)-CH_3$

$CH_3-CH(\overset{+}{O}H_2)-CH_3 + :\ddot{O}H_2 \rightleftharpoons CH_3-CH(OH)-CH_3 + H_3O^+$

(b) The product is isopropyl alcohol because the more stable isopropyl carbocation is produced in the first step. The formation of *n*-propyl alcohol would require the production of the less stable *n*-propyl carbocation.

7.5

$CH_3CH_2CH_2CH_2CH=CH_2 + Hg^+OAc \rightarrow CH_3CH_2CH_2CH_2-\overset{+}{C}H-CH_2-HgOAc$

$CH_3CH_2CH_2CH_2-\overset{+}{C}H-CH_2(HgOAc) + :\ddot{O}(H)-CH_2CH_3 \rightarrow CH_3CH_2CH_2CH_2-CH(\overset{+}{O}(H)-CH_2CH_3)-CH_2-HgOAc$

$\xrightarrow{-H^+} CH_3CH_2CH_2CH_2-CH(OCH_2CH_3)-CH_2-HgOAc \xrightarrow[OH^-]{NaBH_4} CH_3CH_2CH_2CH_2-CH(O-CH_2CH_3)-CH_3$

7.6

(a) $CH_3-C(CH_3)=CH-CH_3 + Hg(OAc)_2 + H_2O \xrightarrow{THF} CH_3-C(OH)(CH_3)-CH(CH_3)-HgOAc$

$\xrightarrow[OH^-]{NaBH_4} CH_3-C(OH)(CH_3)-CH_2-CH_3$

(b) 1-methylcyclohexene $+ Hg(OAc)_2 + H_2O \xrightarrow{THF}$ 1-methyl-1-hydroxy-2-(HgOAc)cyclohexane (CH₃, OH; HgOAc) $\xrightarrow[OH^-]{NaBH_4}$ 1-methylcyclohexanol (CH₃, OH)

(c) cyclohexene $+ Hg(OAc)_2 + CH_3OH \xrightarrow{THF}$ 1-(OCH₃)-2-(HgOAc)cyclohexane $\xrightarrow[OH^-]{NaBH_4}$ methoxycyclohexane (OCH₃)

(d) $CH_3-C(CH_3)_2-CH=CH_2 + Hg(OAc)_2 + CH_3OH \xrightarrow{THF} CH_3-C(CH_3)_2-CH(OCH_3)-CH_2HgOAc$

$\xrightarrow[OH^-]{NaBH_4} CH_3-C(CH_3)_2-CH(OCH_3)-CH_3$

7.7

(a) $3CH_3CH_2CH=CH_2 + \frac{1}{2}(BH_3)_2 \longrightarrow (CH_3CH_2CH_2CH_2)_3B \xrightarrow[OH^-]{H_2O_2}$

$3CH_3CH_2CH_2CH_2OH + H_3BO_3$

(b) $3CH_3-C(CH_3)=CH-CH_3 + \frac{1}{2}(BH_3)_2 \longrightarrow (CH_3-CH(CH_3)-CH(CH_3))_3-B \xrightarrow[OH^-]{H_2O_2}$

$3CH_3-CH(CH_3)-CH(OH)-CH_3 + H_3BO_3$

(c)

3 (1-methylcyclohexene) $+ \frac{1}{2}(BH_3)_2 \longrightarrow$ (trans-2-methylcyclohexyl, H, CH_3)$_3$B $\longrightarrow$ trans-2-methylcyclohexanol (H, OH, H, CH_3) $+ H_3BO_3$

7.8

(a) $3CH_3-C(CH_3)=CH_2 + \frac{1}{2}(BH_3)_2 \longrightarrow (CH_3\ CH(CH_3)\ CH_2)_3\ B$

$\xrightarrow[heat]{CH_3COOD} 3CH_3\ CH(CH_3)\ CH_2D + (CH_3COO)_3B$

(b) (methylenecyclohexane)$=CH_2 + \frac{1}{2}(BH_3)_2 \longrightarrow$ (cyclohexyl(H)$-CH_2$)$_3$B $\xrightarrow[heat]{CH_3COOD}$

3 (cyclohexyl)$-CH_2D + (CH_3COO)_3B$

(c)

3 (1-methylcyclopentene) $+ \frac{1}{2}(BH_3)_2 \longrightarrow$ (cyclopentyl: CH_3, H, H, B—) $\xrightarrow[heat]{CH_3COOD}$ (cyclopentyl: CH_3, H, H, D) $+ (CH_3COO)_3B$

(d)

$$3\ \text{(1-methylcyclopentene)} + \tfrac{1}{2}(BD_3)_2 \longrightarrow \text{(trialkylborane, H, D, } CH_3\text{, B—)} \xrightarrow[\text{heat}]{CH_3COOD} \text{(H, D, D, } CH_3\text{ cyclopentane)} + (CH_3COO)_3B$$

7.9

$$\text{cyclopentene} + Cl_2 \rightleftarrows \text{(π-complex with :Cl:δ+—:Cl:δ−)} \rightleftarrows \text{(bridged complex :Cl:δ+—:Cl:δ−)} \longrightarrow \text{(chloronium ion :Cl:}^+)$$

$$\text{(chloronium ion :Cl:}^+) + H\text{—}\ddot{O}\text{—}H \longrightarrow \text{(}\overset{+}{O}H_2\text{, Cl)} \longrightarrow \text{(OH, Cl)} + H_3O^+$$

7.10

(a) Nucleophilic substitution by an S_N2 mechanism in which the H_2O molecule is the nucleophile. The acid catalyzes the reaction by protonating the epoxide oxygen to make the leaving group a $-CH_2OH$ rather than the very basic $-CH_2O^-$ which would be a very poor leaving group.

(b) The release in ring strain upon opening the ring helps to counteract the inhibiting effect of the poor leaving group $-CH_2O^-$, therefore the strongly nucleophilic OH^- ion is able to bring about the reaction.

7.11

(a) $CH_3CH_2CH{=}CHCH_3$

(b) $CH_3\overset{CH_3}{\overset{|}{C}}{=}\overset{CH_3}{\overset{|}{C}}CH_3$

(c) $CH_3CH_2\underset{CH_3}{\underset{|}{C}}HCH{=}CH_2$

(d) cyclohexene

7.12

The initiation reaction (step 1) has no competitor and all of the subsequent steps must occur rapidly. The reaction disallowed for step 2,

$$R\text{—}\ddot{\underset{..}{O}}\cdot + HBr \not\longrightarrow R\text{—}\ddot{\underset{..}{O}}\text{—}Br + H\cdot \qquad \Delta H^\circ \cong 39\ \text{kcal/mole}$$

must have an $E_{act} > 39$ kcal/mole. Thus it cannot compete with step (2),

$$R{-}\ddot{\underset{..}{O}}\cdot + HBr \longrightarrow R{-}\ddot{\underset{..}{O}}{-}H + Br\cdot \qquad \Delta H^\circ \cong -13 \text{ kcal/mole}$$

which has a much lower E_{act}.

7.13

Chain-initiating steps

(a) (1) $R{-}\ddot{\underset{..}{O}}{-}\ddot{\underset{..}{O}}{-}R \xrightarrow{\text{heat}} 2\,R{-}\ddot{\underset{..}{O}}\cdot$

(2) $R{-}\ddot{\underset{..}{O}}\cdot + H{-}CCl_3 \longrightarrow R{-}\ddot{\underset{..}{O}}H + \cdot CCl_3$

Chain-propagating steps

(3) $CH_3CH_2CH_2CH{=}CH_2 + \cdot CCl_3 \longrightarrow CH_3CH_2CH_2\dot{C}H{-}CH_2CCl_3$

(4) $CH_3CH_2CH_2\dot{C}HCH_2CCl_3 + H{-}CCl_3 \longrightarrow CH_3CH_2CH_2CH_2CH_2CCl_3 + \cdot CCl_3$

then (3), (4), (3), (4), etc.

(b) (1) $R{-}\ddot{\underset{..}{O}}{-}\ddot{\underset{..}{O}}{-}R \xrightarrow{\text{heat}} 2\,R{-}\ddot{\underset{..}{O}}\cdot$

(2) $R{-}\ddot{\underset{..}{O}}\cdot + CH_3CH_2{-}\ddot{\underset{..}{S}}{-}H \longrightarrow R{-}\ddot{\underset{..}{O}}H + CH_3CH_2{-}\ddot{\underset{..}{S}}\cdot$

Chain-propagating steps

(3) $CH_3\overset{CH_3}{\overset{|}{C}}{=}CH_2 + \cdot\ddot{\underset{..}{S}}CH_2CH_3 \longrightarrow CH_3\overset{CH_3}{\overset{|}{\dot{C}}}{-}CH_2{-}\ddot{\underset{..}{S}}{-}CH_2CH_3$

(4) $CH_3\overset{CH_3}{\overset{|}{\dot{C}}}CH_2SCH_2CH_3 + HSCH_2CH_3 \longrightarrow CH_3\overset{CH_3}{\overset{|}{C}}HCH_2SCH_2CH_3 + \cdot\ddot{\underset{..}{S}}CH_2CH_3$

then (3), (4), (3), (4), etc.

(c) (1) $R{-}\ddot{\underset{..}{O}}{-}\ddot{\underset{..}{O}}{-}R \xrightarrow{\text{heat}} 2\,R{-}\ddot{\underset{..}{O}}\cdot$

(2) $R{-}\ddot{\underset{..}{O}}\cdot + Cl{-}CCl_3 \longrightarrow R{-}\ddot{\underset{..}{O}}{-}Cl + \cdot CCl_3$

Chain-propagating steps

(3) $CH_3CH_2\overset{CH_3}{\overset{|}{C}}{=}CH_2 + \cdot CCl_3 \longrightarrow CH_3CH_2\overset{CH_3}{\overset{|}{\dot{C}}}{-}CH_2CCl_3$

(4) $CH_3CH_2\overset{CH_3}{\overset{|}{\dot{C}}}CH_2CCl_3 + CCl_4 \longrightarrow CH_3CH_2\underset{Cl}{\underset{|}{\overset{CH_3}{\overset{|}{C}}}}CH_2CCl_3 + \cdot CCl_3$

then (3), (4), (3), (4), etc.

7.14

(a) $CH_3CH_2CH_2CHClCH_3$, (b) $CH_3CH_2CH_2CHBrCH_2Br$,

(c) $CH_3CH_2CH_2CHOHCH_3$, (d) $CH_3CH_2CH_2\overset{OSO_2OH}{\overset{|}{C}}HCH_3$,

(e) same as (c) , (f) $CH_3CH_2CH_2CHBrCH_3$,

(g) same as (c) , (h) $CH_3CH_2CH_2CH_2CH_2OH$,

(i) $CH_3CH_2CH_2\underset{|}{\overset{OCH_3}{C}}HCH_3$, (j) $CH_3CH_2CH_2CHDCH_2D$,

(k) $CH_3CH_2CH_2CH{=}CH_2$, (l) $CH_3CH_2CH_2\overset{OH}{\overset{|}{C}}HCH_2OH$,

(m) same as (l) , (n) $CH_3CH_2CH_2\overset{O}{\overset{\|}{C}}{-}OH + CO_2$

(o) $CH_3CH_2CH_2\overset{O}{\overset{\|}{C}}H + H\overset{O}{\overset{\|}{C}}H$, (p) $CH_3CH_2CH_2CH_2CH_3$,

(q) $CH_3CH_2CH_2CH_2CH_2Br$

7.15

(a) Cl , (b) Br Br , (c) OH ,

(d) OSO_2OH , (e) same as (c) , (f) Br ,

(g) OH , (h) same as (g) , (i) OCH_3 ,

(j) D D , (k) , (l) OH OH , (m) same as (l) ,

(n) $\overset{O}{\overset{\|}{C}}{-}OH$ $\overset{O}{\overset{\|}{C}}{-}OH$, (o) $\overset{O}{\overset{\|}{C}}H$ $\overset{O}{\overset{\|}{C}}H$, (p) ,

(q) same as (f)

7.16

(a) $CH_3CH_2\underset{\underset{Br}{|}}{C}HCH_2Br$ (b) OH OH *cis* (c) Same as (b)

(d) cyclohexane-1,2-diol (OH, OH) *trans* (e) $CH_3CH_2-O-SO_2-OH$ (f) CH_3CH_2OH

(g) $(CH_3CH(CH_3)CH_2)_3B$ (h) $CH_3CH(CH_3)CH_2OH$ (i) $(CH_3)HC=CH(CH_3)$ (trans) (major product)

(j) cyclopentene (k) $CH_2=CHCH_2CH_2CH_3$ (l) $CH_3C(CH_3)(Br)CH(CH_3)_2$

(m) $CH_3CH(Cl)CH_2CH_2CH_2CH_3$ (n) $BrCH_2CH_2CH_2CH_2CH_2CH_3$

(o) CH_3CH_2CH (–O–, –O–O–) $CHCH_2CH_3$ (ozonide) (p) 2 CH_3CH_2CHO

(q) $2CH_3CH_2C(=O)-O^-$ (r) ozonide of 1-methylcyclopentene (CH_3) (s) $CH_3C(=O)CH_2CH_2CH_2CHO$

(t) $CH_3C(=O)CH_2CH_2CH_2C(=O)-OH$ (u) 1-methylcyclopentanol (CH_3, OH) (v) methylcyclopentane ($-CH_3$)

(w) $CH_3C(CH_3)(I)CH_2CH_2CH_3$ (x) cyclopentane with CH_3 and CH_3, Cl (y) cyclopentane with HO, CH_3 and Cl

7.17

(a) $CH_3CH=CH_2 + H_3O^+ \rightleftarrows CH_3\overset{+}{C}HCH_3 + H_2O$

$(CH_3)_2\overset{+}{C}H + CH_2=CHCH_3 \rightarrow CH_3CH(CH_3)CH_2\overset{+}{C}HCH_3$

$CH_3CH(CH_3)CH(H)-\overset{+}{C}H-CH(H)_2 + :\ddot{O}H_2$ —a→ $CH_3CH(CH_3)CH=CHCH_3$; —b→ $CH_3CH(CH_3)CH_2CH=CH_2$

(b) $CH_3CH(CH_3)CH=CHCH_3$ (more substituted alkene)

7.18

$$CH_3CH{=}CHCH_3 + HCl \rightleftharpoons CH_3CH_2\overset{+}{C}HCH_3 + :\ddot{\underset{..}{Cl}}:^-$$

$$CH_3CH_2\overset{+}{C}HCH_3 + :\ddot{O}(H){-}CH_2CH_3 \longrightarrow CH_3CH_2\underset{\overset{|}{^{+}:O(H)-CH_2CH_3}}{C}HCH_3 \xrightarrow{-H^+} CH_3CH_2\underset{\overset{|}{OCH_2CH_3}}{C}HCH_3$$

7.19

The order of reactivity parallels the order of stability of the carbocations produced by the attack of H^+ on each alkene.

$$\underset{3^\circ}{R{-}\overset{+}{C}(R){-}CH_3} > \underset{2^\circ}{R{-}\overset{+}{C}H{-}CH_3} > \underset{1^\circ}{\overset{+}{C}H_2{-}CH_3}$$

7.20

(a) [chair cyclohexane: D*, CH_3, OD] + [chair cyclohexane: D*, OD, CH_3] (* D may also be axial)

(b) [chair cyclohexane: H, D, CH_3, D] (c) [chair cyclohexane: H, OH, CH_3, D]

7.21

$$\left(2CH_3\overset{O}{\overset{\|}{C}}CH_3\right),\ 4\left(O{=}CHCH_2CH_2\overset{CH_3}{\overset{|}{C}}{=}O\right),\quad O{=}CHCH_2CH_2CH{=}O$$

7.22

$$CH_3\underset{\underset{CH_3}{|}}{C}{=}CH_2 > CH_3CH{=}CH_2 > CH_2{=}CH_2$$

The order is the same as the order of stability of the carbocations formed by protonation of the alkenes.

$$CH_3\underset{\underset{CH_3}{|}}{\overset{+}{C}}{-}CH_3 > CH_3\overset{+}{C}H{-}CH_3 > \overset{+}{C}H_2{-}CH_3$$

7.23

(a) $CH_3CH{=}CHCH_3 + H^+ \rightleftharpoons CH_3\overset{+}{C}HCH_2CH_3$

(*cis* or *trans*)

$$CH_3-\overset{+}{C}H-CH(H)-CH_3 \longrightarrow trans\text{-}CH_3CH{=}CHCH_3 + H^+ \;;\; cis\text{-}CH_3CH{=}CHCH_3 + H^+$$

$$CH_3-CH_2-\overset{+}{C}H-CH_2(H) \longrightarrow CH_3CH_2CH{=}CH_2 + H^+$$

The most stable (most substituted) alkene is formed in greatest amount; i.e., 2-butenes > 1-butene, and *trans*-2-butene > *cis*-2-butene.

(b) 1-Butene, on protonation, gives the same intermediate carbocation: $CH_3CH_2\overset{+}{C}HCH_3$.

(c) The carbocation, $CH_3CH_2\overset{+}{C}HCH_3$, cannot easily rearrange to the branched chain compound because to do so would require the formation of an intermediate primary carbocation, $\overset{+}{C}H_2CH(CH_3)CH_3$.

7.24

$$CH_3-CH(OH)-C(CH_3)_2-CH_3 + H-Cl \rightleftharpoons CH_3-CH(\overset{+}{O}H_2)-C(CH_3)_2-CH_3 + :\ddot{C}l:^-$$

$$\longrightarrow CH_3-\overset{+}{C}H-C(CH_3)_2-CH_3 + H_2O$$

$$CH_3-\overset{+}{C}H-C(CH_3)_2-CH_3 \longrightarrow CH_3-CH(CH_3)-\overset{+}{C}(CH_3)-CH_3 \xrightarrow{:\ddot{C}l:^-} CH_3-CH(CH_3)-C(Cl)(CH_3)-CH_3$$

7.25

4-methylcyclohexene (cyclohexene ring with CH_3)

7.26

(a) $CH_3CH_2CH_3$ (excess) $+ Br_2 \xrightarrow[\text{light}]{CCl_4}$ $CH_3CH_2CH_2Br$ + $CH_3CH(Br)CH_3$ $\xrightarrow[CH_3CH_2OH]{KOH}$ $CH_3CH{=}CH_2$

(b) $CH_3CH{=}CH_2$ (above) $+ HBr \xrightarrow[\text{inhibitor}]{\text{free-radical}} CH_3CH(Br)CH_3$

(c) $CH_3CH{=}CH_2$ (from (a)) + $HBr \xrightarrow{\text{Peroxides}} CH_3CH_2CH_2Br$

(d) $CH_3\overset{CH_3}{\overset{|}{C}}HCH_3 + Br_2 \xrightarrow[\text{light}]{\text{heat}} CH_3\underset{\underset{Br}{|}}{\overset{CH_3}{\overset{|}{C}}}CH_3 \xrightarrow[CH_3CH_2OH]{KOH} CH_3\overset{CH_3}{\overset{|}{C}}{=}CH_2$

(excess)

(e) $CH_3\overset{CH_3}{\overset{|}{C}}{=}CH_2$ (from (d)) + $H_2O \xrightarrow{H_3O^+} CH_3\underset{\underset{OH}{|}}{\overset{CH_3}{\overset{|}{C}}}CH_3$

(f) $CH_3CH_2CH_2CH_2Cl + KOH \xrightarrow{CH_3CH_2OH} CH_3CH_2CH{=}CH_2 \xrightarrow[\text{CCl}_4,\ \text{dark}]{Cl_2} CH_3CH_2\underset{\underset{Cl}{|}}{C}HCH_2Cl$

(g) $CH_3CH_2Br + KOH \xrightarrow{CH_3CH_2OH} CH_2{=}CH_2 \xrightarrow{Br_2\ +\ H_2O} \underset{\underset{OH}{|}}{C}H_2\underset{\underset{Br}{|}}{C}H_2$

(h) $CH_3\overset{CH_3}{\overset{|}{C}}{=}CH_2$ (from (d)) + $HBr \xrightarrow{\text{Peroxides}} CH_3\overset{CH_3}{\overset{|}{C}}HCH_2Br \xrightarrow{Li} CH_3\overset{CH_3}{\overset{|}{C}}HCH_2Li \xrightarrow{CuI} (CH_3\overset{CH_3}{\overset{|}{C}}HCH_2)_2CuLi \xrightarrow{CH_3\overset{CH_3}{\overset{|}{C}}HCH_2Br} CH_3\overset{CH_3}{\overset{|}{C}}HCH_2CH_2\overset{CH_3}{\overset{|}{C}}HCH_3$

(i)

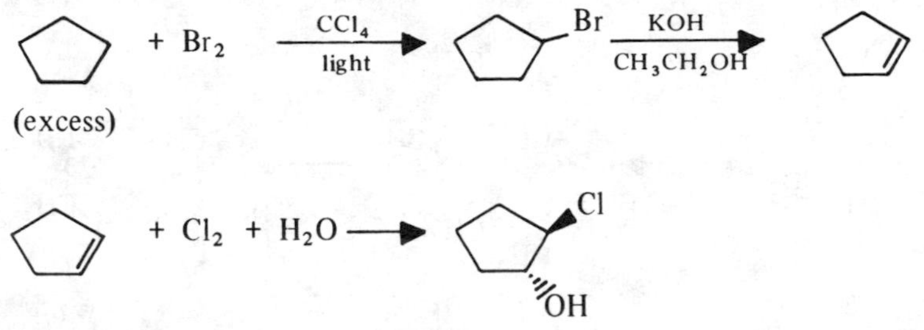

(j) $CH_3CH_2CH_2CH_2Br \xrightarrow[CH_3CH_2OH]{KOH} CH_3CH_2CH{=}CH_2 \xrightarrow[\text{free radical inhibitor}]{HBr} CH_3CH_2\underset{\underset{Br}{|}}{C}HCH_3$

(k) $CH_3CH_2CH_2CH_2Cl \xrightarrow[CH_3CH_2OH]{KOH} CH_3CH_2CH{=}CH_2 \xrightarrow{HCl} CH_3CH_2\underset{\underset{Cl}{|}}{C}HCH_3 \xrightarrow{Li} CH_3CH_2\overset{CH_3}{\overset{|}{C}}HLi \xrightarrow{CuI} \left(CH_3CH_2\overset{CH_3}{\overset{|}{C}}H\right)_2CuLi \xrightarrow{CH_3CH_2\overset{CH_3}{\overset{|}{C}}HCl} CH_3CH_2\overset{CH_3}{\overset{|}{C}}H{-}\overset{CH_3}{\overset{|}{C}}HCH_2CH_3$

(l) $CH_3\underset{\underset{CH_3}{|}}{\overset{CH_3}{\overset{|}{C}}}{-}OH \xrightarrow[\text{heat}]{H_2SO_4} CH_3\overset{CH_3}{\overset{|}{C}}{=}CH_2 \xrightarrow{(BH_3)_2} (CH_3\overset{CH_3}{\overset{|}{C}}HCH_2)_3B$

$$(CH_3\underset{|}{\overset{CH_3}{C}}HCH_2)_3B \xrightarrow[OH^-]{H_2O_2} CH_3\overset{CH_3}{\underset{|}{C}}HCH_2OH$$

7.27

(a) $CH_3CH_2CH_2CH{=}CH_2 + Br_2 \longrightarrow CH_3CH_2CH_2\underset{Br}{\underset{|}{C}}HCH_2Br$

MW=70.12 MW=159.8

159.8 g Br_2 will react with 70.12 g pentene. Therefore ~ 16 g Br_2 will react with 7.0 g pentene.

(b) Since bromine and an alkene react in equimolar proportions:

$$\frac{3.20\ g}{160\ g/mole} = 0.02 \text{ mole } Br_2 = 0.02 \text{ mole alkene}$$

$$2.24 \text{ g alkene} = (0.02 \text{ mole})(\text{Mol. Wt.})$$

$$\therefore \text{Mol. Wt.} = \frac{2.24\ g}{0.02\ mole} = 112 \text{ g/mole alkene}$$

7.28

Rewriting the starting compound, we can better see the required reaction:

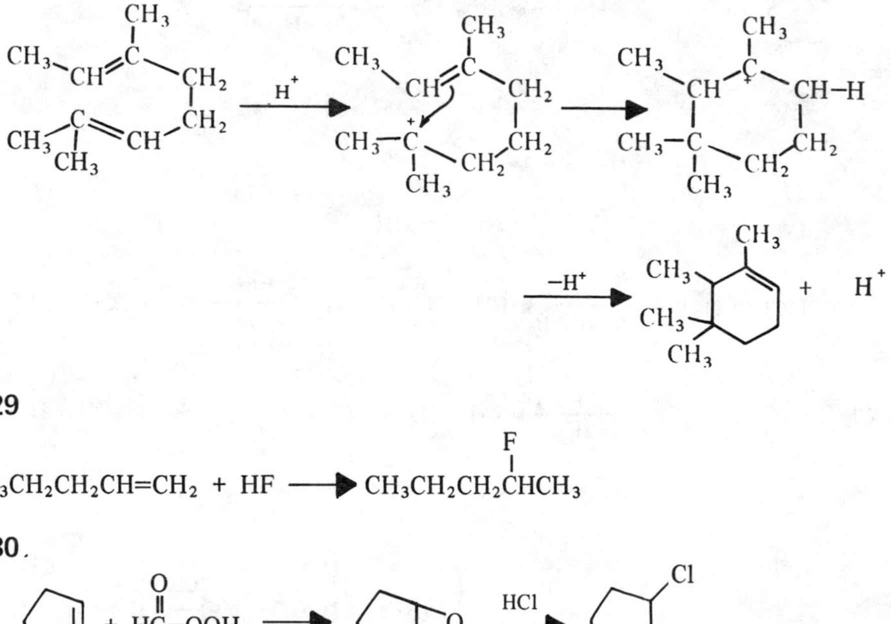

$CH_3CH{=}C(CH_3)CH_2CH_2CH{=}C(CH_3)_2 \xrightarrow{H^+}$ [carbocation $CH_3-\overset{+}{C}(CH_3)-CH_2-$...] $\longrightarrow$ [cyclized cation] $\xrightarrow{-H^+}$ 1,2,2,6-... trimethylcyclohexene (CH_3, CH_3, CH_3, CH_3) + H^+

7.29

$$CH_3CH_2CH_2CH{=}CH_2 + HF \longrightarrow CH_3CH_2CH_2\overset{F}{\overset{|}{C}}HCH_3$$

7.30.

(a) 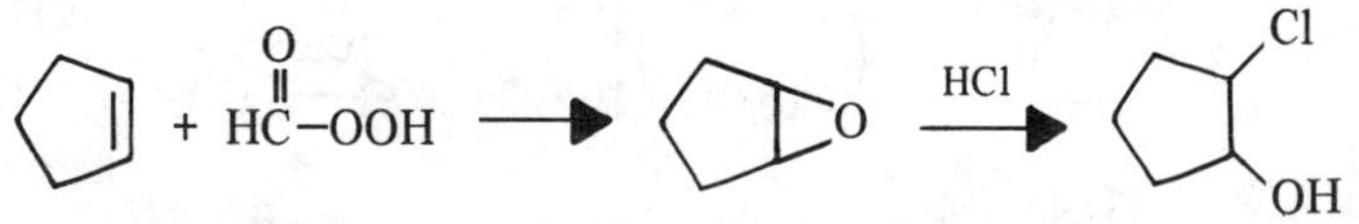 cyclopentene + $H\overset{O}{\overset{\|}{C}}-OOH \longrightarrow$ cyclopentene oxide $\xrightarrow{HCl}$ 2-chlorocyclopentanol (Cl, OH)

(b) *trans*-product because the Cl^- attacks anti to the epoxide and an inversion of configuration occurs.

$$\text{epoxide} \xrightarrow{HCl} \text{protonated epoxide} + :\ddot{Cl}:^- \longrightarrow \text{trans-2-chlorocyclopentanol}$$

7.31

$$CH_3CH(CH_3)CH_2CH_2CH_2CH_2-CH(-O-)CH-CH_2CH_2CH_2CH_2CH_2CH_2CH_2CH_2CH_2CH_3$$

(cis epoxide: both H on the same side)

7.32

$$CH_3\overset{CH_3}{\overset{|}{C}}=CHCH_2CH_2\overset{CH_3}{\overset{|}{C}}=CHCH_2CH_2\overset{CH_2}{\overset{\|}{C}}CH=CH_2$$

7.33

In this case the more highly substituted alkene is less stable. See page 249 for an explanation.

7.34

(a) $$CH_3\overset{CH_3}{\overset{|}{C}}HCH_2Br + KOH \xrightarrow{\text{ethanol}} CH_3\overset{CH_3}{\overset{|}{C}}=CH_2 + KBr + H_2O$$

(b) By dehydrobromination as shown in (a) above, and then by adding HBr to the resulting alkene, isobutyl bromide can be converted into tert-butyl bromide:

$$CH_3\overset{CH_3}{\overset{|}{C}}=CH_2 + HBr \xrightarrow{CCl_4} CH_3\overset{CH_3}{\underset{Br}{\overset{|}{\underset{|}{C}}}}CH_3 \quad \text{(Markovnikov's rule)}$$

7.35

The intermediate I is competitively attacked by Cl^-, Br^- and H_2O.

$$\text{cyclopentene} + Br_2 \rightleftarrows \text{cyclopentene}\cdots :Br:-:Br: \longrightarrow \text{cyclopentene}\cdots :Br:\delta+ - :Br:\delta-$$

$$\text{I (bromonium ion, } :Br:^+) + :\ddot{Cl}:^- \text{ (from NaCl)} \longrightarrow \text{trans-1-bromo-2-chlorocyclopentane}$$

I

7.36

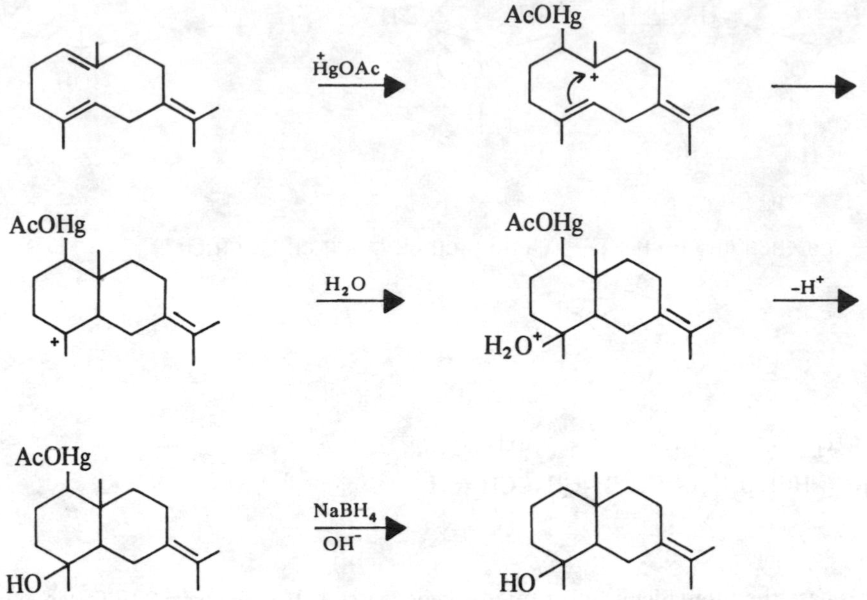

$\xrightarrow{\overset{+}{H}gOAc}$ (AcOHg-substituted carbocation) $\longrightarrow$

AcOHg ... (+) $\xrightarrow{H_2O}$ AcOHg ... H_2O^+ $\xrightarrow{-H^+}$

AcOHg ... HO $\xrightarrow[OH^-]{NaBH_4}$ HO ...

7.37

(a) $H{-}SH \xrightarrow{h\nu} H\cdot + \cdot SH$ **Chain-initiating step**

$R{-}CH{=}CH_2 + \cdot SH \longrightarrow R\dot{C}H{-}CH_2SH$

$R{-}\dot{C}HCH_2SH + H{-}SH \longrightarrow RCH_2CH_2SH$

} **Chain-propagating steps**

(b) $R\dot{C}HCH_2SH + RCH_2CH_2SH \longrightarrow RCH_2CH_2SH + RCH_2CH_2S\cdot$

$RCH{=}CH_2 + \cdot SCH_2CH_2R \longrightarrow R\dot{C}HCH_2SCH_2CH_2R$

$R\dot{C}HCH_2SCH_2CH_2R + RCH_2CH_2SH \longrightarrow (RCH_2CH_2)_2S + RCH_2CH_2S\cdot$

7.38

One dimer (the major product) gives the following products on ozonolysis.

$$CH_3{-}\overset{CH_3}{\underset{CH_3}{C}}{-}CH_2{-}\underset{CH_3}{C}{=}CH_2 \xrightarrow[(2)\ Zn,\ H_2O]{(1)\ O_3} CH_3{-}\overset{CH_3}{\underset{CH_3}{C}}{-}CH_2{-}\overset{O}{\overset{\|}{C}}{-}CH_3 + H{-}\overset{O}{\overset{\|}{C}}{-}H$$

The other dimer gives different products.

$$CH_3{-}\overset{CH_3}{\underset{CH_3}{C}}{-}CH{=}\underset{CH_3}{C}{-}CH_3 \xrightarrow[(2)\ Zn,\ H_2O]{(1)\ O_3} CH_3{-}\overset{CH_3}{\underset{CH_3}{C}}{-}C{\overset{O}{\underset{H}{\lessgtr}}} + CH_3{-}\overset{O}{\overset{\|}{C}}{-}CH_3$$

By isolating and identifying the products of each reaction, Whitmore and his students were able to deduce the structures of the diisobutylenes.

7.39

The isomers are propene tetramers formed by an acid-catalyzed reaction:

$$\mathrm{CH_3CH{=}CH_2 + H_3PO_4 \rightleftharpoons CH_3\overset{+}{C}HCH_3 + H_2PO_4^-}$$

$$\mathrm{CH_3\underset{\stackrel{|}{CH_3}}{C}H^+ + CH_2{=}CHCH_3 \longrightarrow CH_3\underset{\stackrel{|}{CH_3}}{C}HCH_2\underset{\stackrel{|}{CH_3}}{C}H^+}$$

$$\mathrm{CH_3\underset{\stackrel{|}{CH_3}}{C}HCH_2\underset{\stackrel{|}{CH_3}}{C}H^+ + CH_2{=}CHCH_3 \longrightarrow CH_3\underset{\stackrel{|}{CH_3}}{C}HCH_2\underset{\stackrel{|}{CH_3}}{C}HCH_2\underset{\stackrel{|}{CH_3}}{C}H^+}$$

$$\mathrm{CH_3\underset{\stackrel{|}{CH_3}}{C}HCH_2\underset{\stackrel{|}{CH_3}}{C}HCH_2\underset{\stackrel{|}{CH_3}}{C}H^+ + CH_2{=}CHCH_3 \longrightarrow CH_3\underset{\stackrel{|}{CH_3}}{C}HCH_2\underset{\stackrel{|}{CH_3}}{C}HCH_2\underset{\stackrel{|}{CH_3}}{C}HCH_2\overset{+}{C}HCH_3}$$

$$\mathrm{CH_3\underset{\stackrel{|}{CH_3}}{C}HCH_2\underset{\stackrel{|}{CH_3}}{C}HCH_2\underset{\stackrel{|}{CH_3}}{C}HCH_2\overset{+}{C}HCH_3 \xrightarrow{-H+} CH_3\underset{\stackrel{|}{CH_3}}{C}HCH_2\underset{\stackrel{|}{CH_3}}{C}HCH_2\underset{\stackrel{|}{CH_3}}{C}HCH{=}CHCH_3}$$

$$+$$

$$\mathrm{CH_3\underset{\stackrel{|}{CH_3}}{C}HCH_2\underset{\stackrel{|}{CH_3}}{C}HCH_2\underset{\stackrel{|}{CH_3}}{C}HCH_2CH{=}CH_2}$$

7.40

The halogen does not go to "the less hydrogenated carbon" as predicted by the original version of Markovnikov's rule, but the reaction does proceed through the *formation of the more stable carbocation* as required by the modern form of the rule. Of the two paths below, the reaction follows path (1) because the carbocation formed in the path (1) is more stable. This is true even though the carbocation is a 1° cation and the carbocation in path (2) is a 2° cation. The positive charge of the carbocation formed in (2) is located on a carbon that is directly attached to a highly electron-withdrawing group, the CF_3- group. (The group is highly electron withdrawing because of the combined electronegativities of the three fluorines.) The electron-withdrawing CF_3- group intensifies the positive charge of $CF_3\overset{+}{C}HCH_3$ and makes it highly unstable.

$$\text{(1)}\ \mathrm{CF_3CH{=}CH_2 \xrightarrow{H^+} CF_3CH_2\overset{+}{C}H_2 \xrightarrow{Cl^-} CF_3CH_2CH_2Cl}$$

more stable
even though 1°

$$\text{(2)}\ \mathrm{CF_3CH{=}CH_2 \xrightarrow[\times]{H^+} CF_3\overset{+}{C}HCH_3 \xrightarrow{Cl^-} CF_3\underset{\stackrel{|}{Cl}}{C}HCH_3}$$

less stable
even though 2°

(not formed)

We examine this reaction again in Sect. 12.9B.

7.41

X = Br is the only case in which both propagation steps are exothermic (ΔH° is negative). Each of the other reactions has one step with a high E_{act}:

$\dot{X}$ = F, $E_{act} \geqslant +$ 38 kcal/mole (2nd step)
$\dot{X}$ = Cl, $E_{act} \geqslant +$ 5 kcal/mole (2nd step)
$\dot{X}$ = I, $E_{act} \geqslant +$ 7 kcal/mole (1st step)

CHAPTER EIGHT
Stereochemistry

8.1

Chiral (a) screw, (e) foot, (f) ear, (g) shoe, (h) spiral staircase
Achiral (b) plain spoon, (c) fork, (d) cup

8.2

(b) Yes. (c) No. (d) No.

8.3

(a) Yes. (b) Yes. (c) No. (d) No.

8.4

(a) 1-Chloropropane, (c) 2-methyl-1-chloropropane, (d) 2-methyl-2-chloropropane, (f) 1-chloropentane, and (h) 3-chloropentane are all achiral.

(b) I: Cl, H, Br, I; II: Cl, Br, H, I

(c) I: CH_3, H, Br, CH_2CH_3; II: CH_3, Br, H, CH_2CH_3

(g) I: CH_3, H, Cl, $CH_2CH_2CH_3$; II: CH_3, Cl, H, $CH_2CH_2CH_3$

8.5

(a) 1. X, H, H, H 2. H, X, X, H; H, X, Y, H 3. H, Z, X, Y; H, X, Z, Y

(b) 1. One

2. Two: H---X, C, H---X and X---H, C, H---X, or H---X, C, H---Y and H---X, C, Y---H

3. Three:

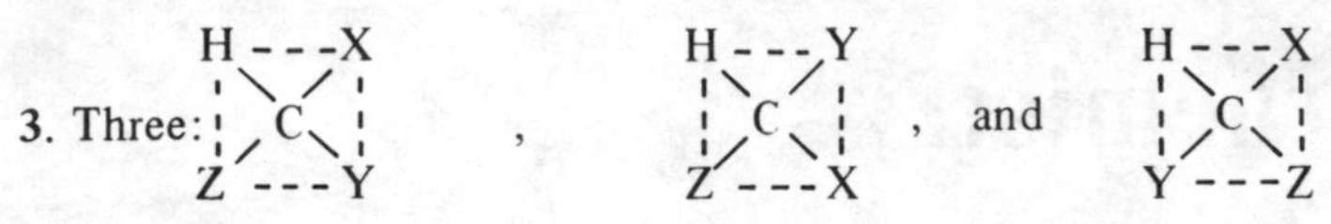

(c) 1. One

2. Three:

or

3. Six:

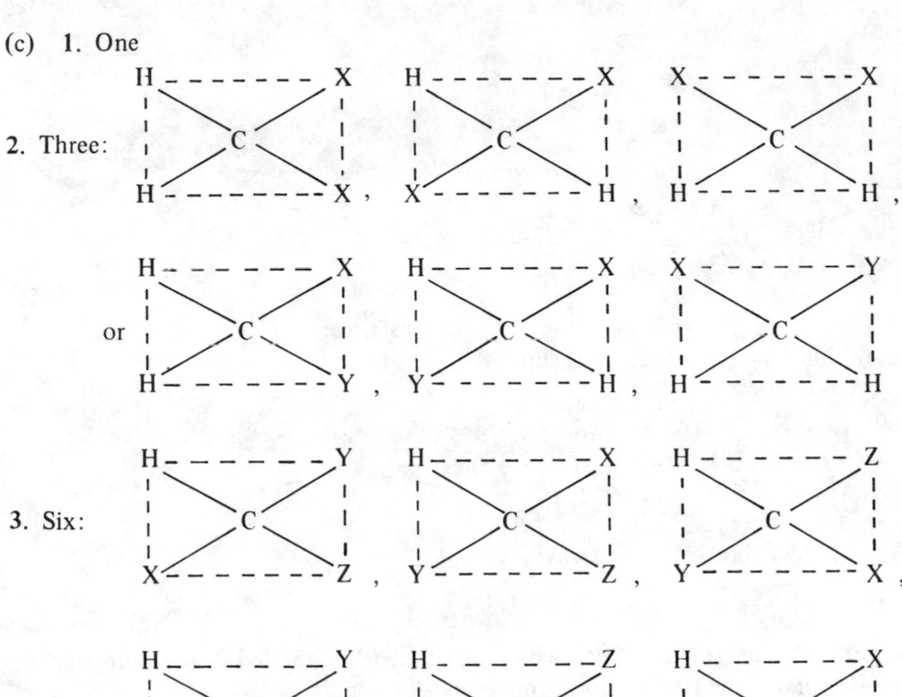

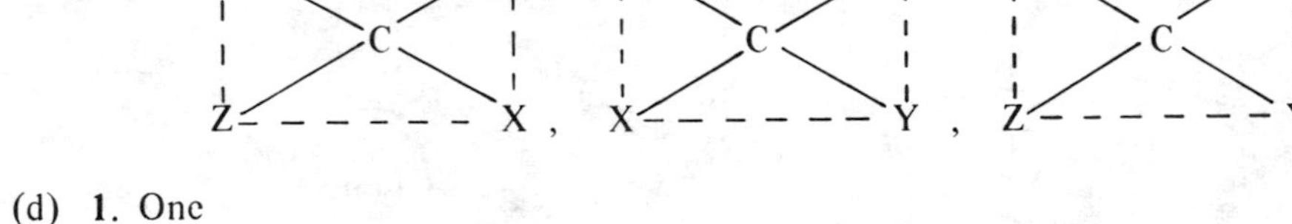

(d) 1. One

2. Two or three:

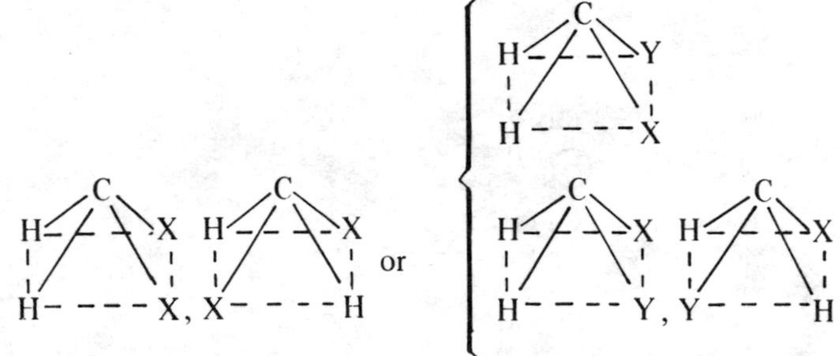

3. Six:

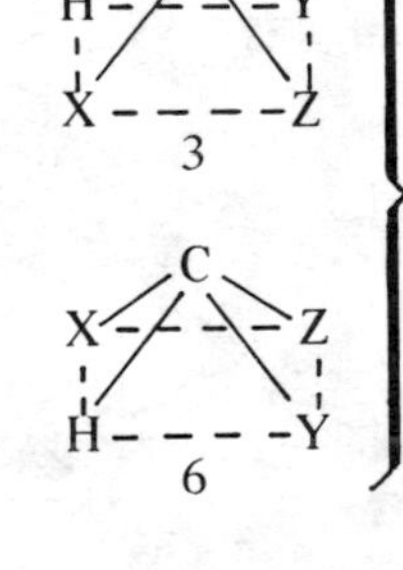

enantiomeric pairs

8.6

(b) plain spoon, (c) fork, (d) cup all possess a plane of symmetry.

8.7

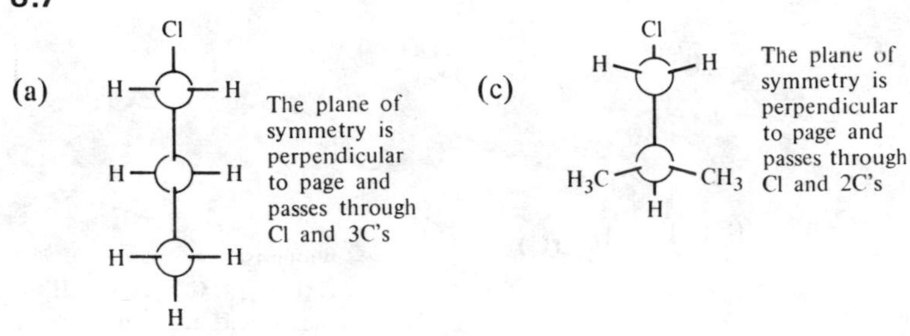

(a) The plane of symmetry is perpendicular to page and passes through Cl and 3C's

(c) The plane of symmetry is perpendicular to page and passes through Cl and 2C's

(d) A vertical plane perpendicular to page passes through Cl, tertiary C, and CH_3 at bottom

(f) A plane perpendicular to page passes through Cl and 5C's

(h) A plane perpendicular to page passes through Cl, C, and H

8.8

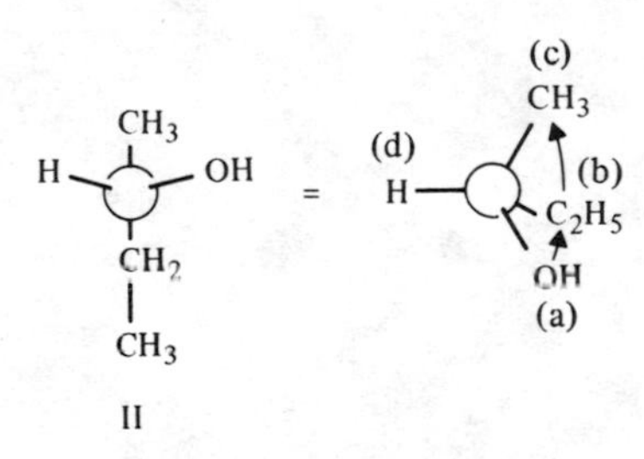

From priority a to b to c, the direction is counterclockwise, therefore II is S–2–butanol

8.9

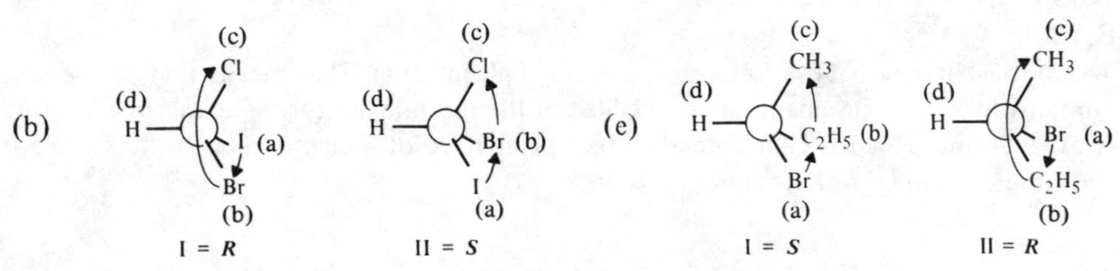

(g) (d) H—C(—C_3H_7 (b))(—CH_3 (c))(—Cl (a))   (d) H—C(—CH_3 (c))(—Cl (a))(—C_3H_7 (b))

I = *S*   II = *R*

8.10

Phenyl (—C_6H_5) is treated as though it were a ring in which each ring carbon carries duplicated (C) atoms and H atoms,

which is:
C then C,C,C, then C,C,H
then C,C,H either way
around the ring

—C≡C—H is treated as though it were —C(C)(C)—C(C)(C)—H

which is:
C, then C,C,C, then C,C,H
then nothing.

—$C(CH_3)_3$ is —C(CH_3)(CH_3)—C(H)(H)—H

which is:
C, then C,C,C, then H,H,H

8.11

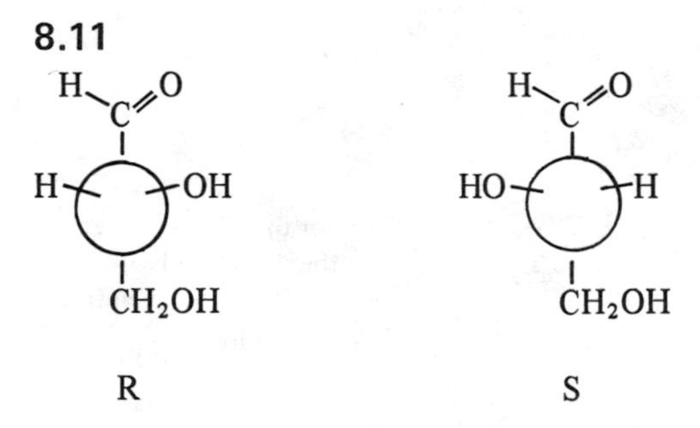

8.12

(a) *R* (b) *R* (c) *R*

8.13

The optical purity is 50% (see previous paragraph in text). That means that the sample contains 50% of the *S*-enantiomer and 50% of the racemic mixture. The racemic mixture is 50% *S*- and 50% *R*. Therefore the total percentage of *S*-enantiomer in the sample is 75%, the percentage of *R*-enantiomer is 25%.

8.14

(a) $CH_3CH_2CH_2CHCH_2CH_3$ (Racemic modification)
with CH_3 on the CH carbon

(b) $CH_3CH_2CHOHCH_3$ (Racemic modification) (c) Same as (b)

8.15

(a) Diastereomers (b) Diastereomers (c) Diastereomers

(d)

	1	2	3	4
1		enantiomers	diastereomers	diastereomers
2	enantiomers		diastereomers	diastereomers
3	diastereomers	diastereomers		enantiomers
4	diastereomers	diastereomers	enantiomers	

(e) Yes. (f) No.

8.16

(a) **A** alone would be optically active.

(b) **B** alone would be optically active.

(c) **C** would not be optically active because it is a meso compound.

(d) An equimolar mixture of **A** and **B** would not be optically active because it is a racemic modification.

8.17

(1) **C** or **D**. (2) **A**. (3) **B**.

8.18

(a)

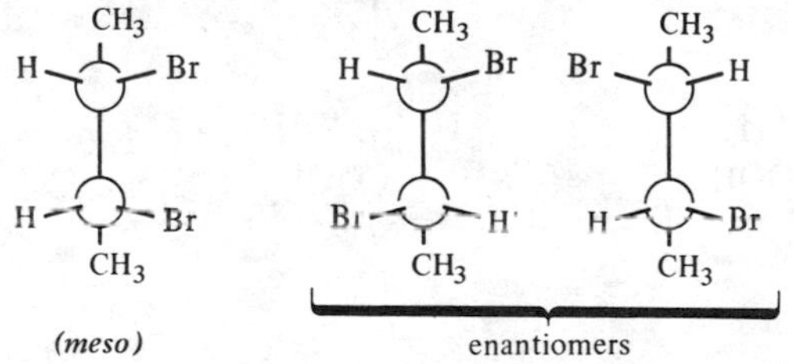

CH_3; H, Br / H, Br; CH_3 — *(meso)*

CH_3; H, Br / Br, H; CH_3 and CH_3; Br, H / H, Br; CH_3 — enantiomers

(b)

CH_3; H, Cl / H, Br; CH_3 and CH_3; Cl, H / Br, H; CH_3 — enantiomers

CH_3; Cl, H / H, Br; CH_3 and CH_3; H, Cl / Br, H; CH_3 — enantiomers

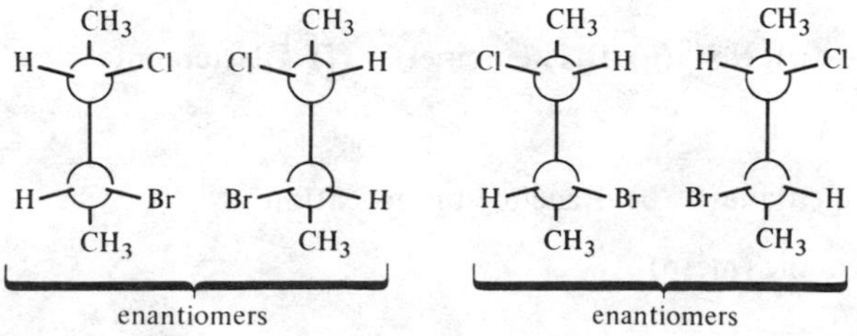

(c)

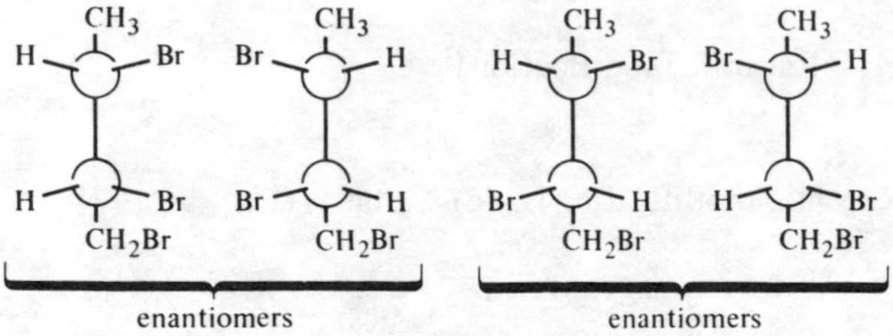

CH_3 H Br | CH_3 Br H | CH_3 H Br | CH_3 Br H

H Br CH_2Br | Br H CH_2Br | Br H CH_2Br | H Br CH_2Br

enantiomers enantiomers

(d)

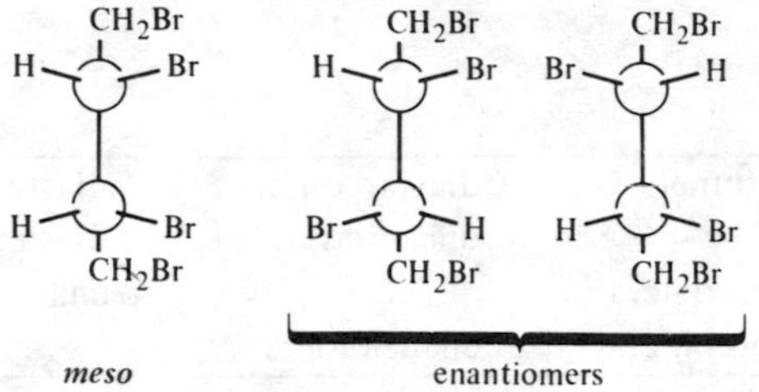

CH_2Br H Br | CH_2Br H Br | CH_2Br Br H

H Br CH_2Br | Br H CH_2Br | H Br CH_2Br

meso enantiomers

(e)

CH_3 H Cl | CH_3 H Cl | CH_3 H Cl | CH_3 Cl H

H Cl | Cl H | H Cl | Cl H

H Cl CH_3 | H Cl CH_3 | Cl H CH_3 | H Cl CH_3

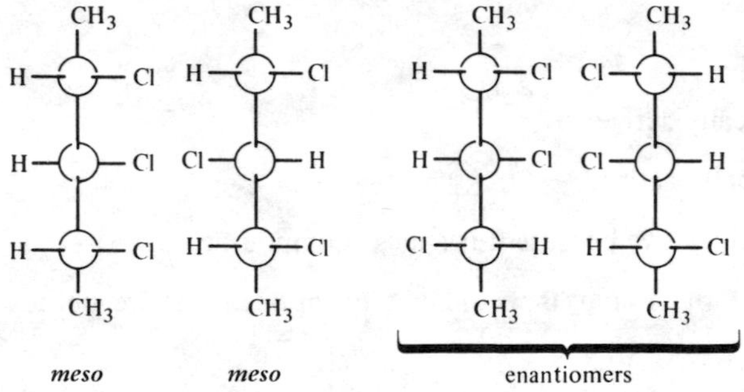

meso *meso* enantiomers

8.19

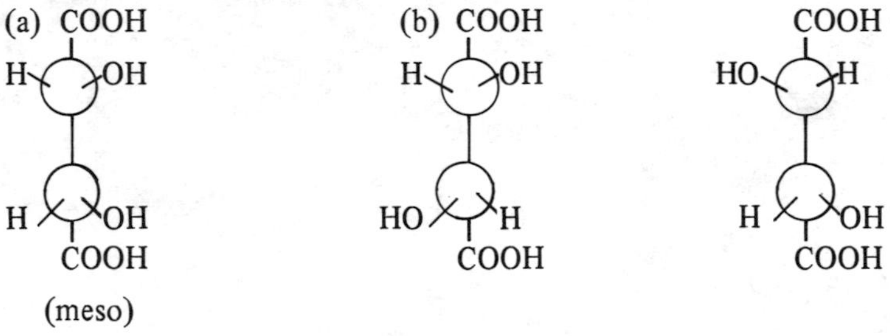

(a) COOH H OH H OH COOH (meso)

(b) COOH H OH HO H COOH

COOH HO H H OH COOH

(c) No (d) A racemic modification

8.20

(a) No (b) Yes (c) No (d) No (e) Diastereomers (f) Diastereomers

8.21

(a) *Trans*-1,2-dibromocyclopentanes (b) Racemic modification

(c) *Cis*-1,2-dibromocyclopentane (meso)

8.22

B: (2S,3S)-2,3-Butanediol, **C**: (2S,3R)-2,3-butanediol

8.23

(a) meso is (2S,3R)-2,3-dibromobutane, the two enantiomers are (2S,3S)-2,3-dibromobutane and (2R,3R)-2,3-dibromobutane.

(b) I: (2S,3R)-2-chloro-3-bromobutane
II: (2R,3S)-2-chloro-3-bromobutane
III: (2R,3R)-2-chloro-3-bromobutane
IV: (2S,3S)-2-chloro-3-bromobutane

8.24

(a) 2, 3-Dibromobutane (racemic modification):

$H_3C(H)C{=}C(H)CH_3$ + Br_2 → bromonium ion (CH$_3$, H$_3$C, H, H, :Br:$^+$) + :Br:$^-$

:Br:$^-$ + bromonium ion (CH$_3$, H$_3$C, H, H, Br:$^+$) → a: CH$_3$, H, Br / Br, H, CH$_3$; b: CH$_3$, Br, H / H, Br, CH$_3$

(b) *Meso* -2, 3-dibromobutane:

$H_3C(H)C{=}C(H)CH_3$ + Br_2 → bromonium ion (CH$_3$, H, H, CH$_3$, Br:$^+$) + enantiomer + :Br:$^-$

:Br:$^-$ + bromonium ion (CH$_3$, H, H, CH$_3$, Br:$^+$) → a: CH$_3$, H, Br / H, Br, CH$_3$; b: CH$_3$, Br, H / Br, H, CH$_3$

These molecules are identical

The enantiomer gives the same result in this step as well.

8.25

(a) H–C(CH$_3$)–OH / H–C(CH$_3$)–OH (*meso*)

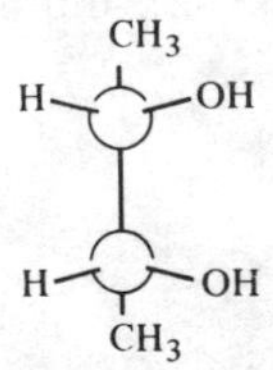

(b) H–C(CH$_3$)–OH / HO–C(CH$_3$)–H + HO–C(CH$_3$)–H / H–C(CH$_3$)–OH (racemic modification)

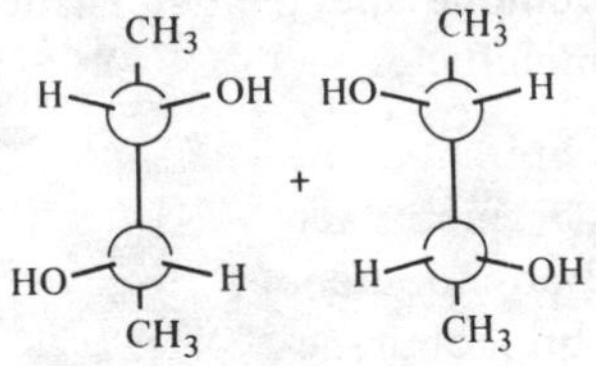

8.26

(a) (b) *(R)*-(−)-glyceric acid

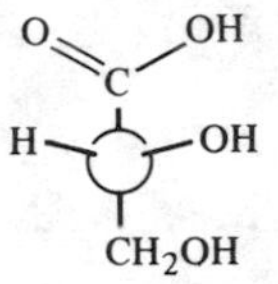

(a)(d) *(S)*-(−)-3-bromo-2-hydroxypropanoic acid

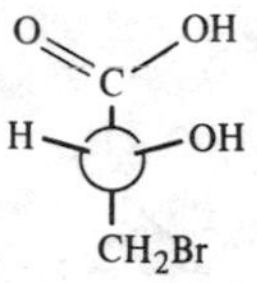

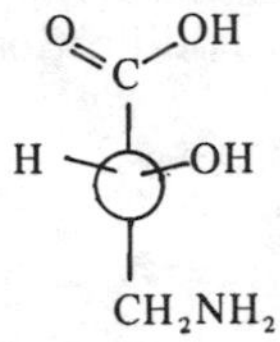

CH_2NH_2 (*R*)-(+)-isoserine

(e) *(R)*-(−)-lactic acid

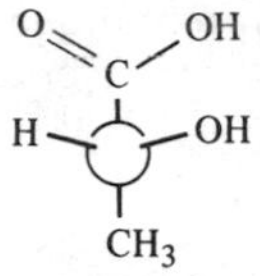

8.27

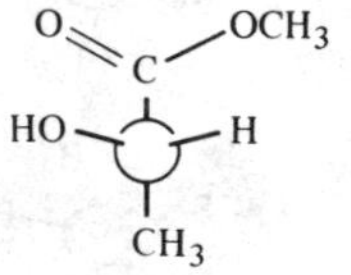

(S)-(−)-methyl lactate

8.28

We know that the reaction of secondary alkyl halides with OH^- is S_N2 and therefore involves inversion of configuration at the carbon atom that bears the bromine atom. We could relate the configurations of the bromide and the alcohol by allowing a pure enantiomer, say (−)-2-bromobutane, to react with $NaOH/H_2O$. If we obtain the (+)-alcohol, which we know has the *S*-configuration, then we would know that (−)-2-bromobutane has the *R* configuration.

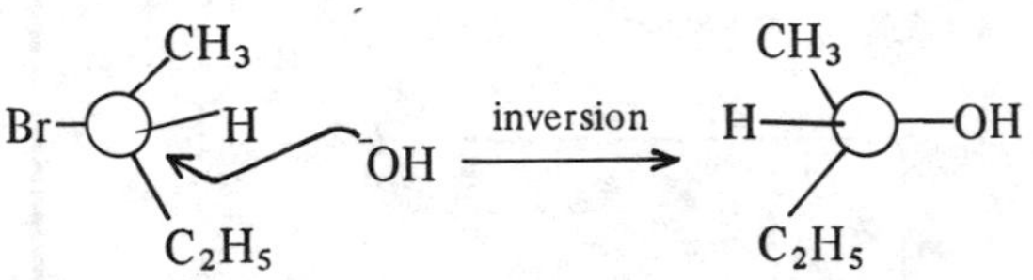

R-(−)-2-bromobutane $\xrightarrow{\text{inversion}}$ *S*-(+)-2-butanol

8.29

The overall process involves racemization. A possible mechanism involves the formation of an achiral carbocation:

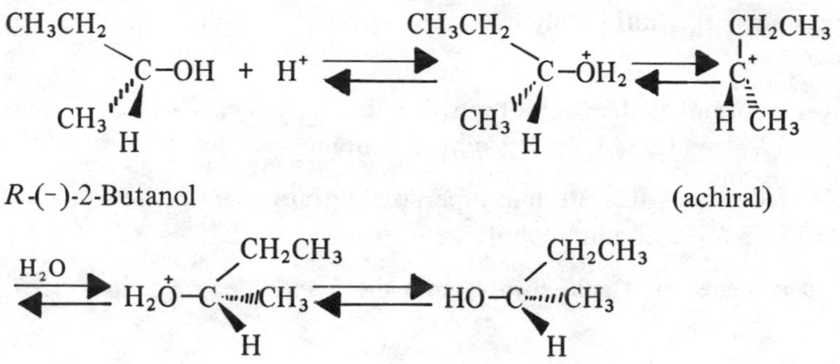

$$CH_3CH_2(CH_3)(H)C{-}OH + H^+ \rightleftharpoons CH_3CH_2(CH_3)(H)C{-}\overset{+}{O}H_2 \rightleftharpoons CH_3CH_2\overset{+}{C}(H)CH_3$$

R-(−)-2-Butanol (achiral)

$$\underset{H_2O}{\rightleftharpoons} H_2\overset{+}{O}{-}C(CH_2CH_3)(H)CH_3 \rightleftharpoons HO{-}C(CH_2CH_3)(H)CH_3$$

S-(+)-2-Butanol

Once formed the carbocation can revert to R-(−)-2-butanol or become S-(+)-2-butanol. After continued heating a racemate will be produced.

8.30

(a) COOH; HO–C–H; H–C–OH; COOH

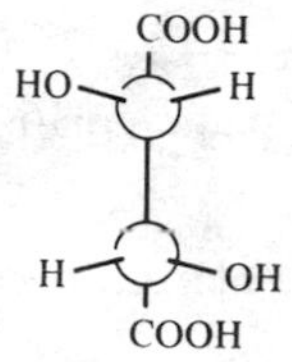

(−)–tartaric acid

(b) COOH; H–C–OH; H–C–OH; COOH

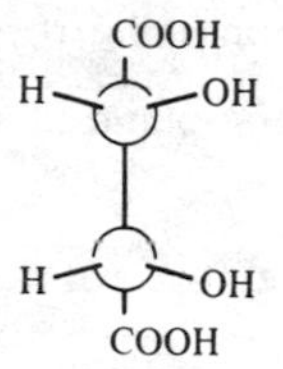

(*meso*)–tartaric acid

(c) No, *meso*-tartaric acid would *not* be optically active.

8.31

They are diastereomers.

8.32

(a) (*Z*) - 1-Bromo-1-chloro-1-butene
(b) (*Z*) - 2-Bromo-1-chloro-1-iodopropene
(c) (*E*) - 3-Ethyl-4-methyl-2-pentene
(d) (*E*) - 1-Chloro-1-fluoro-2-methyl-1-butene

8.33

(a) Isomers are different compounds that have the same molecular formula. C_2H_6O: CH_3CH_2OH and CH_3OCH_3

(b) Structural isomers are isomers that differ because their atoms are joined in a different order. C_4H_{10}: $CH_3CH_2CH_2CH_3$ and $CH_3CH(CH_3)CH_3$

(c) Stereoisomers are isomers that differ only in the arrangement of their atoms in space: *cis*- and *trans* -2-butene.

(d) Diastereomers are stereoisomers that are not mirror reflections of each other: *cis*- and *trans* -2-butene, or (2 *S*, 3 *S*)- and (2 *S*, 3 *R*)- 2, 3-dibromobutane.

(e) Enantiomers are stereoisomers that are non-superposable mirror reflections of each other: (2 *S*, 3 *S*)- and (2 *R*, 3 *R*)- 2, 3-dibromobutane.

(f) A meso compound is made up of achiral molecules that contain chiral centers: (2 *S*, 3 *R*)- 2, 3-dibromobutane.

(g) A racemic modification is an equimolar mixture of a pair of enantiomers.

(h) A plane of symmetry is an imaginary plane that bisects a molecule in such a way that the two halves of the molecule are mirror reflections of each other. (See Fig. 8.7.)

(i) A chiral center is any tetrahedral atom that has four different groups attached to it.

(j) A chiral molecule is one that is not superposable on its mirror reflection.

(k) An achiral molecule is superposable on its mirror reflection.

(l) Optical activity is the rotation of the plane of polarization of plane polarized light by a substance placed in the light path.

(m) A dextrorotatory substance is one that rotates the plane of polarization of plane polarized light in a clockwise direction.

(n) A reaction occurs with retention of configuration when all the groups around the chiral atom retain the same relative configuration after the reaction that they had before the reaction.

8.34

(a) Enantiomers (b) Same (c) Enantiomers (d) Diastereomers (e) Same

(f) Structural isomers (g) Same (h) Diastereomers (i) Same

(j) Enantiomers (k) Same (l) Enantiomers (m) Same (n) Structural isomers (o) Same (p) Diastereomers (q) Enantiomers

8.35

(a)

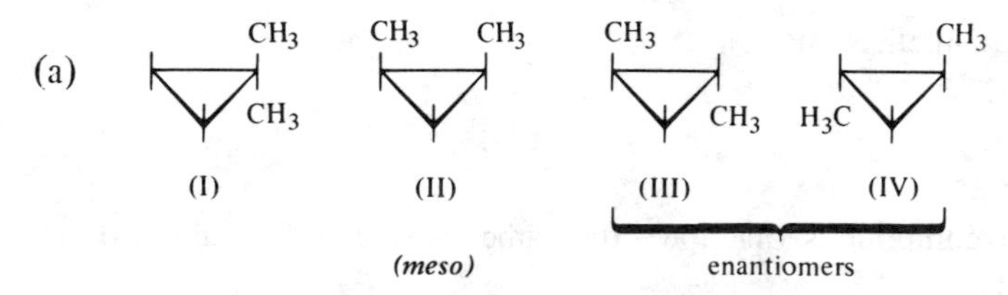

(b) III and IV (c) II (d) Three: I, II, and a mixture of III and IV. (e) None, since the only chiral molecules are **III** and **IV**, and they would be obtained in the same amounts as a racemic modification.

8.36

(a) 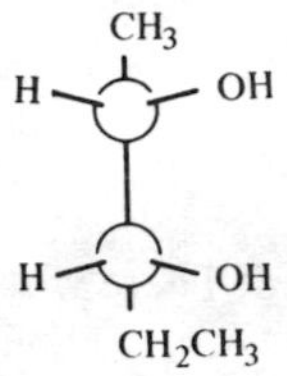 and enantiomer, (b) 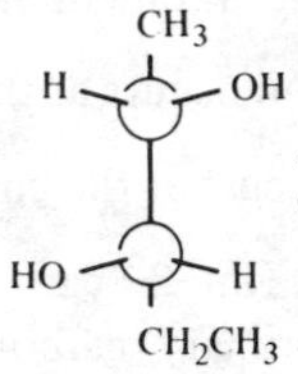 and enantiomer, (c) 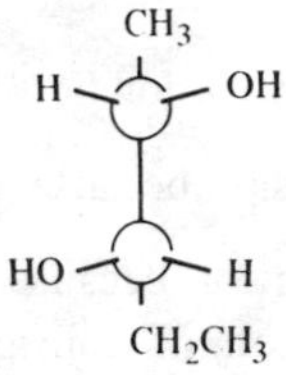 and enantiomer,

(d) 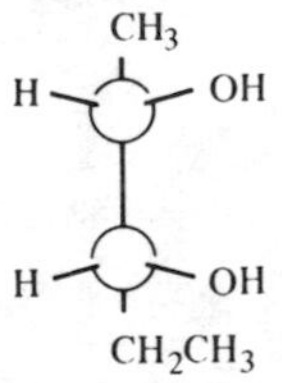 and enantiomer, (e) 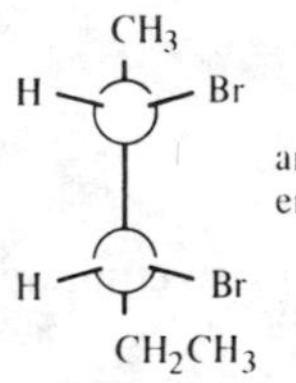 and enantiomer, (f) 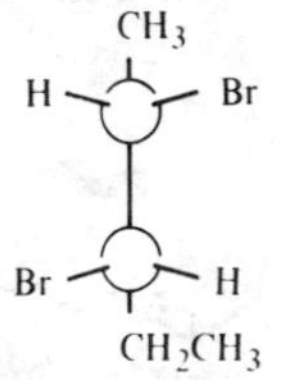 and enantiomer,

8.37

(a) 2*S*, 3*R* - (the enantiomer is 2*R*,3*S*) (b) 2 *S*, 3 *S* - (the enantiomer is 2 *R*, 3 *R* -)

(c) Same as (b) (d) Same as (a) (e) 2 *S*, 3 *R* - (the enantiomer is 2 *R*, 3 *S* -)

(f) 2 *S*, 3 *S* - (the enantiomer is 2 *R*, 3 *R* -)

8.38

(a) 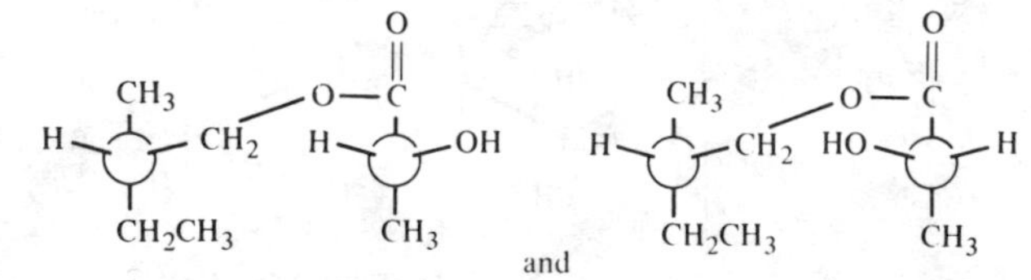 (*S*) (*R*) and (*S*) (*S*)

(b) They are diasteromers.

(c) The boiling points of these esters *will* be different. If there is a large enough difference in boiling points, then separation by fractional distillation will be possible.

(d) Yes.

(e) After separation of the diastereomeric esters by fractional distillation, they could each be hydrolyzed to yield the separate enantiomeric acids.

8.39

(a) Retention because the chiral carbon is not involved.

(b) Retention because the chiral carbon is not involved.

(c) Inversion because an S_N2 reaction occurs at the chiral carbon.

(d) Racemization because proton transfer from H_3O^+ to alcohol is followed by formation of an achiral carbocation (see the answer to Problem 8.29).

(e) Racemization because an S_N1 reaction occurs at the chiral carbon.

(f) Inversion because an S_N2 reaction occurs at the chiral carbon.

(g) Retention because the chiral carbon (the 3° carbon) is not involved. There would be an inversion of configuration at the primary carbon of the CH_2Cl group when it is attacked by OH^-. We would not be able to detect this inversion of configuration because this carbon is not chiral.

8.40

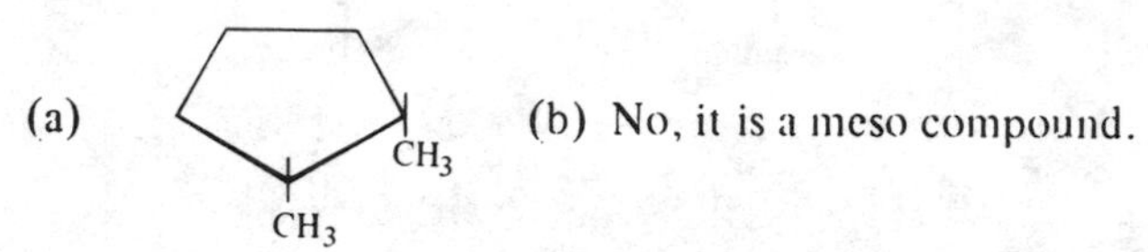

8.41

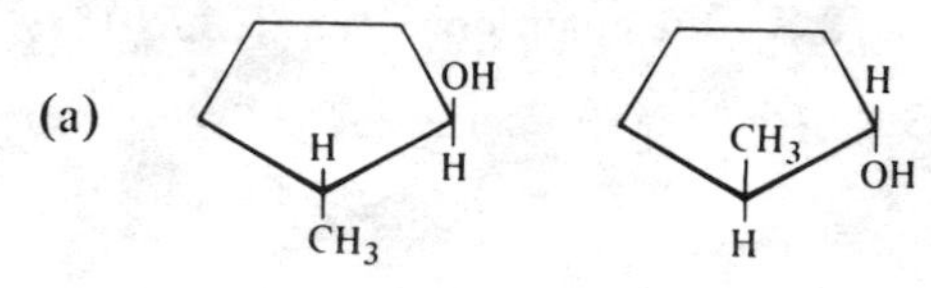

(b) CH_3 — H + $(BH_3)_2$ → [B— H H CH_3 and H CH_3 B— H]

The BH_3 group may attack the double bond from either side of the ring

→ B— H H CH_3 + H CH_3 B— H

H_2O_2, OH^-

OH H H CH_3 H CH_3 OH H

8.42

(a) Four

(b) $CH_3(CH_2)_5$—CH(OH)(H)—CH_2—CH=CH—$(CH_2)_7COOH$ (cis) + enantiomer

$CH_3(CH_2)_5$—CH(OH)(H)—CH_2—CH=CH—$(CH_2)_7COOH$ (trans) + enantiomer

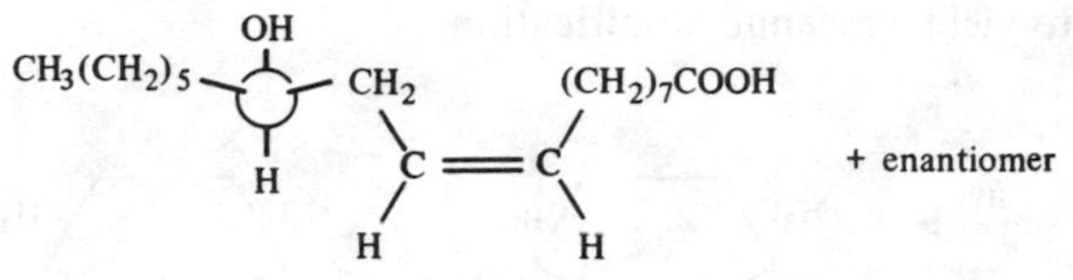

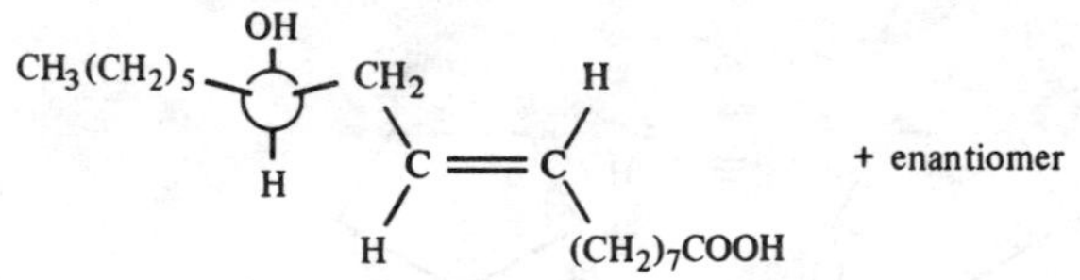

8.43

Hydroxylations by $KMnO_4$ are *syn* hydroxylations (cf p. 265). Thus, maleic acid must be the *cis*-dicarboxylic acid:

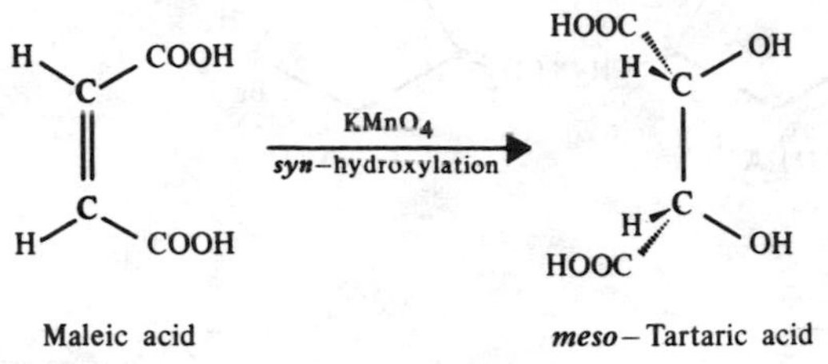

$$\text{Maleic acid} \xrightarrow[\textit{syn}\text{-hydroxylation}]{KMnO_4} \textit{meso}\text{-Tartaric acid}$$

Maleic acid — *meso*–Tartaric acid

Fumaric acid must be the *trans*-dicarboxylic acid:

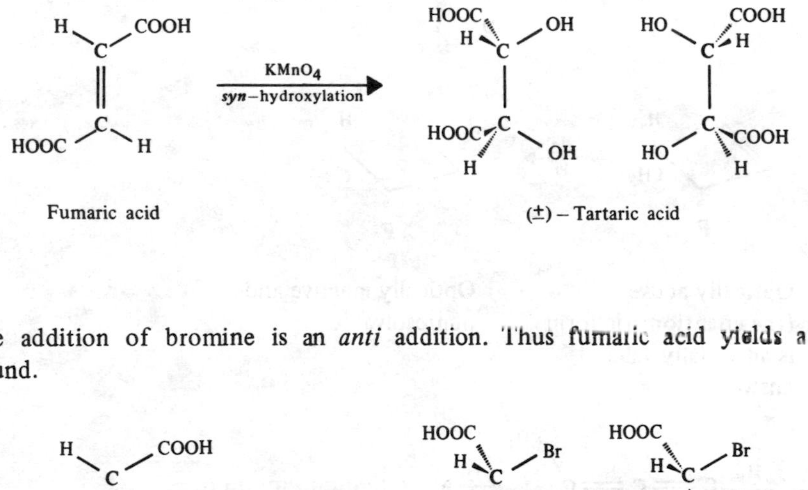

$$\text{Fumaric acid} \xrightarrow[\textit{syn}\text{-hydroxylation}]{KMnO_4} (\pm)\text{-Tartaric acid}$$

Fumaric acid — (±) – Tartaric acid

8.44

(a) The addition of bromine is an *anti* addition. Thus fumaric acid yields a *meso* compound.

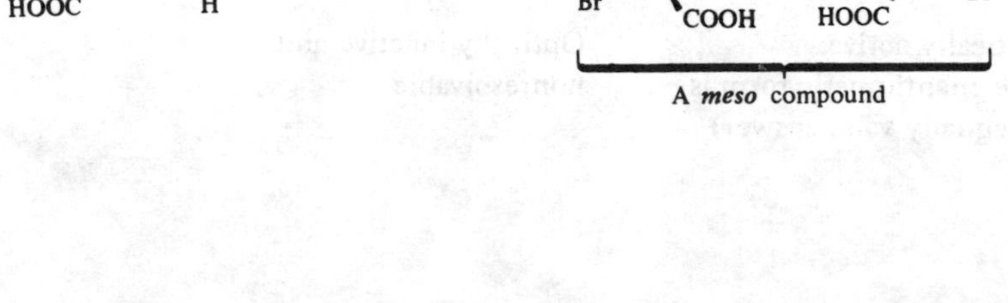

$$\text{Fumaric acid} + Br_2 \xrightarrow[\text{addition}]{\textit{anti}} \text{HOOC(H)(Br)C—C(Br)(H)COOH}$$

A *meso* compound

(b) Maleic acid adds bromine to yield a racemic modification.

8.45

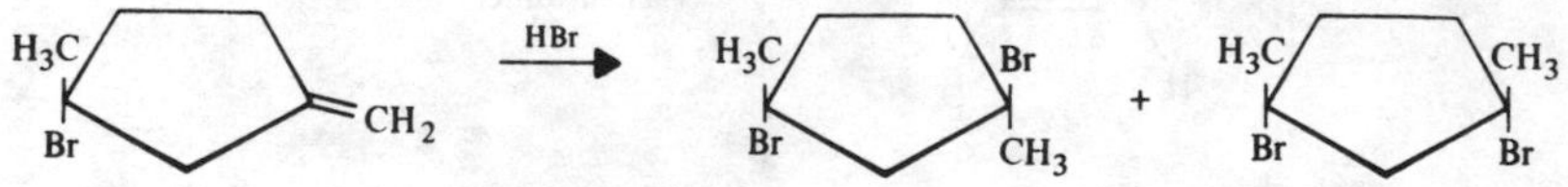

(+) A $\xrightarrow{HBr}$ B (optically active) + C (a *meso* compound)

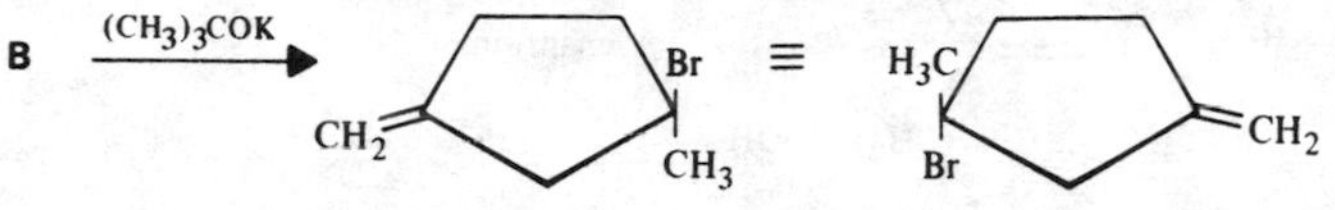

B $\xrightarrow{(CH_3)_3COK}$ (+) A ≡ (+) A

C $\xrightarrow{(CH_3)_3COK}$ (+) A + (−) A

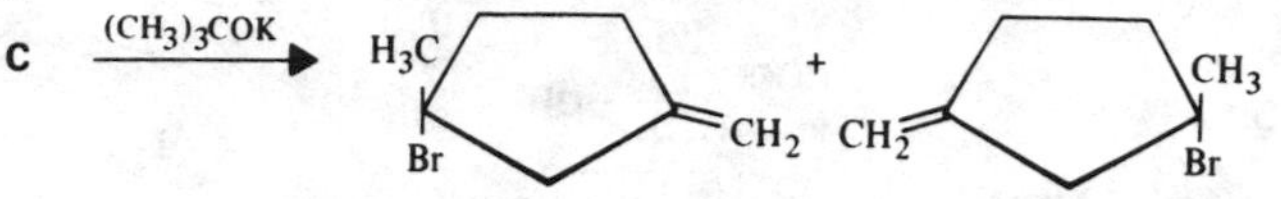

A $\xrightarrow{(CH_3)_3COK}$ D $\xrightarrow{(1)\ O_3,\ (2)\ Zn,\ H_2O}$ + 2HCH(=O)

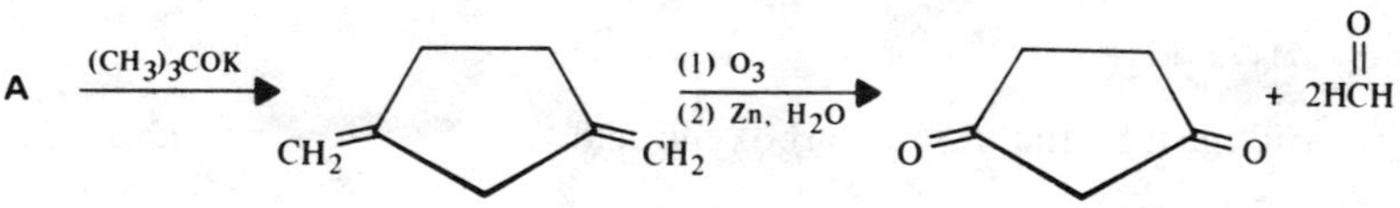

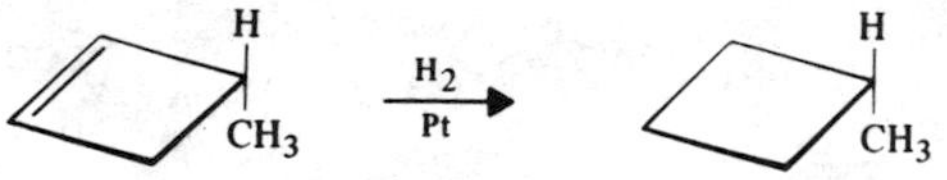

8.46

(a) E $\xrightarrow[Pt]{H_2}$ F

E: Optically active (the enantiomeric form is an equally valid answer)

F: Optically inactive and nonresolvable

(b) $CH_3CH_2(H)C{=}C{=}C(CH_3)H$ $\xrightarrow[Pt]{H_2}$ $CH_3CH_2CH_2CH_2CH_2CH_3$

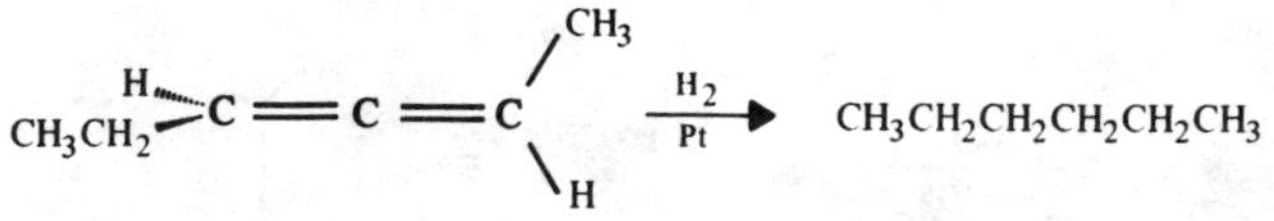

G: Optically active (the enantiomeric form is an equally valid answer)

H: Optically inactive and nonresolvable

8.47

That **I** and **J** rotate plane-polarized light in the same direction tells us that **I** and **J** are not enantiomers of each other. Thus, the following are possible structures for **I**, **J** and **K**. (The enantiomers of **I**, **J** and **K** would form another set of structures, and other answers are possible as well.)

$CH_3CH(CH_3)$—, CH_3, H, $CH=CH_2$ (Newman projection)

I

Optically active

H_2, Pt →

$CH_3CH(CH_3)$—, CH_3, H, CH_2CH_3 (Newman projection)

K

Optically active

$CH_2=C(CH_3)$—, CH_3, H, CH_2CH_3 (Newman projection)

J

Optically active

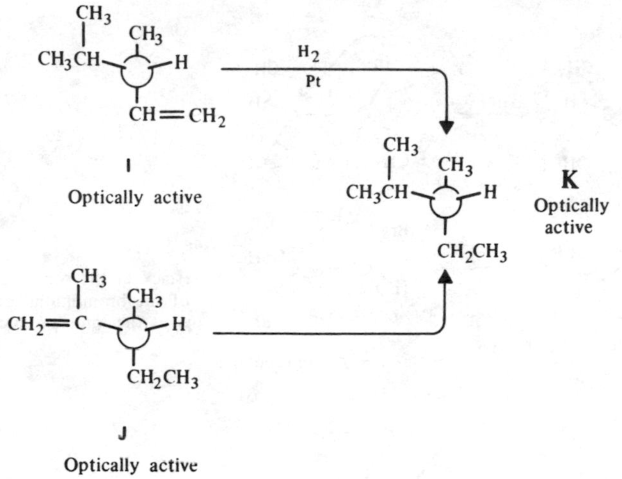

8.48

Possible structures are:

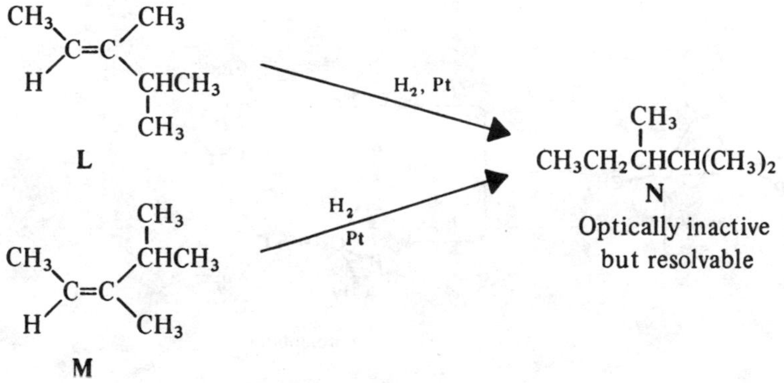

$CH_3(H)C=C(CH_3)CH(CH_3)_2$

L

H_2, Pt →

$CH_3CH_2\overset{CH_3}{CH}CH(CH_3)_2$

N

Optically inactive but resolvable

$CH_3(H)C=C(CH_3)CH(CH_3)_2$

M

H_2, Pt →

(Other answers are possible as well.)

8.49

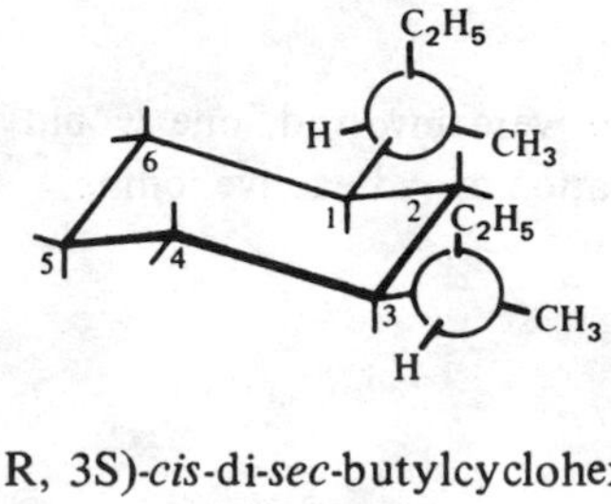

(Chair cyclohexane, carbons numbered 1–6, with sec-butyl groups at C-1 and C-3: H, CH_3, C_2H_5 substituents shown in Newman projections)

The (1R, 3S)-*cis*-di-*sec*-butylcyclohexane shown above is optically inactive because it is a *meso*-compound. Since the *sec*-butyl groups are of opposite chirality, it has a plane of symmetry perpendicular to the ring and passing through atoms 2 and 5.

*8.50

The reactions proceed through the formation of bromonium ions identical to those formed in the bromination of *trans*- and *cis*-2-butene (see problem 8.24).

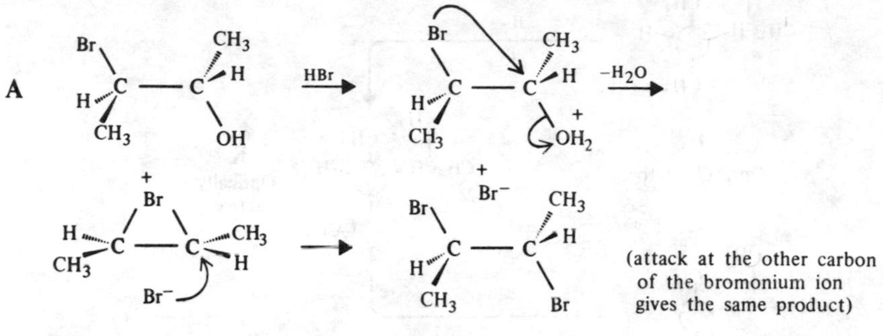

(attack at the other carbon of the bromonium ion gives the same product)

meso–2,3–dibromobutane

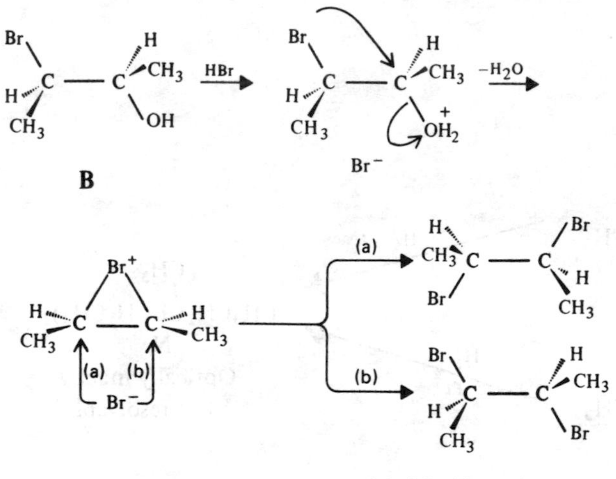

($\pm$)–2,3–dibromobutane

*8.51

(a) If every substitution *involves an inversion*, then racemization will be complete when only *half* the substrate has incorporated radioactive iodine. (At this point there will be an equimolar mixture of the two enantiomers.) Thus, the rate of racemization *should be twice the rate of incorporation of radioactive iodine*.

(b) If an achiral intermediate such as a carbocation were involved, one would expect the rate of racemization to equal the rate of incorporation of radioactive iodine.

C

Special Topic

C.1

The reaction will proceed through the most stable free radical that can be produced. Head-to-head polymerization, as shown, will lead to a primary radical,

$$\underset{CH_3}{R-CH_2-\underset{|}{CH}\cdot} + \underset{CH_3}{\underset{|}{CH}=CH_2} \longrightarrow R-CH_2-\underset{CH_3}{\underset{|}{CH}}-\underset{CH_3}{\underset{|}{CH}}-CH_2\cdot,$$

which is less stable than the secondary radical that is produced by head-to-tail polymerization:

$$R-CH_2-\underset{CH_3}{\underset{|}{CH}}\cdot + CH_2=\underset{CH_3}{\underset{|}{CH}} \longrightarrow R-CH_2-\underset{CH_3}{\underset{|}{CH}}-CH_2-\underset{CH_3}{\underset{|}{CH}}\cdot$$

C.2

(a) $$nCH_2=\underset{F}{\underset{|}{CH}} \xrightarrow[\text{peroxide}]{\text{organic}} \left(CH_2-\underset{F}{\underset{|}{CH}}\right)_n$$

(b) $$nCF_2=\underset{Cl}{\underset{|}{CF}} \xrightarrow[\text{peroxide}]{\text{organic}} \left(CF_2-\underset{Cl}{\underset{|}{CF}}\right)_n$$

(c) $$nCF_2=\underset{CF_3}{\underset{|}{CF}} + mCH_2=CF_2 \xrightarrow[\text{peroxide}]{\text{organic}} \left(CF_2-\underset{CF_3}{\underset{|}{CF}}\right)_n\left(CH_2-CF_2\right)_m$$

Note that the units are randomly ordered, and not necessarily joined to their own kind as shown.

C.3

Polymerization will occur to produce the most stable carbocation possible. The scheme shown in this problem involves formation of the primary carbocations,

$$\underset{CH_3}{\overset{CH_3}{\underset{|}{\overset{|}{CH}}}}-CH_2^+\ ,\quad \underset{CH_3}{\overset{CH_3}{\underset{|}{\overset{|}{CH}}}}-CH_2-\underset{CH_3}{\overset{CH_3}{\underset{|}{\overset{|}{C}}}}-CH_2^+\ ,\ \text{etc.}$$

instead of the tertiary carbocations,

$$\begin{array}{c} \quad CH_3 \qquad\quad CH_3 \\ \quad | \qquad\qquad\; | \\ CH_3-C-CH_2-C^{+} \\ \quad | \qquad\qquad\; | \\ \quad CH_3 \qquad\quad CH_3 \end{array}$$

C.4

(a) By proton transfer from water to the strongly basic carbanion,

$$R-CH_2-\underset{\displaystyle CN}{\underset{|}{CH}}:^{-} + H-\underset{\displaystyle H}{\underset{|}{\ddot{O}}}: \longrightarrow R-CH_2-\underset{\displaystyle CN}{\underset{|}{CH_2}} + :\ddot{\underset{..}{O}}H^{-}$$

(b)

$$\left(CH_2\underset{\displaystyle R}{\underset{|}{CH}}\right)_n CH_2\underset{\displaystyle R}{\underset{|}{CH}}:^{-} + (m+1)\underset{O}{CH_2-CH_2} \longrightarrow$$

$$\left(CH_2\underset{\displaystyle R}{\underset{|}{CH}}\right)_{n+1}\left(CH_2-CH_2-O\right)_m CH_2-CH_2-\ddot{\underset{..}{O}}:^{-}$$

$$\xrightarrow{H_2O} \left(CH_2\underset{\displaystyle R}{\underset{|}{CH}}\right)_{n+1}\left(CH_2-CH_2-O\right)_m CH_2-CH_2-OH$$

In this polymer, each chain consists of a long uninterrupted segment of the first repeating unit, $\left(CH_2\underset{\displaystyle R}{\underset{|}{CH}}\right)_{n+1}$, followed by a long uninterrupted segment of the second repeating unit, $\left(CH_2-CH_2-O\right)_m$.

C.5

(a)

Above chain	H	CH_3	H	CH_3	H	OCH_3–C=O	H	CH_3	H	OCH_3–C=O	H	OCH_3–C=O	H	OCH_3–C=O	H	CH_3
Below chain	H	O=C–OCH_3	H	O=C–OCH_3	H	CH_3	H	O=C–OCH_3	H	CH_3	H	CH_3	H	CH_3	H	O=C–OCH_3

(b)

Above chain	H	CH_3	H	OCH_3–C=O	H	CH_3	H	OCH_3–C=O	H	CH_3	H	OCH_3–C=O	H	CH_3
Below chain	H	O=C–OCH_3	H	CH_3	H	O=C–OCH_3	H	CH_3	H	O=C–OCH_3	H	CH_3	H	O=C–OCH_3

(c)

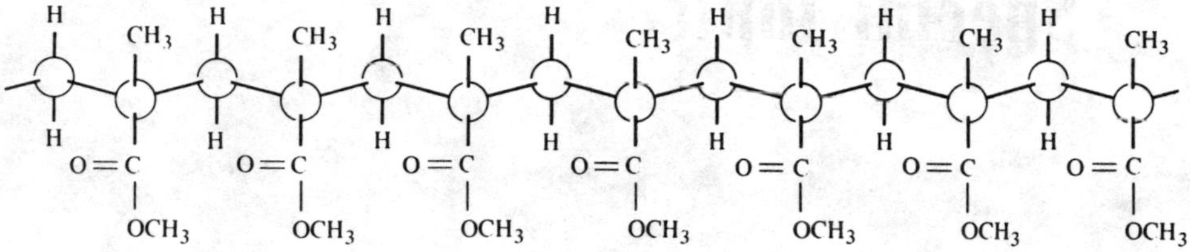

H CH3 H CH3 H CH3 H CH3 H CH3 H CH3 H CH3

H H H H H H H

O=C O=C O=C O=C O=C O=C O=C

OCH3 OCH3 OCH3 OCH3 OCH3 OCH3 OCH3

D

Special Topic

D.1

The singlet methylene reacts with the double bond at the same rate from either side as shown:

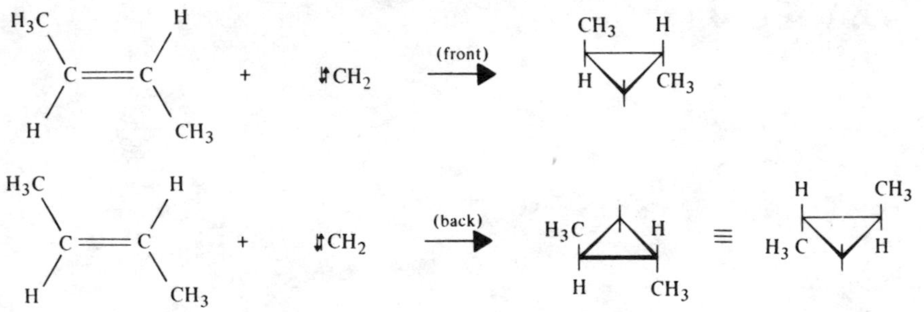

$$H_3C(H)C{=}C(H)CH_3 + {:}CH_2 \xrightarrow{\text{(front)}} \text{cyclopropane (CH}_3\text{, H / H, CH}_3\text{)}$$

$$H_3C(H)C{=}C(H)CH_3 + {:}CH_2 \xrightarrow{\text{(back)}} \text{cyclopropane (H}_3\text{C, H / H, CH}_3\text{)} \equiv \text{cyclopropane (H, CH}_3\text{ / H}_3\text{C, H)}$$

D.2

(a) cyclopropane (CH_3, H / CH_2CH_3, H) + cyclopropane (CH_3H_2C, H / CH_3, H) (racemic modification)

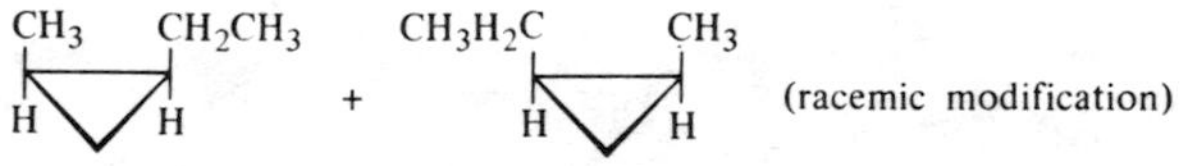

(b) cyclopropane (CH_3, H / H, CH_2CH_3) + cyclopropane (H, H_3C / CH_2CH_3, H) (racemic modification)

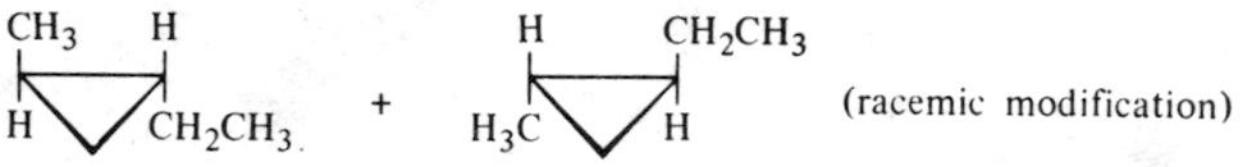

(c) cyclopropane (CH_3, H / CH_2CH_3, H) + cyclopropane (H, H_3C / H, CH_2CH_3) — racemic modification — and cyclopropane (CH_3, H / H, CH_2CH_3) + cyclopropane (H, H_3C / CH_2CH_3, H) — racemic modification

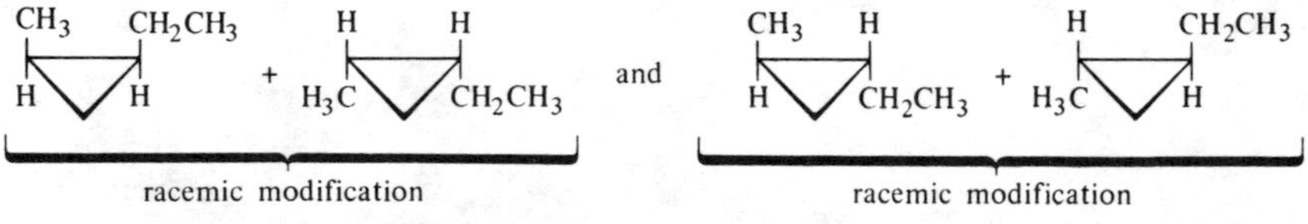

(d) Same mixture as in (c).

D.3

Recombination of radicals should yield ethane, butane, hexane, isobutane, 2-methylpentane, 2,3-dimethylbutane.

D.4

(a) 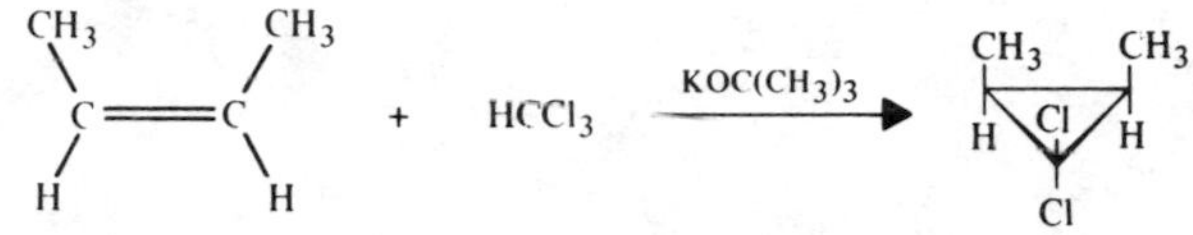

(b) 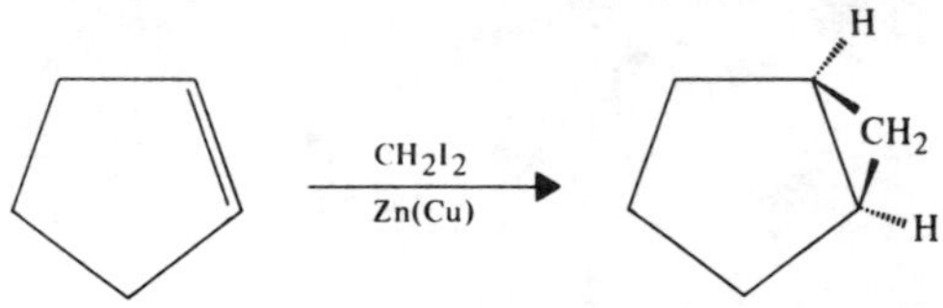

(c) 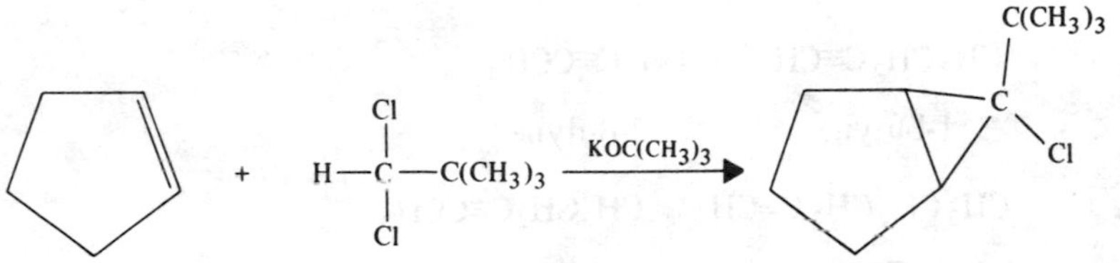

(d) 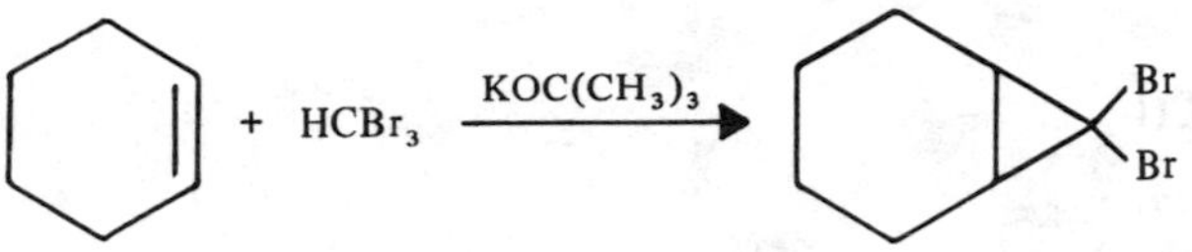

D.5

(a) The triplet state has lower energy than the singlet state because its two electrons are in different orbitals and therefore there is less repulsion between them. (b) In CCl_2 the singlet state is a resonance hybrid of the structures

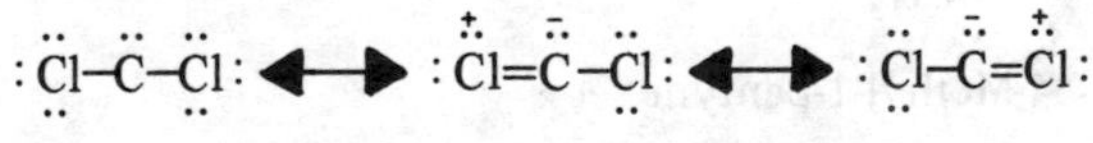

This stabilizes the singlet state.

Corresponding resonance structures of the triplet form of CCl_2 are not possible.

CHAPTER NINE
Alkynes

9.1

(a) C_4H_6 : $CH_3CH_2C{\equiv}CH$, $CH_3C{\equiv}CCH_3$

1-Butyne 2-Butyne

(b) C_5H_8 : $CH_3CH_2CH_2C{\equiv}CH$ $CH_3CH_2C{\equiv}CCH_3$

1-Pentyne 2-Pentyne

$CH_3CH(CH_3)C{\equiv}CH$

3-Methyl-1-butyne

(c) $CH_3CH_2CH_2CH_2C{\equiv}CH$ $CH_3CH_2CH_2C{\equiv}CCH_3$

1-Hexyne 2-Hexyne

$CH_3CH_2C{\equiv}CCH_2CH_3$

3-Hexyne

$CH_3CH(CH_3)CH_2C{\equiv}CH$

4-Methyl-1-pentyne

$CH_3C{\equiv}CCH(CH_3)CH_3$

4-Methyl-2-pentyne

$HC{\equiv}CCH(CH_3)CH_2CH_3$

3-Methyl-1-pentyne

$HC{\equiv}CC(CH_3)_2CH_3$

3,3-Dimethyl-
1-Butyne

9.2

(a) $HC{\equiv}CH + :\ddot{N}H_2^- \rightleftharpoons HC{\equiv}C:^- + :NH_3$

stronger acid, stronger base; weaker base, weaker acid

(No appreciable amount of reactants are present at equilibrium.)

(b) $CH_2{=}CH_2 + :\ddot{N}H_2^- \rightleftarrows CH_2{=}\ddot{C}H^- + :NH_3$ (No appreciable amount of products are present at equilibrium.)

weaker acid — weaker base — stronger base — stronger acid

(c) $CH_3CH_3 + :\ddot{N}H_2^- \rightleftarrows CH_3\ddot{C}H_2^- + :NH_3$ (No appreciable amount of products are present at equilibrium.)

weaker acid — weaker base — stronger base — stronger acid

(d) $HC{\equiv}C:^- + CH_3CH_2\ddot{O}H \rightleftarrows HC{\equiv}CH + CH_3CH_2\ddot{O}:^-$ (No appreciable amount of reactants are present at equilibrium.)

stronger base — stronger acid — weaker acid — weaker base

(e) $HC{\equiv}C:^- + H{-}\ddot{O}{:}{-}H \rightleftarrows HC{\equiv}CH + :\ddot{O}H^-$ (No appreciable amount of reactants are present at equilibrium.)

stronger base — stronger acid — weaker acid — weaker base

9.3

$CH_3CH(CH_3){-}CH(CH_3){-}OH$ from the Sia_2BH.

9.4

(a) $CH_3CH_2C{\equiv}CH \xrightarrow[HgSO_4,\ H_2SO_4]{H_2O} CH_3CH_2\overset{O}{\overset{\|}{C}}CH_3$

(b) cyclopentyl$-C{\equiv}CH + Sia_2BH \xrightarrow{0^\circ}$ cyclopentyl$-CH{=}CH{-}BSia_2 \xrightarrow[OH^-]{H_2O_2}$ cyclopentyl$-CH_2\overset{O}{\overset{\|}{C}}H$

(c) $3CH_3C{\equiv}CCH_3 + \frac{1}{2}(BD_3)_2 \xrightarrow{0^\circ}$ $\left[(CH_3)(D)C{=}C(CH_3)\right]_3B$ (cis) $\xrightarrow[0^\circ]{CH_3COOD}$ $(CH_3)(D)C{=}C(CH_3)(D)$ (cis: both CH_3 on the same side, both D on the same side)

or $CH_3C{\equiv}C{-}CH_3 + D_2 \xrightarrow{Ni_2B(P\text{-}2)}$ $(CH_3)(D)C{=}C(CH_3)(D)$ (cis)

(d) $CH_3CH_2C{\equiv}CH + Sia_2BH \longrightarrow$ $(CH_3CH_2)(H)C{=}C(H)(BSia_2)$ $\xrightarrow{CH_3COOD}$ $(CH_3CH_2)(H)C{=}C(H)(D)$

9.5

$$CH_3-\underset{CH_3}{\overset{CH_3}{C}}-C{\equiv}CH + NaNH_2 \longrightarrow CH_3-\underset{CH_3}{\overset{CH_3}{C}}-C{\equiv}C^{-}:\ Na^{+} + NH_3$$

$$\xrightarrow{CH_3CH_2Br} CH_3-\underset{CH_3}{\overset{CH_3}{C}}-C{\equiv}C-CH_2-CH_3$$

A reaction between $CH_3CH_2C{\equiv}C^{-}:\overset{+}{Na}$ and $CH_3-\underset{CH_3}{\overset{CH_3}{C}}-Br$ would result in elimination to produce $CH_2{=}\underset{CH_3}{C}-CH_3 + CH_3CH_2C{\equiv}CH$.

9.6

(a) $C_6H_{11}-C{\equiv}CH \xrightarrow[HgSO_4,\ H_2SO_4]{H_2O} C_6H_{11}-\overset{O}{\overset{\|}{C}}-CH_3$

(b) $C_6H_{11}-\overset{O}{\overset{\|}{C}}-CH_3 \xrightarrow[0^\circ]{PCl_5} C_6H_{11}-\underset{Cl}{\overset{Cl}{C}}-CH_3 \xrightarrow[\text{mineral oil, heat}]{NaNH_2} C_6H_{11}-C{\equiv}CH$

$C_6H_{11}-C{\equiv}CH + Sia_2BH \longrightarrow C_6H_{11}-CH{=}CH-BSia_2 \xrightarrow[OH^-]{H_2O_2} C_6H_{11}-CH_2\overset{O}{\overset{\|}{C}}H$

9.7

The acetylenic carbon is *sp*-hybridized, whereas the ethylenic carbon is sp^2-hybridized. Because the acetylenic carbon has a greater amount of *s*-character, it is more electronegative than the sp^2-hybridized ethylenic carbon. The polarity of $-\overset{|}{\underset{|}{C}}\rightarrow C\equiv$ is thus greater than the polarity of $-\overset{|}{\underset{|}{C}}\rightarrow\overset{|}{C}=$.

9.8

(a) $CH_3\underset{CH_3}{C}HC{\equiv}CCH_2CH_3$

2-Methyl-3-hexyne

(b) cyclic: $C{\equiv}C$ joined through $CH_2-CH_2-CH_2-CH_2-CH_2-CH_2$ ring

Cyclooctyne

(c) $HC{\equiv}CCH_2CH_2CH_2CH_2CH_3$

1-Heptyne

9.9

(a) 3-Methyl-1-butyne
(b) 2,2-Dimethyl-3-hexyne
(c) 3-Nonyne
(d) 3-Hexyne
(e) 2,2,5,5-Tetramethyl-3-hexyne
(f) 2,5-Dimethyl-3-hexyne
(g) 2-Hexyne
(h) 4-Methyl-2-hexyne
(i) 2,7-Dimethyl-4-octyne
(j) 1-Octyne

9.10

a, j

9.11

(a) d, g (b) None

9.12

(a) $\underset{\text{(coke)}}{3\,C} + \underset{\text{(lime)}}{CaO} \xrightarrow{2500°} CaC_2 + CO$

$CaC_2 + 2H_2O \xrightarrow[\text{temperature}]{\text{room}} HC{\equiv}CH + Ca(OH)_2$

(b) $HC{\equiv}CH + H_2 \xrightarrow{Ni_2B} CH_2{=}CH_2$

(c) $HC{\equiv}CH + NaNH_2 \xrightarrow[NH_3]{\text{liquid}} HC{\equiv}C:^- \overset{+}{Na} + NH_3$

$\underset{\text{(excess)}}{CH_4} + Br_2 \xrightarrow[\text{heat}]{\text{light}} CH_3Br + HBr$

$HC{\equiv}C:^- \overset{+}{Na} + CH_3Br \longrightarrow HC{\equiv}C{-}CH_3 + \overset{+}{Na}\overset{-}{Br}$

(d) $CH_3C{\equiv}CH + H_2 \xrightarrow{Ni_2B} CH_3CH{=}CH_2$

(e) $CH_3C{\equiv}CH + H_2O \xrightarrow[H_2SO_4]{HgSO_4} CH_3\overset{O}{\overset{\|}{C}}CH_3$

(f) $CH_3C{\equiv}CH + NaNH_2 \xrightarrow[NH_3]{\text{liquid}} CH_3C{\equiv}C:^- \overset{+}{Na} + NH_3$

$\xrightarrow{CH_3Br} CH_3C{\equiv}CCH_3 + \overset{+}{Na}Br^-$

(g) $CH_2{=}CH_2 + HBr \longrightarrow CH_3CH_2Br \xrightarrow{HC{\equiv}\bar{C}:\overset{+}{Na}} HC{\equiv}CCH_2CH_3 + \overset{+}{Na}Br^-$

(h) $CH_3C{\equiv}CCH_3 + H_2O \xrightarrow[H_2SO_4]{HgSO_4} CH_3\overset{O}{\overset{\|}{C}}CH_2CH_3$

(i) See (g) above.

(j) $CH_3C{\equiv}C:^- \overset{+}{Na} + CH_3CH_2Br \longrightarrow CH_3C{\equiv}CCH_2CH_3 + \overset{+}{Na}Br^-$

(k) $CH_2{=}CH_2 + H_2O \xrightarrow{H_2SO_4} CH_3CH_2OH$

(l) $CH_2{=}CH_2 + Br_2 \xrightarrow[\text{dark}]{CCl_4} CH_2BrCH_2Br$

(m) $CH_3C{\equiv}CH + 2HCl \longrightarrow CH_3{-}C(Cl)_2{-}CH_3$

(n) $CH_3C{\equiv}CCH_3 + H_2 \xrightarrow{Ni_2B}$ $(CH_3)(H)C{=}C(CH_3)(H)$ (cis: both CH_3 on same side)

(o) $CH_3C{\equiv}CCH_3 \xrightarrow[-78^\circ]{Li\ +\ C_2H_5NH_2}$ $(CH_3)(H)C{=}C(H)(CH_3)$ (trans)

(p) $CH_3CH_2C{\equiv}CH + H_2 \xrightarrow{Ni_2B} CH_3CH_2CH{=}CH_2$

(q) $CH_3CH_2CH{=}CH_2 + HBr \xrightarrow[\text{inhibitor}]{\text{peroxide}} CH_3CH_2CHBrCH_3$

(r) $CH_3CH_2CH{=}CH_2 + HBr \xrightarrow[\text{peroxide}]{} CH_3CH_2CH_2CH_2Br$

(s) $CH_3CH_2CH{=}CH_2 + (BH_3)_2 \longrightarrow (CH_3CH_2CH_2CH_2)_3B \xrightarrow[OH^-]{H_2O_2}$ $CH_3CH_2CH_2CH_2OH$

(t) $CH_3CH_2CH{=}CH_2 + Hg(OAc)_2 \longrightarrow CH_3CH_2CH(HgOAc)CH_3 \xrightarrow[OH^-]{NaBH_4} CH_3CH_2CH(OH)CH_3$

or $CH_3CH_2CH{=}CH_2 + H_2O \xrightarrow{H_2SO_4} CH_3CH_2CH(OH){-}CH_3$

(u) $(H_3C)(H)C{=}C(H)(CH_3)$ (trans) $+ Br_2 \xrightarrow[\text{dark}]{CCl_4}$ [Newman-type projection: upper carbon CH_3, H, Br; lower carbon H, Br, CH_3]

(v) $(H_3C)(H)C{=}C(CH_3)(H)$ (cis) $+ Br_2 \xrightarrow[\text{dark}]{CCl_4}$ [projection: upper carbon CH_3, H, Br; lower carbon Br, H, CH_3] + enantiomer

(w) $CH_3C{\equiv}CCH_3 + HCl \xrightarrow[CH_3COOH]{Cl^-,\ 25°}$ $\underset{Cl}{\overset{CH_3}{}}\!\!>C{=}C<\!\!\underset{CH_3}{\overset{H}{}}$

(x) $CH_3CH{=}CH_2 + HBr \xrightarrow{\text{peroxide}} CH_3CH_2CH_2Br$

(y) $CH_3CH_2C{\equiv}CH + NaNH_2 \xrightarrow[NH_3]{\text{liquid}} CH_3CH_2C{\equiv}C:^-\ \overset{+}{Na} + NH_3$

$\qquad \xrightarrow{CH_3CH_2Br} CH_3CH_2C{\equiv}CCH_2CH_3$

(z) $CH_3CH_2C{\equiv}CH + HBr \xrightarrow{\text{peroxide}} CH_3CH_2CH{=}CHBr$

9.13

(a) $\underset{Br}{\overset{CH_3CH_2CH_2}{}}\!\!>C{=}C<\!\!\underset{H}{\overset{Br}{}}$ (b) $\underset{Cl}{\overset{CH_3CH_2CH_2}{}}\!\!>C{=}CH_2$ (c) $CH_3CH_2CH_2\overset{Cl}{\underset{Cl}{C}}CH_3$

(d) $CH_3CH_2CH_2CH{=}CHBr$ (e) $CH_3CH_2CH_2\overset{O}{\overset{\|}{C}}CH_3$ (f) $CH_3CH_2CH_2CH{=}CH_2$

(g) $CH_3CH_2CH_2CH{=}CH_2$ (h) $CH_3CH_2CH_2CH_2\overset{O}{\overset{\|}{C}}H$ (i) $CH_3CH_2CH_2C{\equiv}C^{-}:\overset{+}{Na}$

(j) $CH_3CH_2CH_2C{\equiv}CCH_3$ (k) $CH_3CH_2CH_2C{\equiv}C{-}C{\equiv}CCH_2CH_2CH_3$

(l) $CH_3CH_2CH_2C{\equiv}CAg$ (m) $CH_3CH_2CH_2C{\equiv}CCu$

9.14

(a) $\underset{Cl}{\overset{CH_3CH_2}{}}\!\!>C{=}C<\!\!\underset{CH_2CH_3}{\overset{H}{}}$ (b) $CH_3CH_2\overset{Cl}{\underset{Cl}{C}}CH_2CH_2CH_3$

(c) $\underset{Br}{\overset{CH_3CH_2}{}}\!\!>C{=}C<\!\!\underset{CH_2CH_3}{\overset{Br}{}}$ (d) $CH_3CH_2\overset{Br}{\underset{Br}{C}}{-}\overset{Br}{\underset{Br}{C}}CH_2CH_3$

(e) $\underset{H}{\overset{CH_3CH_2}{}}\!\!>C{=}C<\!\!\underset{H}{\overset{CH_2CH_3}{}}$ (f) Same as (e)

(g) $\underset{H}{\overset{CH_3CH_2}{}}\!\!>C{=}C<\!\!\underset{CH_2CH_3}{\overset{H}{}}$ (h) $CH_3CH_2\overset{O}{\overset{\|}{C}}CH_2CH_2CH_3$

(i) No reaction (j) No reaction (k) Same as (e)

(l) Same as (h) (m) No reaction

(n) $CH_3CH_2CH_2CH_2CH_2CH_3$ (o) $2CH_3CH_2COOH$ (p) $2CH_3CH_2COOH$

(q) No reaction

9.15

(a) $CH_3CH_2CH_2CH{=}CH_2 + Br_2 \longrightarrow CH_3CH_2CH_2CHBrCH_2Br \xrightarrow[\text{liq NH}_3]{2NaNH_2} CH_3CH_2CH_2C{\equiv}CH$

(b) $CH_3CH_2CH_2CH_2CH_2Cl \xrightarrow[CH_3CH_2OH]{KOH} CH_3CH_2CH_2CH{=}CH_2$

then proceed as in (a) above.

(c) $CH_3CH_2CH_2CH{=}CHCl \xrightarrow[NH_3]{NaNH_2} CH_3CH_2CH_2C{\equiv}CH$

(d) $CH_3CH_2CH_2CH_2CHCl_2 \xrightarrow[NH_3]{2NaNH_2} CH_3CH_2CH_2C{\equiv}CH$

(e) $HC{\equiv}CH \xrightarrow[\text{liq. NH}_3]{NaNH_2} HC{\equiv}C:^- Na^+ \xrightarrow{CH_3CH_2CH_2Br} HC{\equiv}CCH_2CH_2CH_3$

9.16

(a) Propyne is soluble in cold, concentrated H_2SO_4; propane is not. Other tests are Br_2/CCl_4 and $KMnO_4/H_2O$.

(b) $Ag(NH_3)_2^+\ \bar{O}H$ gives a precipitate with propyne, not with propene.

(c) Dilute $KMnO_4$ oxidizes 1-bromopropene and not 2-bromopropane.

(d) $Ag(NH_3)_2^+\ \bar{O}H$ gives a precipitate with 1-butyne, not with 2-bromo-2-butene.

(e) Sodium fusion followed by acidification with dilute HNO_3 and addition of $AgNO_3$ gives a AgBr precipitate with 2-bromo-2-butene, not with 2-butyne.

(f) Br_2/CCl_4 is decolorized by 2-butyne, not by *n*-butyl alcohol.

(g) $AgNO_3/C_2H_5OH$ gives a AgBr precipitate wth 2-bromobutane, not with 2-butyne.

(h) Br_2/CCl_4 is decolorized by $CH_3C{\equiv}CCH_2OH$, not by $CH_3CH_2CH_2CH_2OH$.

(i) Br_2/CCl_4 is decolorized by $CH_3CH{=}CHCH_2OH$, not by $CH_3CH_2CH_2CH_2OH$.

(In many cases above other tests are possible.)

9.17

(a) A = $CH_3CH_2CH_2C{\equiv}CH$, B = $CH_3CH_2C{\equiv}CCH_3$, C = (cyclopentene)

(b) Yes, B may also be $CH_3CH{=}CH{-}CH{=}CH_2$ or $CH_2{=}CH{-}CH_2{-}CH{=}CH_2$.

C may also be (methylcyclobutene)$-CH_3$, (methylcyclobutene)$-CH_3$, or (methylenecyclobutane, $=CH_2$) or (ethylcyclopropene, CH_2CH_3) , etc.

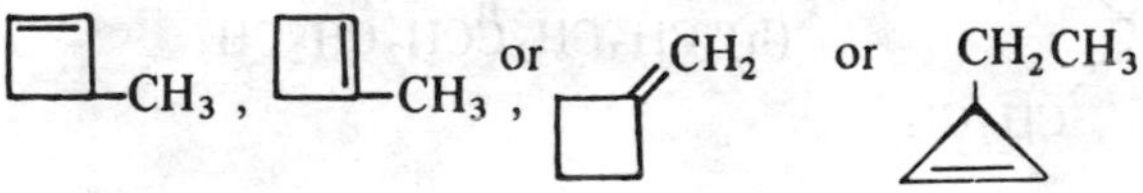

(c) B = $CH_3CH_2C{\equiv}CCH_3$

(d)  (cyclopentene)

9.18

(a) $CH_3\overset{CH_3}{\overset{|}{C}}HC{\equiv}CH + HCl\ (1\ mole) \longrightarrow CH_3\overset{CH_3}{\overset{|}{C}}H\underset{Cl}{\underset{|}{C}}{=}CH_2$

(b) $CH_3\overset{CH_3}{\overset{|}{C}}HC{\equiv}CH \xrightarrow[\text{quinoline}]{H_2/Pd,BaSO_4} CH_3\overset{CH_3}{\overset{|}{C}}HCH{=}CH_2 \xrightarrow[\text{peroxides}]{HBr} CH_3\overset{CH_3}{\overset{|}{C}}HCH_2CH_2Br$

(c) Product of (a) $\xrightarrow{H_2/Ni} CH_3\overset{CH_3}{\overset{|}{C}}H\underset{Cl}{\underset{|}{C}}HCH_3$

(d) Product of (a) $\xrightarrow[\text{dark}]{Cl_2/CCl_4} CH_3\overset{CH_3}{\overset{|}{C}}H{-}\overset{Cl}{\overset{|}{\underset{Cl}{\underset{|}{C}}}}{-}CH_2Cl$

(e) Product of (a) $\xrightarrow[\text{peroxide inhibitor}]{HBr} CH_3\overset{CH_3}{\overset{|}{C}}H{-}\overset{Cl}{\overset{|}{\underset{Br}{\underset{|}{C}}}}{-}CH_3$

(f) $CH_3\overset{CH_3}{\overset{|}{C}}HC{\equiv}CH \xrightarrow[\text{then}\ H^+]{KMnO_4/OH^-,} CH_3\overset{CH_3}{\overset{|}{C}}HCOOH + CO_2$

9.19

(a) $CH_3CH_2CH_2C{\equiv}CH + D_2 \xrightarrow{Ni_2B}$ $(CH_3CH_2CH_2)(D)C{=}C(H)(D)$ [$CH_3CH_2CH_2$ and H on top; D and D on bottom]

(b) $CH_3CH_2CH_2C{\equiv}CH + Sia_2BH \longrightarrow$ $(CH_3CH_2CH_2)(H)C{=}C(H)(BSia_2)$ [$CH_3CH_2CH_2$ and H on top; H and $BSia_2$ on bottom] $\xrightarrow{CH_3COOD}$ $(CH_3CH_2CH_2)(H)C{=}C(H)(D)$ [$CH_3CH_2CH_2$ and H on top; H and D on bottom]

(c) $CH_3CH_2CH_2C{\equiv}CH + DCl \longrightarrow$ $(CH_3CH_2CH_2)(Cl)C{=}C(D)(H)$ [$CH_3CH_2CH_2$ and D on top; Cl and H on bottom]

(d) $CH_3CH_2CH_2C{\equiv}CH + Sia_2BD \longrightarrow$ $(CH_3CH_2CH_2)(D)C{=}C(H)(BSia_2)$ [$CH_3CH_2CH_2$ and H on top; D and $BSia_2$ on bottom] $\xrightarrow[OH^-]{H_2O_2} CH_3CH_2CH_2\underset{D}{\underset{|}{C}}H\overset{O}{\overset{\|}{C}}H$

9.20

The syntheses involve combinations of chlorination, dehydrochlorination, and hydrogenation as follows:

(a)-(b) $H{-}C{\equiv}C{-}H \xrightarrow[Cl^-]{Cl_2}$ (H)(Cl)C=C(Cl)(H) $\xrightarrow[Cat.]{H_2}$ $Cl{-}CH_2CH_2{-}Cl$

trans-1,2-Dichloroethene; 1,2-Dichloroethane

(c)-(e) $H{-}C{\equiv}C{-}H + 2Cl_2 \longrightarrow CHCl_2{-}CHCl_2$

1,1,2,2-Tetrachloroethane

$\xrightarrow[(-HCl)]{base} CHCl{=}CCl_2 \xrightarrow[Cat.]{H_2} CH_2ClCHCl_2$

1,1,2-Trichloroethene; 1,1,2-Trichloroethane

(f)-(g) $CHCl{=}CCl_2 \xrightarrow{Cl_2} CHCl_2CCl_3 \xrightarrow[(-HCl)]{base} CCl_2{=}CCl_2$

1,1,1,2,2-Pentachloroethane; 1,1,2,2-Tetrachloroethene

9.21

(a) $(BH_3)_2 \rightleftharpoons 2BH_3$

$CH_3C(CH_3){=}CHCH_3 + H{-}BH{-}H \longrightarrow CH_3CH(CH_3){-}CH(CH_3){-}BH_2$ (SiaBH$_2$)

$CH_3C(CH_3){=}CHCH_3 + H{-}B(CH(CH_3)CH(CH_3)CH_3){-}H \longrightarrow (CH_3CH(CH_3){-}CH(CH_3))_2BH$

Sia_2BH

(b) Steric hindrance. It is difficult to place three bulky groups around the boron atom.

9.22

$HC{\equiv}C{-}C{\equiv}C(H){C}{=}C{=}C(H)(CH{=}CH)_2CH_2COOH$

$HC{\equiv}C{-}C{\equiv}C(H){C}{=}C{=}C((CH{=}CH)_2CH_2COOH)(H)$

9.23

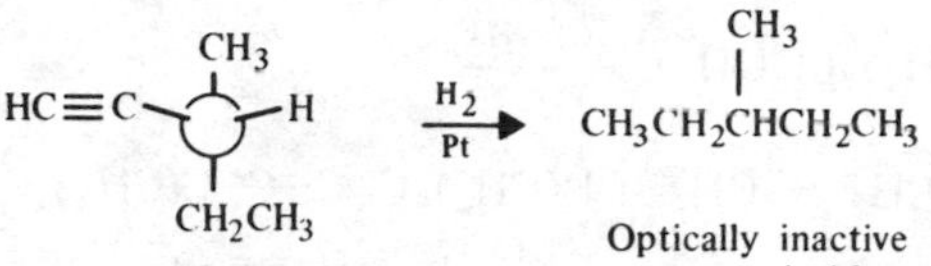

$$HC{\equiv}C\text{–}C(CH_3)(H)(CH_2CH_3) \xrightarrow[Pt]{H_2} CH_3CH_2CH(CH_3)CH_2CH_3$$

D

Optically active (the other enantiomer is an equally valid answer)

Optically inactive nonresolvable

9.24

Ordinary alkenes *are* more reactive toward electrophilic reagents. But, the alkenes obtained from the addition of an electrophilic reagent to an alkyne have at least one electronegative atom (Cl, Br, etc.) attached to a carbon of the double bond.

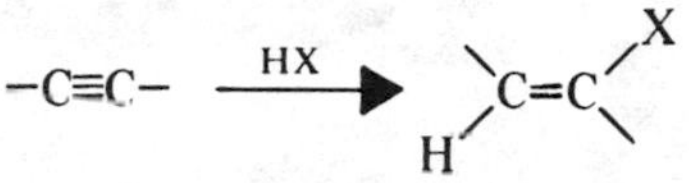

$$-C{\equiv}C- \xrightarrow{HX} (H)C{=}C(X)$$

or

$$-C{\equiv}C- \xrightarrow{X_2} (X)C{=}C(X)$$

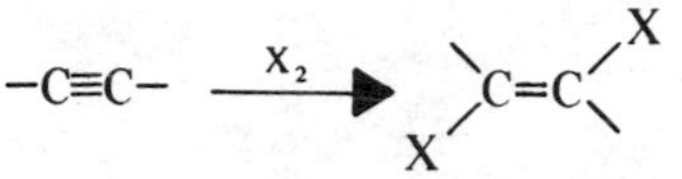

These alkenes are less reactive than alkynes toward electrophilic addition because the electronegative group makes the double bond "electron poor."

9.25

(a) That hydrogenation of erythrogenic acid produces the unbranched product, $CH_3(CH_2)_{16}COOH$, tells us that the carbon chain of erythrogenic acid is unbranched. That five moles of hydrogen are absorbed tells us that the carbon chain of erythrogenic acid contains combinations of two triple bonds and one double bond, one triple bond and three double bonds, or five double bonds.

(b) $CH_2{=}CHCH_2C{\equiv}C{-}C{\equiv}C{-}(CH_2)_{10}COOH$

or

$CH_3CH{=}CHC{\equiv}C{-}C{\equiv}C(CH_2)_{10}COOH$

or

$CH_2{=}CHCH{=}CHCH{=}CHC{\equiv}C(CH_2)_9COOH$

or

$CH_3C{\equiv}CCH_2CH{=}CHCH_2CH{=}CHCH_2CH{=}CH(CH_2)_5COOH$

etc.

(c) Either of the compounds shown below.

$CH_2{=}CH(CH_2)_4C{\equiv}C{-}C{\equiv}C(CH_2)_7COOH$

or

$CH_2{=}CHC{\equiv}C(CH_2)_4C{\equiv}C(CH_2)_7COOH$

(d) Here we have an oxidative coupling of two different alkynes.

$$CH_2{=}CH(CH_2)_4C{\equiv}CH \ + \ HC{\equiv}C(CH_2)_7COOH \xrightarrow[O_2]{CuCl,\ NH_3}$$

$$CH_2{=}CH(CH_2)_4C{\equiv}C{-}C{\equiv}C(CH_2)_4CH{=}CH_2 \ + \ CH_2{=}CH(CH_2)_4C{\equiv}C{-}C{\equiv}C(CH_2)_7COOH \ +$$

Nonacidic product — Erythrogenic acid

$$HOOC(CH_2)_7C{\equiv}C{-}C{\equiv}C(CH_2)_7COOH$$

Dicarboxylic acid

CHAPTER TEN

Conjugated Unsaturated Systems: Visible - Ultraviolet Spectroscopy

10.1

(a) $^{14}CH_2{=}CHCH_2X$ and $CH_2{=}CH^{14}CH_2X$

(b) The reaction proceeds through the resonance stabilized free radical,

$$^{14}CH_2{=}CH{-}\dot{C}H_2 \longleftrightarrow {}^{14}\dot{C}H_2{-}CH{=}CH_2 \text{ or } {}^{14}\overset{\delta\cdot}{C}H_2 \mathrel{\overset{\cdots}{-}} CH \mathrel{\overset{\cdots}{-}} \overset{\delta\cdot}{C}H_2$$

Thus attack on X_2 can occur by the carbon at either end of the chain since they are equivalent.

(c) 50:50

10.2

(a)

$$\underset{\textbf{D}}{\overset{4}{C}H_3{-}\overset{+}{\overset{3}{C}H}{-}\overset{2}{C}H{=}\overset{1}{C}H_2} \longleftrightarrow \underset{\textbf{E}}{\overset{4}{C}H_3{-}\overset{3}{C}H{=}\overset{2}{C}H{-}\overset{+}{\overset{1}{C}H_2}} \text{ or } \underset{\textbf{F}}{\overset{4}{C}H_3{-}\underset{\delta+}{\overset{3}{C}H} \mathrel{\overset{\cdots}{-}} \overset{2}{C}H \mathrel{\overset{\cdots}{-}} \underset{\delta+}{\overset{1}{C}H_2}}$$

(b) We know that the allyl cation is almost as stable as a tertiary carbocation. Here we find not only the resonance stabilization of an allyl cation but also the additional stabilization that arises from contributor **D** in which the plus charge is on a secondary carbon.

(c) $CH_3{-}\underset{}{\overset{Cl}{\overset{|}{C}H}}{-}CH{=}CH_2$ and $CH_3{-}CH{=}CH{-}CH_2{-}Cl$

10.3

(a) *Cis*-1,3-pentadiene, *trans,trans*-2,4-hexadiene, *cis,trans*-2,4-hexadiene, and 1,3-cyclohexadiene are conjugated dienes.

(b) 1,4-Cyclohexadiene is an isolated diene.

(c) 1-Penten-4-yne is an isolated enyne.

10.4

(a) $CH_3CH_2\underset{Cl}{\underset{|}{C}H}CH{=}CHCH_3$ and $CH_3CH_2CH{=}CH\underset{Cl}{\underset{|}{C}H}CH_3$

(b) The most stable cation is a hybrid of equivalent forms: $CH_3\overset{}{\underset{+}{C}H}CH{=}CHCH_3 \longleftrightarrow$

$\overset{+}{CH_3CH{=}CHCHCH_3}$. Thus 1,4 and 1,2- addition yield the same product,

$$CH_3\underset{\underset{Cl}{|}}{C}HCH{=}CHCH_3$$

10.5

(a) Addition of the proton gives the resonance hybrid of

$$CH_3{-}\overset{+}{C}H{-}CH{=}CH_2 \longleftrightarrow CH_3{-}CH{=}CH{-}\overset{+}{C}H_2$$

$$\text{I} \qquad\qquad\qquad \text{II}$$

The inductive effect of the methyl group in I stabilizes the positive charge on the adjacent carbon. Such stabilization of the positive charge does not occur in II. Because I contributes more heavily to the resonance hybrid than does II, C-2 bears a greater positive charge and reacts faster with the bromide ion.

(b) In the 1,4-addition product, the double bond is more highly substituted than in the 1,2-addition product, hence it is the more stable alkene.

10.6

(a) $CH_2{=}\overset{\overset{CH_3}{|}}{C}{-}\dot{C}H_2 \longleftrightarrow \cdot CH_2{-}\overset{\overset{CH_3}{|}}{C}{=}CH_2$

(b) $CH_2{=}CH{-}\underset{+}{C}H{-}CH{=}CH_2 \longleftrightarrow \overset{+}{C}H_2{-}CH{=}CH{-}CH{=}CH_2 \longleftrightarrow$

$$CH_2{=}CH{-}CH{=}CH{-}\overset{+}{C}H_2$$

(c) 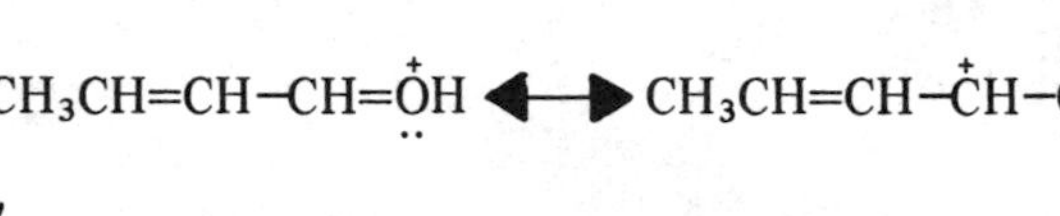 ↔ ↔

(d) 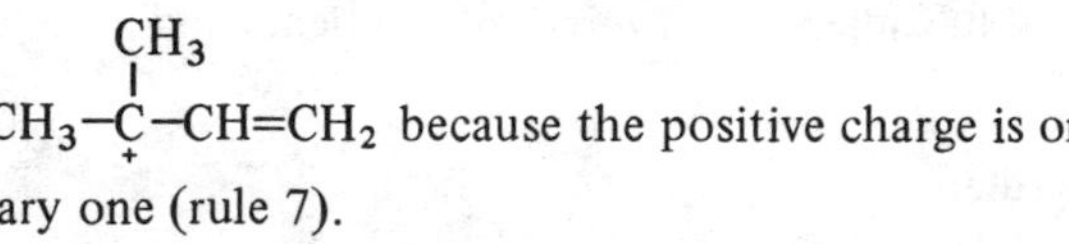 ↔ ↔

(e) $CH_3CH{=}CH{-}CH{=}\overset{+}{\ddot{O}}H \longleftrightarrow CH_3CH{=}CH{-}\overset{+}{C}H{-}\ddot{O}H \longleftrightarrow CH_3\overset{+}{C}H{-}CH{=}CH{-}\ddot{O}H$

10.7

(a) $CH_3{-}\underset{+}{\overset{\overset{CH_3}{|}}{C}}{-}CH{=}CH_2$ because the positive charge is on a tertiary carbon rather than a primary one (rule 7).

(b) 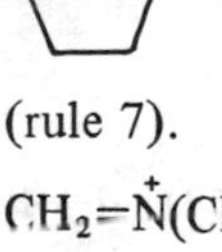 CH_2 + because the positive charge is on a secondary carbon rather than a primary one (rule 7).

(c) $CH_2{=}\overset{+}{N}(CH_3)_2$ because all atoms have a complete octet (rule 9), and there are more covalent bonds (rule 8).

(d) $CH_3{-}\overset{\overset{\displaystyle O}{\|}}{C}{-}OH$ because it has no charge separation (rule 10).

(e) $CH_2{=}CH\dot{C}HCH{=}CH_2$ because the radical is on a secondary carbon rather than a primary one (rule 7).

10.8

In resonance structures, the positions of the nuclei must remain the same for all structures (rule 2). The keto and enol forms shown not only differ in the positions of their electrons, they also differ in the position of one of the hydrogen atoms. In the enol form it is attached to an oxygen; in the keto form it has been moved so that it is attached to a carbon.

10.9

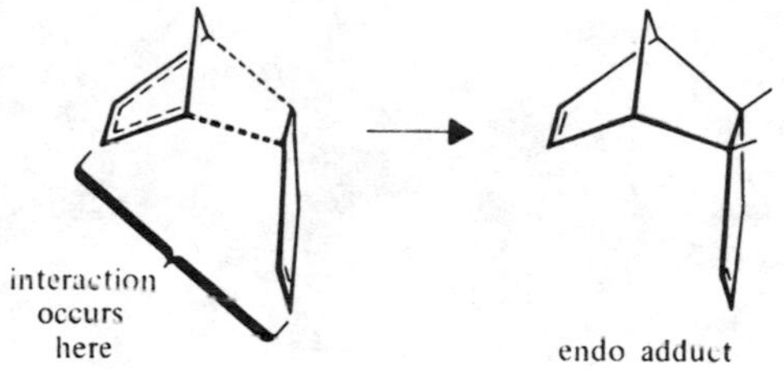

interaction occurs here

endo adduct

10.10

(a) H_3C, H_3C, H, $COCH_3$ + H_3C, H_3C, $COCH_3$, H

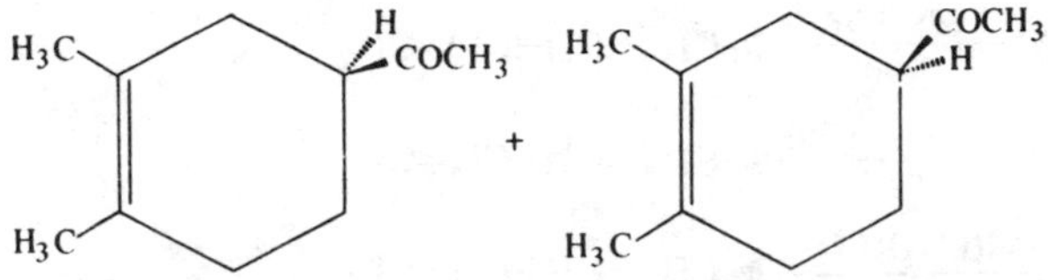

(b) H_3C, H_3C, H, O, $COCH_3$, $COCH_3$, H, O

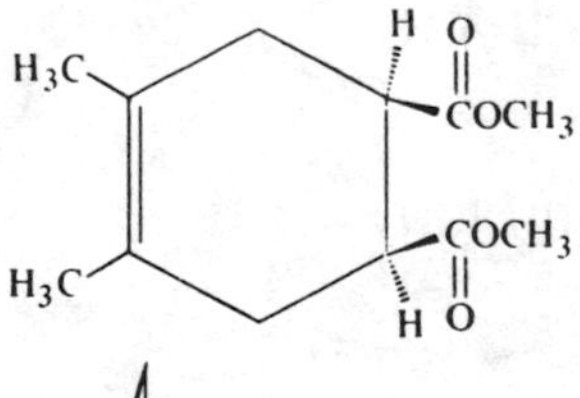

(c) O, O, O (major product)

\+

O, O, O (minor product)

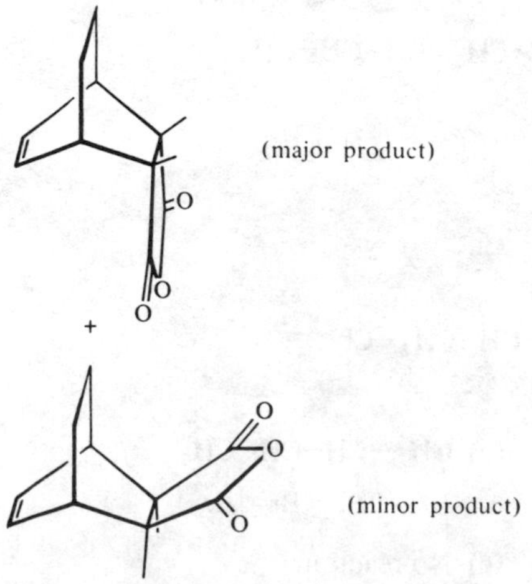

10.11

Cyclopentadiene + $CH_3OC(=O)-CH=CH-C(=O)OCH_3$ (trans; H and $COCH_3$ on one carbon, CH_3OC and H on the other)

10.12

Cyclopentadiene + $CH_3O-C(=O)-C\equiv C-C(=O)-OCH_3$

10.13

(a) $BrCH_2CH_2CH_2CH_2Br \xrightarrow[(CH_3)_3COH]{(CH_3)_3COK} CH_2{=}CH{-}CH{=}CH_2$

(b) $HOCH_2CH_2CH_2CH_2OH \xrightarrow[\text{heat}]{\text{conc.}\ H_2SO_4} CH_2{=}CH{-}CH{=}CH_2$

(c) $CH_2{=}CH{-}CH_2CH_2{-}OH \xrightarrow[\text{heat}]{\text{conc.}\ H_2SO_4} CH_2{=}CH{-}CH{=}CH_2$

(d) $CH_2{=}CH{-}CH_2CH_2{-}Cl \xrightarrow[(CH_3)_3COH]{(CH_3)_3COK} CH_2{=}CH{-}CH{=}CH_2$

(e) $CH_2{=}CH{-}CH(Cl){-}CH_3 \xrightarrow[(CH_3)_3COH]{(CH_3)_3COK} CH_2{=}CH{-}CH{=}CH_2$

(f) $CH_2{=}CH{-}CH(OH){-}CH_3 \xrightarrow[\text{heat}]{\text{conc.}\ H_2SO_4} CH_2{=}CH{-}CH{=}CH_2$

(g) $HC{\equiv}C{-}CH{=}CH_2 + H_2 \xrightarrow[\text{quinoline}]{\text{Pd. BaSO}_4} CH_2{=}CH{-}CH{=}CH_2$

10.14

$CH_2{=}C(CH_3){-}C(CH_3){=}CH_2$

10.15

(a) $Cl{-}CH_2CH(Cl)CH{=}CH_2 + Cl{-}CH_2{-}CH{=}CH{-}CH_2{-}Cl$

(b) $CH_2(Cl){-}CH(Cl){-}CH(Cl){-}CH_2(Cl)$

(c) $CH_2(Br){-}CH(Br){-}CH(Br){-}CH_2(Br)$

(d) $CH_3{-}CH_2{-}CH_2{-}CH_3$

(e) No reaction

(f) $Cl{-}CH_2{-}CH(OH){-}CH{=}CH_2$ + $Cl{-}CH_2{-}CH{=}CH{-}CH_2{-}OH$

(g) $4CO_2$ (Note: $KMnO_4$ oxidizes HOOC−COOH to $2CO_2$)

(h) $CH_3{-}CH(OH){-}CH{=}CH_2$ + $CH_3{-}CH{=}CH{-}CH_2OH$

10.16

(a) $CH_2{=}CH{-}CH_2{-}CH_3$ + NBS (N-bromosuccinimide: CH_2–C(=O)–NBr–C(=O)–CH_2 ring) $\xrightarrow{CCl_4}$ $CH_2{=}CH{-}CH(Br){-}CH_3$ (+ $CH_2(Br){-}CH{=}CH{-}CH_3$) $\xrightarrow[(CH_3)_3COH]{(CH_3)_3COK}$ $CH_2{=}CH{-}CH{=}CH_2$

(NBS)

(b) $CH_2{=}CH{-}CH_2CH_2CH_3$ + NBS $\xrightarrow{CCl_4}$ $CH_2{=}CH{-}CH(Br)CH_2CH_3$

(+ $CH_2(Br)CH{=}CHCH_2CH_3$) $\xrightarrow[(CH_3)_3COH]{(CH_3)_3COK}$ $CH_2{=}CH{-}CH{=}CH{-}CH_3$

(c) $CH_3CH_2CH_2CH_2OH$ $\xrightarrow[heat]{conc.\ H_2SO_4}$ $CH_3CH_2CH{=}CH_2$ $\xrightarrow{[as\ in\ (a)]}$ $CH_2{=}CH{-}CH{=}CH_2$ $\xrightarrow[heat]{Br_2}$ $CH_2(Br){-}CH{=}CH{-}CH_2(Br)$

(d) $CH_3{-}CH{=}CH{-}CH_3$ + NBS $\xrightarrow{CCl_4}$ $CH_3{-}CH{=}CH{-}CH_2{-}Br$

(e) cyclopentane (excess) + Br_2 $\xrightarrow[heat]{light}$ bromocyclopentane $\xrightarrow[(CH_3)_3COH]{(CH_3)_3COK}$ cyclopentene $\xrightarrow[CCl_4]{NBS}$ 3-bromocyclopentene (Br)

(f) 3-bromocyclopentene (Br) $\xrightarrow[(CH_3)_3COH]{(CH_3)_3COK}$ cyclopentadiene (same as cyclopentadiene)

10.17

$$R{-}\ddot{\underset{..}{O}}{-}\ddot{\underset{..}{O}}{-}R \xrightarrow[or\ heat]{light} 2R{-}\ddot{\underset{..}{O}}\cdot$$

$$R{-}\ddot{\underset{..}{O}}\cdot + H{-}\ddot{\underset{..}{Br}}: \longrightarrow R{-}\ddot{\underset{..}{O}}{-}H + \cdot\ddot{\underset{..}{Br}}:$$

$$CH_2{=}CH{-}CH{=}CH_2 + \cdot\ddot{\underset{..}{Br}}: \longrightarrow \left[CH_2{=}CH{-}\dot{C}H{-}CH_2Br \longleftrightarrow \dot{C}H_2{-}CH{=}CH{-}CH_2Br\right]$$

$$\xrightarrow{HBr} CH_2{=}CH{-}\underset{H}{CH}{-}\underset{Br}{CH_2} + \underset{H}{CH_2}{-}CH{=}CH{-}\underset{Br}{CH_2} + \cdot\ddot{Br}:$$

(*cis* and *trans*)

10.18

(a) $Ag(NH_3)_2OH$ gives a precipitate with 1-butyne only.

(b) 1,3-Butadiene decolorizes bromine solution; *n*-butane does not.

(c) CrO_3/H_2SO_4 oxidizes the alcohol. The solution changes from orange to green. No color change with 1,3-butadiene.

(d) $AgNO_3$ in C_2H_5OH gives a AgBr precipitate with $CH_2{=}CHCH_2CH_2Br$. No reaction with 1,3-butadiene.

(e) $AgNO_3$ in C_2H_5OH gives a AgBr precipitate with $BrCH_2CH{=}CHCH_2Br$ (it is an allylic bromide), but not with $CH_3\underset{Br}{CH}{=}\underset{Br}{C}HCH_3$ (a vinyl bromide).

10.19

(a) Because a highly resonance-stabilized free radical is formed:

$$CH_2{=}CH{-}\dot{C}H{-}CH{=}CH_2 \longleftrightarrow \dot{C}H_2{=}CH{-}CH{=}CH{-}\dot{C}H_2 \longleftrightarrow \dot{C}H_2{-}CH{=}CH{-}CH{=}CH_2$$

(b) Because the carbanion is more stable:

$$CH_2{=}CH{-}\ddot{C}H{-}CH{=}CH_2 \longleftrightarrow CH_2{=}CH{-}CH{=}CH{-}\ddot{C}H_2 \longleftrightarrow \ddot{C}H_2{-}CH{=}CH{-}CH{=}CH_2$$

i.e., we can write more resonance structures of nearly equal energies.

10.20

$$CH_2{=}\overset{CH_3}{C}{-}CH{=}CH_2 \xrightarrow{H^+} \left[CH_3\overset{CH_3}{\underset{+}{C}}{-}CH{=}CH_2 \longleftrightarrow CH_3{-}\overset{CH_3}{C}{=}CH{-}\overset{+}{C}H_2\right] \quad \text{I}$$

$$\longrightarrow \left[CH_2{=}\overset{CH_3}{C}{-}\underset{+}{CH}{-}CH_3 \longleftrightarrow \overset{+}{C}H_2{-}\overset{CH_3}{C}{=}CH{-}CH_3\right] \quad \text{II}$$

The resonance hybrid, I, has the positive charge, in part, on the tertiary carbon; in II, the positive charge is on primary and secondary carbons only. Therefore hybrid I is more stable, and will be the intermediate carbocation. 1,4-addition to I gives

$$CH_3{-}\overset{CH_3}{C}{=}CH{-}CH_2Cl$$

10.21

(a) butadiene + $CH_2{=}CH{-}CH{=}O$

(d) $CH_3{-}CH{=}CH{-}CH{=}CH_2$ + maleic anhydride (H–C=C–H with two C=O joined through O)

(b) $(CH_3)_2C=CH_2$-type diene $CH_2=C(CH_3)-C(CH_3)=CH_2$ + $H(CH_3OOC)C=C(COOCH_3)H$ (e) butadiene + $CH_2=CH-CN$

(c) $CH_2=C(CH_3)-C(CH_3)=CH_2$ + $H(CH_3OOC)C=C(H)COOCH_3$

10.22

Neither compound can assume the s-*cis*-conformation. 1,3-Butadiyne is linear, and (cyclohexene)$=CH_2$ is forced into the s-*trans* conformation by the requirements of the ring.

10.23

(a) cyclohexadiene with two $C(=O)-OCH_3$ groups (b) cyclohexadiene with two CF_3 groups

10.24

The formula, C_6H_8, tells us that **A** and **B** have six hydrogens less than an alkane. This unsaturation may be due to three double bonds, one triple bond and one double bond, or combinations of two double bonds and a ring, or one triple bond and a ring. Since both **A** and **B** react with two moles of H_2 to yield cyclohexane, they are either cyclohexyne or cyclohexadienes. The absorption maximum of 256 nm for **A** tells us that it is conjugated. **B**, with no absorption maximum beyond 200, possesses isolated double bonds. We can rule out cyclohexyne because of ring strain caused by the requirement of linearity of the $-C{\equiv}C-$ system. Therefore **A** is 1,3-cyclohexadiene; **B** is 1,4-cyclohexadiene.

10.25

In the dimer, the remaining double bonds are isolated:

CH_2 CH_2

The rate of dimerization can be followed by observing the rate of disappearance of the absorption maximum at 239 nm that is due to the conjugated double bonds in cyclopentadiene (see Table 10.3). The product does not absorb at 239 nm.

10.26

All three compounds have an unbranched five-carbon chain. The formula, C_5H_6, suggests that they have one double bond and one triple bond. **D**, **E**, and **F** must differ, therefore, in the way the multiple bonds are distributed in the chain. **E** and **F** have a terminal $-C{\equiv}CH$ (reaction with $Ag(NH_3)_2^+OH^-$). The absorption maximum near 230 nm for **D** and **E** suggests that in these compounds, the multiple bonds are conjugated. The structures are:

$CH_3-C{\equiv}C-CH{=}CH_2$ (**D**) $HC{\equiv}C-CH{=}CH-CH_3$ (**E**) $HC{\equiv}C-CH_2-CH{=}CH_2$ (**F**)

10.27

The *endo* adduct is less stable than the *exo*, but is produced at a faster rate at 25°. At 90°. equilibrium is established, and the more stable *exo* adduct predominates.

10.28

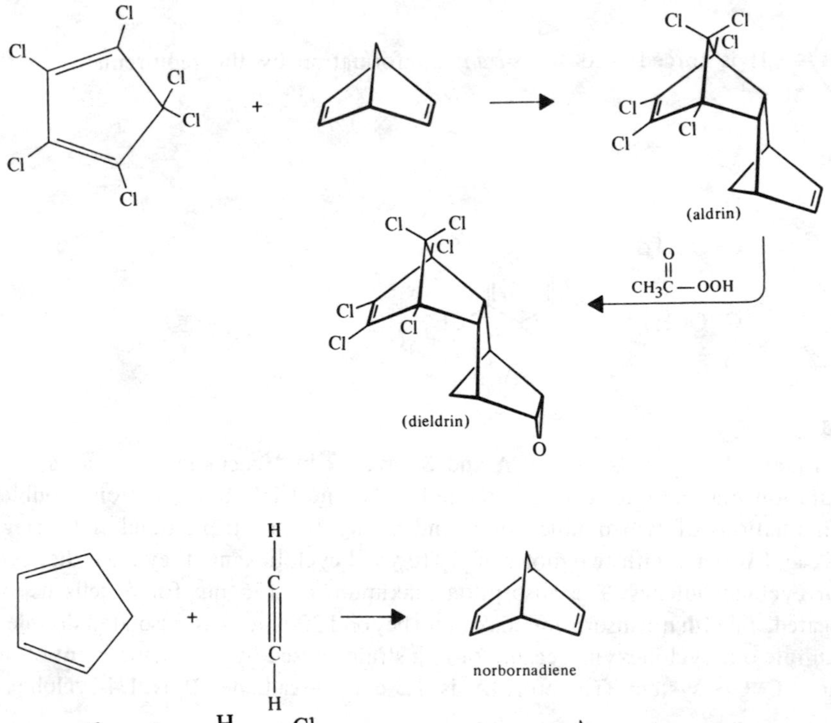

Cl Cl Cl Cl Cl Cl + → (aldrin)

$CH_3\overset{O}{\overset{\|}{C}}-OOH$ → (dieldrin)

10.29

(a) cyclopentadiene + $H-C{\equiv}C-H$ → norbornadiene

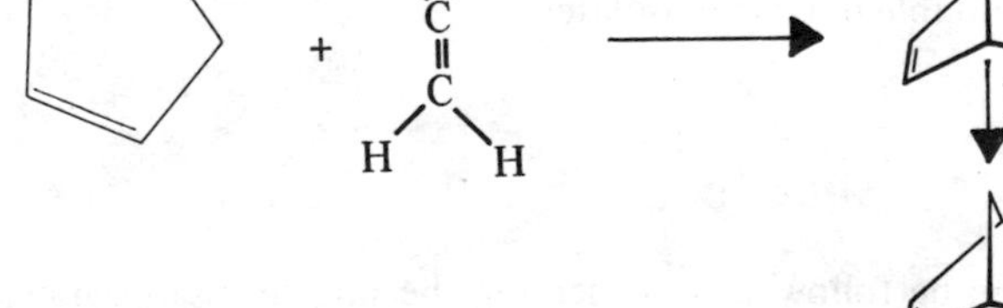

(b) cyclopentadiene + $H(Cl)C{=}CH_2$ → (chloronorbornene) $\xrightarrow{NaOC_2H_5/C_2H_5OH}$ norbornadiene

10.30

$\xrightarrow{Cl_2\ (dark)}$

Note: The other double bond is less reactive because of the presence of the two chlorine substituents.

chlordan

$\xrightarrow{\text{allylic chlorination}}$

heptachlor

10.31

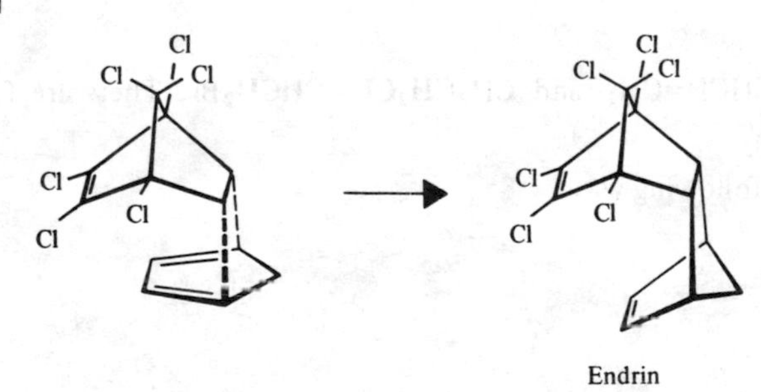

Endrin

10.32

Protonation of the alcohol leads to an allylic cation that can react with a chloride ion at either carbon-1 or carbon-3.

$$CH_3CH{=}CHCH_2OH \xrightarrow{H^+} CH_3CH{=}CHCH_2{-}\overset{\overset{\displaystyle H}{|}}{O^+}{-}H \xrightarrow{-H_2O}$$

$$CH_3CH{=}CHCH_2^+ \longleftrightarrow CH_3\underset{+}{C}HCH{=}CH_2 \xrightarrow{Cl^-}$$

$$CH_3CH{=}CHCH_2Cl + CH_3\underset{\underset{\displaystyle Cl}{|}}{C}HCH{=}CH_2$$

10.33

(1) $CH_2{=}CH{-}CH{=}CH_2 + Cl_2 \longrightarrow ClCH_2{-}\overset{+}{C}H{-}CH{=}CH_2$

$\updownarrow$

$ClCH_2{-}CH{=}CH{-}\overset{+}{C}H_2$

$ClCH_2{-}\overset{\delta+}{C}H{\overset{\dots}{-}}CH{\overset{\dots}{-}}\overset{\delta+}{C}H_2$

(2) $ClCH_2{-}\overset{\delta+}{C}H{\overset{\dots}{-}}CH{\overset{\dots}{-}}\overset{\delta+}{C}H_2 \xrightarrow[(-H^+)]{CH_3OH} ClCH_2{-}\underset{OCH_3}{\underset{|}{C}H}{-}CH{=}CH_2$

$+ \; ClCH_2{-}CH{=}CH{-}CH_2OCH_3$

10.34

A six-membered ring cannot accommodate a triple bond because of the strain that would be introduced.

1,2-Dibromocyclohexane $\xrightarrow{2\ RO^-Na^+}$ cyclohexyne (ring of CH_2, CH_2, CH_2, $C{\equiv}C$, CH_2) ✗ too highly strained

→ 1,3-cyclohexadiene

10.35

The products are $CH_3CH_2\underset{Br}{\underset{|}{C}H}CH{=}CH_2$ and $CH_3CH_2CH{=}CHCH_2Br$. They are formed from an allylic radical in the following way:

$Br_2 \longrightarrow 2\ Br\cdot$
(from NBS)

$Br\cdot + CH_3CH_2CH_2CH{=}CH_2 \longrightarrow CH_3CH_2\dot{C}HCH{=}CH_2$

$\updownarrow \qquad + HBr$

$CH_3CH_2CH{=}CH\dot{C}H_2$

$CH_3CH_2\overset{\delta\cdot}{C}H{\overset{\dots}{-}}CH{\overset{\dots}{-}}\overset{\delta\cdot}{C}H_2 + Br_2 \longrightarrow CH_3CH\underset{Br}{\underset{|}{C}H}CH{=}CH_2$

$+ \; CH_3CH_2CH{=}CHCH_2Br + Br\cdot$

10.36

(a) The same carbocation is produced in the dissociation step:

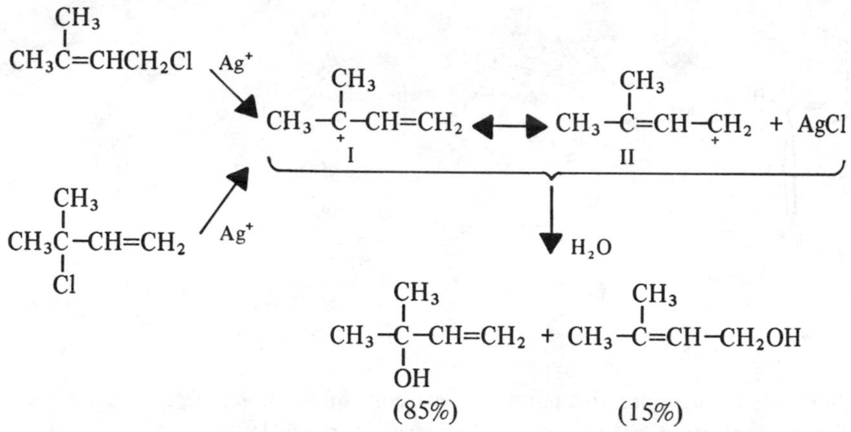

$$CH_3\overset{CH_3}{\overset{|}{C}}=CHCH_2Cl \xrightarrow{Ag^+} CH_3-\underset{+}{\overset{CH_3}{\overset{|}{C}}}-CH=CH_2 \longleftrightarrow CH_3-\overset{CH_3}{\overset{|}{C}}=CH-\underset{+}{CH_2} + AgCl$$

I II

$$CH_3\underset{\underset{Cl}{|}}{\overset{CH_3}{\overset{|}{C}}}-CH=CH_2 \xrightarrow{Ag^+} \text{(I} \longleftrightarrow \text{II)}$$

$$\downarrow H_2O$$

$$CH_3-\underset{\underset{OH}{|}}{\overset{CH_3}{\overset{|}{C}}}-CH=CH_2 + CH_3-\overset{CH_3}{\overset{|}{C}}=CH-CH_2OH$$

(85%) (15%)

(b) Structure I contributes more than II to the resonance hybrid of the carbocation (rule 7). Therefore the hybrid carbocation has a larger positive charge on the tertiary carbon than on the primary carbon. Reaction of the carbocation with water will therefore occur more frequently at the tertiary carbon.

10.37

(a) Propyne. (b) Base $(:B^-)$ removes a proton leaving the anion whose resonance structures are shown:

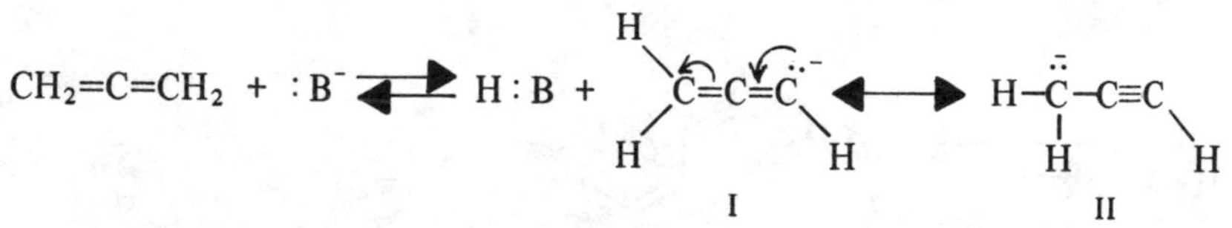

$$CH_2=C=CH_2 + :B^- \rightleftharpoons H:B + \underset{\text{I}}{H_2C=C=\ddot{C}^-H} \longleftrightarrow \underset{\text{II}}{H-\underset{\underset{H}{|}}{\ddot{C}}-C\equiv C-H}$$

Reaction with H : B may then occur at the CH_2 carbanion. The overall reaction is

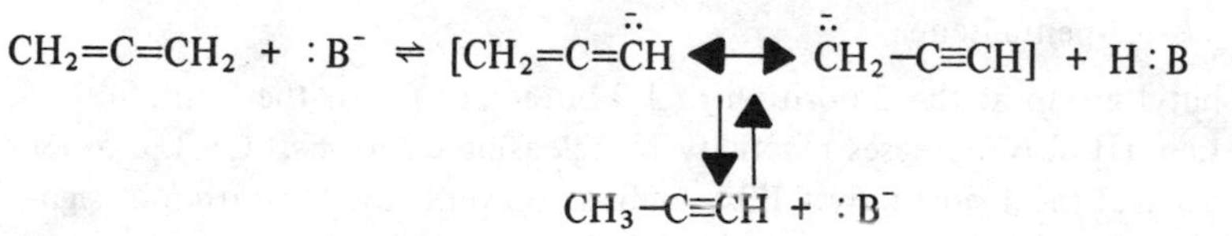

$$CH_2=C=CH_2 + :B^- \rightleftharpoons [CH_2=C=\ddot{C}H \longleftrightarrow \ddot{C}H_2-C\equiv CH] + H:B$$

$$\updownarrow$$

$$CH_3-C\equiv CH + :B^-$$

10.38

The first crystalline solid is the Diels-Alder adduct below, mp 125°,

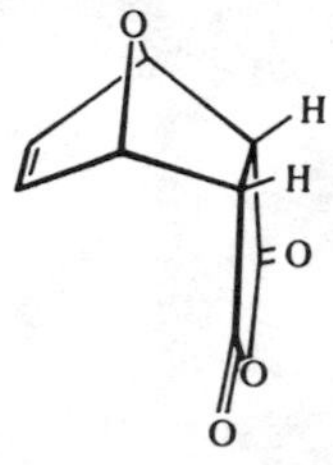

On melting, this adduct undergoes a reverse Diels-Alder reaction yielding furan (which vaporizes) and maleic anhydride, mp 56°.

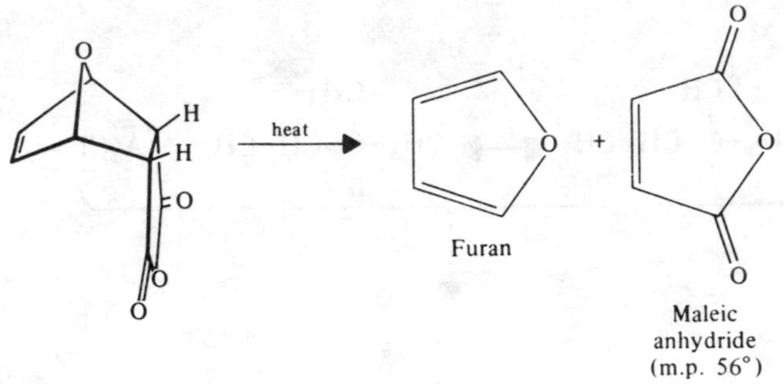

heat

Furan

Maleic anhydride (m.p. 56°)

10.39

(a) *trans*-1,3-Pentadiene assumes the required *s-cis* conformation readily. With *cis*-1,3-pentadiene the concentration of the *s-cis* conformation is much lower because of steric hindrance presented by the internal methyl group.

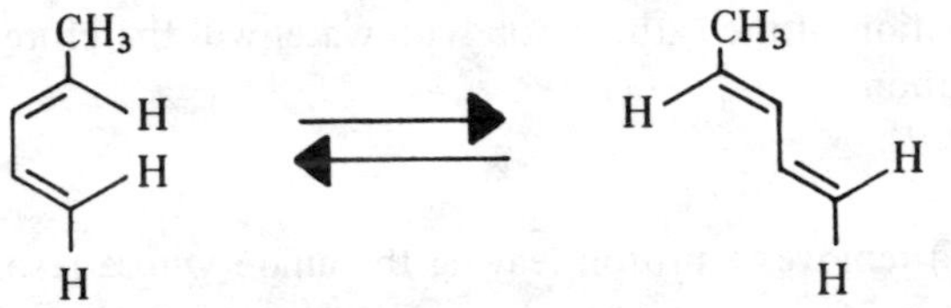

trans-1,3-Pentadiene

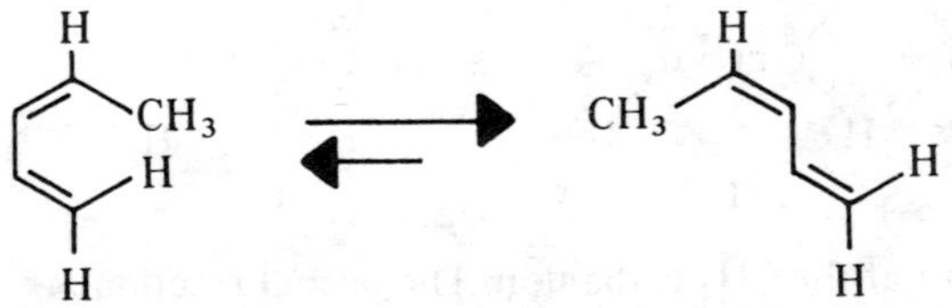

cis-1,3-pentadiene

(b) The large *tert*-butyl group at the 2-position of 1,3-butadiene favors the formation of the *s-cis* conformation. (It also increases reactivity by releasing electrons.) On the other hand, a *tert*-butyl group at the 1-position of 1,3-butadiene prevents the formation of significant concentrations of the *s-cis* conformation and thus no Diels-Alder reactions can take place.

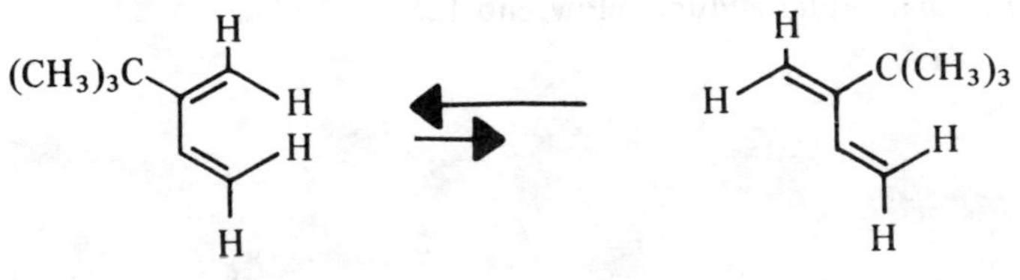

2-*tert*-Butyl-1,3-butadiene

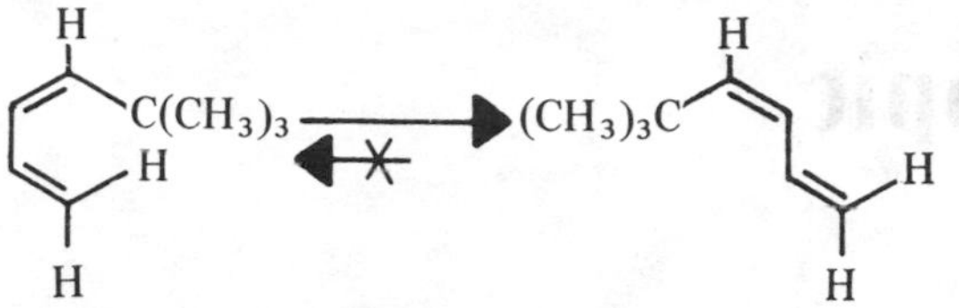

Cis-5,5-dimethyl-1,3-hexadiene
(*Cis*-1-*tert*-butyl-1,3-butadiene)

10.40

The following reaction takes place:

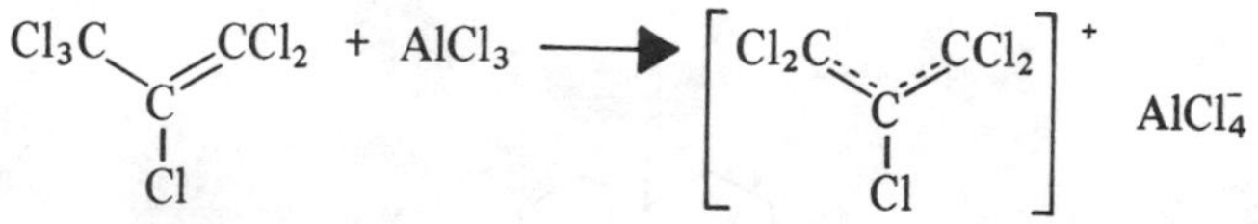

The salt forms because the organic cation is an allylic cation and is, therefore, unusually stable. It is also stabilized by resonance structures involving nonbonding electron pairs of the chlorine atoms such as the second structure below.

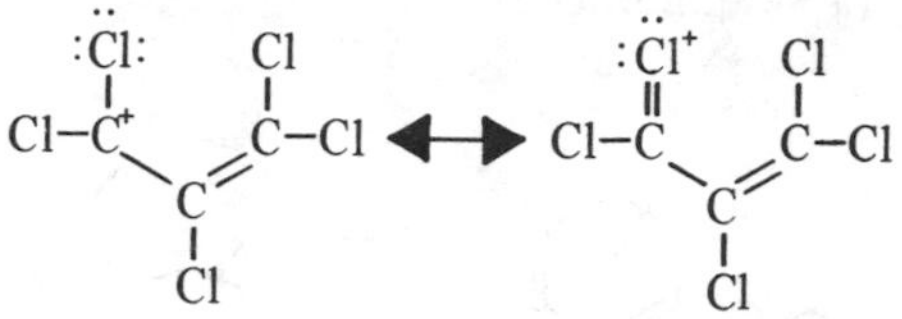

E

Special Topic

E.1

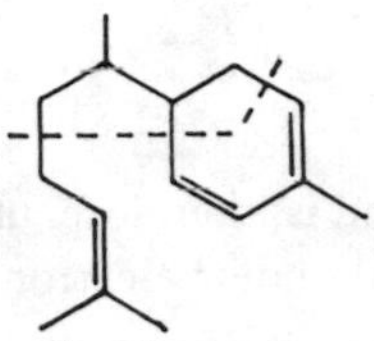

Zingiberene
(a sesquiterpene)

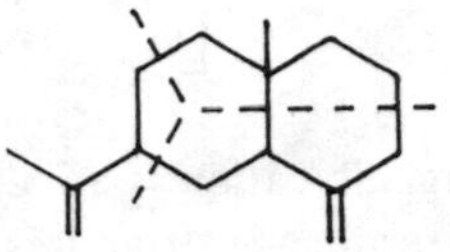

β-Selinene
(a sesquiterpene)

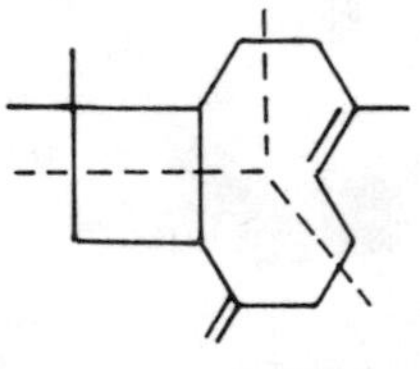

caryophyllene
(a sesquiterpene)

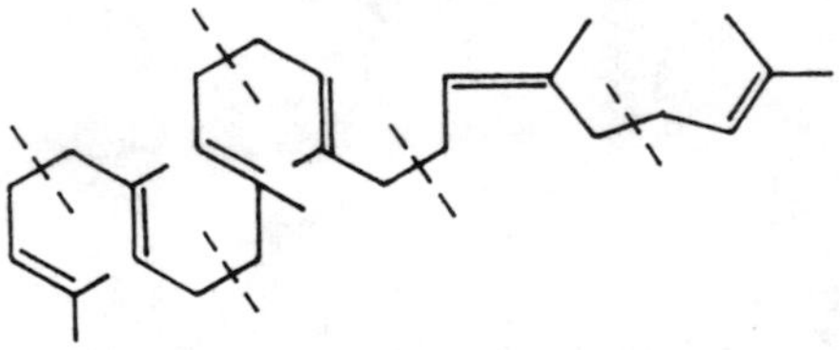

squalene
(a triterpene)

E.2

(a) Myrcene $\xrightarrow[(2)\ Zn,\ H_2O]{(1)\ O_3}$ HCH(=O) + $OHC{-}CH_2{-}CH_2{-}C(=O){-}CHO$ + HCH(=O) + $CH_3C(=O)CH_3$

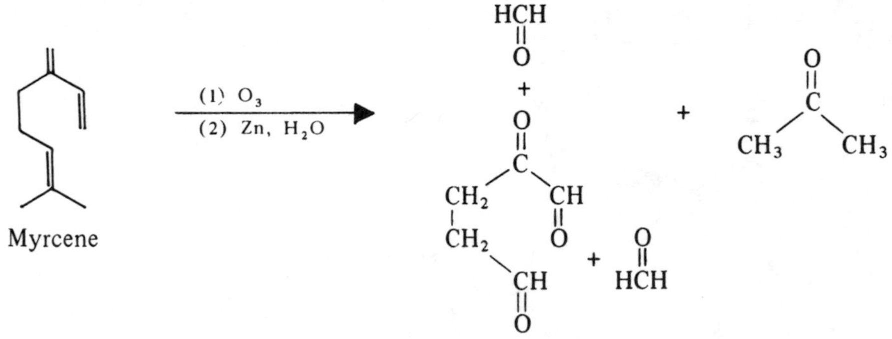

(b) Limonene $\xrightarrow[(2)\ Zn,\ H_2O]{(1)\ O_3}$ $OHC{-}CH_2{-}CH(CH_2CH_2C(=O)CH_3){-}C(=O)CH_3$ + HCH(=O)

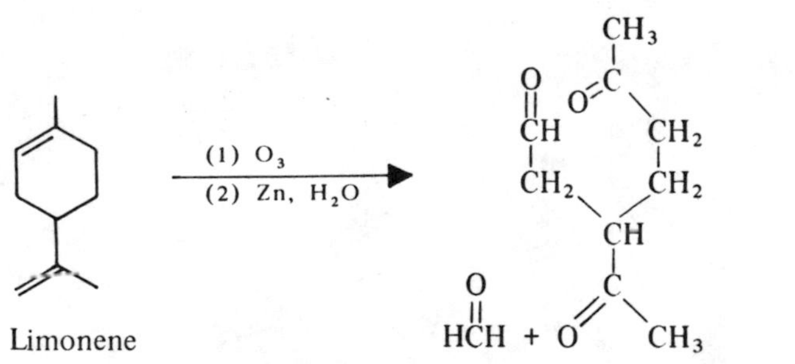

(c) α-Farnesene (See p. 423)

$$\xrightarrow[(2)\ Zn,\ H_2O]{(1)\ O_3} CH_3\overset{O}{\overset{\|}{C}}CH_3 + H\overset{O}{\overset{\|}{C}}CH_2CH_2\overset{O}{\overset{\|}{C}}CH_3$$

$$+ H\overset{O}{\overset{\|}{C}}CH_2\overset{O}{\overset{\|}{C}}H + H\overset{O}{\overset{\|}{C}}-\overset{O}{\overset{\|}{C}}CH_3 + H\overset{O}{\overset{\|}{C}}H$$

(d) Geraniol (See p. 423)

$$\xrightarrow[(2)\ Zn,\ H_2O]{(1)\ O_3} CH_3\overset{O}{\overset{\|}{C}}CH_3 + H\overset{O}{\overset{\|}{C}}CH_2CH_2\overset{O}{\overset{\|}{C}}CH_3$$

$$+ H\overset{O}{\overset{\|}{C}}CH_2OH$$

(e) Squalene (See p. 423)

$$\xrightarrow[(2)\ Zn,\ H_2O]{(1)\ O_3} 2CH_3\overset{O}{\overset{\|}{C}}CH_3 + H\overset{O}{\overset{\|}{C}}CH_2CH_2\overset{O}{\overset{\|}{C}}H$$

$$+ 4CH_3\overset{O}{\overset{\|}{C}}CH_2CH_2\overset{O}{\overset{\|}{C}}H$$

E.3

(a) 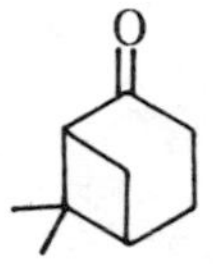 O + CO_2

(c) 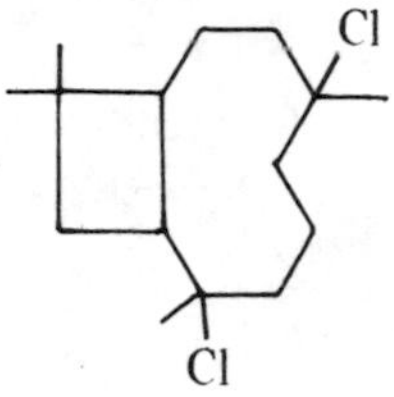 Cl Cl

(b)

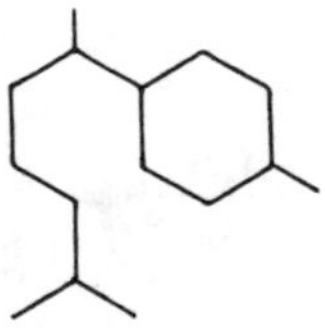

(d) 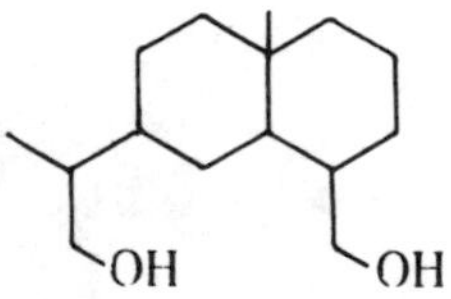 OH OH

E.4

Br_2 in CCl_4 or $KMnO_4$ in H_2O. Either reagent would give a positive result with geraniol and a negative result with menthol.

E.5

(a)

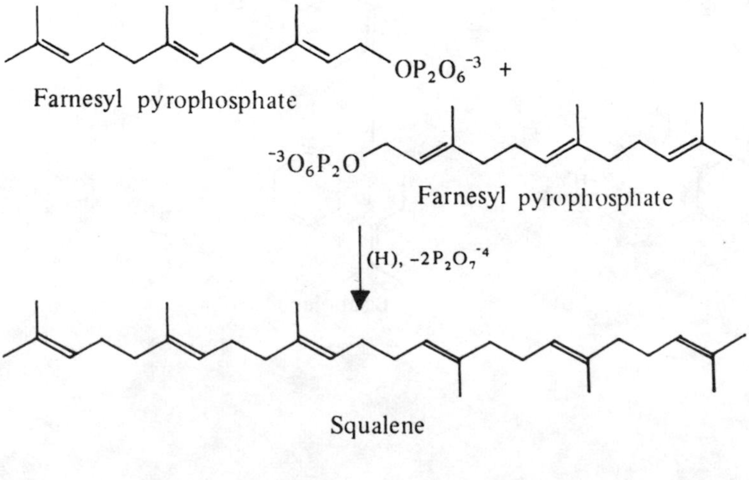

$OP_2O_6^{-3}$ +

Farnesyl pyrophosphate

$^{-3}O_6P_2O$

Farnesyl pyrophosphate

(H), $-2P_2O_7^{-4}$

Squalene

(b) $OP_2O_6^{-3}$ +

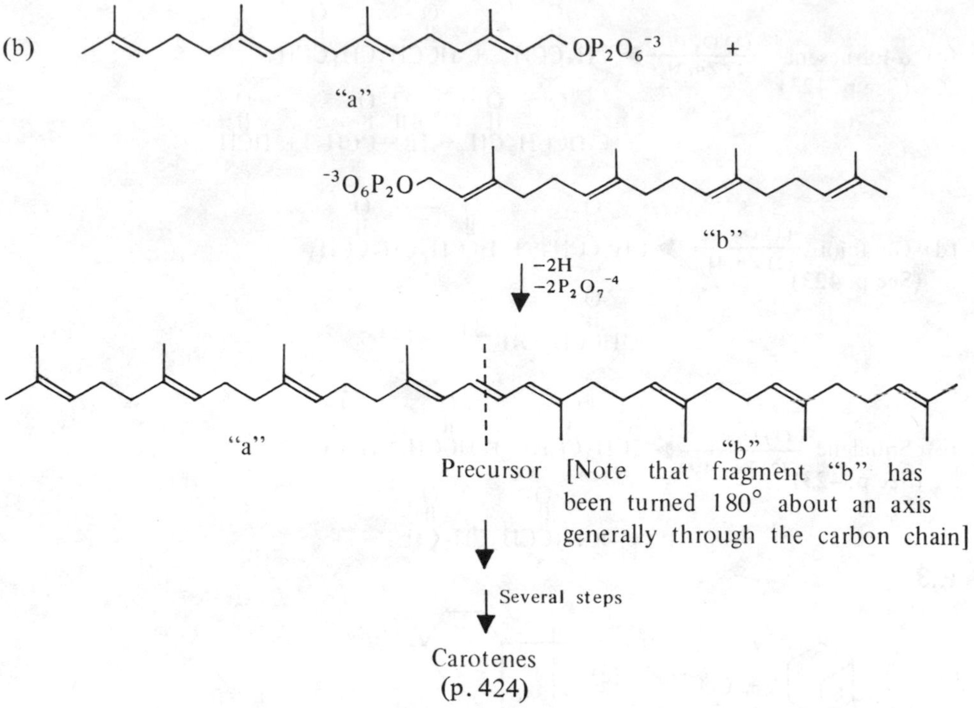

"a"

$^{-3}O_6P_2O$

"b"

$-2H$
$-2P_2O_7^{-4}$

"a"

"b"

Precursor [Note that fragment "b" has been turned 180° about an axis generally through the carbon chain]

Several steps

Carotenes
(p. 424)

E.6

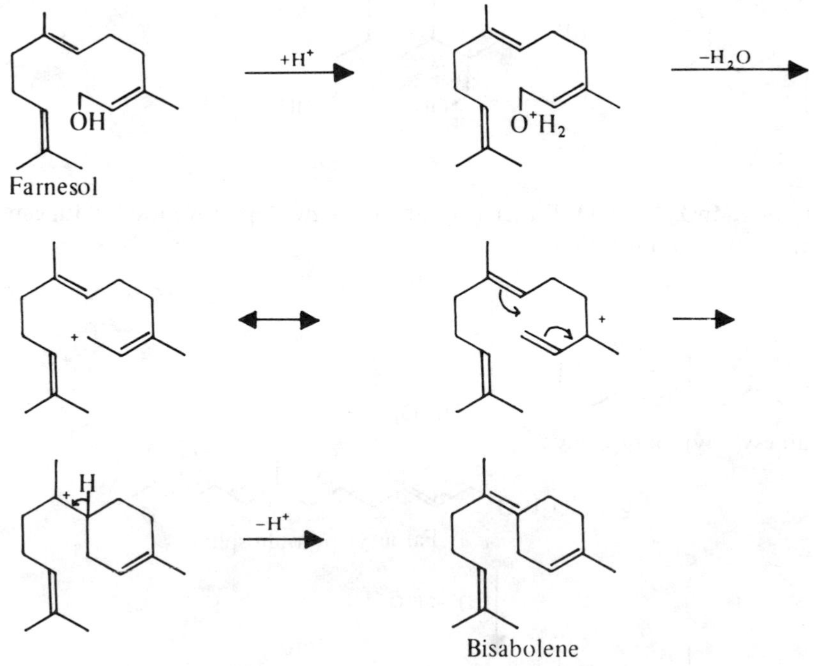

OH

Farnesol

$+H^+$

O^+H_2

$-H_2O$

H

$-H^+$

Bisabolene

E.7

A reverse Diels-Alder reaction takes place.

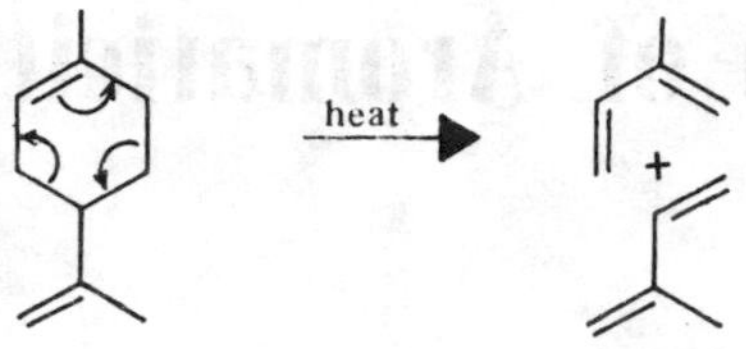

heat

E.8

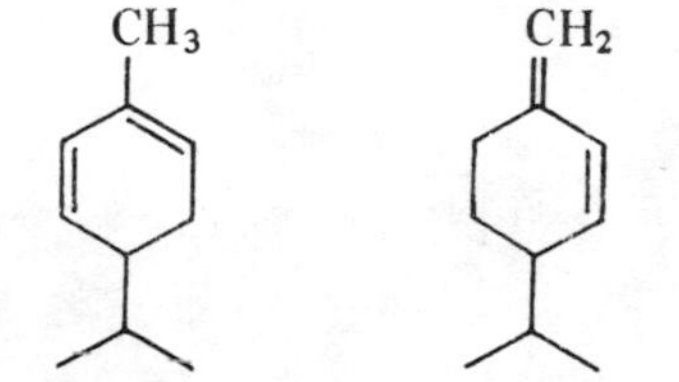

CH_3 CH_2

α-Phellandrene β-Phellandrene

Note: on permanganate oxidation, the $=CH_2$ group of β-phellandrene is converted to CO_2 and thus is not detected in the reaction.

E.9

The Diels-Alder reaction requires that the diene units assume an *s-cis* conformation (p. 395). Vitamin A can do this easily; however, for Neovitamin A steric hindrance is considerable and thus the concentration of the *s-cis* conformation is very small.

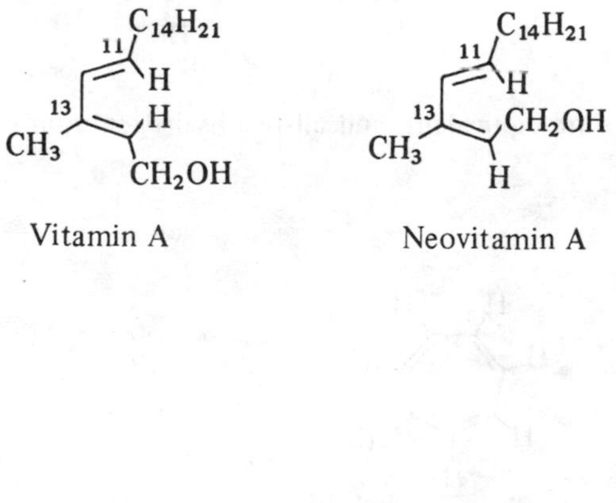

$C_{14}H_{21}$ 11 H 13 H CH_3 CH_2OH

Vitamin A

$C_{14}H_{21}$ 11 H 13 CH_2OH CH_3 H

Neovitamin A

CHAPTER ELEVEN

Aromatic Compounds I: The Phenomenon of Aromaticity

11.1

(a) None (b) None

11.2

Resonance structures may differ only in the positions of the electrons. In the two 1,3,5-cyclohexatrienes shown, the carbons are in different positions; therefore they cannot be resonance structures.

11.3

(a)

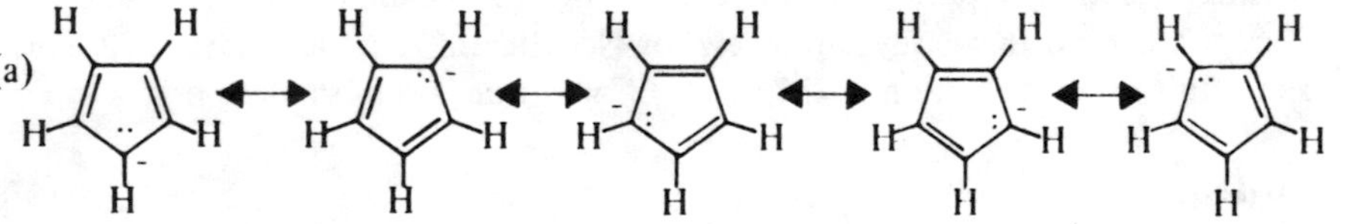

(b) Yes, all of the five resonance structures are equivalent, and all five hydrogen atoms are equivalent.

11.4

(a) ClO_4^- ⟶ ClO_4^- + $HC(C_6H_5)_3$

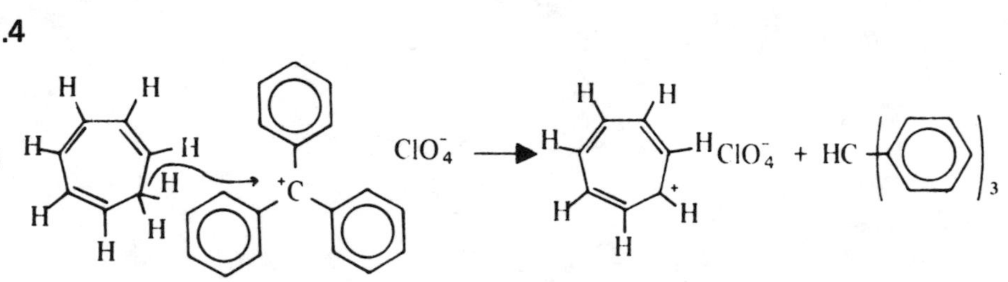

(b) Triphenylmethane (See (a) above).

(c) ClO_4^-

11.5

Tropylium bromide is ionic and has the structure,  + Br^-. The ring is aromatic.

11.6

(a) $+CH(CH=CH_2)(CH=CH-CH=CH_2)$

(b) Cycloheptatrienyl cation is aromatic.

11.7

(a) $+CH(CH=CH_2)_2$ —(π electron energy increases)→ cyclopentadienyl cation $+ H_2$

The π electron energy of cyclopentadienyl cation is higher than that of the open chain counterpart.

(b) $+CH_2-CH=CH_2$ —(π electron energy decreases)→ cyclopropenyl cation $+ H_2$

The π electron energy of cyclopropenyl cation is lower than that of the open chain counterpart.

(c) $4n+2 = 2$ when $n = 0$. Hückel's rule predicts that cyclopropenyl cation is aromatic.

(d) The π electron energy of cyclopropenyl anion is higher than that of the open chain counterpart.

11.8

(a)

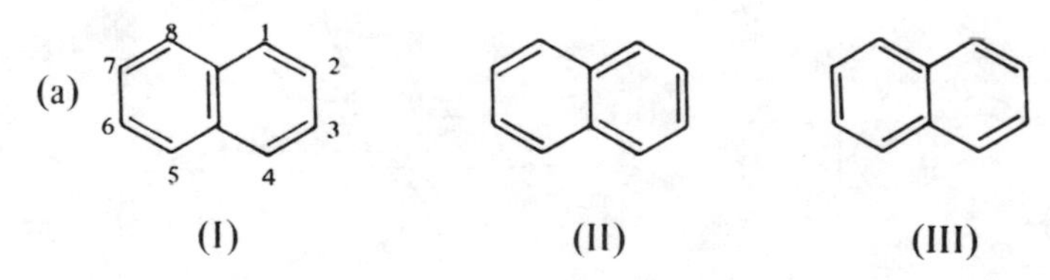

(b) Two of the structures (I and III) have a double bond between the C_1-C_2 carbons, whereas only structure II has a double bond between the C_2-C_3 carbons. Assuming that the three structures contribute nearly equally, the C_1-C_2 bond should be more like a double bond and therefore should be shorter than the C_2-C_3 bond.

11.9

$(C_6H_5)C=C(C_6H_5)$ ring with $C^+-\ddot{O}:^-$

III

III is a more important contributor to the resonance hybrid of I than a corresponding ionic structure of II is to the hybrid of II. III is an important contributor to the hybrid of diphenylcyclopropenone because it resembles the aromatic cyclopropenyl cation (Cf problem 11.7); i.e., the ring in structure III has 2 π electrons and is a $4n + 2$ system where $n = 0$.

11.10

7, 6, 5, 1 N, 8, 2, N 4, N 9, 3, H

1,3,7 are pyridine type nitrogens; 9 is a pyrrole type nitrogen

11.11

(a) O_2N– –SO_3H (b) CH_3, Cl (c) Cl, Cl (d) NO_2, NO_2

(e) OCH_3, Br (f) COOH, NO_2 (g) OH, I (h) $COCH_3$, Cl

(i) Br (j) Cl (k) NO_2 (l) NO_2, N

(m) N, H, CH_3 (n) NO_2, Cl, Cl (o) CH_2Br, NO_2 (p) NH_2, Cl

(q) COOH, Br, Br, NO_2 (r) CH_3, CH_3, CH_3 (s) COOH, OH (t) CH=CH_2

(u)

(v) OH

(w) CH_3, O_2N, NO_2, NO_2

(x)

(y)

(z)

11.12

(a) Cl, Cl, Cl

1,2,3-Trichlorobenzene | 1,2,4-Trichlorobenzene | 1,3,5-Trichlorobenzene

(b) NO_2, Br, Br

2,3-Dibromo-1-nitrobenzene | 2,4-Dibromo-1-nitrobenzene | 1,4-Dibromo-2-nitrobenzene | 1,3-Dibromo-2-nitrobenzene

1,2-Dibromo-4-nitrobenzene | 3,5-Dibromo-1-nitrobenzene

(c)

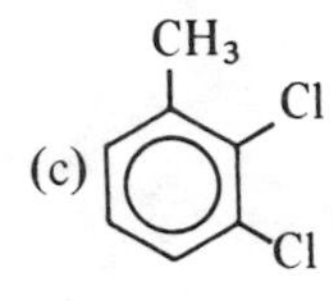

2,3-Dichlorotoluene | 2,4-Dichlorotoluene | 2,5-Dichlorotoluene | 2,6-Dichlorotoluene

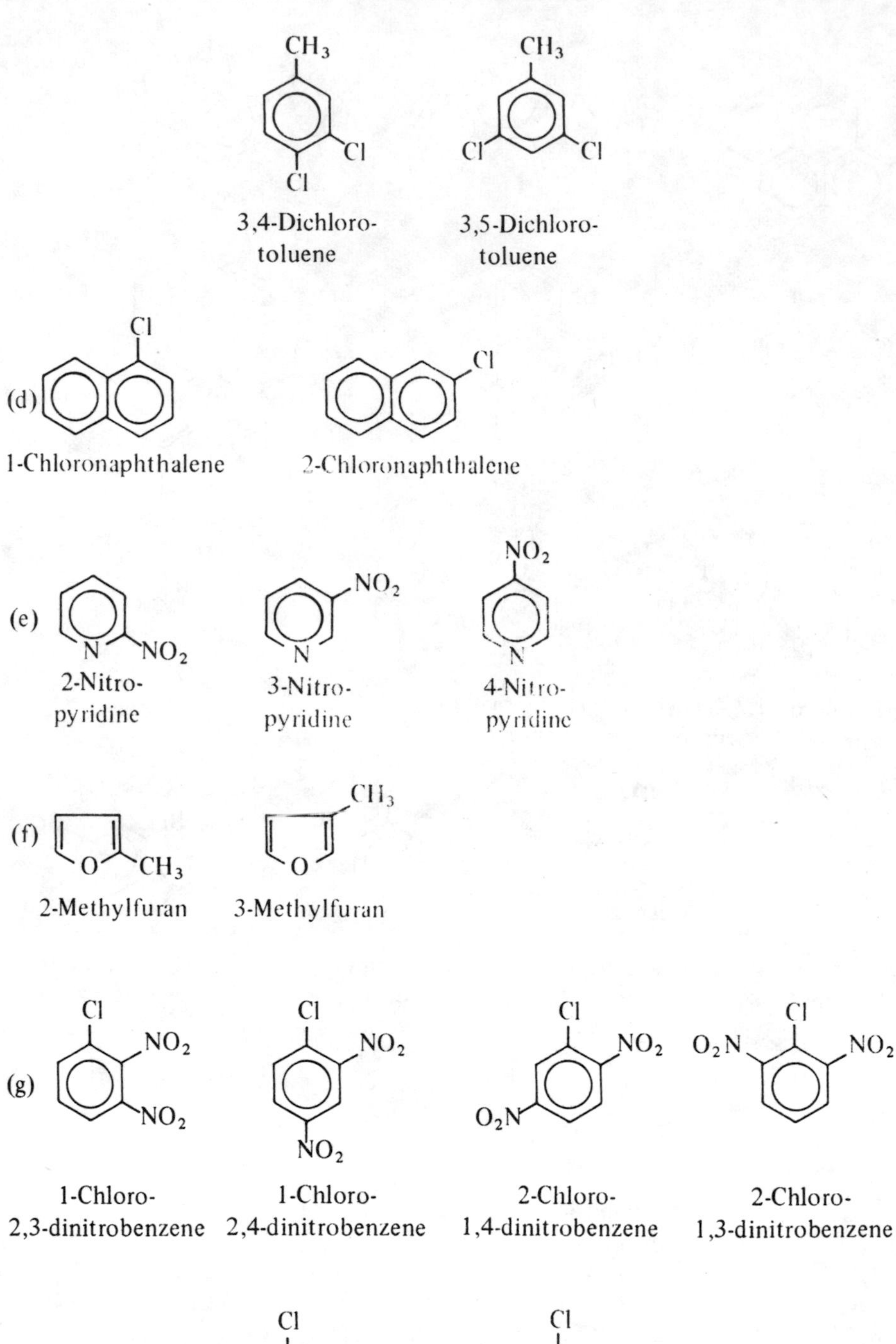
CH3
Cl
Cl
3,4-Dichloro-
toluene
CH3
Cl
Cl
3,5-Dichloro-
toluene
(d)
Cl
1-Chloronaphthalene
Cl
2-Chloronaphthalene
(e)
N
NO2
2-Nitro-
pyridine
NO2
N
3-Nitro-
pyridine
NO2
N
4-Nitro-
pyridine
(f)
O
CH3
2-Methylfuran
CH3
O
3-Methylfuran
(g)
Cl
NO2
NO2
1-Chloro-
2,3-dinitrobenzene
Cl
NO2
NO2
1-Chloro-
2,4-dinitrobenzene
Cl
NO2
O2N
2-Chloro-
1,4-dinitrobenzene
Cl
O2N
NO2
2-Chloro-
1,3-dinitrobenzene
Cl
NO2
NO2
4-Chloro-
1,2-dinitrobenzene
Cl
O2N
NO2
1-Chloro-
3,5-dinitrobenzene

(h)

CH_3 CH_3 Cl — 1-Chloro-2,3-dimethylbenzene

CH_3 CH_3 Cl — 4-Chloro-1,2-dimethylbenzene

CH_3 Cl CH_3 — 2-Chloro-1,3-dimethylbenzene

CH_3 CH_3 Cl — 1-Chloro-2,4-dimethylbenzene

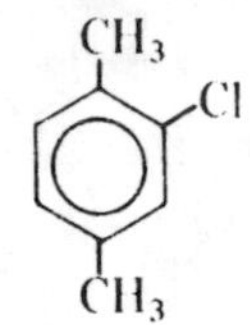

1-Chloro-3,5-dimethylbenzene

CH_3 Cl CH_3 — 2-Chloro-1,4-dimethylbenzene

(i)

CH_3 OH — o-Cresol

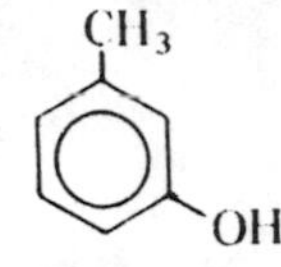

m-Cresol

CH_3 OH — p-Cresol

11.13

(a)

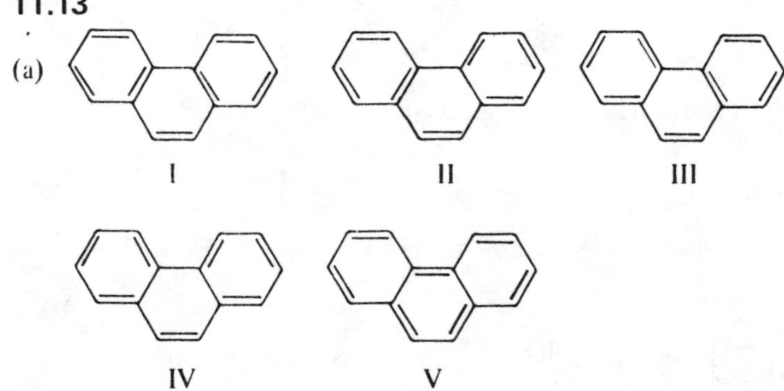

(b) The 9,10 bond should be close to that of a double bond, 1.33Å, since in four of the five contributors it is a double bond.

(c) Almost that of an actual double bond.

(d) Bromine adds to the 9,10 double bond because of its large double bond character and because addition disrupts only one of three aromatic rings.

11.14

(a)

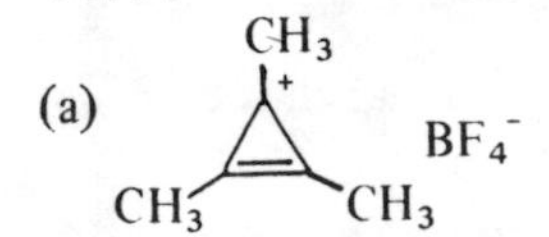

(b) The trimethylcyclopropenyl cation is aromatic.

11.15

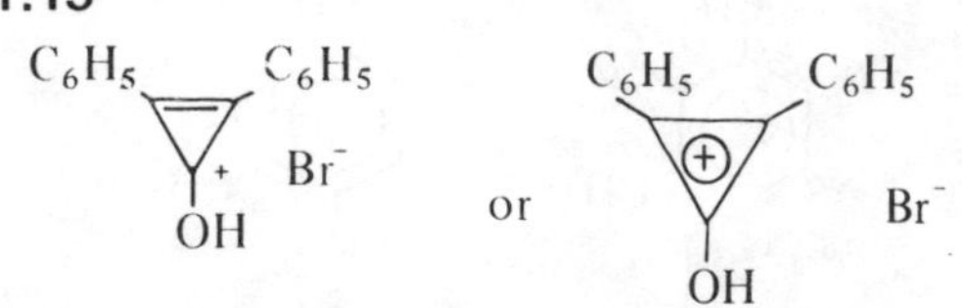

11.16

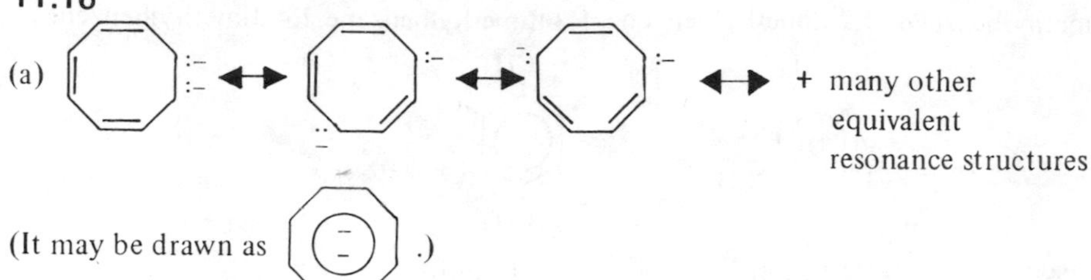

(It may be drawn as .)

(b) It has $4n + 2 = 10$ π electrons ($n = 2$), and is therefore aromatic; i.e., it obeys Huckel's rule.

11.17

(a) Cyclononatetraenyl anion: Li^+

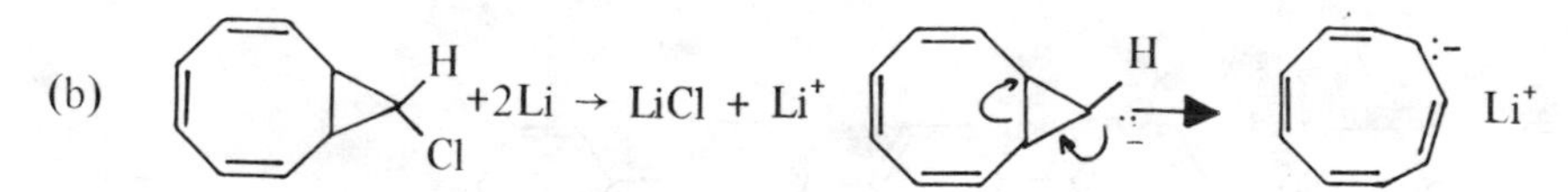

The product anion has 10 π-electrons and is therefore aromatic.

11.18

(a) + H_2 ⟶ $\Delta H° = (-49.8) - (-55.4) = +5.6$ kcal/mole

(b) 2 ⟶ + $\Delta H° = -5.6 + (-55.4 - (-28.6))$
$= -5.6 - 26.8 = -32.4$ kcal/mole

(c) Reaction (a) is endothermic, and the competing reaction (reaction (b)) is exothermic. Therefore we must conclude that reaction (a) is unlikely under conditions (i.e., the presence of a catalyst) that will allow equilibrium to be established.

11.19

(a) Would not be aromatic. It is a monocyclic system of 12 π electrons and thus does not obey Huckel's rule.

(b) Would not be aromatic; it is not a conjugated system.

(c) Would not be aromatic; it is an 8 π electron monocyclic system and thus does not obey Huckel's rule.

(d) Would not be aromatic; it is a 16 π electron monocyclic system and thus does not obey Huckel's rule.

(e) Would be aromatic because of resonance structures (below) that consist of a cycloheptatrienyl cation and cyclopentadienyl anion.

etc.

(f) Would be aromatic; it is a planar monocyclic system of 14 π electrons. (We count only two electrons of the triple bond because only two are in p-orbitals that overlap with those of the double bonds on either side.)

(g) Would be aromatic; it is a planar monocyclic system of 10 π electrons.

(h) Would be aromatic; it is a planar monocyclic system of 10 π electrons. (The bridging $-CH_2-$ groups allows the ring system to be planar.)

11.20

Resonance contributors that involve the carbonyl group of **I** resemble the *aromatic* cycloheptatrienyl cation and thus stabilize **I**. Similar contributors to the hybrid of **II** resemble the *antiaromatic* cyclopentadienyl cation (see Prob. 11.7) and thus destabilize **II**.

(a)

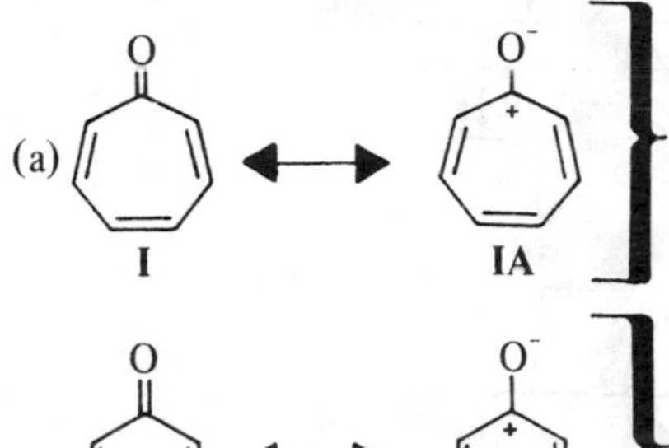

Contributors like this are exceptionally stable because they resemble an aromatic compound. They therefore make large stabilizing contributions to the hybrid.

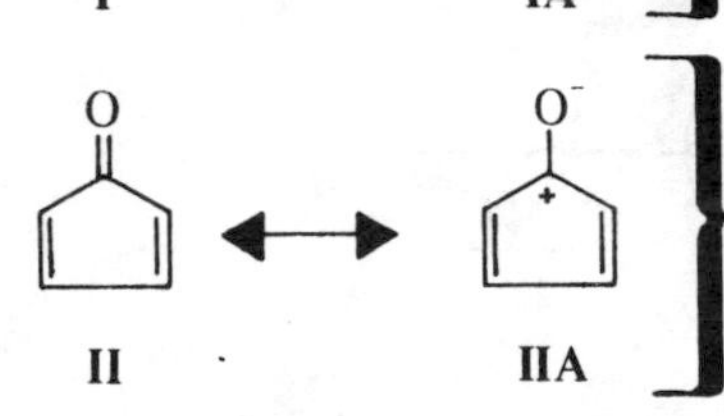

Contributors like this are exceptionally unstable because they resemble an antiaromatic compound. Any contribution they make to the hybrid is destabilizing.

(b)

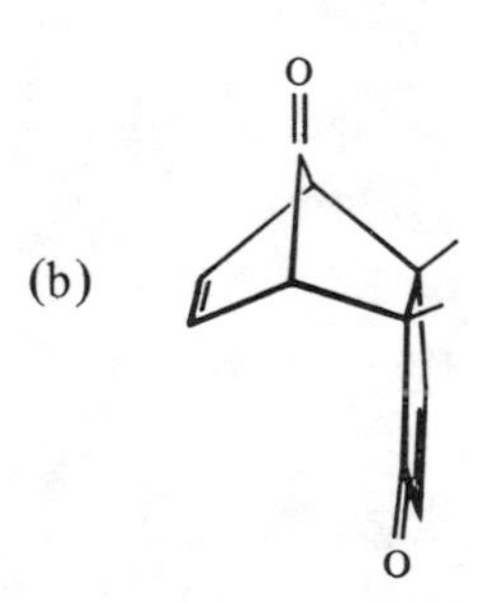

11.21

C is [14]-annulene. The steps in its synthesis are the following:

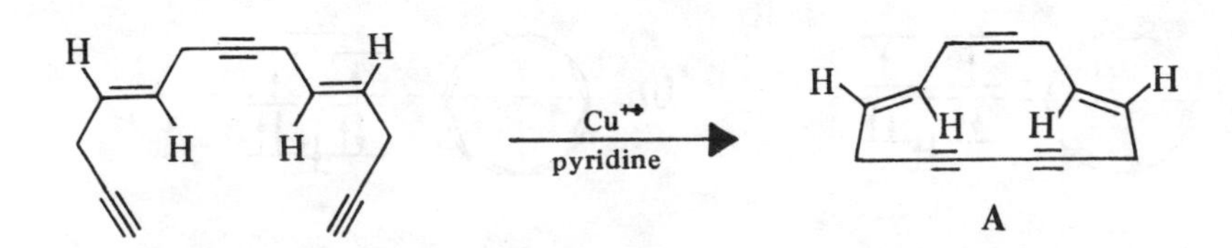

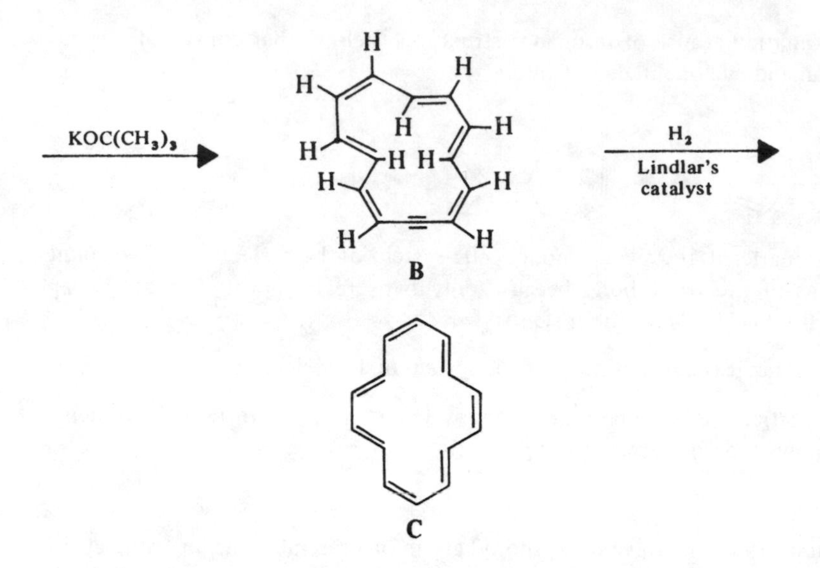

11.22

(a)

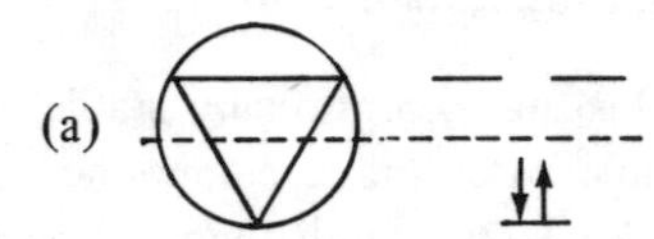

(b)

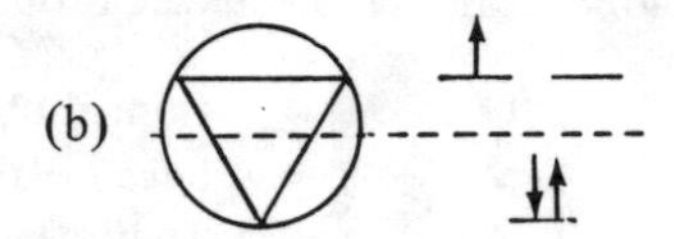

(c)

(d)

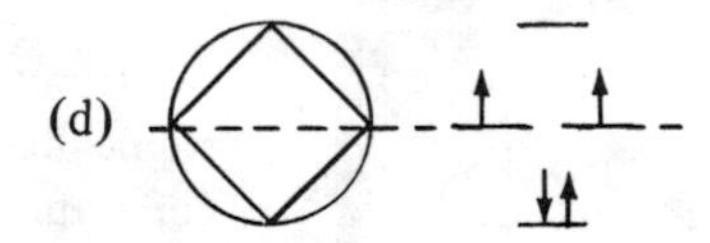

(e)

(f)

(g)

(h)

(i)

(j)

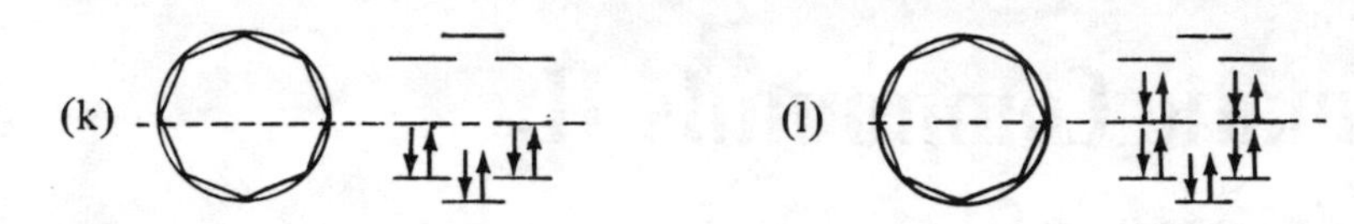

(m) b, c, d, f, i, j, (n) a, e, g, h, k, l, (o) same as for (n),

(p) yes

CHAPTER TWELVE

Aromatic Compounds II: Reactions of Aromatic Compounds with Electrophiles

12.1

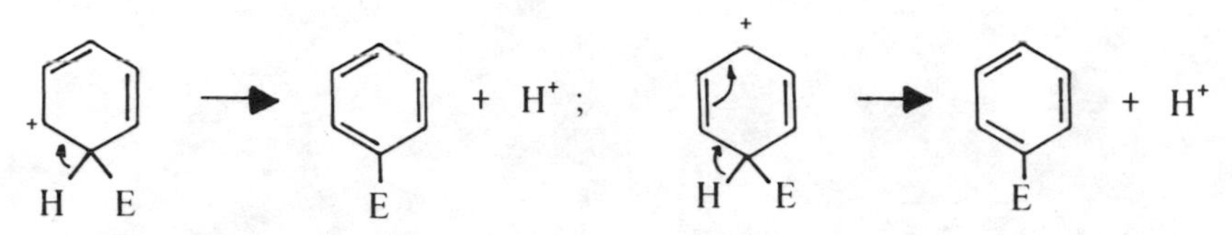

12.2

Oxidation of the Fe by X_2 generates the ferric salt:

$$2Fe + 3X_2 \longrightarrow 2FeX_3$$

12.3

$$H\text{–}\ddot{O}\text{–}NO_2 + H\text{–}\ddot{O}\text{–}NO_2 \longrightarrow H\text{–}\overset{+}{\ddot{O}}(H)\text{–}NO_2 + NO_3^-$$

$$H\text{–}\overset{+}{\ddot{O}}(H)\text{–}NO_2 + HONO_2 \longrightarrow NO_2^+ + H_3O^+ + NO_3^-$$

12.4

(a)

$$C_6H_5SO_3H + D_2O \rightleftarrows C_6H_5SO_3^- + D_2HO^+$$

$$C_6H_5SO_3^- + D_3O^+ \rightleftarrows C_6H_5^+(D)(SO_3^-) + D_2O$$

$$C_6H_5^+(D)(SO_3^-) \longrightarrow C_6H_5D + :\ddot{O}=S(=\ddot{O}:)=\ddot{O}:$$

(b) $:\ddot{Br}\text{–}\ddot{Br}\text{–}\overset{-}{Fe}Br_3$ ($\delta+$ $\delta+$) is the electrophile.

$$C_6H_5SO_3^- + :\ddot{B}r-Br-FeBr_3 \longrightarrow [C_6H_5(SO_3^-)(Br)]^+ + FeBr_4^-$$

$$[C_6H_5(Br)SO_3^-]^+ \longrightarrow C_6H_5Br + SO_3$$

12.5

(a)

$$\text{cyclohexene} \underset{}{\overset{HF}{\rightleftharpoons}} C_6H_{11}^+$$

$$C_6H_6 + C_6H_{11}^+ \longrightarrow [C_6H_6-C_6H_{11}]^+ \longrightarrow C_6H_5-C_6H_{11} + H^+$$

(b)

$$C_6H_{11}OH + BF_3 \rightleftharpoons C_6H_{11}\overset{+}{O}H-\overset{-}{B}F_3 \rightleftharpoons C_6H_{11}^+ + HO\overset{-}{B}F_3$$

$$C_6H_6 + C_6H_{11}^+ \longrightarrow [C_6H_6-C_6H_{11}]^+ \longrightarrow C_6H_5-C_6H_{11} + H^+$$

12.6

$$(CH_3CO)_2O + AlCl_3 \rightleftharpoons CH_3C(=O)-\overset{+}{O}(-\overset{-}{A}lCl_3)-C(=O)CH_3 \longrightarrow CH_3\overset{+}{C}=\ddot{O}: + CH_3\overset{O}{\overset{\|}{C}}O\overset{-}{A}lCl_3$$

$$(CH_3CO)_2O + AlCl_3 \longrightarrow CH_3C(=\overset{+}{O}\overset{-}{A}lCl_3)-O-C(=O)CH_3 \longrightarrow CH_3\overset{+}{C}=\ddot{O}: + CH_3\overset{O}{\overset{\|}{C}}O\overset{-}{A}lCl_3$$

12.7

40% ortho, 40% meta, 20% para.

12.8

(a) At the lower temperature the proportions are determined by the relative reaction rates. At the higher temperature the proportions are determined by the relative stabilities; i.e., the more stable isomer predominates even if it is produced at the slower rate because all steps are reversible.

(b) p-Toluenesulfonic acid.

12.9

(a) CH_2CH_3, Cl + CH_2CH_3, Cl (b) CF_3, Cl (c) $\overset{+}{N}(CH_3)_3\ Cl^-$, Cl (d) $COOCH_3$, Cl

12.10

(a) $Cl{-}CH_2{-}CH_2{-}\overset{+}{N}(CH_3)_3\overset{-}{Cl}$. This is anti-Markovnikov addition because the intermediate carbocation formed by Markovnikov addition would have positive charges on adjacent atoms, i.e.,

$$CH_3{-}\overset{+}{C}H{-}\overset{+}{N}(CH_3)_3\overset{-}{Cl}$$

The anti-Markovnikov carbocation is $\overset{+}{C}H_2{-}CH_2{-}\overset{+}{N}(CH_3)_3\overset{-}{Cl}$. Although this is a primary carbocation, it is more stable because of the greater separation of charges.

(b) Slower because it requires formation of an intermediate primary cation with two positive charges.

(c) The positive charge of the trimethylammonium ion makes it a powerful electron-withdrawing group.

(d) $^+N(CH_3)_3$, H, E $\longleftrightarrow$ $^+N(CH_3)_3$, H, E $\longleftrightarrow$ $^+N(CH_3)_3$, H, E

None of these structures is especially unstable, whereas in *ortho* or *para* attack at least one structure would be.

12.11

(a) The electronic influence of the trimethylammonium ion on the ring is an inductive effect that causes *meta* orientation and it is diminished by every methylene group that separates it from the ring.

(b) The increasing number of chlorines makes the methyl group increasingly electron-withdrawing, and hence increasingly *meta*-directing.

12.12

12.13

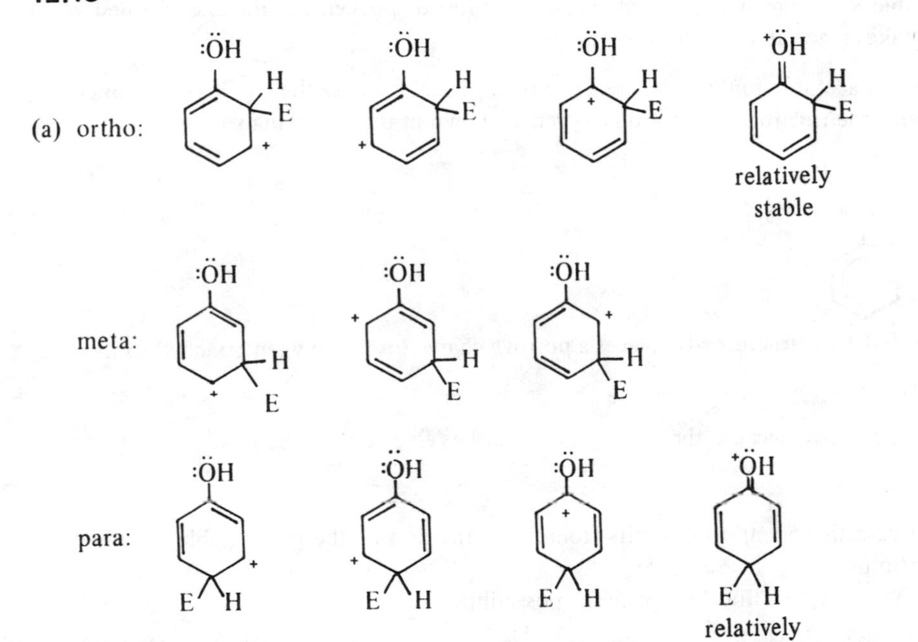

(b) The electron-releasing ability of the –OH group through resonance increases the electron density of the ring, and it stabilizes the positive charge of the intermediate carbocation.

(c) An exceptionally stable structure (above) contributes to the intermediate carbocation only when attack is *ortho* or *para*.

(d,e) More reactive because the negatively charged $-\bar{O}$ group of the phenoxide ion is an even more powerful electron-releasing group than the $-OH$ group of phenol.

12.14

(a)

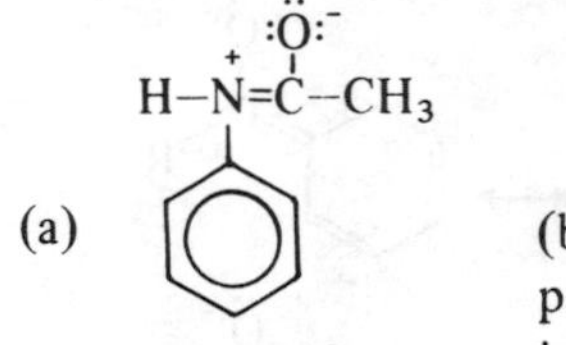

(b) The $-\overset{O}{\overset{\|}{C}}-CH_3$ group competes with the ring for the electron pair on N, therefore stabilization of the intermediate carbocation is less effective than in aniline.

(c)

$H-\overset{+}{N}(=\text{ring})-\overset{\ddot{O}:}{\overset{\|}{C}}-CH_3$ (ring carbon bearing H and E) **A**

$H-\overset{+}{N}(=\text{ring})-\overset{\ddot{O}:}{\overset{\|}{C}}-CH_3$ (para ring carbon bearing E and H) **B**

Yes, resonance accounts for an electron release from nitrogen to the ring, and exceptionally stable structures (**A** and **B** above) contribute to the sigma complexes formed when attack takes place at an *ortho* or *para* carbon.

(d) Phenyl acetate should be less reactive than phenol because the $-COCH_3$ group competes with the ring for electrons on oxygen as shown in the following structure.

$:\overset{+}{O}=\overset{:\ddot{O}:^-}{\overset{|}{C}}-CH_3$ (O attached to benzene ring)

Notice that this structure also places a positive charge on the oxygen attached to the ring.

(e) Ortho-para

(f) More reactive because the $-\ddot{\underset{..}{O}}-$ group furnishes electrons to the ring.

12.15

In each case the orientation results from the formation of the more stable intermediate carbocation.

With 3,3,3-tri fluoropropene the possibilities are:

$$CF_3-CH=CH_2 + H^+ \longrightarrow CF_3-\overset{+}{C}H-CH_3 \quad \text{I (less stable)}$$

$$CF_3-CH=CH_2 + H^+ \longrightarrow CF_3-CH_2-\overset{+}{C}H_2 \quad \text{II (more stable)} \xrightarrow{Cl^-} CF_3-CH_2-CH_2Cl$$

Although II is a primary carbocation and I is a secondary carbocation, II is more stable because in it the positive charge is separated from the highly electron-withdrawing CF_3 group by an intervening carbon.

With chloroethene the possibilities are:

$$:\ddot{\underset{..}{Cl}}-CH=CH_2 + H^+ \longrightarrow :\ddot{\underset{..}{Cl}}-CH_2-\overset{+}{C}H_2 \quad \text{III (less stable)}$$

$$:\ddot{\underset{..}{Cl}}-CH=CH_2 + H^+ \longrightarrow :\ddot{\underset{..}{Cl}}-\overset{+}{C}H-CH_3 \longleftrightarrow :\overset{+}{\underset{..}{Cl}}=CH-CH_3 \quad \text{IV (more stable)}$$

$$:\ddot{\underset{..}{Cl}}-\overset{+}{C}H-CH_3 \xrightarrow{Cl^-} Cl-\underset{\underset{Cl}{|}}{C}H-CH_3$$

Here carbocation IV is more stable than III because of resonance involving an electron pair of the chlorine atom.

12.16

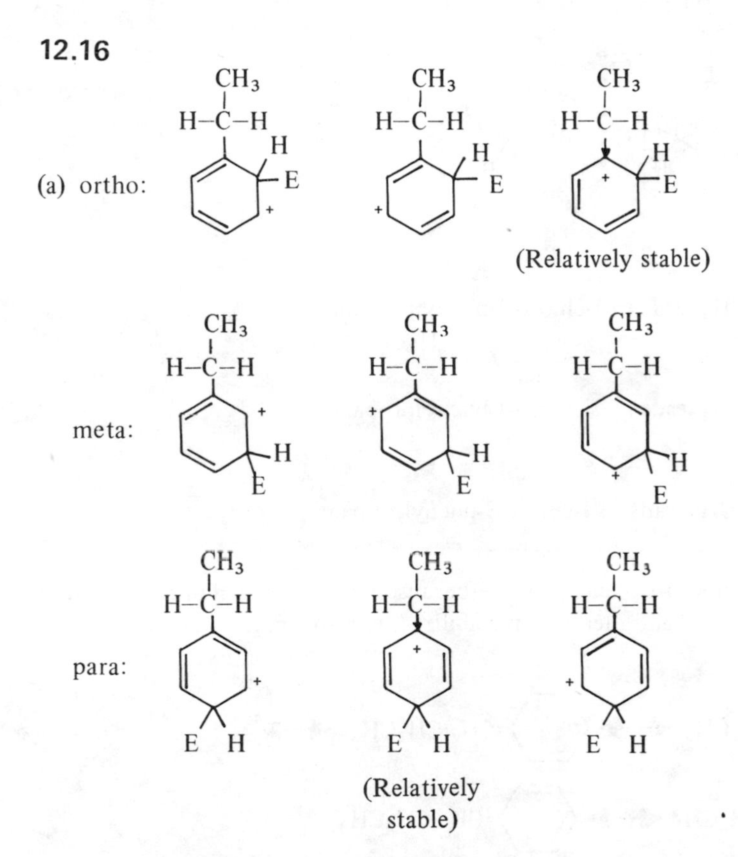

(b) Yes, substitution at an ortho or para position yields a carbocation that is stabilized by the contribution of a relatively stable structure. The relatively stable structures (above) have a positive charge on the ring carbon that bears the ethyl group.

(c) Because the ethyl group is electron-releasing it stabilizes the intermediate sigma complexes.

12.17

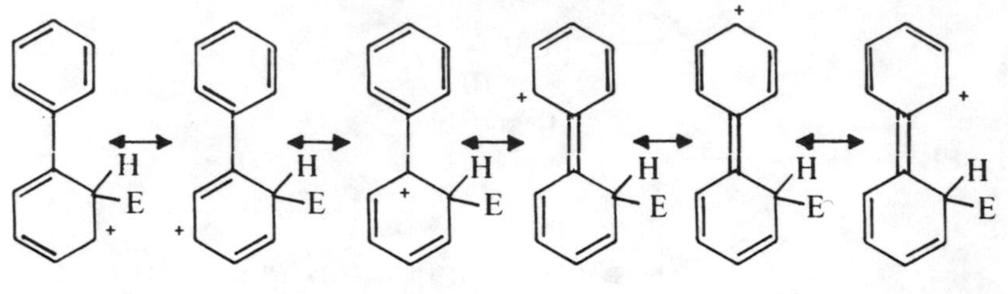

The carbocation above has the positive charge delocalized over both rings and thus it is relatively stable. Similar structures can be drawn for the sigma complex formed when substitution takes place at a para position. However, when electrophilic attack takes place at the meta position it produces a carbocation whose positive charge cannot be delocalized over both rings:

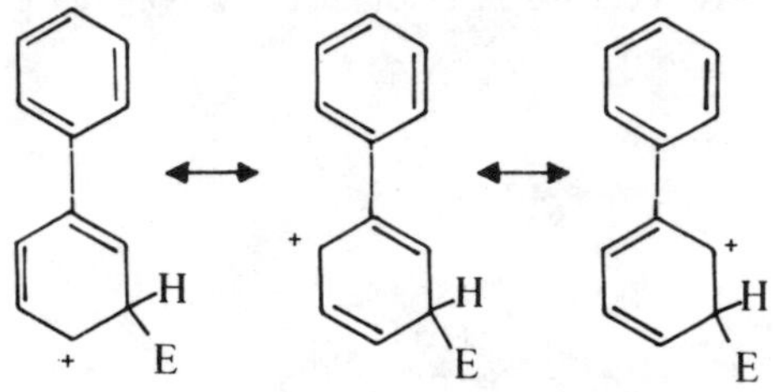

12.18

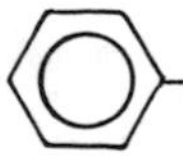

C_6H_5–$\dot{C}HCH_2CH_3$ leads to 1-chloro-1-phenylpropane
(I)

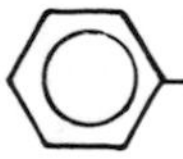

C_6H_5–$CH_2\dot{C}HCH_3$ leads to 2-chloro-1-phenylpropane
(II)

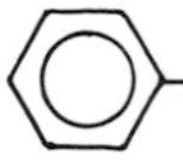

C_6H_5–$CH_2CH_2\dot{C}H_2$ leads to 1-chloro-3-phenylpropane
(III)

The major product is 1-chloro-1-phenylpropane because I is the most stable free-radical because it is a benzylic radical and therefore is stabilized by resonance.

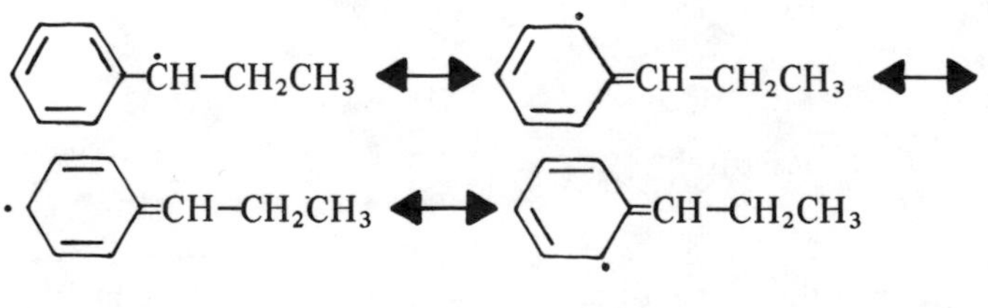

12.19

(a) 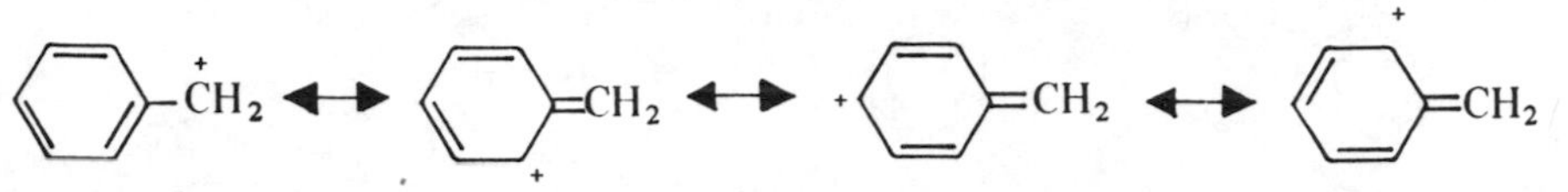

(b) The benzyl cation is stabilized by resonance.

(c) Yes, because the benzyl cation is a hybrid of the structures given in (a).

(d) Over the ortho and para ring carbons and the benzylic carbon.

(e) π_1, π_2, π_3 (see Figure 12.4).

(f) Since π_4 is vacant, molecular orbital theory predicts the same thing as resonance theory—that the positive charge is delocalized over the benzylic carbon and the ortho and para ring carbons.

12.20

(a) Anti addition gives:

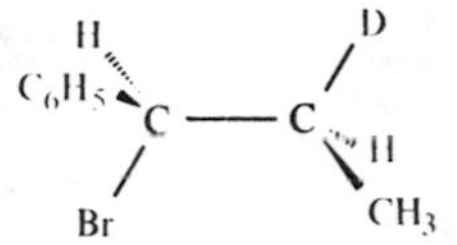

+

enantiomer

Syn-addition gives:

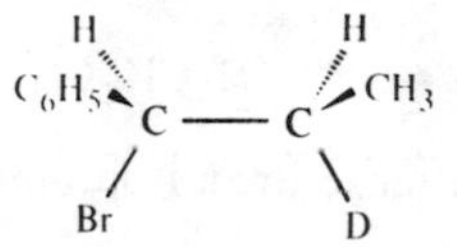

+

enantiomer

(b) Yes, anti-addition and syn-addition give diastereomeric products.

12.21

(a) 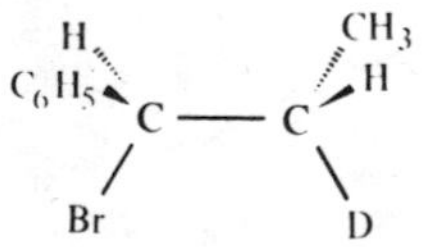

+

enantiomer

(b) 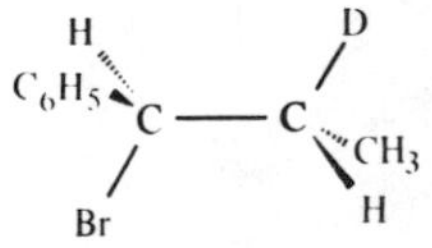

+

enantiomer

12.22

Chlorinate the ring first. If we were to introduce the side-chain double bond first, chlorination of the ring would result in addition of chlorine to the side-chain double bond.

12.23

(a) 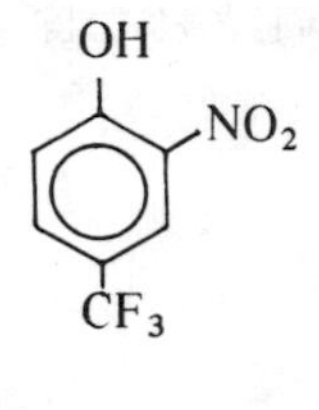

(b) 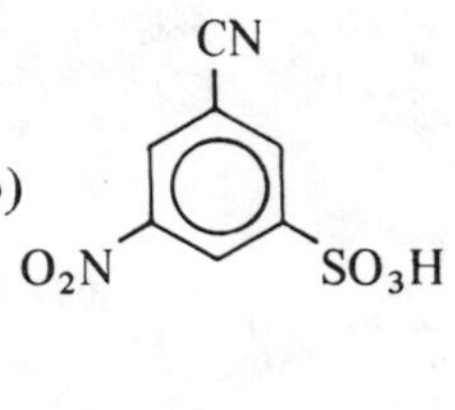

(c) 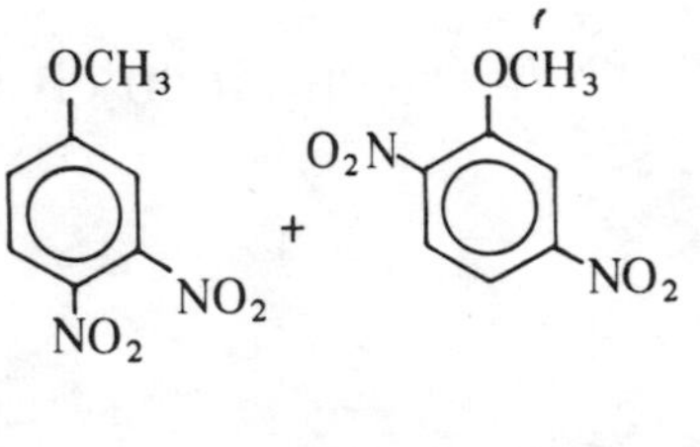

(d) 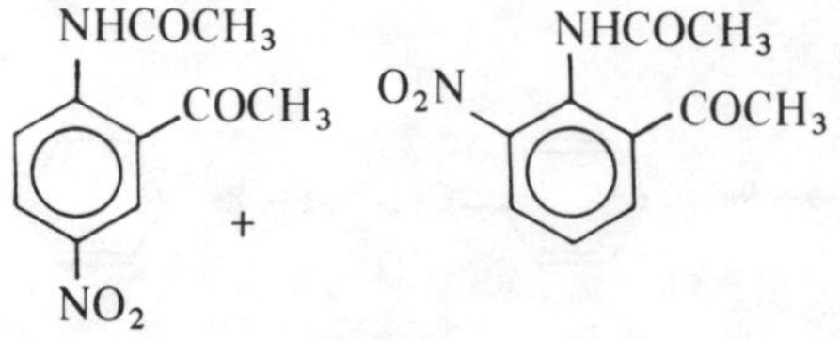 NHCOCH3, COCH3, NO2 + O2N, NHCOCH3, COCH3

(e) 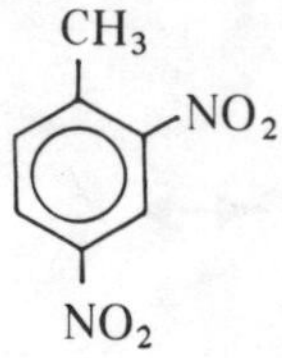 CH3, NO2, NO2

(f) 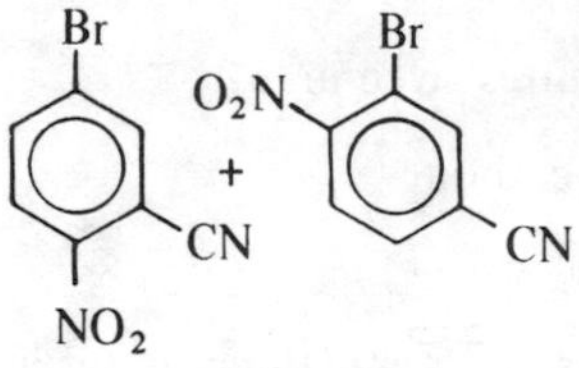 Br, CN, NO2 + Br, O2N, CN

(g) 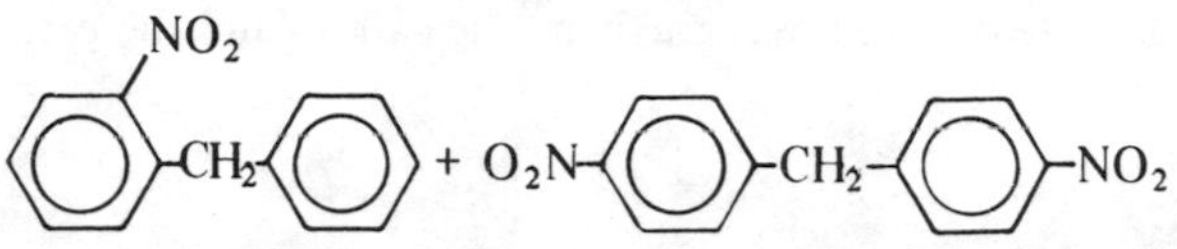 NO_2-C_6H_4-CH_2-C_6H_5 + O_2N-C_6H_4-CH_2-C_6H_4-NO_2

12.24

$$CH_3CH_2CH_2{-}OH + BF_3 \rightleftharpoons CH_3CH_2\overset{+}{C}H_2 + HO\overset{-}{B}F_3$$

The *n*-propyl cation can rearrange to an isopropyl cation:

$$CH_3CH_2\overset{+}{C}H_2 \xrightarrow[\text{Shift}]{\text{Hydride}} CH_3\overset{+}{C}HCH_3$$

Both cations can then attack the benzene ring.

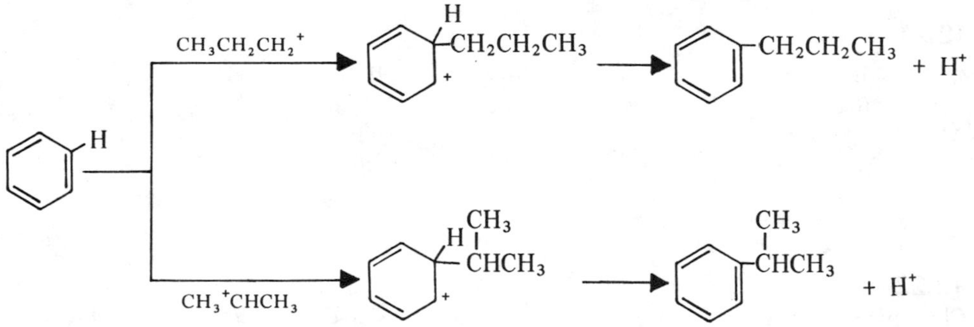

Benzene (H) $\xrightarrow{CH_3CH_2CH_2^+}$ [H, $CH_2CH_2CH_3$, +] $\longrightarrow$ C_6H_5-$CH_2CH_2CH_3$ + H^+

Benzene (H) $\xrightarrow{CH_3\overset{+}{C}HCH_3}$ [H, CH_3, $CHCH_3$, +] $\longrightarrow$ C_6H_5-$CH(CH_3)CH_3$ + H^+

12.25

(a) 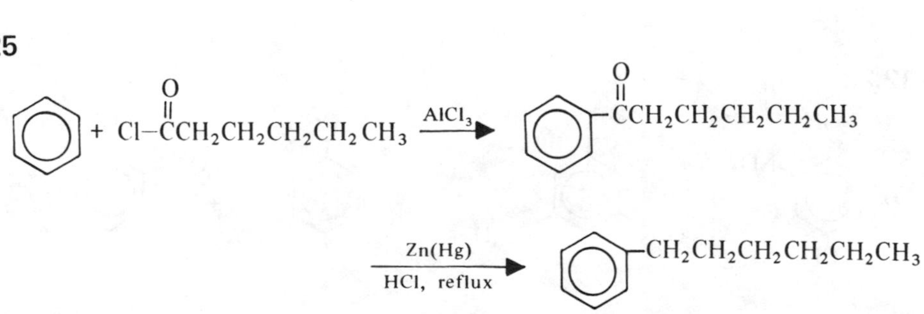 $C_6H_6 + Cl{-}\overset{O}{\overset{\|}{C}}CH_2CH_2CH_2CH_2CH_3 \xrightarrow{AlCl_3} C_6H_5{-}\overset{O}{\overset{\|}{C}}CH_2CH_2CH_2CH_2CH_3$

$$\xrightarrow[\text{HCl, reflux}]{\text{Zn(Hg)}} C_6H_5{-}CH_2CH_2CH_2CH_2CH_2CH_3$$

(b) $C_6H_6 + Cl-C(=O)-CH(CH_3)CH_3 \xrightarrow{AlCl_3} C_6H_5-C(=O)-CH(CH_3)CH_3 \xrightarrow[\text{HCl, reflux}]{\text{Zn(Hg)}} C_6H_5-CH_2CH(CH_3)CH_3$

(c) benzene + phthalic anhydride $\xrightarrow{AlCl_3}$ o-benzoylbenzoic acid $\xrightarrow[\text{HCl, reflux}]{\text{Zn(Hg)}}$

o-benzylbenzoic acid (HO–C(=O)) $\xrightarrow[\text{heat}]{SOCl_2}$ o-benzylbenzoyl chloride (Cl–C(=O)) $\xrightarrow[CS_2]{AlCl_3}$ anthrone (H, H; O)

12.26

The acyl group deactivates the ring; therefore, an acylbenzene is less reactive than benzene itself.

12.27

(a) $C_6H_5OCH_3 + Br_2 \xrightarrow{FeBr_3}$ o-Bromoanisole + p-Bromoanisole

$C_6H_5OCH_3 + HNO_3 \xrightarrow{H_2SO_4}$ o-nitroanisole + p-nitroanisole

$C_6H_5OCH_3 + SO_3 \xrightarrow{H_2SO_4}$ o-methoxybenzenesulfonic acid (SO_3H) + p-methoxybenzenesulfonic acid (SO_3H)

Reactions are faster than the corresponding reactions of benzene.

(b) $C_6H_5CHF_2 + Br_2 \xrightarrow{FeBr_3}$ *m*-$BrC_6H_4CHF_2$

m-Bromobenzal difluoride

$C_6H_5CHF_2 + HNO_3 \xrightarrow{H_2SO_4}$ *m*-$O_2NC_6H_4CHF_2$

m-Nitrobenzal difluoride

$C_6H_5CHF_2 + SO_3 \xrightarrow{H_2SO_4}$ *m*-$HO_3SC_6H_4CHF_2$

m-Difluoromethyl-benzenesulfonic acid

Reactions are slower than corresponding reactions of benzene.

(c) $C_6H_5CH_2CH_3 + Br_2 \xrightarrow{FeBr_3}$ *o*-$BrC_6H_4CH_2CH_3$ + *p*-$BrC_6H_4CH_2CH_3$

o-Bromoethyl-benzene   *p*-Bromoethyl-benzene

nitration ⟶ *o*-nitroethylbenzene and *p*-nitroethylbenzene

sulfonation ⟶ *o*-ethylbenzenesulfonic acid and *p*-ethylbenzenesulfonic acid.

Reactions are faster than corresponding reactions of benzene.

(d) $C_6H_5NO_2 + Br_2 \xrightarrow{FeBr_3}$ *m*-$BrC_6H_4NO_2$

m-Bromonitrobenzene

nitration ⟶ *m*-dinitrobenzene

sulfonation ⟶ *m*-nitrobenzenesulfonic acid

Reactions are slower than corresponding reactions of benzene.

(e) chlorobenzene + Br_2 $\xrightarrow{FeBr_3}$ *o*-Bromochlorobenzene + *p*-bromochlorobenzene

nitration ⟶ *o*-nitrochlorobenzene + *p*-nitrochlorobenzene

sulfonation ⟶ *o*-chlorobenzenesulfonic acid + *p*-chlorobenzenesulfonic acid

Reactions are slower than corresponding reactions of benzene.

(f) $C_6H_5SO_3H$ + Br_2 $\xrightarrow{FeBr_3}$ *m*-Bromobenzenesulfonic acid

nitration ⟶ *m*-nitrobenzene sulfonic acid

sulfonation ⟶ *m*-benzenedisulfonic acid

Reactions are slower than corresponding reactions of benzene.

(g) $C_6H_5COOC_2H_5$ + Br_2 $\xrightarrow{FeBr_3}$ ethyl *m*-Bromobenzoate*

$C_6H_5COOC_2H_5$ + HNO_3 $\xrightarrow{H_2SO_4}$ ethyl *m*-Nitrobenzoate*

$C_6H_5COOC_2H_5$ + SO_3 $\xrightarrow{H_2SO_4}$ ethyl *m*-Sulfobenzoate*

Reactions are slower than corresponding reactions of benzene.

*Rules for naming the starred compounds in this problem have not been given in the text.

(h) C_6H_5–O–C_6H_5 + Br_2 $\xrightarrow{FeBr_3}$ *o*-bromophenoxybenzene + *p*-bromophenoxybenzene

nitration ⟶ *o*-nitrophenoxybenzene + *p*-nitrophenoxybenzene sulfonation ——▶ *o*-phenoxybenzenesulfonic acid + *p*-phenoxybenzenesulfonic acid

Reactions are faster than corresponding reactions of benzene.

(i) Biphenyl + Br_2 $\xrightarrow{FeBr_3}$ 2-Bromobiphenyl* + 4-Bromobiphenyl*

nitration ——▶ 2-nitrobiphenyl + 4-nitrobiphenyl

sulfonation ——▶ *o*-phenylbenzenesulfonic acid + *p*-phenylbenzenesulfonic acid

Reactions are faster than corresponding reactions of benzene.

(j) $C_6H_5C(CH_3)_3$ + Br_2 $\xrightarrow{FeBr_3}$ *o*-Bromo-*tert*-butylbenzene + *p*-Bromo-*tert*-butylbenzene

nitration ——▶ *o*-nitro-*tert*-butylbenzene + *p*-nitro-*tert*-butylbenzene

sulfonation ——▶ *o*-*tert*-butylbenzenesulfonic acid + *p*-*tert*-butylbenzenesulfonic acid

Reactions are faster than corresponding reactions of benzene.

(k) C_6H_5F + Br_2 $\xrightarrow{FeBr_3}$ *o*-Bromofluorobenzene + *p*-Bromofluorobenzene

nitration ——▶ *o*-nitrofluorobenzene + *p*-nitrofluorobenzene

sulfonation ——▶ *o*-fluorobenzenesulfonic acid + *p*-fluorobenzenesulfonic acid

Reactions are slower than corresponding reactions of benzene.

(l) $C_6H_5\overset{O}{\overset{\|}{C}}CH_2CH_3$ + Br_2 $\xrightarrow{FeBr_3}$ $m\text{-}BrC_6H_4\overset{O}{\overset{\|}{C}}\text{-}CH_2CH_3$

Ethyl *m*-bromophenyl ketone

nitration ——▶ Ethyl *m*-nitrophenyl ketone*

sulfonation ——▶ Ethyl *m*-sulfophenyl ketone*

Reactions are slower than corresponding reactions of benzene.

(m) Benzonitrile (CN) + Br_2 $\xrightarrow{FeBr_3}$ *m*-Bromobenzonitrile

nitration ⟶ *m*-nitrobenzonitrile
sulfonation ⟶ *m*-cyanobenzenesulfonic acid

Reactions are slower than corresponding reactions of benzene.

(n) Phenyl acetate ($O{-}\overset{O}{\overset{\|}{C}}CH_3$) + Br_2 $\xrightarrow{FeBr_3}$ *o*-Bromophenyl acetate + *p*-Bromophenyl acetate

nitration ⟶ *o*-nitrophenyl acetate + *p*-nitrophenyl acetate

Phenyl acetate ($\overset{O}{\overset{\|}{O C}}CH_3$) + SO_3 $\xrightarrow{H_2SO_4}$ *o*-Acetoxybenzenesulfonic acid* (SO_3H) + *p*-Acetoxybenzenesulfonic acid* (SO_3H)

Reactions are faster than corresponding reactions of benzene.

(o) Benzamide ($CONH_2$) + Br_2 $\xrightarrow{FeBr_3}$ *m*-Bromobenzamide

nitration ⟶ *m*-nitrobenzamide

Benzamide ($CONH_2$) + SO_3 $\xrightarrow{H_2SO_4}$ *m*-Carbamoylbenzenesulfonic acid* (SO_3H)

Reactions are slower than corresponding reactions of benzene.

(p) 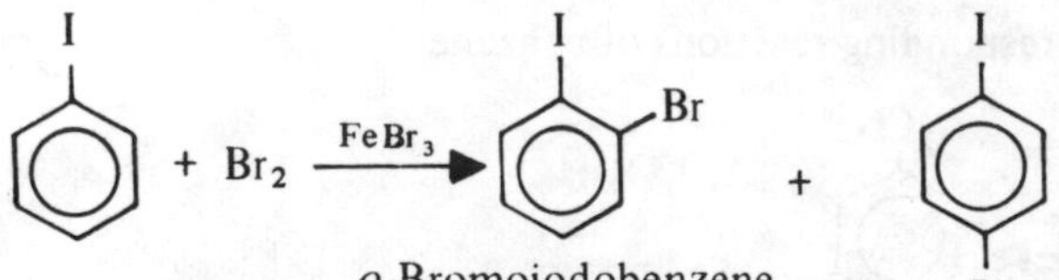 Iodobenzene + Br_2 $\xrightarrow{FeBr_3}$ *o*-Bromoiodobenzene + *p*-Bromoiodobenzene

nitration ⟶ *o*-nitroiodobenzene + *p*-nitroiodobenzene

sulfonation ⟶ *o*-iodobenzenesulfonic acid + *p*-iodobenzenesulfonic acid

Reactions are slower than corresponding reactions of benzene.

12.28

(a) 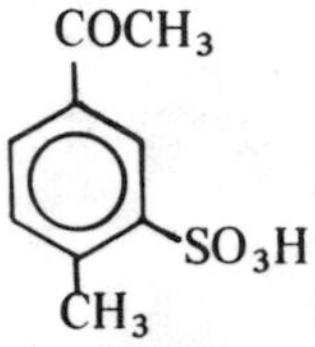 benzene ring with $COCH_3$, SO_3H, CH_3

(b) 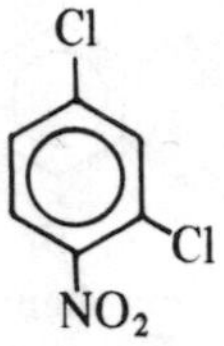 benzene ring with Cl, Cl, NO_2

(c) 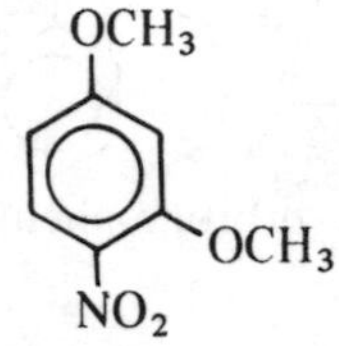 benzene ring with OCH_3, OCH_3, NO_2

(d) 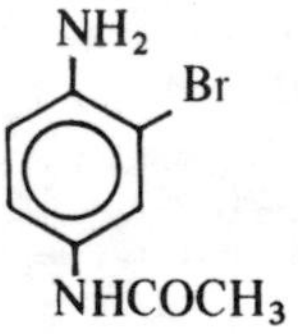 benzene ring with NH_2, Br, $NHCOCH_3$

(e) 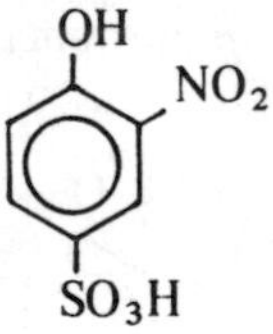 benzene ring with OH, NO_2, SO_3H

(f) 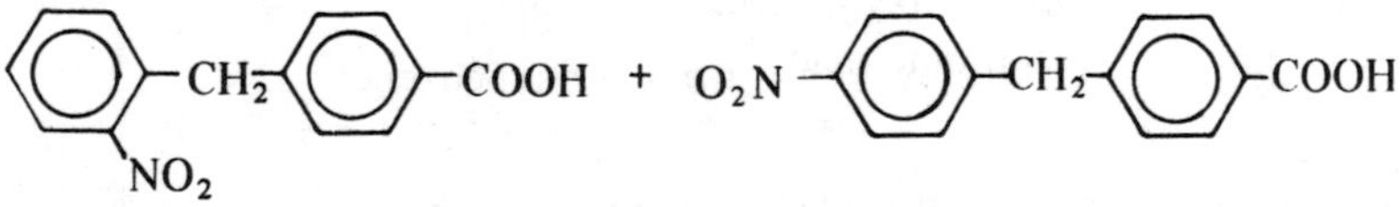 o-O_2N-C_6H_4-CH_2-C_6H_4-COOH + O_2N-C_6H_4-CH_2-C_6H_4-COOH

(g) 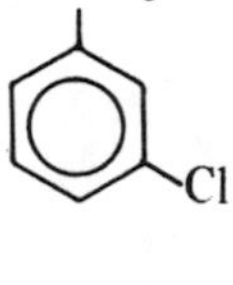 benzene ring with CCl_3, Cl

12.29

(a) 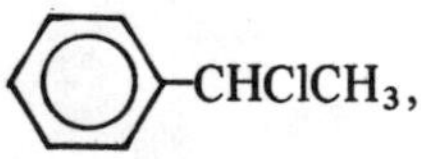 C_6H_5-$CHClCH_3$, (b) C_6H_5-$CH{=}CHCH_3$, (c) C_6H_5-$CH{=}CHCH_2CH_3$,

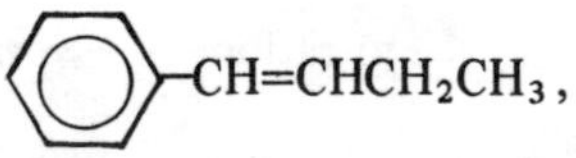

(d) 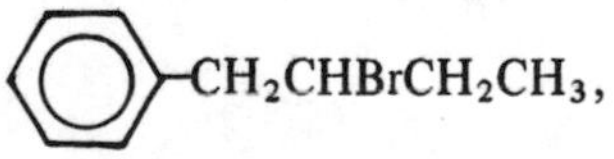 C_6H_5-$CH_2CHBrCH_2CH_3$, (e) 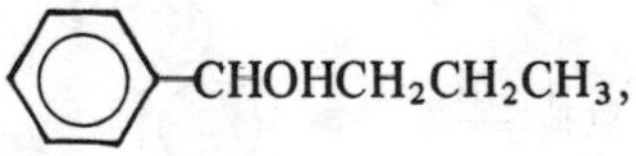 C_6H_5-$CHOHCH_2CH_2CH_3$,

(f) 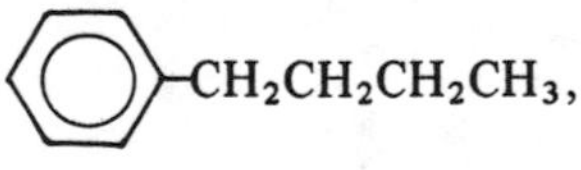 C_6H_5-$CH_2CH_2CH_2CH_3$, (g) 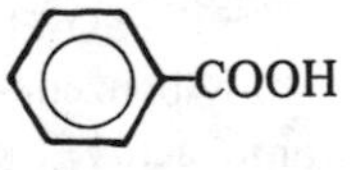 C_6H_5-COOH

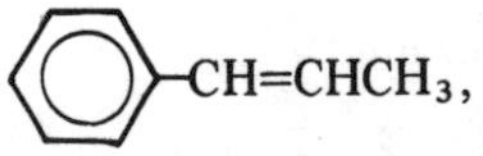

12.30

(a) CH_3 ; $\xrightarrow[(2)H_3O^+]{(1)KMnO_4, OH^-, heat}$ COOH ; $\xrightarrow[FeCl_3]{Cl_2}$ COOH, Cl

(b) NH_2 ; $\xrightarrow[base]{CH_3COCl}$ $NHCOCH_3$; $\xrightarrow[FeBr_3]{Br_2}$ $NHCOCH_3$, Br ; $\xrightarrow{H^+, H_2O}$ NH_2, Br

(c) $NHCOCH_3$; $\xrightarrow[H_2SO_4]{SO_3}$ $NHCOCH_3$, SO_3H ; $\xrightarrow[FeBr_3]{Br_2}$ $NHCOCH_3$, Br, SO_3H ; $\xrightarrow[H_2SO_4, heat]{57\%}$ NH_2, Br

(d) $NHCOCH_3$ + HNO_3 $\xrightarrow{H_2SO_4}$ $NHCOCH_3$, NO_2 ; $\xrightarrow[FeBr_3]{Br_2}$ $NHCOCH_3$, Br, NO_2 ; $\xrightarrow[H_2O]{H^+}$ NH_2, Br, NO_2

(e) $NHCOCH_3$; $\xrightarrow[FeBr_3]{Br_2}$ $NHCOCH_3$, Br ; $\xrightarrow[H_2SO_4]{HNO_3}$ $NHCOCH_3$, NO_2, Br ; $\xrightarrow[H_2O]{H^+}$ NH_2, NO_2, Br

(f) CH_3 + CH_3COCl $\xrightarrow{AlCl_3}$ CH_3, $COCH_3$ + para isomer

(g) CH_3, $\overset{O}{\overset{\|}{C}}CH_3$; $\xrightarrow[(2)H_3O^+]{(1)Br_2, OH^-}$ CH_3, COOH

(h) + I_2 + $AgClO_4$ $\xrightarrow{25^\circ}$ I ; $\xrightarrow[H_2SO_4]{SO_3}$ I, SO_3H + Ortho isomer

(i) Toluene + HNO_3 $\xrightarrow{H_2SO_4}$ p-nitrotoluene (+ Ortho isomer) $\xrightarrow[FeBr_3]{Br_2}$ 2-bromo-4-nitrotoluene

CH_3 ; NO_2 ; CH_3, Br, NO_2

+ Ortho isomer

(j) Toluene $\xrightarrow[FeBr_3]{Br_2}$ p-bromotoluene (+ Ortho isomer) $\xrightarrow[(2) H_3O^+]{(1) KMnO_4, OH^-, heat}$ p-bromobenzoic acid

CH_3 ; CH_3, Br ; COOH, Br

+

Ortho isomer

(k) Benzene + $Cl-\overset{O}{\overset{\|}{C}}CH_2CH_2CH_3$ $\xrightarrow{AlCl_3}$ $C_6H_5\overset{O}{\overset{\|}{C}}CH_2CH_2CH_3$ $\xrightarrow[HCl]{Zn(Hg)}$ $C_6H_5CH_2CH_2CH_2CH_3$

(l) $C_6H_5CH_2CH_2CH_2CH_3$ $\xrightarrow[FeBr_3]{Br_2}$ p-Br$C_6H_4CH_2CH_2CH_2CH_3$ $\xrightarrow{\text{N-Bromo-succinimide}}$ p-Br$C_6H_4\overset{Br}{\overset{|}{C}}H-CH_2CH_2CH_3$ $\xrightarrow{KOH, C_2H_5OH}$ p-Br$C_6H_4CH=CHCH_2CH_3$

(m) Toluene (CH_3) $\xrightarrow[light]{Cl_2\ (excess)}$ $C_6H_5CCl_3$ $\xrightarrow[H_2SO_4]{HNO_3}$ m-nitro(CCl_3)benzene (CCl_3, NO_2)

(n) Toluene (CH_3) + HNO_3 (fuming) $\xrightarrow[heat]{H_2SO_4}$ 2,4,6-trinitrotoluene (CH_3, O_2N, NO_2, NO_2)

12.31

(a) 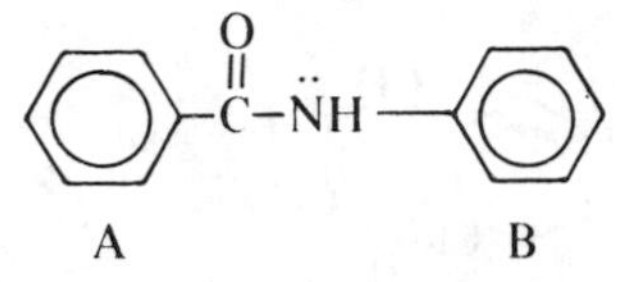 C6H5–C(=O)–N̈H–C6H5 (ring A: benzoyl phenyl; ring B: N-phenyl)

A B

Ring B undergoes electrophilic substitution more readily than ring A.

(b) Resonance structures such as the one below stabilize the intermediate carbocation:

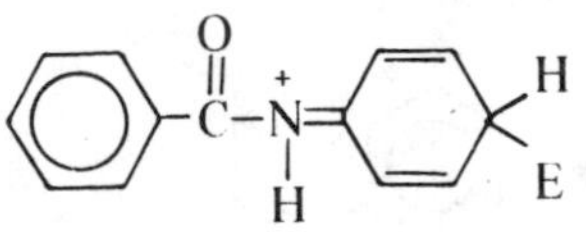

C6H5–C(=O)–N⁺(H)=C6H4(H)(E)

O H E H

12.32

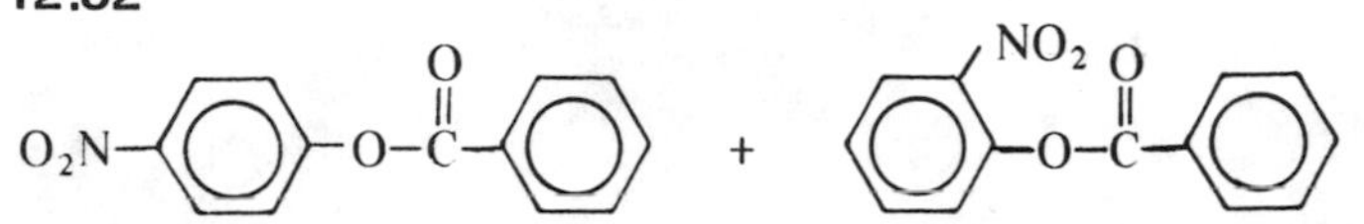

O_2N–C6H4–O–C(=O)–C6H5 + (o-NO_2)C6H4–O–C(=O)–C6H5

12.33

(a) 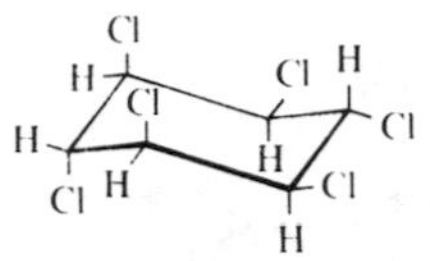 Cl H Cl H Cl H Cl H Cl H Cl H (b) No (c) Lindane is a meso compound.

(d) 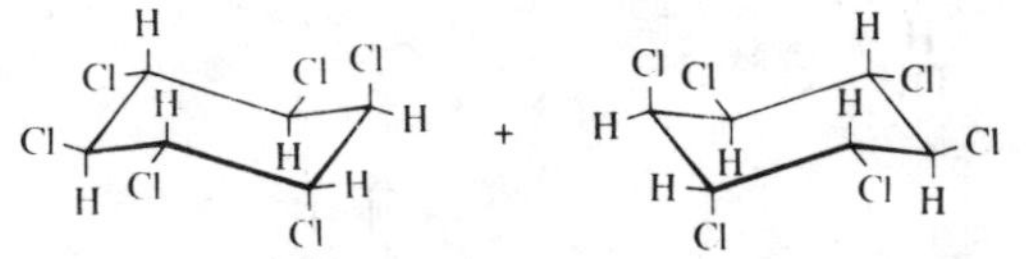 H Cl Cl Cl H H Cl H Cl H H Cl + Cl Cl H Cl H H Cl H H Cl Cl H

12.34

If we consider resonance structures for the ring that undergoes electrophilic attack, two structures are possible for the arenium ion that forms when attack takes place at the 1-position,

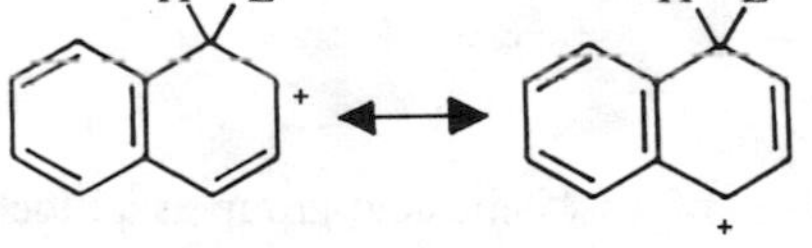

H E + ⟷ H E +

whereas only one is possible when attack takes place at the 2-position,

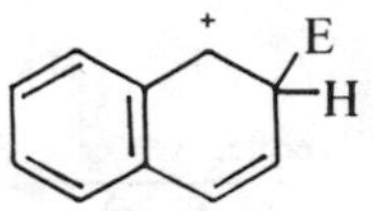

+ E H

Attack at the 1-position, therefore, takes place faster.

12.35

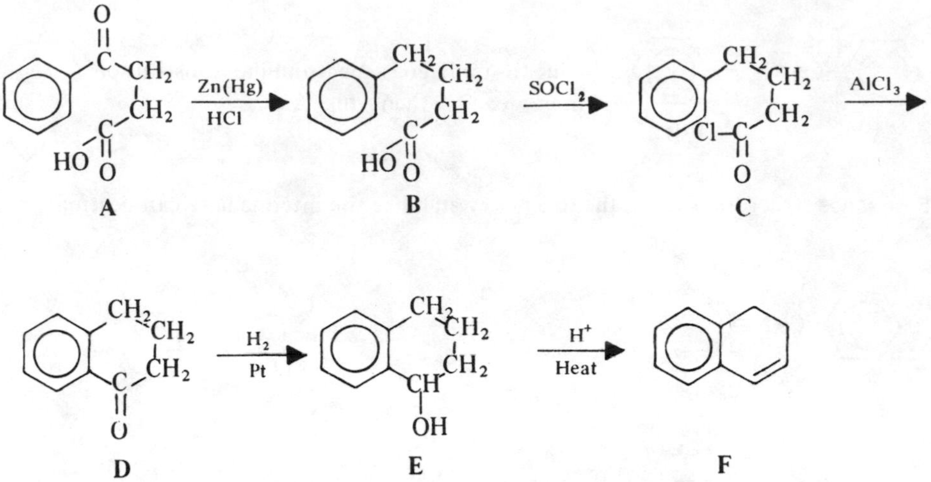

O ‖ C–CH_2–CH_2–C(=O)–HO (A) —Zn(Hg), HCl→ CH_2–CH_2–CH_2–C(=O)–HO (B) —$SOCl_2$→ CH_2–CH_2–CH_2–Cl–C(=O) (C) —$AlCl_3$→

CH_2, CH_2, CH_2, C=O (D) —H_2, Pt→ CH_2, CH_2, CH_2, CH–OH (E) —H^+, Heat→ F

A B C

D E F

12.36

CH_2, C=Ö:, R —$+H^+$ / $-H^+$⇌ CH_2, C^+, R, OH

CH_2, C^+, R, OH ⟶ CH_2 +, C, H, R, OH —$-H^+$→ CH_2, C, R, OH

$+H^+$ ⇅ $-H^+$

R ← $-H^+$ — H, H, C, +, R ← $-H_2O$ — CH_2, R, OH_2 +

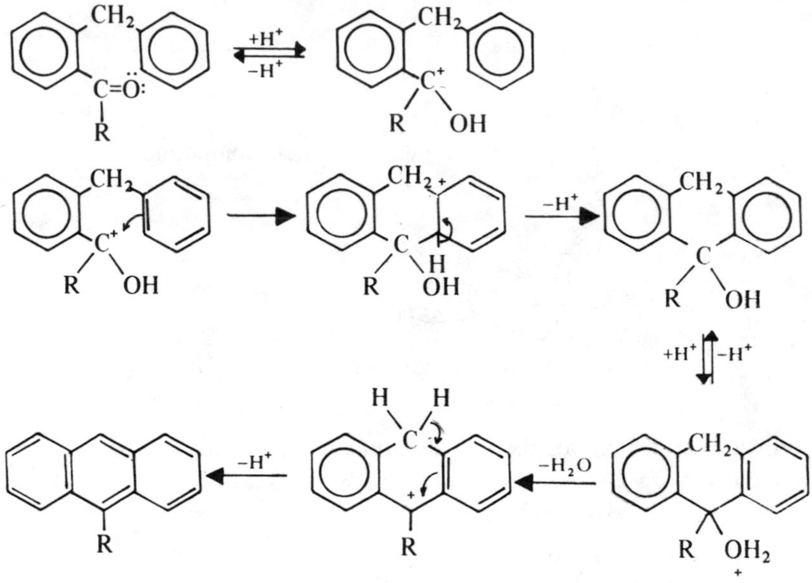

12.37

This problem serves as another illustration of the use of a sulfonic acid group as a blocking groups in a synthetic sequence. Here we are able to bring about nitration between two *meta* substituents.

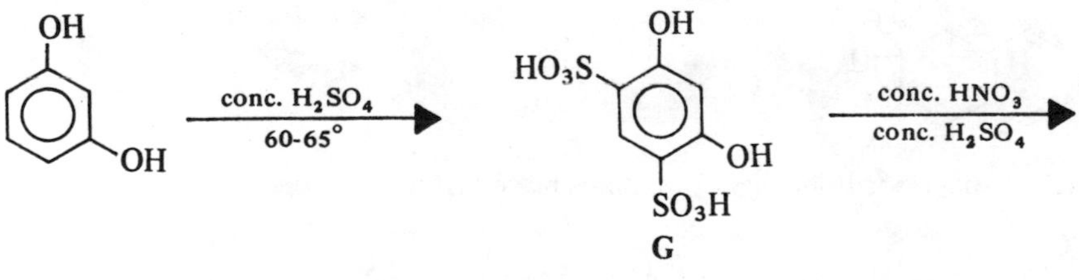

OH, OH —conc. H_2SO_4, 60-65°→ OH, HO_3S, OH, SO_3H (G) —conc. HNO_3, conc. H_2SO_4→

G

$$\underset{\mathbf{H}}{\text{2-nitro-4,6-disulfo-resorcinol } (HO_3S, OH, NO_2, OH, SO_3H)} \xrightarrow[\text{heat}]{H_3O^+/H_2O} \underset{\mathbf{I}}{\text{2-nitroresorcinol } (OH, NO_2, OH)}$$

12.38

$$\text{phenol} \xrightarrow[100^\circ]{H_2SO_4} \text{p-}HOC_6H_4SO_3H \xrightarrow[Fe]{Cl_2} \text{2,6-dichloro-4-}SO_3H\text{-phenol}$$

$$\xrightarrow[\text{heat}]{H_3O^+} \text{2,6-dichlorophenol}$$

12.39

$$CH_3\overset{O}{\overset{\|}{C}}Cl + AlCl_3 \rightleftharpoons CH_3\overset{+}{C}{=}O + AlCl_4^-$$

$$CH_3\overset{+}{C}{=}O + CH_2{=}CHCH_3 \longrightarrow CH_3\overset{O}{\overset{\|}{C}}CH_2\overset{+}{C}HCH_3$$

$$CH_3\overset{O}{\overset{\|}{C}}CH_2\overset{+}{C}HCH_3 \longrightarrow \underset{\mathbf{A}}{CH_3\overset{O}{\overset{\|}{C}}CH{=}CHCH_3} + H^+$$

$$CH_3\overset{O}{\overset{\|}{C}}CH_2\overset{+}{C}HCH_3 \longrightarrow \underset{\mathbf{B}}{CH_3\overset{O}{\overset{\|}{C}}CH_2CH{=}CH_2} + H^+$$

$$CH_3\overset{O}{\overset{\|}{C}}CH_2\overset{+}{C}HCH_3 \xrightarrow{Cl^-} \underset{\mathbf{C}}{CH_3\overset{O}{\overset{\|}{C}}CH_2\underset{Cl}{\underset{|}{C}}HCH_3}$$

12.40

(a)

(1) $C_6H_5CH{=}CH{-}CH{=}CH_2 \xrightarrow{H^+} C_6H_5CH{=}CH{-}\overset{+}{C}H{-}CH_3$

$$\updownarrow$$

$$C_6H_5\overset{+}{C}H{-}CH{=}CH{-}CH_3$$

$$C_6H_5\overset{\delta+}{C}H\text{=--}CH\text{=--}\overset{\delta+}{C}H{-}CH_3$$

(2) $C_6H_5\overset{\delta+}{C}H\text{---}CH\text{---}\overset{\delta+}{C}H-CH_3 \xrightarrow{X^-} C_6H_5CH=CH-\underset{X}{\underset{|}{C}H}-CH_3$

(b) 1,2-Addition (or in this instance actually 3,4-addition).

(c) Yes. The carbocation given in (a) is a hybrid of *secondary allylic and benzylic* contributors and is therefore more stable than any other possibility; for example

$$C_6H_5CH=CH-CH=CH_2 \xrightarrow{H^+} \left.\begin{array}{c} C_6H_5CH_2-\overset{+}{C}H-CH=CH_2 \\ \updownarrow \\ C_6H_5CH_2-CH=CH-\overset{+}{C}H_2 \end{array}\right\} \text{a hybrid of allylic contributors only}$$

(d) Since the reaction produces only *the more stable isomer*—that is, the one in which the double bond is conjugated with the benzene ring—the reaction must be under equilibrium control:

$$C_6H_5\overset{\delta+}{C}H\text{---}CH\text{---}\overset{\delta+}{C}H-CH_3 + Cl^- \longrightarrow C_6H_5-CH=CH-\underset{Cl}{\underset{|}{C}H}-CH_3 \quad \text{actual product}$$

more stable isomer

$$\longrightarrow C_6H_5\underset{Cl}{\underset{|}{C}H}-CH=CH-CH_3 \quad \text{not formed}$$

less stable isomer

12.41

(a) The addition product is the *meso* compound:

C_6H_5, H, Cl / H, Cl, C_6H_5

meso–1,2–dichloro–1,2–diphenylethane

(b) The reaction is a *syn* addition:

C_6H_5, H, C_6H_5, H + Cl–Cl → C_6H_5, H, C_6H_5, H, Clδ+ – Clδ− → C_6H_5, H, Cl, C_6H_5, H, +, Cl^- → C_6H_5, H, Cl, C_6H_5, H, Cl

π–complex — Carbocation–ion pair — ***meso*** compound

(c) A similar reaction starting with *trans*-stilbene would yield a racemic modification of 1,2-dichloro-1,2-diphenylethane: *syn* addition at one face would yield one enantiomer; *syn* addition at the other face would yield the other enantiomer. The racemic modification would be optically inactive, but it could be resolved into the separate (and optically active) enantiomers.

12.42

(a) [fluorene]–CH_2OH $\xrightarrow{H^+}$ [fluorene]–$CH_2\overset{+}{O}H_2$ $\xrightarrow{-H_2O}$

[fluorene]–$\overset{+}{C}H_2$ $\xrightarrow[\text{ment}]{\text{rearrange-}}$ [carbocation with H, H, H shown] $\xrightarrow{-H^+}$ [phenanthrene]

(b) $CH_3{-}\underset{C_6H_5}{\underset{|}{C}}{=}CH_2 \xrightarrow{H^+} CH_3{-}\underset{C_6H_5}{\underset{|}{\overset{+}{C}}}{-}CH_3$

$\xrightarrow{CH_2=\underset{C_6H_5}{\underset{|}{C}}-CH_3}$ [C_6H_5–$C(CH_3)_2$–CH_2–$\overset{+}{C}(CH_3)$–C_6H_5] $\longrightarrow$

[arenium ion: CH_3, CH_3, C_6H_5, H, CH_3] $\xrightarrow{-H^+}$ [indane: CH_3, CH_3, C_6H_5, CH_3]

12.43

[benzene] + $CH_3\underset{CH_3}{\underset{|}{C}H}\left(CH_2\underset{CH_3}{\underset{|}{C}H}\right)_2CH{=}CHCH_3$ and $CH_3\underset{CH_3}{\underset{|}{C}H}\left(CH_2\underset{CH_3}{\underset{|}{C}H}\right)_2CH_2CH{-}CH_2$ $\xrightarrow[35\text{-}45^\circ]{AlCl_3}$

$CH_3\underset{CH_3}{\underset{|}{C}H}\left(CH_2\underset{CH_3}{\underset{|}{C}H}\right)_2CH_2\underset{CH_3}{\underset{|}{C}H}{-}C_6H_5$

and

$CH_3\underset{CH_3}{\underset{|}{C}H}\left(CH_2\underset{CH_3}{\underset{|}{C}H}\right)_2\underset{CH_2CH_3}{\underset{|}{C}H}{-}C_6H_5$ $\xrightarrow[\text{heat}]{H_2SO_4}$

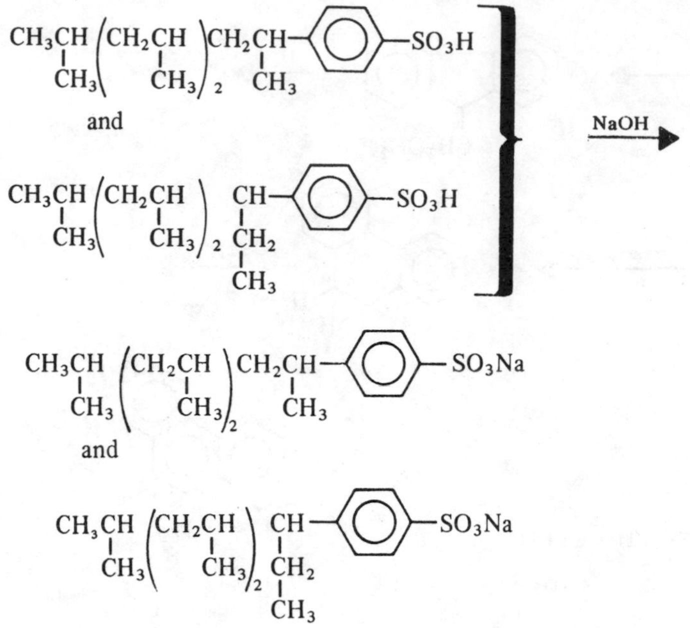

$CH_3CH(CH_3)(CH_2CH(CH_3))_2CH_2CH(CH_3)-C_6H_4-SO_3H$

and

$CH_3CH(CH_3)(CH_2CH(CH_3))_2CH(CH_2CH_3)-C_6H_4-SO_3H$

$\xrightarrow{NaOH}$

$CH_3CH(CH_3)(CH_2CH(CH_3))_2CH_2CH(CH_3)-C_6H_4-SO_3Na$

and

$CH_3CH(CH_3)(CH_2CH(CH_3))_2CH(CH_2CH_3)-C_6H_4-SO_3Na$

12.44

(a) Large *ortho* substituents prevent the two rings from becoming coplanar and if the correct substitution patterns are present, the molecule as a whole will be chiral. Thus enantiomeric forms are possible even though the molecules do not have a chiral carbon. The compound with 2-NO_2, 6-COOH, 2′-NO_2, 6′-COOH is an example.

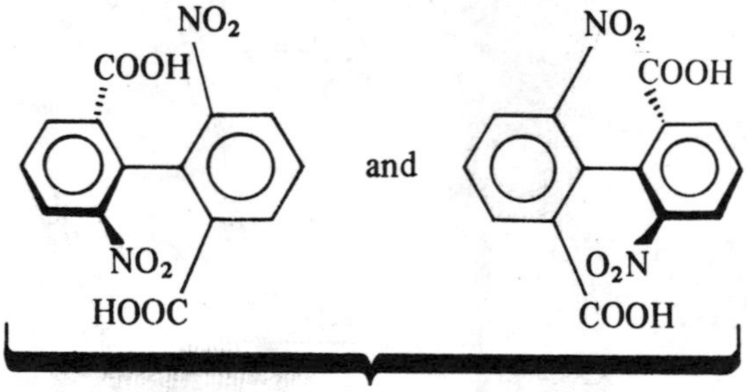

NO_2, COOH, NO_2, HOOC and NO_2, COOH, O_2N, COOH

These molecules are nonsuperposable mirror reflections and, thus, are enantiomers.

(b) Yes

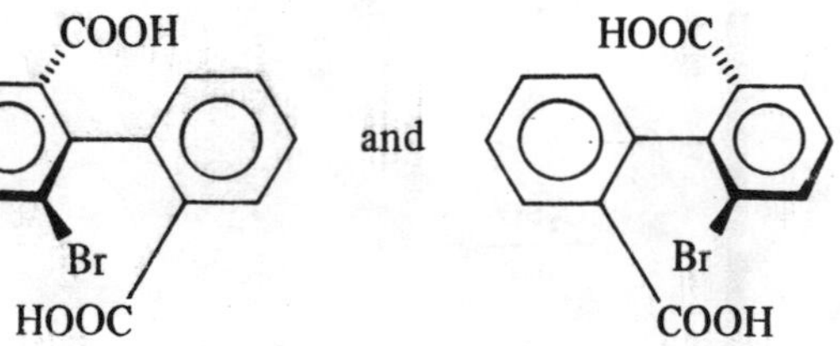

COOH, Br, HOOC and HOOC, Br, COOH

(c) This molecule has a plane of symmetry.

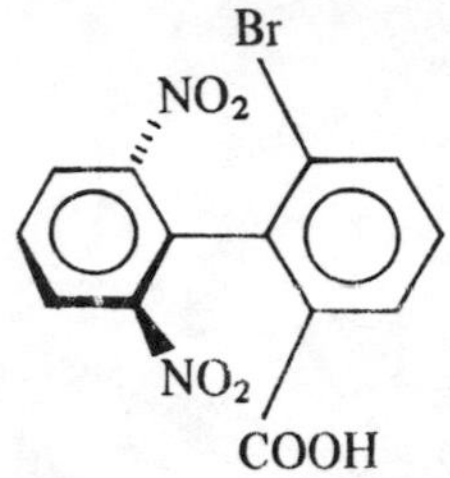

Br
NO_2
NO_2
COOH

The plane of the page is a plane of symmetry.

12.45

A. 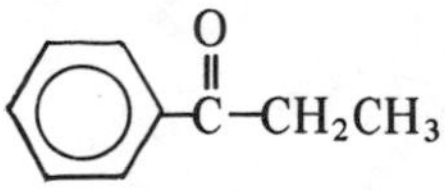 $C_6H_5-\overset{O}{\overset{\|}{C}}-CH_2CH_3$, B. $C_6H_5-CCl_2CH_2CH_3$, C. $C_6H_5-C{\equiv}CCH_3$,

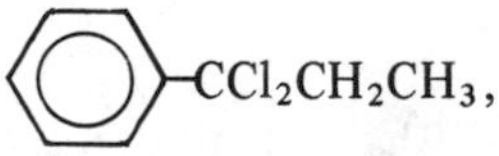

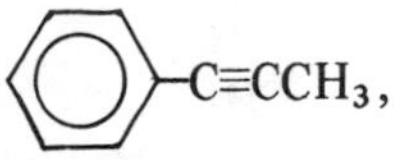

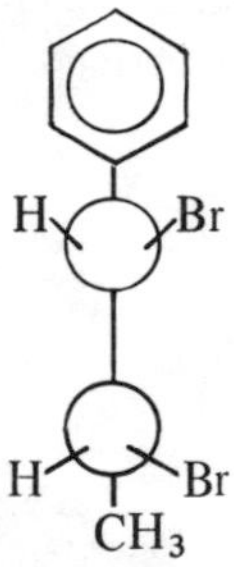

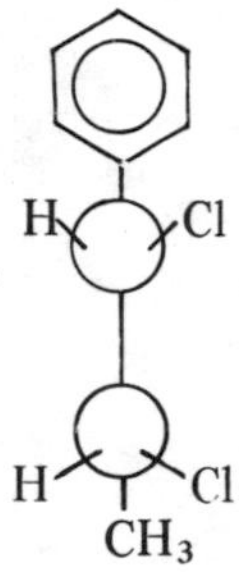

D. 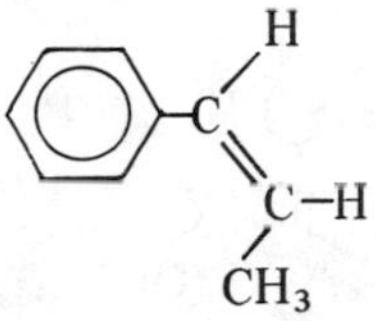 (H, C–H, CH_3), E. 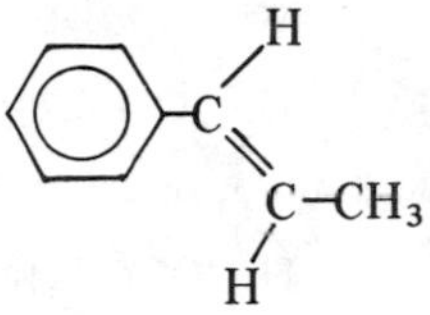 (H, C–CH_3, H), F. 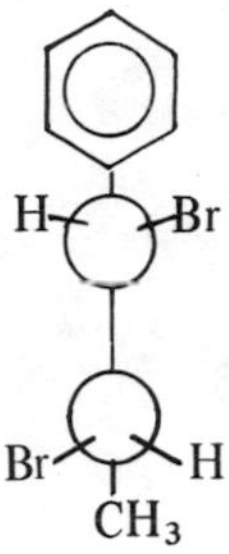 (H, Br; Br, H, CH_3)

G. (H, Br; H, Br, CH_3), H. (H, Cl; H, Cl, CH_3), I. 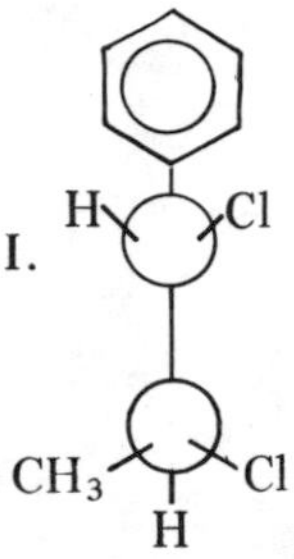 (H, Cl; CH_3, Cl, H)

12.46

(a) 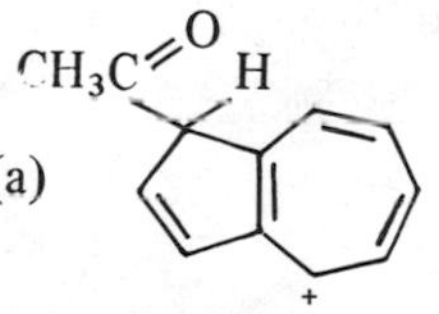 $CH_3C{=}O$, H

(b) This arenium ion is especially stable because its seven-membered ring is an aromatic cation. (c) 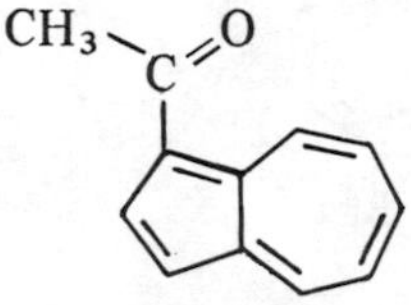 $CH_3-C{=}O$

G

Special Topic

G.1

(a) $C_6H_5CH_2OH \cdots Tl(OOCCF_3)_3$ (b) $C_6H_5CH_2OCH_3 \cdots Tl(OOCCF_3)_3$

(c) $C_6H_5CH_2CH_2O{-}CH_3 \cdots Tl(OOCCF_3)_3$ (d) $C_6H_5CH_2C(OH){=}O \cdots Tl(OOCCF_3)_3$ (e) $C_6H_5CH_2C(OCH_3){=}O \cdots Tl(OOCCF_3)_3$

G.2

In the complex the Tl atom is too far from the ortho position:

$C_6H_5CH_2CH_2\ddot{O}{-}C(CH_3){=}O \cdots Tl(OOCCF_3)_3$

It must therefore react in the usual way at the para (least crowded) position.

G.3

(a) $C_6H_5C(CH_3)_3 \xrightarrow[Tl(OOCCF_3)_3]{Br_2} p\text{-}BrC_6H_4C(CH_3)_3$

(b) $C_6H_5CH_2COOH \xrightarrow[CF_3COOH,\ 25°]{Tl(OOCCF_3)_3} o\text{-}(Tl(OOCCF_3)_2)C_6H_4CH_2COOH \xrightarrow[(DMF)]{CuCN} o\text{-}(CN)C_6H_4CH_2COOH$

(c) OCH_3-benzene $\xrightarrow[\text{(2)}Pb(OOCCH_3)_4, (C_6H_5)_3P \quad \text{(3)}OH^-]{\text{(1)}Tl(OOCCF_3)_3, CF_3COOH, 25°}$ p-HO–C_6H_4–OCH_3

(d) $CH_2CH_2CH_3$-benzene $\xrightarrow[\text{(2)}KI, H_2O]{\text{(1)}Tl(OOCCF_3)_3, CF_3COOH, 25°}$ p-I–C_6H_4–$CH_2CH_2CH_3$

(e) $CH_2CH_2CH_3$-benzene $\xrightarrow[\text{(2)}KI, H_2O]{\text{(1)}Tl(OOCCF_3)_3, CF_3COOH, 73°}$ m-I–C_6H_4–$CH_2CH_2CH_3$

(f) m-xylene (CH_3, CH_3) $\xrightarrow[\text{above}]{\text{same as (d)}}$ 2,4-dimethyliodobenzene (CH_3, CH_3, I)

(g) CH_2CH_3-benzene $\xrightarrow[\text{(2)}CuCN, DMF]{\text{(1)}Tl(OOCCF_3)_3, CF_3COOH, 25°}$ p-CN–C_6H_4–CH_2CH_3

G.4

(a) CH_3-benzene $\xrightarrow[Tl(OOCCF_3)_3]{Br_2}$ p-Br–C_6H_4–CH_3

(b) CH_3-benzene $\xrightarrow[\text{(2)}KI, H_2O]{\text{(1)}Tl(OOCCF_3)_3, CF_3COOH, 25°}$ p-I–C_6H_4–CH_3

(c) CH_3–C_6H_5 $\xrightarrow[\text{(2) CuCN}]{\text{(1) }Tl(OOCCF_3)_3, CF_3COOH, 25°}$ CH_3–C_6H_4–CN

(d) CH_3–C_6H_5 $\xrightarrow[\text{(2) }Pb(OOCCH_3)_4, (C_6H_5)_3P \quad \text{(3) }OH^-]{\text{(1) }Tl(OOCCF_3)_3, CF_3COOH, 25°}$ CH_3–C_6H_4–OH

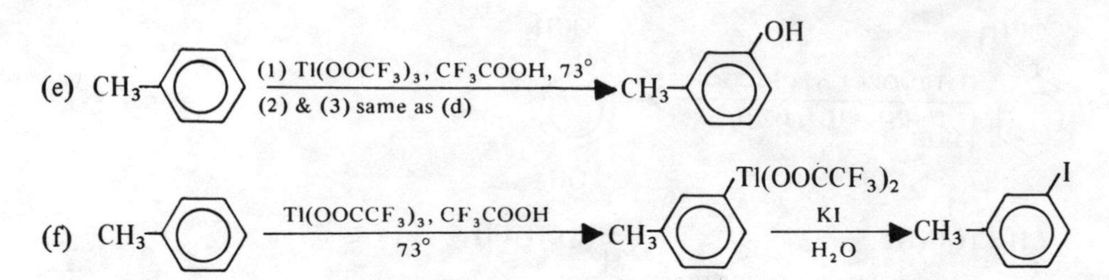
(e) CH3
(1) Tl(OOCF3)3, CF3COOH, 73°
(2) & (3) same as (d)
OH
CH3
(f) CH3
Tl(OOCCF3)3, CF3COOH
73°
Tl(OOCCF3)2
CH3
KI
H2O
I
CH3

CHAPTER THIRTEEN

Physical Methods of Structure Determination, Nuclear Magnetic Resonance Spectroscopy, Infrared Spectroscopy

13.1

The methyl protons of 15,16-dimethyldihydropyrene are highly shielded by the induced field in the center of the aromatic system where the induced field opposes the applied field. (p 532).

13.2

(a) The six protons (hydrogens) of ethane are equivalent:

$$\overset{a}{CH_3}-\overset{a}{CH_3}$$

Ethane gives a single signal in its proton nmr spectrum.

(b) Propane has two different sets of equivalent protons:

$$\overset{a}{CH_3}-\overset{b}{CH_2}-\overset{a}{CH_3}$$

Propane gives two signals.

(c) The six protons of dimethyl ether are equivalent:

$$\overset{a}{CH_3}-O-\overset{a}{CH_3}$$

One signal.

(d) Three different sets of equivalent protons:

$$\overset{a}{CH_3}-\overset{b}{CH_2}-C_6H_4(\overset{c}{H})_4-\overset{b}{CH_2}-\overset{a}{CH_3}$$

Three signals.

(e) Two different sets of equivalent protons:

$$\overset{a}{CH_3}-\overset{O}{\overset{\|}{C}}-O-\overset{b}{CH_3}$$

Two signals.

(f) Three different sets of equivalent protons:

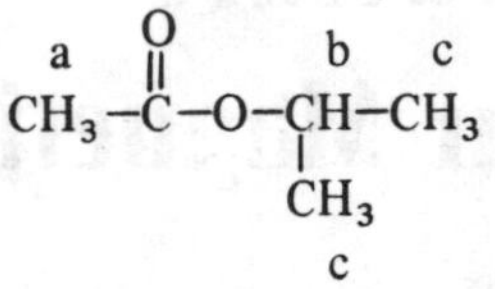

$$\overset{a}{CH_3}-\overset{O}{\overset{\|}{C}}-O-\overset{b}{CH}(\underset{c}{CH_3})-\overset{c}{CH_3}$$

Three signals.

13.3

(a) HO, H, CH$_3$ / H, H, CH$_3$ [replacement by W] → HO, H, CH$_3$ / H, W, CH$_3$ and HO, H, CH$_3$ / W, H, CH$_3$

diastereomers

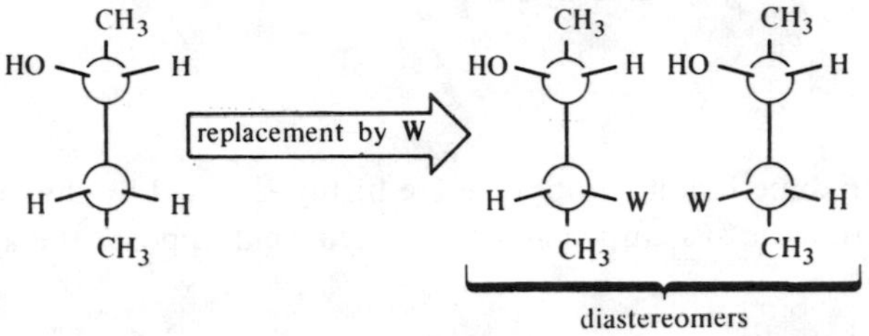

(b) Six,

$$\overset{a}{CH_3}-\overset{b}{H}C(\overset{c}{OH})-\overset{d}{H}C\overset{e}{H}-\underset{f}{CH_3}$$

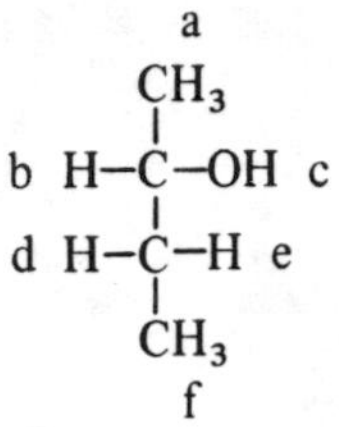

(c) Six signals.

13.4

(a) Two, $\overset{a}{CH_3}-\overset{b}{CH_2}-\overset{b}{CH_2}-\overset{a}{CH_3}$

(b) Three, $\overset{a}{CH_3}-\overset{b}{CH_2}-O-\overset{c}{H}$

(c) Four, $\overset{a}{CH_3}(\underset{b}{H})C=C(\overset{c}{H})(\underset{d}{H})$

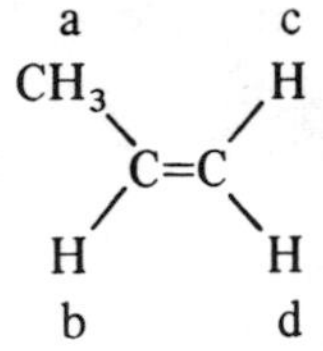

(d) Two, $\overset{a}{CH_3}(\underset{b}{H})C=C(\overset{b}{H})(\underset{a}{CH_3})$

(e) Four, $\overset{a}{CH_3}-\overset{b}{CHBr}-C(\overset{}{H}c)(H\,d)-Br$

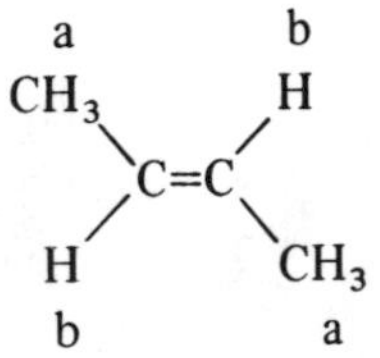

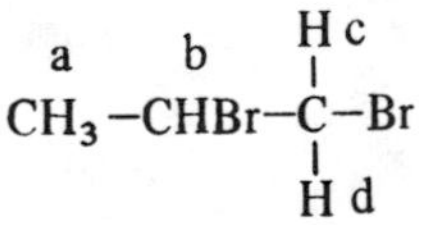

(f) Two,

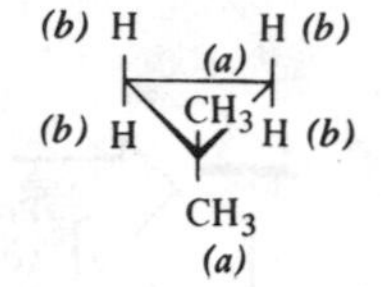

(b) H　H (b)
(a) CH_3
(b) H　H (b)
CH_3
(a)

(g) Three,

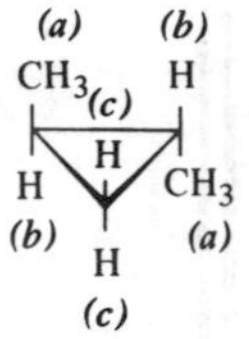

(a) CH_3　(b) H
(c) H
H (b)　CH_3 (a)
H
(c)

(h) Four,

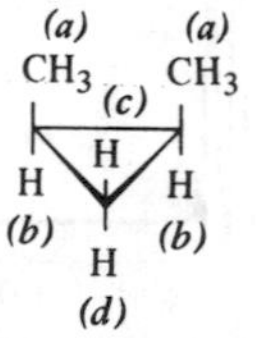

(a) CH_3　(a) CH_3
(c) H
H (b)　H (b)
H
(d)

(i) Six,

$$\overset{a}{CH_3}-\overset{b}{CH_2}-\overset{c}{CH_2}\!\!\diagdown\; C{=}C \;\diagup H\ e$$
H d　H f

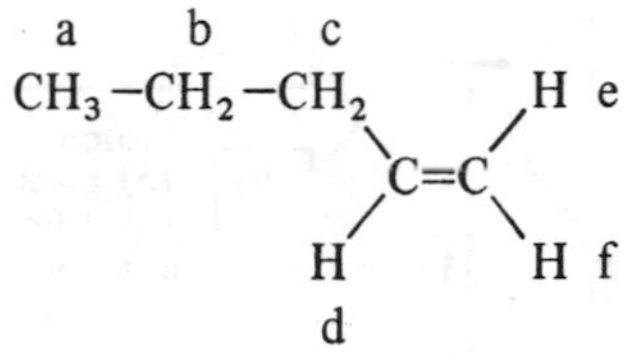

13.5

The proton nmr spectrum of $CHBr_2CHCl_2$ consists of two doublets. The doublet from the proton of the $-CHCl_2$ group should occur at lowest magnetic field strength because the greater electronegativity of chlorine reduces the electron density in the vicinity of the $-CHCl_2$ proton, and consequently, reduces its shielding relative to $-CHBr_2$.

13.6

The determining factors here are the number of chlorine atoms attached to the carbons bearing protons and the deshielding that results from chlorine's electronegativity. In 1,1,2-trichloroethane the proton that gives rise to the triplet is on a carbon that bears two chlorines, and the signal from this proton is downfield. In 1,1,2,3,3-pentachloropropane the proton that gives rise to the triplet is on a carbon that bears only one chlorine; the signal from this proton is upfield.

13.7

The signal from the three equivalent protons designated **a** should be split into a doublet by the proton **b**. This doublet, because of the electronegativity of the attached chlorines, should occur downfield.

$$(\overset{a}{Cl_2CH})_3-\overset{b}{CH}$$

The proton designated **b** should be split into a quartet by the three equivalent protons **a**. The quartet should occur upfield.

13.8

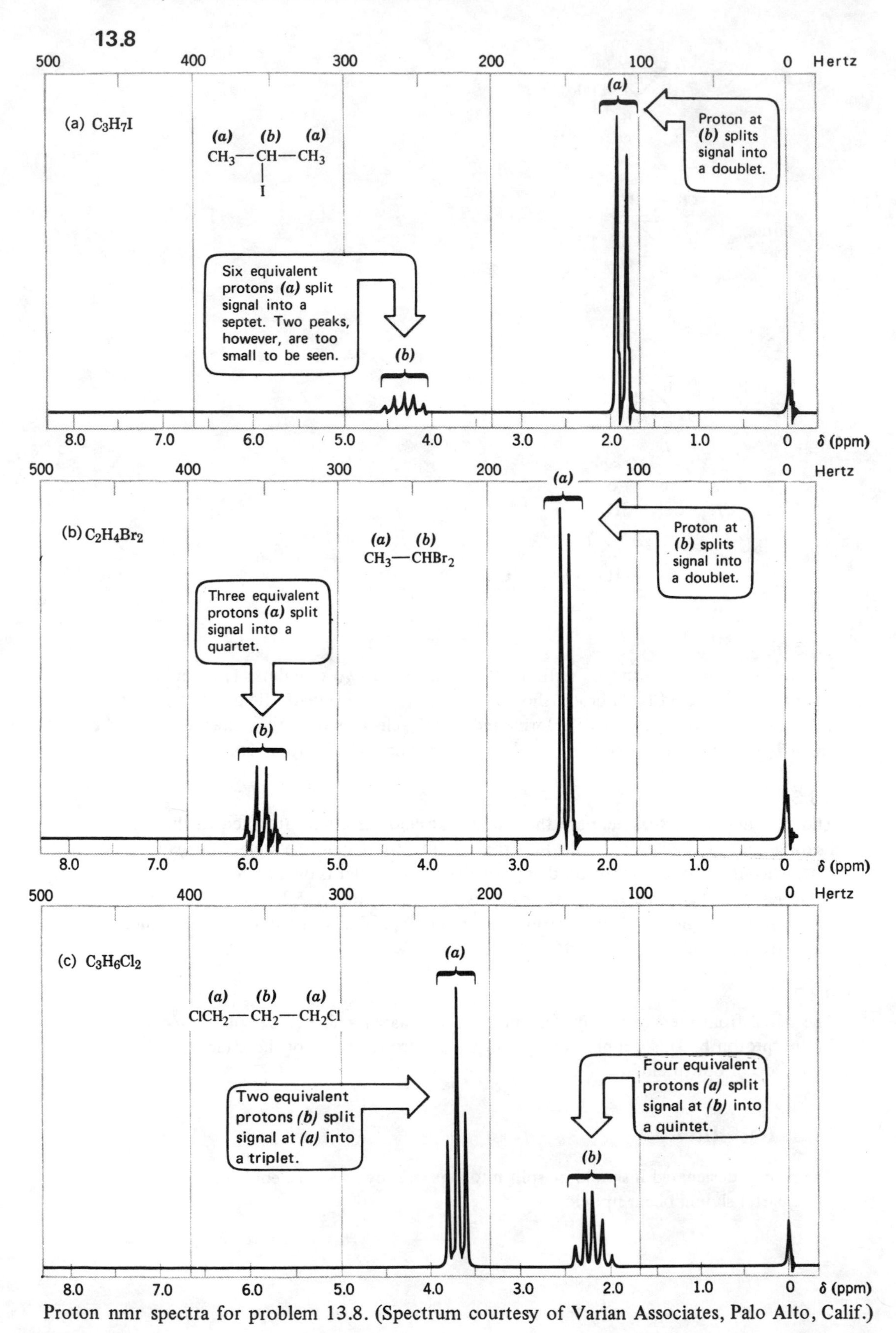

Proton nmr spectra for problem 13.8. (Spectrum courtesy of Varian Associates, Palo Alto, Calif.)

13.9

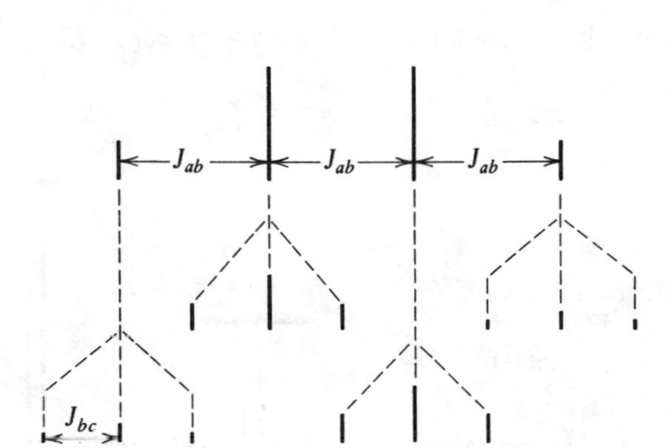

Result: (Nine peaks)

(b) $J_{ab} = J_{bc}$

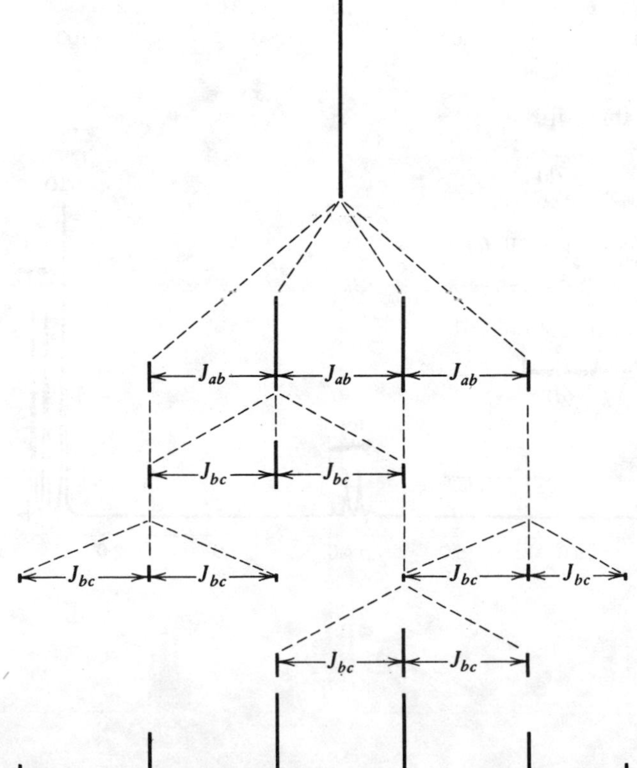

13.10

(a) $C_6H_5CH(CH_3)_2$

(b) $C_6H_5CHCH_3$ with NH_2 on the CH

(c) 

Proton nmr spectra for problem 13.10 are given below and on page 173.

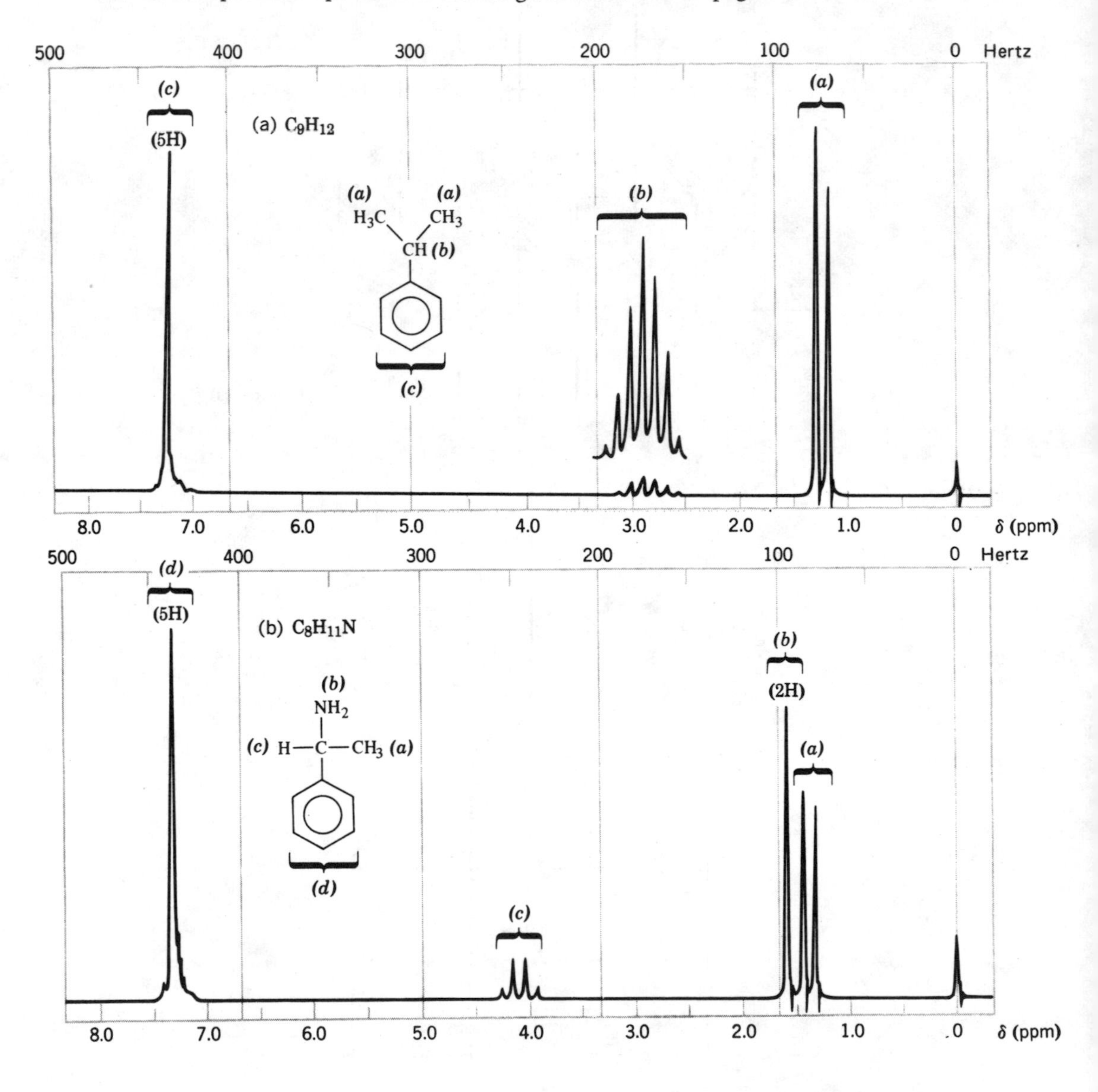

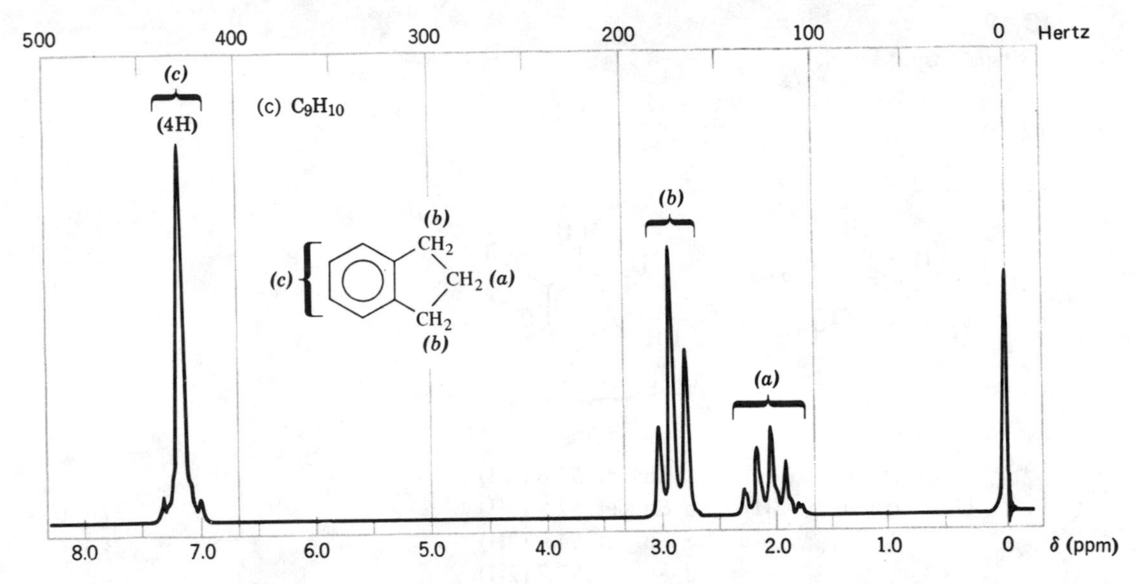

Proton nmr spectra for problem 13.10 (Spectra courtesy of Varian Associates, Palo Alto, Calif.)

13.11

(a)

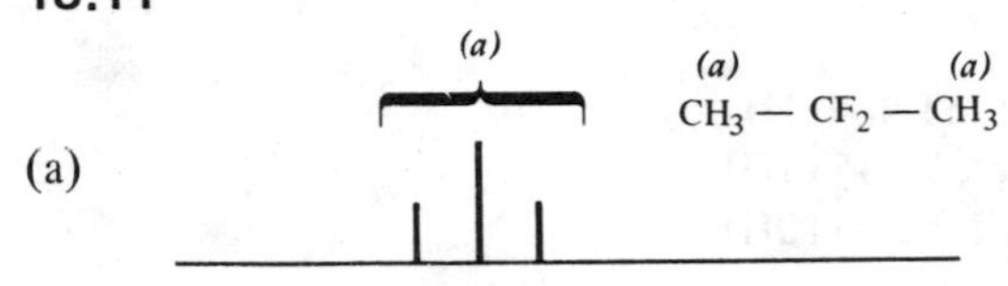

(b)

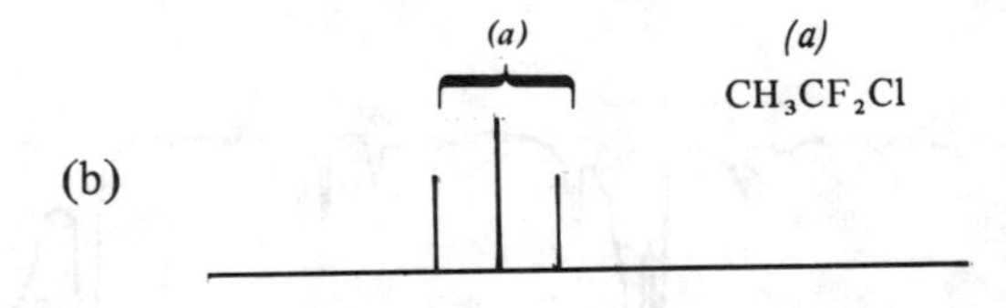

(c)

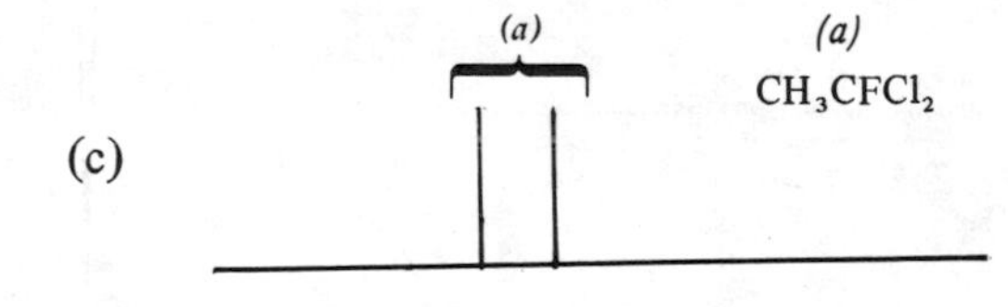

(d)

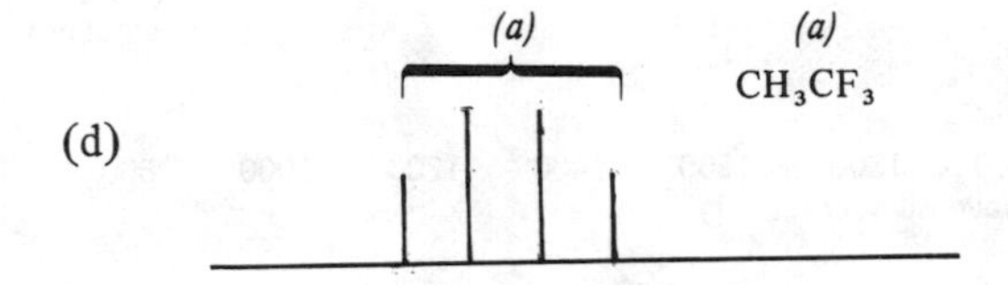

13.12

A single unsplit signal.

13.13

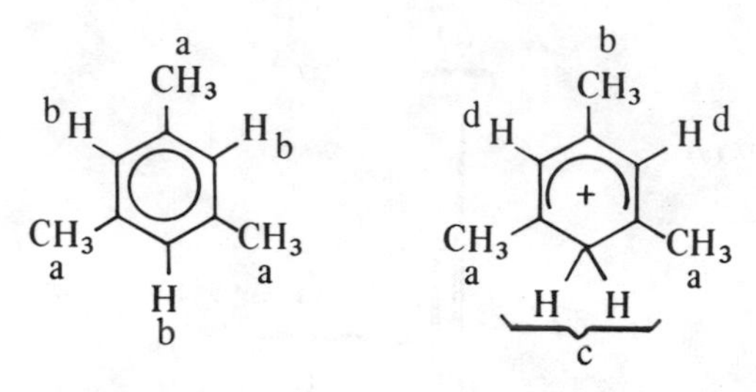

a Singlet, δ2.35 (9H)	a Singlet, δ2.8 (6H)
b Singlet, δ6.70 (3H)	b Singlet, δ2.9 (3H)
	c Singlet, δ4.6 (2H)
	d Singlet, δ7.7 (2H)

13.14

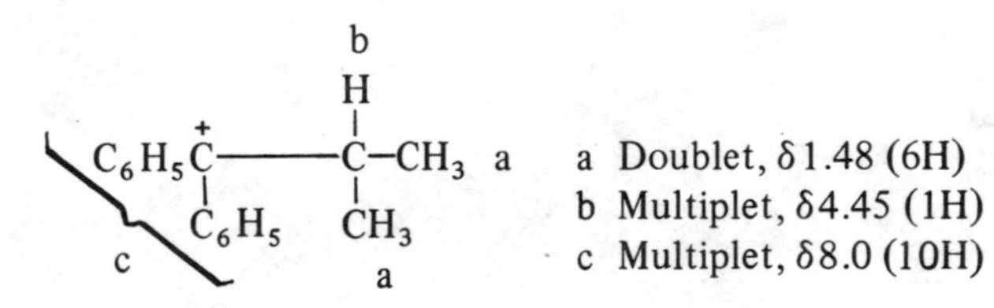

a Doublet, δ1.48 (6H)
b Multiplet, δ4.45 (1H)
c Multiplet, δ8.0 (10H)

13.15

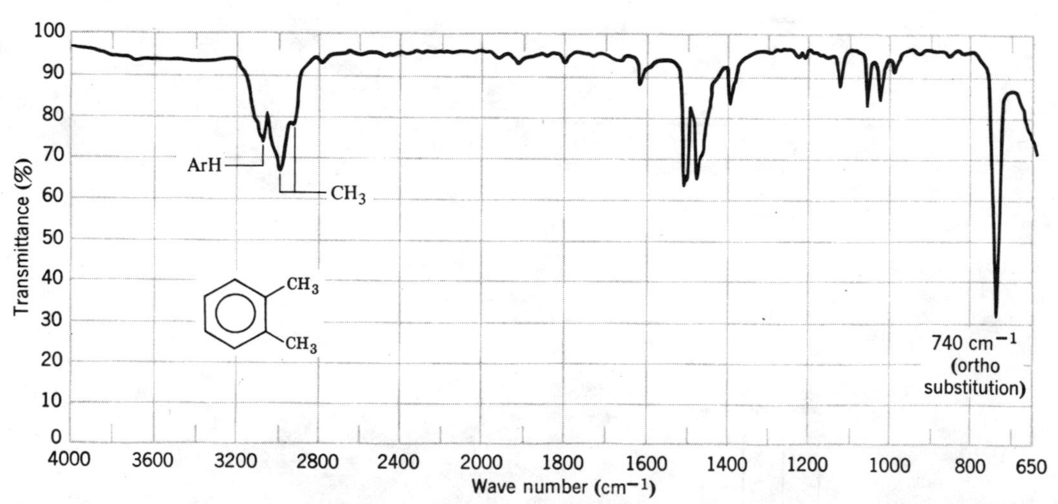

The infrared spectrum of o-xylene.
Similar assignments can be made for m-xylene and p-xylene.

13.16

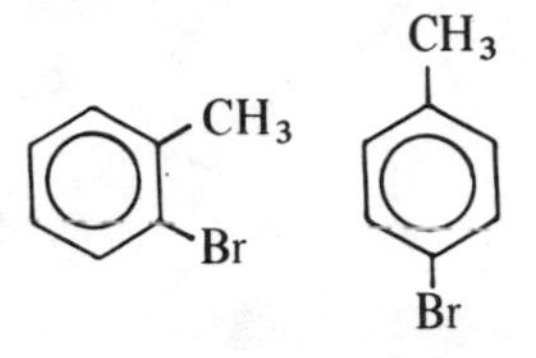

CH_3 / Br — CH_3 / Br — CH_2Br

A B C D

13.17

(a) $(CH_3)_3C{-}OH$ (a = CH_3, b = OH)

a Singlet, δ 1.28 (9H)
b Singlet, δ 1.35 (1H)

(b) $CH_3{-}CH(Br){-}CH_3$ (a = CH_3, b = CH)

a Doublet, δ 1.71 (6H)
b Septet, δ 4.32 (1H)

(c) $CH_3{-}C(=O){-}CH_2{-}CH_3$ (b = CH_3, c = CH_2, a = CH_3)

a Triplet, δ 1.05 (3H)
b Singlet, δ 2.13 (3H)
c Quartet, δ 2.47 (2H)
C=O, 1720 cm^{-1}

(d) $C_6H_5{-}CH_2{-}OH$ (c = C_6H_5, b = CH_2, a = OH)

a Singlet, δ 2.43 (1H)
b Singlet, δ 4.58 (2H)
c Multiplet, δ 7.28 (5H)
O–H, 3200-3600 cm^{-1}

(e) $CH_3{-}CH(CH_3){-}CH_2Cl$ (a = CH_3, b = CH, c = CH_2)

a Doublet, δ 1.04 (6H)
b Multiplet, δ 1.95 (1H)
c Doublet, δ 3.35 (2H)

(f) $C_6H_5{-}CH(C_6H_5){-}C(=O){-}CH_3$ (c = C_6H_5, b = CH, a = CH_3)

a Singlet, δ 2.20 (3H)
b Singlet, δ 5.08 (1H)
c Multiplet, δ 7.25 (10H)
C=O, near 1720 cm^{-1}

(g) $CH_3{-}CH_2{-}CH(Br)COOH$ (a = CH_3, b = CH_2, c = CH, d = COOH)

a Triplet, δ 1.08 (3H)
b Multiplet, δ 2.07 (2H)
c Triplet, δ 4.23 (1H)
d Singlet, δ 10.97 (1H)
O–H, 2500-3000 cm^{-1}

(h) 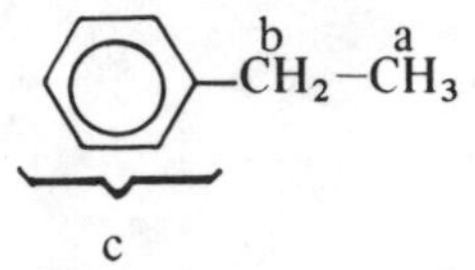 C_6H_5(c)$-CH_2$(b)$-CH_3$(a)

a Triplet, δ1.25 (3H)
b Quartet, δ2.68 (2H)
c Multiplet, δ7.23 (5H)

(i) $\overset{a}{CH_3}-\overset{b}{CH_2}-O-\overset{c}{CH_2}-\overset{d}{COOH}$

a Triplet, δ1.27 (3H)
b Quartet, δ3.66 (2H)
c Singlet, δ4.13 (2H)
d Singlet, δ10.95 (1H)
O–H, 2500-3000 cm^{-1}

(j) 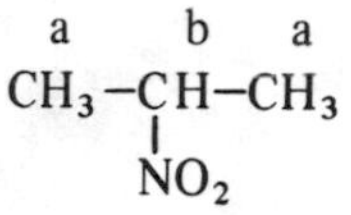 $\overset{a}{CH_3}-\overset{b}{CH}(NO_2)-\overset{a}{CH_3}$

a Doublet, δ1.55 (6H)
b Septet, δ4.67 (1H)

(k) $\overset{a}{CH_3}O-\overset{b}{CH_2}\overset{b}{CH_2}-O\overset{a}{CH_3}$

a Singlet, δ3.25 (6H)
b Singlet, δ3.45 (4H)

(l) 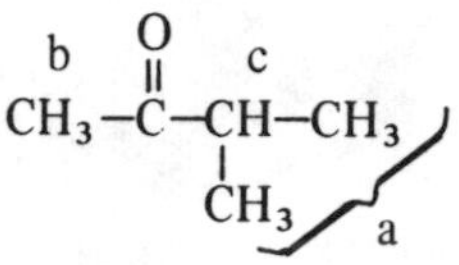 $\overset{b}{CH_3}-\overset{O}{\overset{\|}{C}}-\overset{c}{CH}(CH_3)-CH_3$ (a = both CH_3 on CH)

a Doublet, δ1.10 (6H)
b Singlet, δ2.10 (3H)
c Septet, δ2.50 (1H)
C=O, near 1720 cm^{-1}

(m) 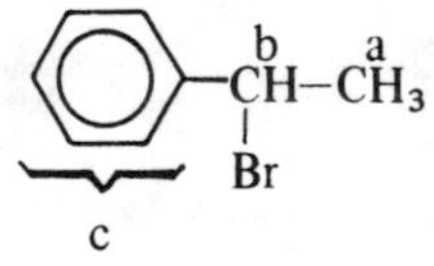 C_6H_5(c)$-\overset{b}{CH}(Br)-\overset{a}{CH_3}$

a Doublet, δ2.0 (3H)
b Quartet, δ5.15 (1H)
c Multiplet, δ7.35 (5H)

13.18

Compound **E** is phenylacetylene, $C_6H_5C{\equiv}CH$. We can make the following assignments in the infrared spectrum:

~ 3300 cm^{-1} , ≡C–H
~ 3030 cm^{-1} , Ar–H
~ 2100 cm^{-1} (weak) , –C≡C–

~ 690 cm^{-1} and ~ 710 cm^{-1} , C_6H_5–

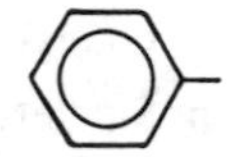

13.19

A pmr signal this far upfield indicates that cyclooctatetraene is a cyclic polyene and is not aromatic.

13.20

Both [14] annulene and dehydro[14] annulene are aromatic as shown by the signals at δ7.78 (10H) and at δ8.0 (10H) respectively. [14] Annulene has four "internal" protons (δ −0.61) and dehydro[14] annulene has only two (δ0.0).

13.21

Compound **F** is *p*-isopropyltoluene.

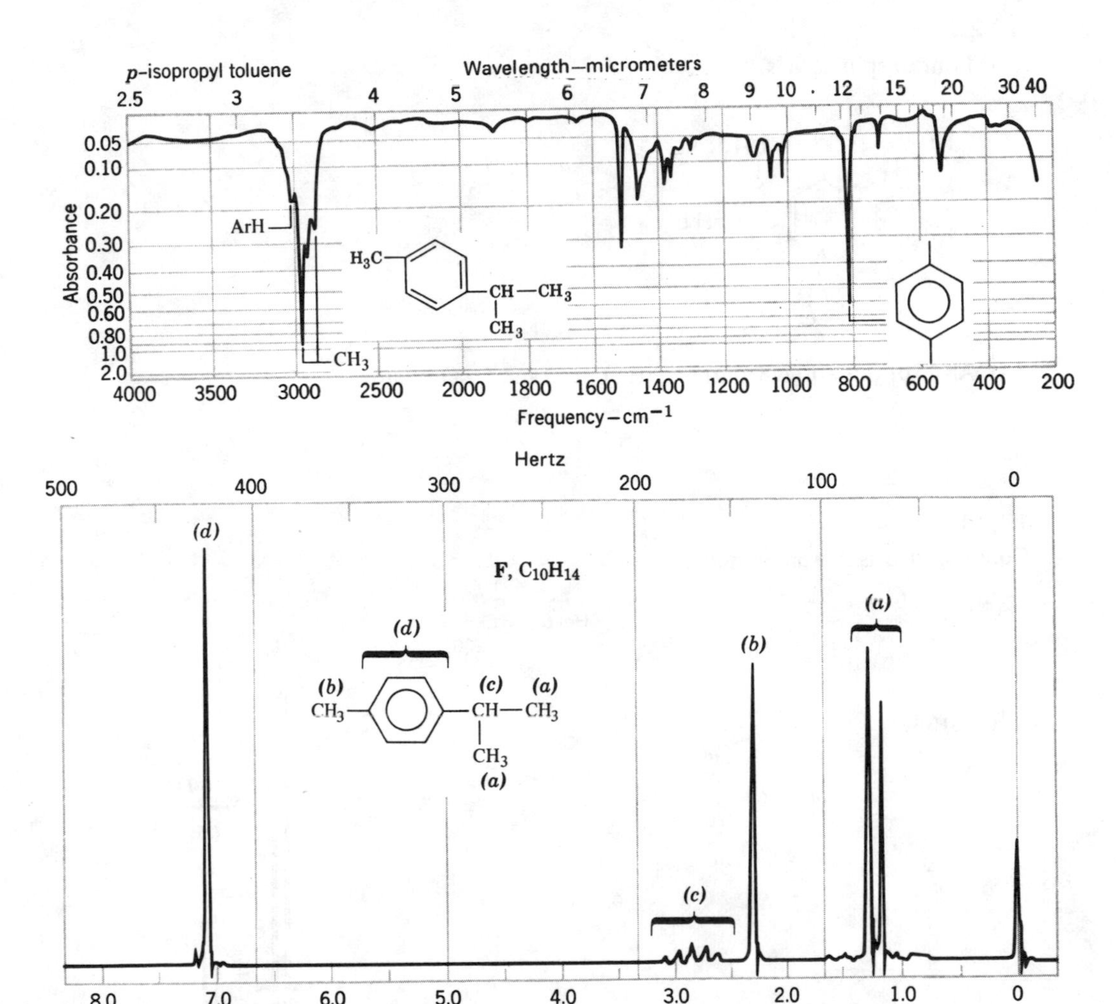

The IR and proton nmr spectra of compound F, problem 13.21 (proton nmr spectrum adapted from Varian Associates, Palo Alto, Calif. IR spectrum adapted from Sadtler Research Laboratories, Philadelphia, Pa.)

13.22

(a) In SbF_5 the carbocations formed initially apparently undergo a complex series of rearrangements to the more stable *tert*-butyl cation.

(b) All of the cations formed intially rearrange to the more stable *tert*-pentyl cation,

$$CH_3CH_2\overset{\overset{\displaystyle CH_3}{|}}{\underset{\underset{\displaystyle CH_3}{|}}{C}}{+}$$

The spectrum of the *tert*-pentyl cation should consist of a singlet (6H), a quartet (2H), and a triplet (3H). The triplet should be most upfield and the quartet most downfield.

13.23

(a) Four unsplit signals,

(b) Absorptions arising from: =C—H , CH_3 , and C=O groups.

13.24

Compound **G** is 2-bromobutane.

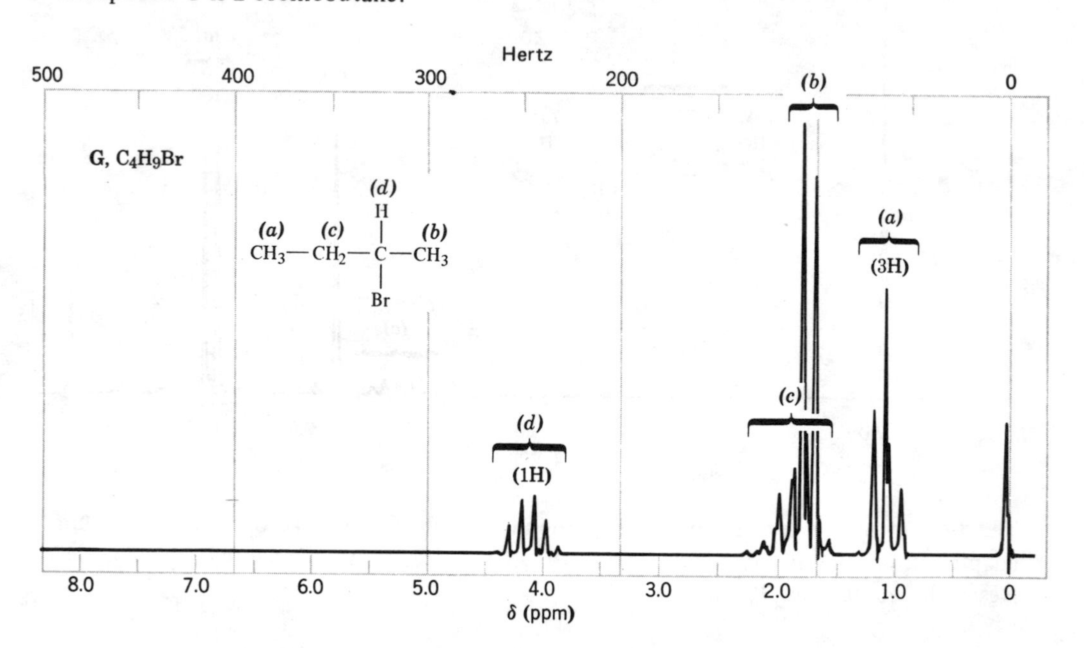

The proton nmr spectrum of compound **G** (problem 13.24). (Spectrum courtesy of Varian Associates, Palo Alto, Calif.)

Compound **H** is 2,3-dibromopropene.

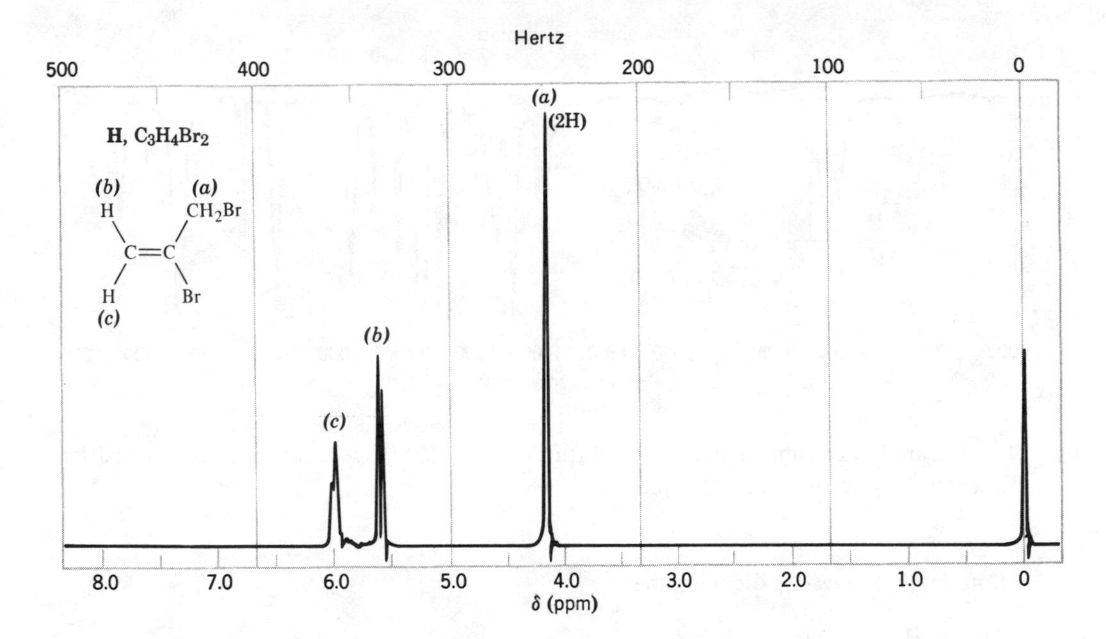

The proton nmr spectrum of compound **H** (problem 13.24). (Spectrum courtesy of Varian Associates, Palo Alto, Calif.).

13.25

Compound **I** is *p*-methoxytoluene.

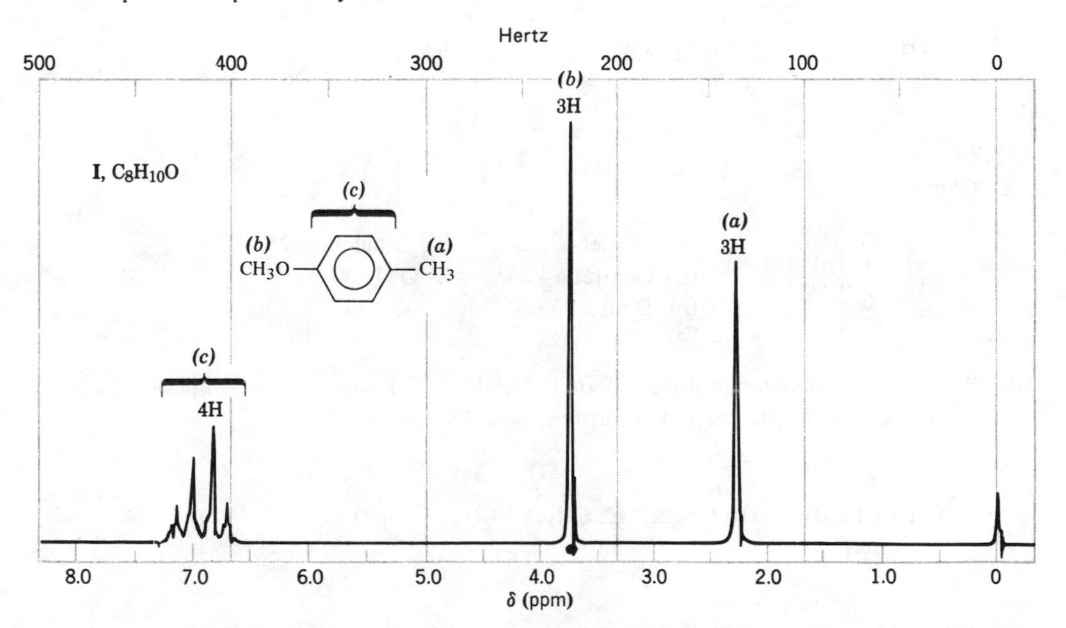

The proton nmr spectrum of compound **I** (problem 13.25). (Spectrum courtesy of Varian Associates, Palo Alto, Calif.)

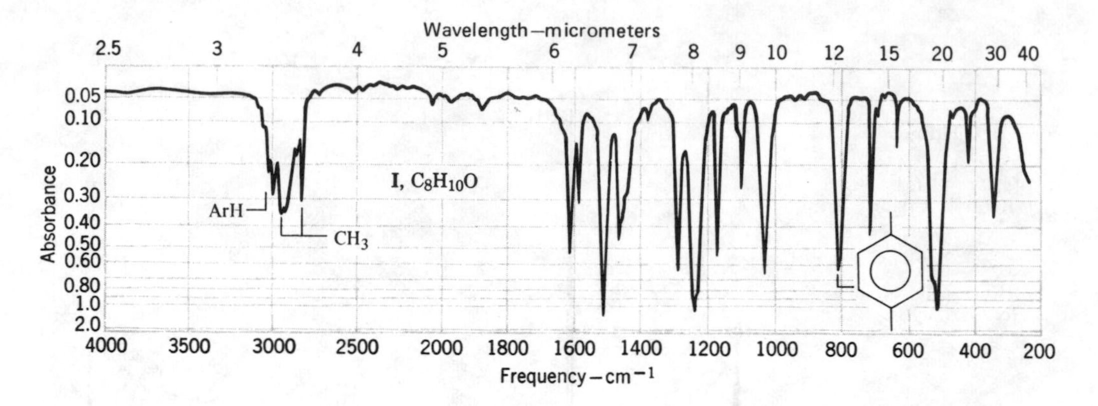

The infrared spectrum of compound **I** (problem 13.25). (Spectrum courtesy of Sadtler Research Laboratories, Philadelphia, Pa.)

13.26

Compound *J* is *cis*-1,2-dichloroethene,

```
H       H
 \     /
  C=C
 /     \
Cl      Cl
```

We can make the following infrared assignments:

3125 cm^{-1}, alkene C–H stretching
1625 cm^{-1}, C=C stretching
695 cm^{-1}, out-of-plane bending of *cis* double bond.

13.27

(a) Compound *K* is,

```
       O
 (a)   ‖  (b)  (c)
 CH3–C–CH–CH3
           |
           OH (d)
```

(a) Singlet δ2.15
(b) Quartet δ4.25
(c) Doublet δ1.35
(d) Singlet δ3.75
C=O, 1720 cm^{-1}

(b) When the compound is dissolved in D_2O, the –OH proton (d) is replaced by a deuteron and thus the proton nmr absorption peak disappears.

```
     O                                 O
     ‖                                 ‖
CH3CCHCH3  +  D2O  ⇌  CH3CCHCH3  +  DHO
       |                             |
       OH                            OD
```

13.28

Compound **L** is allylbenzene,

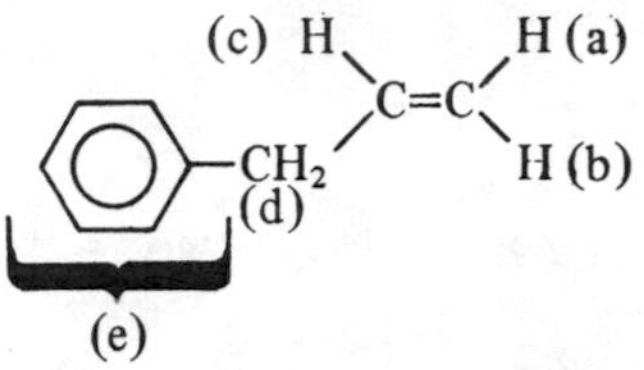

(c) H, H (a), C=C, C_6H_5–CH_2 (d), H (b), (e)

(d)	Doublet	δ3.1 (2H)
(a) or (b)	Multiplet	δ4.8
(a) or (b)	Multiplet	δ5.1
(c)	Multiplet	δ5.8
(e)	Multiplet	δ7.1 (5H)

The following infrared assignments can be made.

3035 cm^{-1}, C–H stretching of benzene ring
3020 cm^{-1}, C–H stretching of $-CH=CH_2$ group
2925 cm^{-1} and 2853 cm^{-1}, C–H stretching of $-CH_2-$ group
1640 cm^{-1}, C=C stretching
990 cm^{-1} and 915 cm^{-1}, C–H bendings of $-CH=CH_2$ group
740 cm^{-1} and 695 cm^{-1}, C–H bendings of $-C_6H_5$ group

The ultraviolet absorbance maximum at 255 nm is indicative of a benzene ring that is not conjugated with a double bond.

13.29

Run the spectrum with the spectrometer operating at a different magnetic field strength (i.e., at 30 MHz or at 100 MHz). If the peaks are two singlets the distance between them–*when measured in Hertz*–will change because chemical shifts *expressed in Hertz* are proportional to the strength of the applied field (see p. 530). If, however, the two peaks represent a doublet then the distance that separates them, expressed in Hertz, will not change because this distance represents the magnitude of the coupling constant and coupling constants are independent of the applied magnetic field (p. 540).

13.30

Compound **M** is *m*-ethyltoluene. We can make the following assignments in the pmr spectrum.

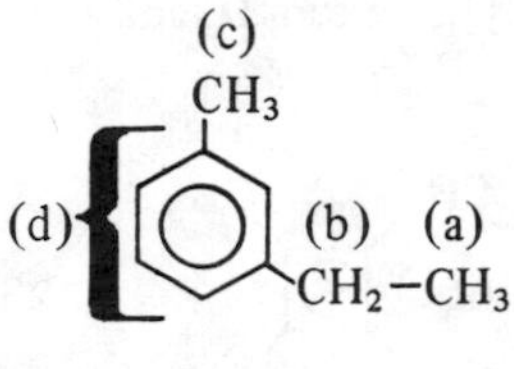

(c) CH_3, (d) ring H's, (b) CH_2–CH_3 (a)

(a) Triplet, δ1.3

(b) Quartet, δ2.6

(c) Singlet, δ2.4

(d) Multiplet, δ7.1

Meta substitution is indicated by the strong peaks at 690 cm^{-1} and 780 cm^{-1} in the infrared spectrum.

13.31

Compound **N** is $C_6H_5CH{=}CHOCH_3$. The absence of absorption peaks due to O–H or C=O stretching in the infrared spectrum of **N** suggests that the oxygen atom is present as part of an ether linkage. The (5H) pmr multiplet at δ7.3 strongly suggests the presence of a monosubstituted benzene ring; this is confirmed by the strong peaks at ~690 cm^{-1} and ~770 cm^{-1} in the infrared spectrum.

We can make the following assignments in the proton nmr spectrum:

$$\overset{(a)}{C_6H_5}-\overset{(b)}{CH}=\overset{(c)}{CH}-O\overset{(d)}{CH_3}$$

(a) Multiplet δ7.3

(c) or (b) Doublet δ6.05

(b) or (c) Doublet δ5.15

(d) Singlet δ3.7

*13.32

That the proton nmr spectrum shows only one signal indicates that all 12 protons of the carbocation are equivalent, and suggests very strongly that what is being observed is the bromonium ion:

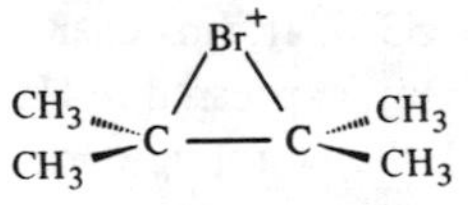

$$(CH_3)_2C \overset{Br^+}{\frown} C(CH_3)_2$$

While this experiment does not prove that bromonium ions are intermediates in alkene additions, it does show that bromonium ions are capable of existence and thus makes postulating them as intermediates more plausible.

*13.33

In the presence of SbF_5, **I** dissociates first to the cyclic allylic cation, **II**, and then to the aromatic dication, **III**.

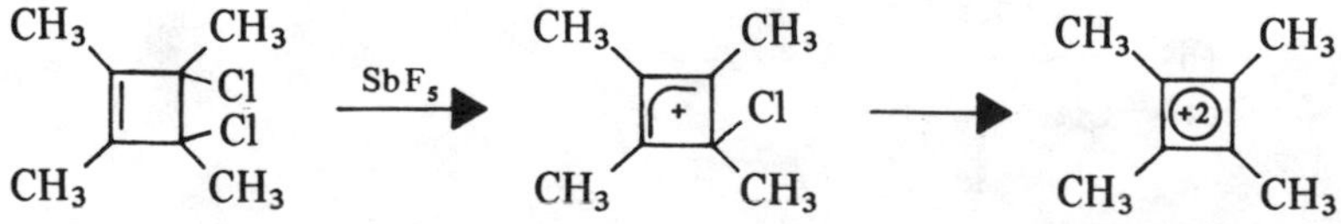

$$\text{I} \xrightarrow{SbF_5} \text{II} \;(SbF_5Cl^-) \longrightarrow \text{III} + 2SbF_5Cl^-$$

I: tetramethyl-dichlorocyclobutene (CH_3, CH_3, Cl, Cl, CH_3, CH_3)

II: CH_3, CH_3, Cl, CH_3, CH_3 — SbF_5Cl^-

III: CH_3, CH_3, CH_3, CH_3 (+2)

III
A 2π electron aromatic dication

*13.34

The vinylic protons of *p*-chlorostyrene should give a spectrum approximately like the following:

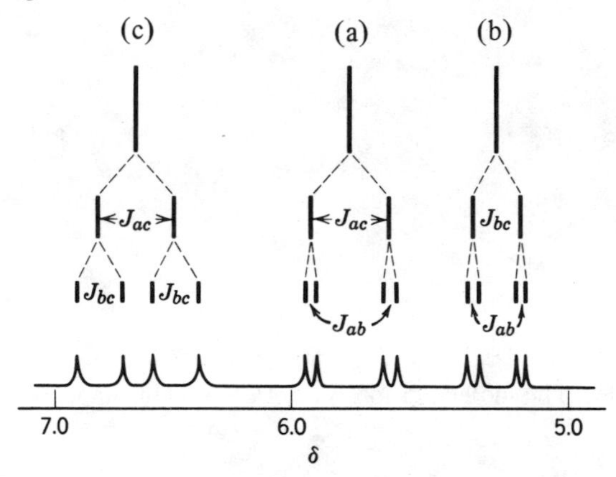

H

Special Topic

H.1

The compound is methane, CH_4. The molecular ion is at *m/e* 16. (This happens also to be the base peak.)

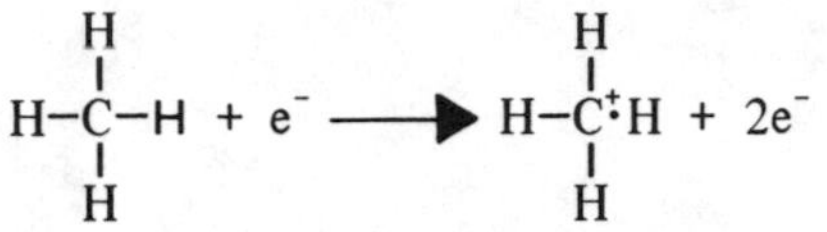

$$\begin{array}{c}H\\|\\H-C-H\\|\\H\end{array} + e^- \longrightarrow \begin{array}{c}H\\|\\H-C^{\dot{+}}H\\|\\H\end{array} + 2e^-$$

m/e 16
$M^{\dot{+}}$

The peaks at *m/e* 15, 14, 13, and 12 are caused by successive losses of hydrogen atoms.

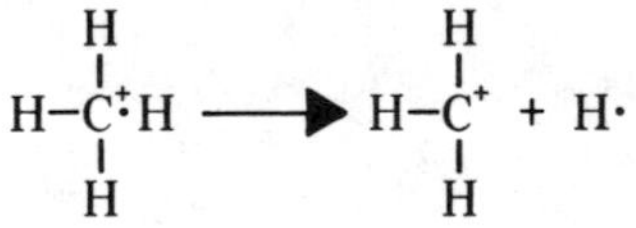

$$\begin{array}{c}H\\|\\H-C^{\dot{+}}H\\|\\H\end{array} \longrightarrow \begin{array}{c}H\\|\\H-C^{+}\\|\\H\end{array} + H\cdot$$

m/e 15

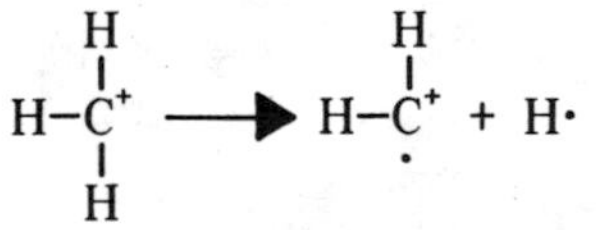

$$\begin{array}{c}H\\|\\H-C^{+}\\|\\H\end{array} \longrightarrow \begin{array}{c}H\\|\\H-\underset{\cdot}{C}^{+}\end{array} + H\cdot$$

m/e 14

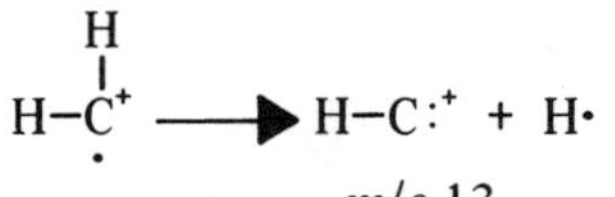

$$\begin{array}{c}H\\|\\H-\underset{\cdot}{C}^{+}\end{array} \longrightarrow H-C\colon^{+} + H\cdot$$

m/e 13

$$H-C\colon^{+} \longrightarrow \cdot C\colon^{+} + H\cdot$$

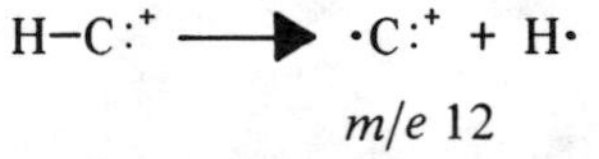

m/e 12

The small peak at *m/e* 17 ($M^{\dot{+}}$ + 1) comes mainly from methane molecules that contain ^{13}C.

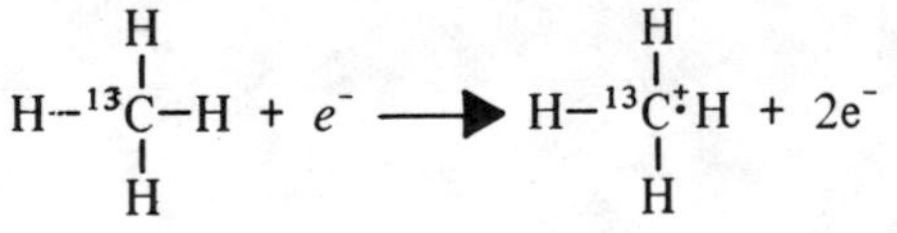

$$\begin{array}{c}H\\|\\H-{}^{13}C-H\\|\\H\end{array} + e^- \longrightarrow \begin{array}{c}H\\|\\H-{}^{13}C^{\dot{+}}H\\|\\H\end{array} + 2e^-$$

m/e 17
($M^{\dot{+}}$ + 1)

H.2

The compound is water.

$$\text{H}-\ddot{\underset{\cdot\cdot}{\text{O}}}-\text{H} + e^- \longrightarrow \text{H}-\overset{\cdot +}{\underset{\cdot\cdot}{\text{O}}}-\text{H} + 2e^-$$

m/e 18 ($M^{\ddagger}$)

$$\text{H}-\overset{\cdot +}{\underset{\cdot\cdot}{\text{O}}}-\text{H} \longrightarrow \text{H}-\ddot{\underset{\cdot\cdot}{\text{O}}}^+ + \text{H}\cdot$$

m/e 17

$$\text{H}-\ddot{\underset{\cdot\cdot}{\text{O}}}^+ \longrightarrow \cdot\ddot{\underset{\cdot\cdot}{\text{O}}}^+ + \text{H}\cdot$$

m/e 16

The peaks at m/e 19 and m/e 20 are due (primarily) to naturally occurring oxygen isotopes.

$$\text{H}-{}^{17}\ddot{\underset{\cdot\cdot}{\text{O}}}-\text{H} + e^- \longrightarrow \text{H}-{}^{17}\overset{\cdot +}{\underset{\cdot\cdot}{\text{O}}}-\text{H} + 2e^-$$

m/e 19 ($M^{\ddagger} + 1$)

$$\text{H}-{}^{18}\ddot{\underset{\cdot\cdot}{\text{O}}}-\text{H} + e^- \longrightarrow \text{H}-{}^{18}\overset{\cdot +}{\underset{\cdot\cdot}{\text{O}}}-\text{H} + 2e^-$$

m/e 20 ($M^{\ddagger} + 2$)

H.3

The compound is methyl fluoride, CH_3F.

$$CH_3-F + e^- \longrightarrow [CH_3F]^{\ddagger} + 2e^-$$

m/e 34 ($M^{\ddagger}$)

$$[CH_3F]^{\ddagger} \longrightarrow [CH_2F]^+ + H\cdot$$

m/e 33

$$[CH_2F]^+ \longrightarrow [CHF]^{\ddagger} + H\cdot$$

m/e 32

$$[CHF]^{\ddagger} \longrightarrow [CF]^+ + H\cdot$$

m/e 31

$$[CH_3F]^{\ddagger} \longrightarrow [F]^+ + CH_3\cdot$$

m/e 19

$$[CH_3F]^{\ddagger} \longrightarrow [CH_3]^+ + F\cdot$$

m/e 15

$$[CH_3]^+ \longrightarrow [CH_2]^{\ddagger} + H\cdot$$

m/e 14

H.4

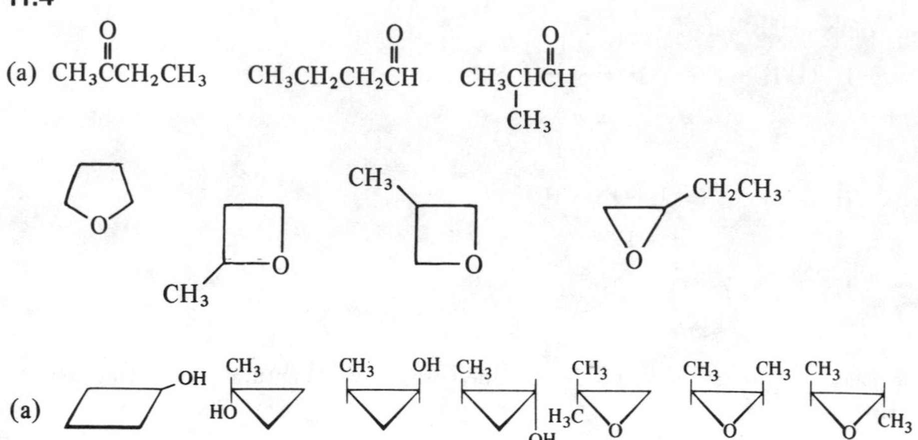

(b) Only the first three. (The peak at 1730 cm^{-1} is due to a C=O group.)

H.5

First we recalculate the intensities of the peaks so as to base them on the $M^{\ddagger}$ peak:

m/e		Intensity % of $M^{\ddagger}$
86 $M^{\ddagger}$	$10.0/10.0 \times 100 =$	100
87	$0.56/10.0 \times 100 =$	5.6
88	$0.04/10.0 \times 100 =$	0.4

A. Since $M^{\ddagger}$ is even, the compound must contain an even number of nitrogen atoms (i.e., 0, 2, 4 etc.)

B. The value of the $M^{\ddagger} + 1$ peak gives the number of carbon atoms

$$\text{Number of Carbon atoms} = 5.6/1.1 \simeq 5$$

The compound must contain no nitrogen atoms because $C_5N_2 = (5 \times 12) + (2 \times 14) = 88$ and the molecular weight of the compound (from the $M^{\ddagger}$ peak) is only 86.

C. The very low value of the $M^{\ddagger} + 2$ peak (0.4%) tells us that the compound does not contain S, Cl, or Br.

D. If the compound were composed only of C and H it would have to be C_5H_{26}:

$$H = 86 - (5 \times 12) = 26$$

But C_5H_{26} is impossible.

However, a formula with one oxygen gives a reasonable number of hydrogens,

$$H = 86 - (5 \times 12) - 16 = 10$$

and thus our compound has the formula $C_5H_{10}O$

H.6

(a) The $M^{+\cdot} + 2$ peak due to $CH_3{}^{37}Cl$ (at *m/e* 52) should be almost one third (32.5%) as large as the $M^{+\cdot}$ peak at *m/e* 50.

(b) The peaks due to $CH_3{}^{79}Br$ and $CH_3{}^{81}Br$ (at *m/e* 94 and *m/e* 96 respectively) should be of nearly equal intensity.

(c) That the $M^{+\cdot}$ and $M^{+\cdot} + 2$ peaks are of nearly equal intensity tells us that the compound contains bromine. C_3H_7Br is therefore a likely molecular formula.

C_3	= 36	C_3	= 36
H_7	= 7	H_7	= 7
^{79}Br	= $\underline{79}$	^{81}Br	= $\underline{81}$
m/e	= 122	*m/e*	= 124

H.7

Recalculating the intensities to base on $M^{+\cdot}$

Peak	*m/e*	% of Base Peak	% of $M^{+\cdot}$
$M^{+\cdot}$	73	86.1	100
$M^{+\cdot} + 1$	74	3.2	3.72
$M^{+\cdot} + 2$	75	0.2	0.23

These best fit the formula C_3H_7NO.

H.8

(a) First recalculating the intensities so as to base them on the $M^{+\cdot}$ peak:

m/e		Intensity % of $M^{+\cdot}$
78 $M^{+\cdot}$	$24/24 \times 100 =$	100
79	$0.8/24 \times 100 =$	3.3
80	$8/24 \times 100 =$	33

A. Since $M^{+\cdot}$ is even the compound contains an even number of nitrogen atoms.

B. Number of Carbon atoms = $(M^{+\cdot} + 1)/1.1 = 3.3/1.1 = 3$

C. The intensity of the $M^{+\cdot} + 2$ peak (33%) tells us that the compound contains one chlorine atom.

D. We use the molecular weight (from the $M^{+\cdot}$ peak) to calculate the number of hydrogens.

$$H = 78 - (3 \times 12) - 35 = 7$$

Thus the formula for the compound is C_3H_7Cl.

(b) CH_3CHCH_3 with Cl on the central carbon:

```
CH3CHCH3
    |
    Cl
```

H.9

(a) A *tert*-butyl cation, $(CH_3)_3C^+$.

$$\left[\begin{array}{c} CH_3 \\ | \\ CH_3-C-CH_3 \\ | \\ CH_3 \end{array}\right]^{+\cdot} \longrightarrow \begin{array}{c} CH_3 \\ | \\ CH_3-C^+ \\ | \\ CH_3 \end{array} + CH_3\cdot$$

m/e 57

H.10

A peak at $M^{+\cdot} - 15$ involves the loss of a methyl radical and the formation of a 1° carbocation.

$$\underset{M^{+\cdot}}{[CH_3CH_2\overset{\overset{CH_3}{|}}{C}HCH_2CH_3]^{+\cdot}} \longrightarrow \underset{M^{+\cdot}-15}{CH_3CH_2\overset{\overset{CH_3}{|}}{C}HCH_2^+} + CH_3\cdot$$

A peak at $M^{+\cdot} - 29$ arises from the loss of an ethyl radical and the formation of a 2° carbocation.

$$\underset{M^{+\cdot}}{[CH_3CH_2\overset{\overset{CH_3}{|}}{C}HCH_2CH_3]^{+\cdot}} \longrightarrow \underset{M^{+\cdot}-29}{CH_3CH_2\overset{\overset{CH_3}{|}}{C}H^+} + CH_3CH_2\cdot$$

Since a 2° carbocation is more stable, the peak at $M^{+\cdot} - 29$ is more intense.

H.11

Both peaks arise from allylic fragmentations

$$^+CH_2-\dot{C}H-CH_2-\underset{\underset{CH_3}{|}}{C}HCH_2CH_3 \longrightarrow \underset{\text{Allyl radical}}{\dot{C}H_2-CH=CH_2} + \underset{m/e\ 57}{^+\underset{\underset{CH_3}{|}}{C}HCH_2CH_3}$$

$$CH_2\overset{+\cdot}{-}CH-CH_2:\underset{\underset{CH_3}{|}}{C}HCH_2CH_3 \longrightarrow \underset{\substack{m/e\ 41 \\ \text{Allyl cation}}}{\overset{+}{C}H_2-CH=CH_2} + \cdot\underset{\underset{CH_3}{|}}{C}HCH_2CH_3$$

H.12

(a) Alcohols undergo rapid cleavage of a carbon-carbon bond next to oxygen because this leads to a resonance-stabilized cation.

$$1^\circ \text{ alcohol } R:CH_2-\overset{+\cdot}{\ddot{O}}H \xrightarrow{-R\cdot} CH_2=\overset{+}{\ddot{O}}H \longleftrightarrow \overset{+}{C}H_2-\ddot{O}H$$

$$2° \text{ alcohol } R{-}\overset{R}{C}H{-}\ddot{O}H^{+\cdot} \xrightarrow{-R\cdot} RCH{=}\overset{+}{O}H \longleftrightarrow R\overset{+}{C}H{-}\ddot{O}H$$

$$3° \text{ alcohol } R{-}\underset{R}{\overset{R}{C}}{-}\ddot{O}H^{+\cdot} \xrightarrow{-R\cdot} R{-}\underset{R}{C}{=}\overset{+}{O}H \longleftrightarrow R{-}\underset{R}{\overset{+}{C}}{-}\ddot{O}H$$

The cation obtained from a tertiary alcohol is the most stable (because of the electron-releasing R groups).

(b) Primary alcohols give a peak at *m/e* 31 due to $CH_2{=}\overset{+}{O}H$.

(c) Secondary alcohols give peaks at *m/e* 45, 59, 73, and so forth, because ions like the following are produced.

$$\underset{m/e\ 45}{CH_3CH{=}\overset{+}{O}H} \qquad \underset{m/e\ 59}{CH_3CH_2CH{=}\overset{+}{O}H} \qquad \underset{m/e\ 73}{CH_3CH_2CH_2CH{=}\overset{+}{O}H}$$

(d) Tertiary alcohols give peaks at *m/e* 59, 73, 87, and so forth, because ions like the following are produced.

$$\underset{m/e\ 59}{CH_3\underset{CH_3}{C}{=}\overset{+}{O}H} \qquad \underset{m/e\ 73}{CH_3CH_2\underset{CH_3}{C}{=}\overset{+}{O}H} \qquad \underset{m/e\ 87}{CH_3CH_2CH_2\underset{CH_3}{C}{=}\overset{+}{O}H}$$

H.13

The spectrum given in Fig. H.12 is that of isopropyl butyl ether. The main clues are the peaks at *m/e* 101 and *m/e* 73 due to the following fragmentations.

$$\left[CH_3{-}\overset{CH_3}{C}H{-}OCH_2CH_2CH_2CH_3\right]^{+\cdot} \xrightarrow{-CH_3\cdot} \underset{m/e\ 101}{CH_3CH{=}\overset{+}{O}CH_2CH_2CH_2CH_3}$$

$$\left[CH_3\overset{CH_3}{C}H{-}O{-}CH_2CH_2CH_2CH_3\right]^{+\cdot} \xrightarrow{-CH_3CH_2CH_2\cdot} \underset{m/e\ 73}{CH_3\overset{CH_3}{C}HO^{+}{=}CH_2}$$

Propyl butyl ether (Fig. H.13) has no peak at *m/e* 101 but has a peak at *m/e* 87 instead.

$$\left[CH_3CH_2CH_2{-}O{-}CH_2CH_2CH_2CH_3\right]^{+\cdot} \xrightarrow{-CH_3CH_2\cdot} \underset{m/e\ 87}{CH_2{=}\overset{+}{O}CH_2CH_2CH_2CH_3}$$

Propyl butyl ether also has a peak at *m/e* 73.

$$\left[CH_3CH_2CH_2{-}O{-}CH_2CH_2CH_2CH_3\right]^{+\cdot} \xrightarrow{-CH_3CH_2CH_2\cdot} \underset{m/e\ 73}{CH_3CH_2CH_2{-}\overset{+}{O}{=}CH_2}$$

[Although it does not help us decide it is interesting to notice that both spectra have intense peaks at *m/e* 43 and *m/e* 57 corresponding to propyl (or isopropyl) and butyl cations formed by carbon-oxygen bond cleavage.]

H.14

The compound is butanal. The peak at *m/e* 44 arises from a McLafferty rearrangement.

$$\left[\text{H-C(=O)-CH}_2\text{-CH}_2\text{-CH}_2\text{-H}\right]^{+\cdot} \longrightarrow \left[\text{H-C(-O-H)=CH}_2\right]^{+\cdot} + \text{CH}_2\text{=CH}_2$$

m/e 72 $M^{+\cdot}$ *m/e* 44 $(M^{+\cdot} - 28)$

The peak at *m/e* 29 arises from a fragmentation producing an acylium ion.

$$\text{H-C(=O}^{+\cdot}\text{)-CH}_2\text{CH}_2\text{CH}_3 \longrightarrow \text{H-C}\equiv\overset{+}{\text{O}} + \text{CH}_3\text{CH}_2\text{CH}_2\cdot$$

m/e 29

H.15

The ion, $CH_2{=}\overset{+}{N}H_2$, produced by the following fragmentation.

$$\text{R:CH}_2\text{-}\overset{+\cdot}{\text{N}}\text{H}_2 \xrightarrow{-\text{R}\cdot} \text{CH}_2\text{=}\overset{+}{\text{N}}\text{H}_2 \longleftrightarrow \overset{+}{\text{C}}\text{H}_2\text{-}\ddot{\text{N}}\text{H}_2$$

m/e 30

H.16

Compound **A** is *tert*-butylamine. Our first clue is the molecular ion at *m/e* 73 (an odd-numbered mass unit) indicating the presence of an odd number of nitrogen atoms. The base peak at *m/e* 58 is our second important clue. It arises from the following fragmentation.

$$\left[\text{CH}_3\text{-C(CH}_3\text{)}_2\text{-NH}_2\right]^{+\cdot} \xrightarrow{-\text{CH}_3\cdot} \text{CH}_3\text{-C(CH}_3\text{)=}\overset{+}{\text{N}}\text{H}_2$$

m/e 58

The pmr spectrum confirms the structure

(a) (b)
$(CH_3)_3C{-}NH_2$

(a) Singlet δ 1.2 (9H)

(b) Singlet δ 1.3 (2H)

H.17

The compound is 2-methyl-2-butanol. Although the molecular ion is not discernible, we are given that it is at *m/e* 88. This gives us the molecular weight of **B** and rules out the possibility of a structure with an odd number of nitrogen atoms.

The infrared absorption (3200-3600 cm^{-1}) suggests the presence of an $-OH$ group.

Two important peaks in the mass spectrum are the intense peaks at *m/e* 59 and *m/e* 73. These correspond to fragmentation reactions that produce resonance-stabilized oxonium ions and strongly suggest that we have a tertiary alcohol [see problem H.12, part (d)].

$$\left[CH_3CH_2\overset{CH_3}{\underset{CH_3}{C}}-OH\right]^{+\cdot} \xrightarrow{-CH_3CH_2\cdot} \overset{CH_3}{\underset{CH_3}{C}}={}^{+}OH \quad m/e\ 59$$

$$\xrightarrow{-CH_3\cdot} CH_3CH_2\underset{CH_3}{C}=\overset{+}{O}H \quad m/e\ 73$$

The peak at *m/e* 70 corresponds to the loss of a molecule of water from the molecular ion and the peak at *m/e* 55 probably arises from a subsequent allylic cleavage

$$\left[CH_3CH_2\overset{CH_3}{\underset{CH_3}{C}}-OH\right]^{+\cdot} \xrightarrow{-H_2O} \left[CH_3CH=\overset{CH_3}{C}-CH_3\right]^{+\cdot} \quad m/e\ 70$$

$$\left[CH_3CH_2\overset{CH_3}{\underset{CH_3}{C}}-OH\right]^{+\cdot} \xrightarrow{-H_2O} \left[CH_3CH_2\overset{CH_3}{C}=CH_2\right]^{+\cdot} \quad m/e\ 70$$

$$\xrightarrow{-CH_3\cdot} \overset{+}{C}H_2-\overset{CH_3}{C}=CH_2 \quad m/e\ 55$$

The pmr spectrum of **B** confirms that it is 2-methyl-2-butanol

$$\underset{(a)}{CH_3}-\underset{(b)}{CH_2}-\overset{\overset{(c)}{CH_3}}{\underset{\underset{(d)}{OH}}{C}}-CH_3\,(c)$$

(a) Triplet, δ0.9 (3H)

(b) Quartet, δ 1.6 (2H)

(c) and (d) Overlapping singlets, δ 1.1 (7H)

H.18

Compound **C** is 3-methyl-1-butanol. Here, (because the compound is a primary alcohol) the molecular ion (*m/e* 88) is small but discernible. Again, the even-numbered mass of the molecular ion rules out a compound with an odd number of nitrogen atoms and the infrared absorption suggests the presence of an −OH group.

An important indication that **C** is a primary alcohol is the peak at *m/e* 31 corresponding to the following fragmentation [see also problem H.12, part (b)].

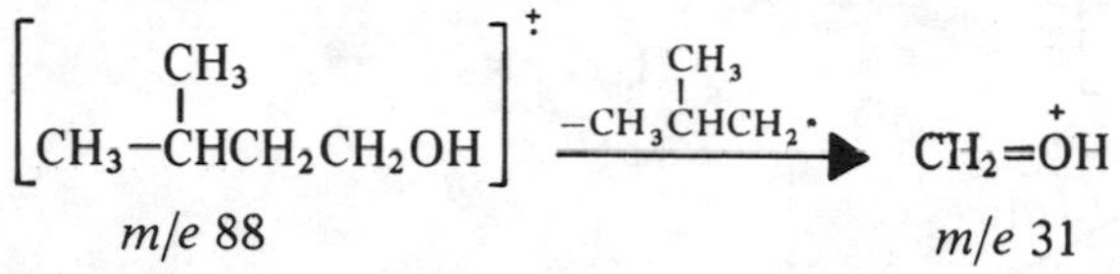

$$\left[CH_3-\underset{}{\overset{CH_3}{\overset{|}{C}}}HCH_2CH_2OH\right]^{\ddagger} \xrightarrow{-CH_3\overset{CH_3}{\overset{|}{C}}HCH_2\cdot} CH_2{=}\overset{+}{O}H$$

$$m/e\ 88 \qquad\qquad\qquad m/e\ 31$$

The peak at *m/e* 70 ($M^{\ddagger}$ - 18) corresponds to the loss of water from the molecular ion.

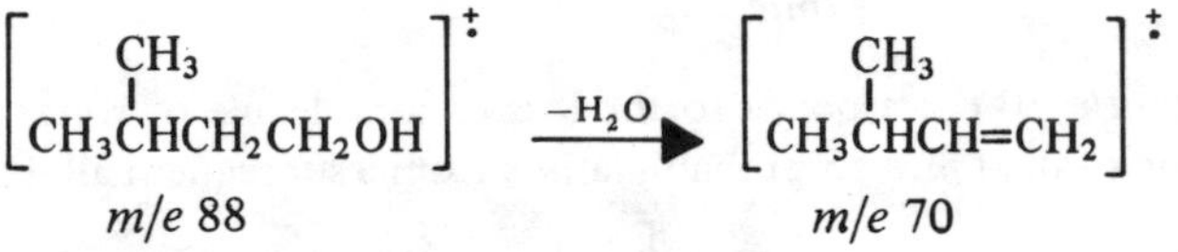

$$\left[CH_3\overset{CH_3}{\overset{|}{C}}HCH_2CH_2OH\right]^{\ddagger} \xrightarrow{-H_2O} \left[CH_3\overset{CH_3}{\overset{|}{C}}HCH{=}CH_2\right]^{\ddagger}$$

$$m/e\ 88 \qquad\qquad\qquad m/e\ 70$$

The peak at *m/e* 55 probably comes from a subsequent allylic cleavage.

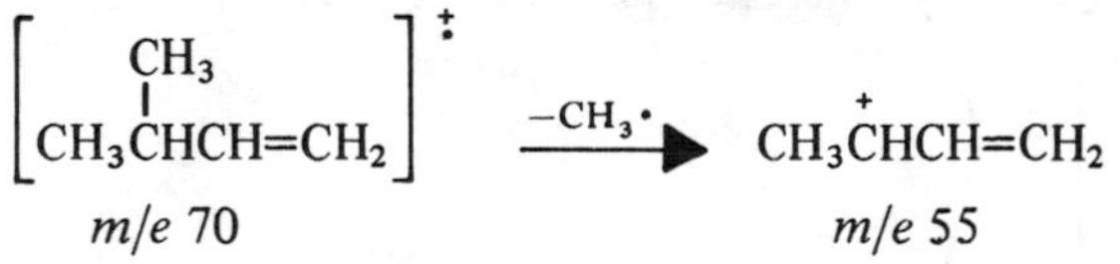

$$\left[CH_3\overset{CH_3}{\overset{|}{C}}HCH{=}CH_2\right]^{\ddagger} \xrightarrow{-CH_3\cdot} CH_3\overset{+}{C}HCH{=}CH_2$$

$$m/e\ 70 \qquad\qquad\qquad m/e\ 55$$

The pmr spectrum is consistent with this structure. We can make the following assignments.

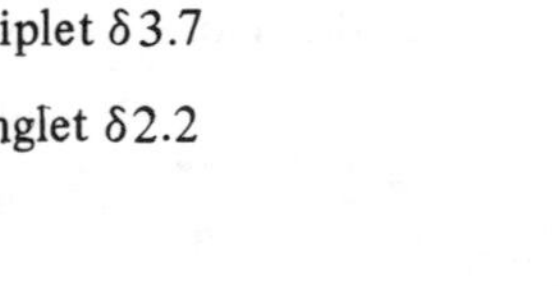

$$\underset{(a)}{CH_3}-\underset{(b)}{\overset{\overset{(a)}{CH_3}}{\overset{|}{C}H}}-\underset{(c)}{CH_2}-\underset{(d)}{CH_2}\underset{(e)}{OH}$$

(a) Doublet, δ0.9

(b) and (c) Multiplet δ 1.5

(d) Triplet δ 3.7

(e) Singlet δ 2.2

H.19

The compound is 2-pentanone. The infrared absorption at 1710 cm^{-1} strongly indicates the presence of a carbonyl group. In the mass spectrum the molecular ion peak at *m/e* 86 gives us the molecular weight and rules out structures with an odd number of nitrogens. A possible formula is $C_5H_{10}O$. (See problem H.5).

The peaks at *m/e* 71 and *m/e* 43 correspond to $M^{+\cdot} - 15$ and $M^{+\cdot} - 43$. Fragmentations of 2-pentanone would produce acylium ions with these mass numbers.

$$\left[CH_3\overset{O}{\overset{\|}{C}}CH_2CH_2CH_3\right]^{+\cdot} \xrightarrow{-CH_3\cdot} \overset{+}{O}{\equiv}CCH_2CH_2CH_3$$

m/e 86 → *m/e* 71

$$\left[CH_3\overset{O}{\overset{\|}{C}}CH_2CH_2CH_3\right]^{+\cdot} \xrightarrow{-CH_3CH_2CH_2\cdot} CH_3C{\equiv}\overset{+}{O}$$

m/e 86 → *m/e* 43

The peak at *m/e* 58 ($M^{+\cdot} - 28$) comes from a McLafferty rearrangement.

$$\left[CH_3{-}C(=O)(CH_2{-}CH_2{-}CH_2{-}H)\right]^{+\cdot} \longrightarrow \left[CH_3{-}C(OH){=}CH_2\right]^{+\cdot} + CH_2{=}CH_2$$

m/e 58

The pmr spectrum confirms our structure.

$$\overset{(a)}{CH_3}\overset{O}{\overset{\|}{C}}\overset{(b)}{CH_2}\overset{(c)}{CH_2}\overset{(d)}{CH_3}$$

(a) Singlet, δ 2.2

(b) Triplet, δ 2.4

(c) Multiplet, δ 1.6

(d) Triplet, δ 0.9

H.20

The compound is bromobenzene. That the compound contains bromine is indicated by the $M^{+\cdot}$ and $M^{+\cdot} + 2$ peaks of nearly equal intensity at *m/e* 156 and *m/e* 158. The peak at *m/e* 77 (the base peak) strongly suggests the presence of a benzene ring.

$$C_6H_5Br^{+\cdot} \xrightarrow{-Br\cdot} C_6H_5^{+}$$

m/e 77

Putting these facts together with the molecular weight (156) leads us to only one logical conclusion.

H.21

Compound **F** is 2-bromoethyl phenyl ether, $C_6H_5OCH_2CH_2Br$. The presence of bromine is indicated by the $M^{+\cdot}$ and $M^{+\cdot} + 2$ peaks of nearly equal intensity at *m/e* 200 and *m/e* 202. The presence of a benzene ring is indicated by the peak at *m/e* 77.

m/e 77

If we subtract the masses of these two large fragments (i.e., Br = 79 and C_6H_5 = 77) from the mass of the molecular ion (200) we are left with a mass difference of 44. One way that we can account for this mass difference is with C_2H_4O.

$$
\begin{aligned}
C_6H_5 &= 77 \\
Br &= 79 \\
C_2H_4O &= \underline{44} \\
MW &= 200
\end{aligned}
$$

Thus we can tentatively assume that the compound has a molecular formula C_8H_9BrO.

The complex multiplet (5H) at δ7.0 in the pmr spectrum confirms our assignment of a benzene ring, C_6H_5-. The triplets at δ2.5 and at δ4.2 suggest mutually coupled $-CH_2-$ groups. Thus we are led to the following as a possible structure.

$C_6H_5-OCH_2CH_2Br$

The peak in the mass spectrum at *m/e* 107 strongly suggests that this assignment is correct.

$$[C_6H_5-OCH_2CH_2Br]^{+\cdot} \longrightarrow C_6H_5-O^+{=}CH_2 + \cdot CH_2Br$$

The peak at *m/e* 94 comes from the following rearrangement.

$$[C_6H_5-O-CH_2-CHBr(H)]^{+\cdot} \longrightarrow [C_6H_6O]^{+\cdot} + CH_2{=}CHBr$$

m/e 94

The small peaks of nearly equal intensity at *m/e* 93 and *m/e* 95 come from $CH_2{=}Br^+$ fragments in which the bromines are ^{79}Br and ^{81}Br.

CHAPTER FOURTEEN

Organic Halides and Organometallic Compounds

14.1

(a) $R\text{-}CH_3\underset{\displaystyle Br}{\underset{|}{C}}HCH_2CH_3 + NaOH \xrightarrow[S_N2]{H_2O} S\text{-}CH_3\underset{\displaystyle OH}{\underset{|}{C}}HCH_2CH_3 + NaBr$

(b) trans-1-bromo-4-methylcyclohexane (substituents: CH_3, H; H, Br) $+ NaOH \xrightarrow[S_N2]{H_2O}$ cis-4-methylcyclohexanol (substituents: CH_3, H; OH, H) $+ NaBr$

(c) $C_6H_5\underset{\displaystyle Br}{\underset{|}{C}}HCH_2CH_3 + NaOC_2H_5 \xrightarrow{C_2H_5OH} C_6H_5CH{=}CHCH_3 + NaCl + C_2H_5OH$

(d) $C_6H_5\overset{\displaystyle Br}{\overset{|}{\underset{\displaystyle Br}{\underset{|}{C}}}}CH_3 + 2NaNH_2 \xrightarrow{NH_3} C_6H_5C{\equiv}CH + 2NaBr + 2NH_3$

(e) $C_6H_5CH_2CH_2\overset{\displaystyle Br}{\overset{|}{\underset{\displaystyle Br}{\underset{|}{C}}}}H + 2NaNH_2 \xrightarrow{NH_3} C_6H_5CH_2C{\equiv}CH + 2NaBr + 2NH_3$

(f) $CH_3CH_2CH_2CH_2CH_2Br + NaOC_2H_5 \xrightarrow{C_2H_5OH} CH_3CH_2CH_2CH{=}CH_2$

$+ NaBr + C_2H_5OH$

14.2

(a) In concentrated base and ethyl alcohol (a relatively nonpolar solvent) the S_N2 reaction is favored. Thus the rate depends on the concentration of both the alkyl halide and $NaOC_2H_5$. Since no carbocation is formed, the only product is

$$CH_3CH{=}CHCH_2OCH_2CH_3.$$

(b) When the concentration of $C_2H_5O^-$ ion is small or zero, the reaction occurs through the S_N1 mechanism. The carbocation that is produced in the first step of the S_N1 mechanism is a resonance hybrid.

$$CH_3CH{=}CHCH_2Cl \rightleftharpoons \left[\begin{array}{c} CH_3CH{=}CH\overset{+}{C}H_2 \\ \updownarrow \\ CH_3\overset{+}{C}H{-}CH{=}CH_2 \end{array}\right] + Cl^-$$

This ion reacts with the nucleophile ($C_2H_5O^-$ or C_2H_5OH) to produce two isomeric ethers

$$CH_3CH{=}CHCH_2{-}OCH_2CH_3 \quad \text{and} \quad CH_3\underset{}{\overset{OCH_2CH_3}{\overset{|}{C}}}HCH{=}CH_2$$

(c) In the presence of water, the first step of the S_N1 reaction occurs. The reverse of this reaction produces two compounds because the positive charge on

$$\left[\begin{array}{c} CH_3CH{=}CH\overset{+}{C}H_2 \\ \updownarrow \\ CH_3\overset{+}{C}H{-}CH{=}CH_2 \end{array}\right] + Cl^- \longrightarrow \begin{array}{c} CH_3CH{=}CHCH_2Cl \\ + \\ CH_3\overset{Cl}{\overset{|}{C}}H{-}CH{=}CH_2 \end{array}$$

the carbocation is distributed over carbons one and three.

14.3

(a) The carbocation that is produced in the S_N1 reaction is exceptionally stable because one resonance contributor is not only allylic but also tertiary.

$$CH_3\overset{CH_3}{\overset{|}{C}}{=}CHCH_2Cl \underset{}{\overset{S_N1}{\rightleftharpoons}} \left[CH_3\overset{CH_3}{\overset{|}{C}}{=}CH\overset{+}{C}H_2 \longleftrightarrow CH_3{-}\underset{+}{\overset{CH_3}{\overset{|}{C}}}{-}CH{=}CH_2 \right]$$

a 3° allylic carbocation

(b) $CH_3\overset{CH_3}{\overset{|}{C}}{=}CHCH_2OH + CH_3\underset{OH}{\underset{|}{\overset{CH_3}{\overset{|}{C}}}}CH{=}CH_2$

14.4

Compounds that undergo reactions by an S_N1 path must be capable of forming relatively stable carbocations. Primary halides of the type, $ROCH_2X$ form carbocations that are stabilized by resonance:

$$R{-}\ddot{\underset{..}{O}}{-}\overset{+}{C}H_2 \longleftrightarrow R{-}\overset{+}{\underset{..}{O}}{=}CH_2$$

14.5

The relative rates are in the order of the relative stabilities of the carbocations:

$$C_6H_5\overset{+}{C}H_2 < C_6H_5\overset{+}{C}HCH_3 < (C_6H_5)_2\overset{+}{C}H < (C_6H_5)_3\overset{+}{C}$$

The solvolysis reaction involves a carbocation intermediate.

14.6

(a) $NO_2{-}C_6H_4{-}OCH_3$ (para)

(b) $C_6H_4(NHCH_3)(NO_2)$ (ortho)

(c) $C_6H_3(NHC_6H_5)(NO_2)_2$ (2,4-dinitro)

14.7

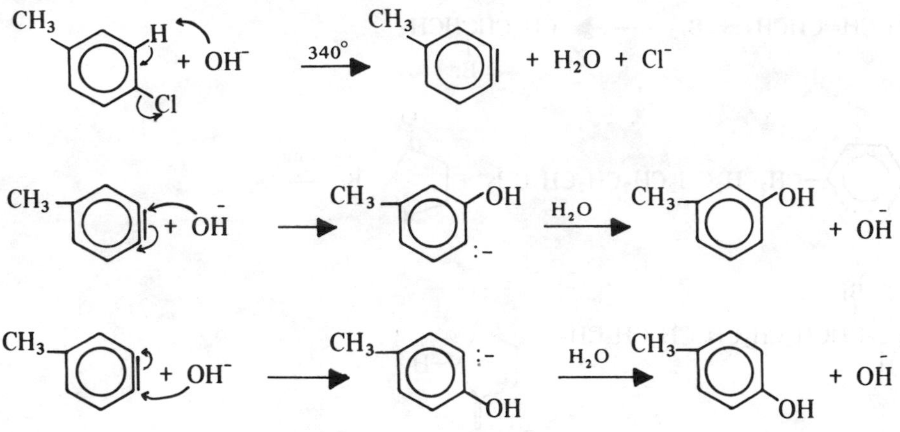

$$\text{3-chloro-toluene (}CH_3\text{, H, Cl)} + OH^- \xrightarrow{340^\circ} CH_3\text{-benzyne} + H_2O + Cl^-$$

$$CH_3\text{-benzyne} + \overset{-}{OH} \longrightarrow CH_3\text{-C}_6H_3(OH)^{:-} \xrightarrow{H_2O} CH_3\text{-C}_6H_4\text{-}OH + \overset{-}{OH}$$

$$CH_3\text{-benzyne} + OH^- \longrightarrow CH_3\text{-C}_6H_3^{:-}(OH) \xrightarrow{H_2O} CH_3\text{-C}_6H_4\text{-}OH + \overset{-}{OH}$$

14.8

Since there are no hydrogens *ortho* to the halogen, elimination cannot take place. (Reaction by a bimolecular displacement is not possible either because the substrate lacks strong electron-withdrawing groups.) Thus the absence of a reaction must be due to the inability of 2-bromo-3-methylanisole to form a benzyne intermediate.

14.9

(a) $CH_3CH_2CH_2CH{=}CH_2 + HBr \xrightarrow{\text{peroxides}} CH_3CH_2CH_2CH_2CH_2Br$

(b) $CH_3CH_2CH_2CH{=}CH_2 + HBr \xrightarrow[\text{inhibitor}]{\text{peroxide}} CH_3CH_2CH_2\underset{|}{\overset{Br}{C}}HCH_3$

(c) $CH_3CH_2\overset{CH_3}{\overset{|}{C}}HCH_3 + Br_2 \xrightarrow[\text{or heat}]{h\nu} CH_3CH_2\overset{CH_3}{\overset{|}{\underset{Br}{\underset{|}{C}}}}CH_3$ (major product)

(d) $CH_3CH_2C{\equiv}CH + HCl\,(1\text{ mole}) \longrightarrow CH_3CH_2\underset{Cl}{\underset{|}{C}}{=}CH_2$

(e) 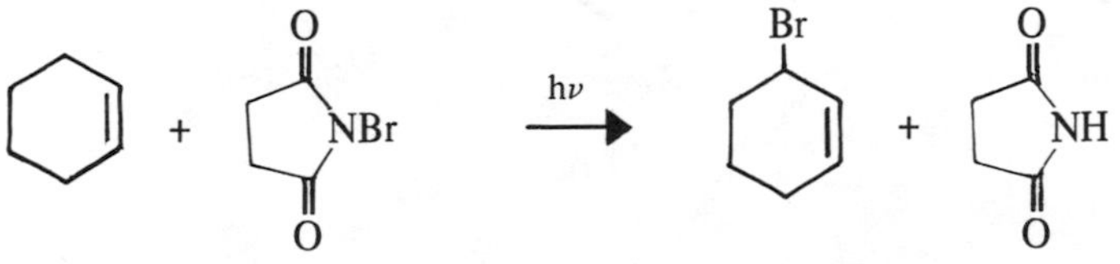 cyclohexene + N-bromosuccinimide $\xrightarrow{h\nu}$ 3-bromocyclohexene + succinimide (NH)

(f) $CH_3CH_2C{\equiv}CH + 2HBr \xrightarrow[\text{inhibitor}]{\text{peroxide}} CH_3CH_2\overset{Br}{\overset{|}{\underset{Br}{\underset{|}{C}}}}CH_3$

(g) $CH_3CH{=}CHCH_3 + Br_2 \longrightarrow CH_3CH(Br)CH(Br)CH_3$

(h) $C_6H_5{-}CH_2CH_2CH_2CH_2CH_2CH_2CH_3$ + N-bromosuccinimide $\xrightarrow{h\nu}$ $C_6H_5{-}CH(Br)CH_2CH_2CH_2CH_2CH_2CH_3$ + succinimide (N–H)

(i) $C_6H_5{-}C_7H_{15} + Br_2 \xrightarrow{FeBr_3} Br{-}C_6H_4{-}C_7H_{15} + HBr$

(j) $C_6H_5{-}CH_3 + Cl_2 \xrightarrow{h\nu} C_6H_5{-}CH_2Cl + HCl$

(k) cyclopentene + HF $\longrightarrow$ cyclopentyl–F

(l) cyclohexene + HI $\longrightarrow$ cyclohexyl–I

14.10

$$CH_3CH(CH_3)CH(OH)CH_3 + HBr \rightleftharpoons CH_3CH(CH_3)CH(\overset{+}{O}H_2)CH_3 + Br^-$$

$$\xrightarrow{-H_2O} CH_3CH(CH_3){-}\overset{+}{C}HCH_3 \longrightarrow CH_3\overset{+}{C}(CH_3){-}CH_2CH_3 \xrightarrow{Br^-} CH_3C(CH_3)(Br){-}CH_2CH_3$$

14.11

The protonated phenol $C_6H_5{-}\overset{+}{O}H_2$ would not dissociate to form the carbocation because that carbocation is highly unstable.

(Highly unstable carbocation.)

14.12

(a) No. (b) The S_N2 reaction of the Cl^- ion with the alkyl chlorosulfite would lead to inversion.

$$RR'CHOH + Cl-S(=O)-Cl \longrightarrow RR'CH-O-S(=O)-Cl + HCl$$

$$HCl + (CH_3)_3N \longrightarrow (CH_3)_3NH^+ + Cl^-$$

$$Cl^- + RR'CH-O-S(=O)-Cl \xrightarrow{\text{inversion}} Cl-CHRR' + {}^-O-S(=O)-Cl$$

$${}^-O-S(=O)-Cl \longrightarrow SO_2 + Cl^-$$

14.13

(a) $\overset{\delta-}{CH_3CH_2CH_2CH_2}:\overset{\delta+}{Li} + H:\ddot{O}H \longrightarrow CH_3CH_2CH_2CH_2-H + Li^+ :\ddot{O}H^-$

(stronger base) (stronger acid) (weaker acid) (weaker base)

(b) $\overset{\delta-}{CH_3CH_2CH_2CH_2}:\overset{\delta+}{Li} + H:\ddot{O}CH_2CH_3 \longrightarrow CH_3CH_2CH_2CH_2-H + Li^+ :\ddot{O}^-CH_2CH_3$

(stronger base) (stronger acid) (weaker acid (weaker base)

14.14

$$(CH_3)_3C-Br + Mg \xrightarrow[35^\circ]{\text{ether}} (CH_3)_3C-MgBr \xrightarrow{D_2O} (CH_3)_3C-D$$

14.15

(a) (1) $CH_3MgBr + CH_3\overset{O}{\overset{\|}{C}}CH_3 \xrightarrow[(2)H_3O^+]{} CH_3-\underset{CH_3}{\overset{CH_3}{C}}-OH$

(2) $2CH_3MgBr\ CH_3\overset{O}{\overset{\|}{C}}-OC_2H_5 \xrightarrow[(2)H_3O^+]{} CH_3-\underset{CH_3}{\overset{CH_3}{C}}-OH$

(b) (1) $CH_3MgBr + CH_3CH_2CH_2\overset{O}{\overset{\|}{C}}H \xrightarrow[(2)H_3O^+]{} CH_3CH_2CH_2\overset{OH}{\overset{|}{C}}HCH_3$

(2) $CH_3CH_2CH_2MgBr + CH_3\overset{O}{\overset{\|}{C}}H \xrightarrow[(2)H_3O^+]{} CH_3CH_2CH_2\overset{OH}{\overset{|}{C}}HCH_3$

(c) (1) $C_6H_5MgBr + CH_3\overset{O}{\overset{\|}{C}}CH_2CH_3 \xrightarrow[(2)H_3O^+]{} C_6H_5\underset{OH}{\underset{|}{\overset{CH_3}{\overset{|}{C}}}}CH_2CH_3$

(2) $CH_3MgBr + C_6H_5\overset{O}{\overset{\|}{C}}CH_2CH_3 \xrightarrow[(2)H_3O^+]{} C_6H_5\underset{OH}{\underset{|}{\overset{CH_3}{\overset{|}{C}}}}CH_2CH_3$

(3) $CH_3CH_2MgBr + C_6H_5\overset{O}{\overset{\|}{C}}CH_3 \xrightarrow[(2)H_3O^+]{} C_6H_5\underset{OH}{\underset{|}{\overset{CH_3}{\overset{|}{C}}}}CH_2CH_3$

(d) (1) $CH_3CH_2CH_2CH_2MgBr + \overset{O}{CH_2{-}CH_2} \xrightarrow[(2)H_3O^+]{} CH_3CH_2CH_2CH_2CH_2CH_2OH$

(2) $CH_3CH_2CH_2CH_2CH_2MgBr + CH_2O \xrightarrow[(2)H_3O^+]{} CH_3CH_2CH_2CH_2CH_2CH_2OH$

14.16

The acetyl group is deactivating, thus the unsubstituted ring of the intermediate, mono-acetylated ferrocene is more reactive than the substituted one.

14.17

(a) $3CH_3CH_2\underset{OH}{\underset{|}{C}}HCH_3 + PBr_3 \longrightarrow CH_3CH_2\underset{Br}{\underset{|}{C}}HCH_3 + H_3PO_3$

(b) $CH_3CH_2CH_2CH_2OH \xrightarrow{PBr_3} CH_3CH_2CH_2CH_2Br \xrightarrow{(CH_3)_3COK}$

$CH_3CH_2CH{=}CH_2 \xrightarrow[\text{(no peroxides)}]{HBr} CH_3CH_2\underset{Br}{\underset{|}{C}}HCH_3$

(c) See (b) above.

(d) $CH_3CH_2C{\equiv}CH \xrightarrow[\text{(no peroxides)}]{HBr} CH_3CH_2\underset{Br}{\underset{|}{C}}{=}CH_2 \xrightarrow[Pt]{H_2} CH_3CH_2\underset{Br}{\underset{|}{C}}HCH_3$

14.18

(a) $CH_3CH_2\underset{OH}{\underset{|}{C}}HCH_3 \xrightarrow{PBr_3} CH_3CH_2\underset{Br}{\underset{|}{C}}HCH_3 \xrightarrow{(CH_3)_3COK} CH_3CH_2CH{=}CH_2$ (+2-butenes)

$\xrightarrow[\text{peroxides}]{HBr} CH_3CH_2CH_2CH_2Br$

(b) $3CH_3CH_2CH_2CH_2OH \xrightarrow{PBr_3} CH_3CH_2CH_2CH_2Br + H_3PO_3$

(c) $CH_3CH_2CH{=}CH_2 \xrightarrow[\text{(peroxides)}]{HBr} CH_3CH_2CH_2CH_2Br$

(d) $CH_3CH_2C{\equiv}CH \xrightarrow[\text{peroxides}]{HBr} CH_3CH_2CH{=}CHBr \xrightarrow[H_2]{Pt} CH_3CH_2CH_2CH_2Br$

14.19

(a) cyclohexanol + $SOCl_2$ → chlorocyclohexane + SO_2 + HCl

(b) cyclohexene + HCl → chlorocyclohexane

(c) 1-methylcyclohexene (CH_3) + HBr $\xrightarrow[\text{peroxides}]{\text{no}}$ 1-bromo-1-methylcyclohexane (Br, CH_3)

(d) 1-methylcyclohexene (CH_3) + HBr $\xrightarrow{\text{peroxides}}$ 1-bromo-2-methylcyclohexane (CH_3, Br)

(e) 1-bromo-1-methylcyclohexane (Br, CH_3) + Mg $\xrightarrow{\text{ether}}$ (MgBr, CH_3)cyclohexane $\xrightarrow{D_2O}$ (D, CH_3)cyclohexane

14.20

(a) $(CH_3)_2CHCH_2OH$ (b) $(CH_3)_2CHCH_2CN$ (c) $(CH_3)_2CHCH_2OC(CH_3)_3$

(d) $(CH_3)_2C{=}CH_2$ (e) $(CH_3)_2CHCH_2{-}C(OH)(CH_3){-}CH_3$ (f) $(CH_3)_2CHCH_2CH(OH)CH_3$

(g) $(CH_3)_2CHCH_2C(OH)(CH_3)CH_2CH(CH_3)_2$ (h) $(CH_3)_2CHCH_2CH_2CH_2OH$

(i) $(CH_3)_2CHCH_2CH_2OH$ (j) $(CH_3)_2CHCH_3$ (k) $(CH_3)_2CHCH_3 + CH_3C{\equiv}CLi$

14.21

(a) CH_3CH_3 (b) CH_3CH_2D (c) $C_6H_5CH(OH)CH_2CH_3$

(d) $C_6H_5-C(OH)(CH_2CH_3)-C_6H_5$ (e) $C_6H_5-C(OH)(CH_2CH_3)-CH_2CH_3$ (f) $C_6H_5-C(OH)(CH_3)-CH_2CH_3$

(g) CH_3CH_3 + $CH_3CH_2C{\equiv}C-CH(OH)CH_3$ (h) CH_3CH_3 + cyclopentadienyl-MgBr

(i) $(CH_3CH_2)_2Hg$ + $2MgBrCl$ (j) $(CH_3CH_2)_2Cd$ (k) $(CH_3CH_2)_3P$

14.22

(a) $(CH_3)_2CHCH(OH)CH_2CH_2CH_3$ (b) $(CH_3)_2CHC(OH)(CH_3)CH_2CH_2CH_3$

(c) $CH_3CH_2CH_3$ + $CH_3CH_2CH_2C{\equiv}C-C(OH)(CH_3)-CH_3$ (d) $CH_3CH_2CH_3$

(e) $CH_3CH_2CH_2CH_2CH{=}CH_2$ (f) 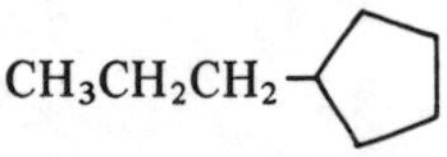 $CH_3CH_2CH_2$-cyclopentyl

(g) 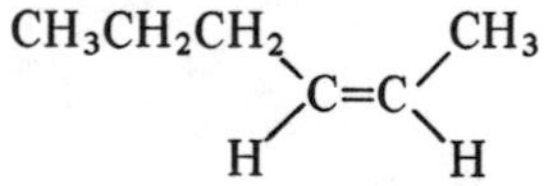 $CH_3CH_2CH_2(H)C{=}C(H)CH_3$ (h) $CH_3CH_2CH_2CH_3$

(i) $CH_3CH_2CH_2D$ (j) $(CH_3CH_2CH_2)_4Si$ (k) $(CH_3CH_2CH_2)_2Zn$

14.23

(a) (1) $CH_3CH_2MgBr + (CH_3)_2C{=}O \xrightarrow{(2)\ H_3O^+} CH_3CH_2C(CH_3)_2-OH$

(2) $CH_3MgBr + CH_3CH_2C(CH_3){=}O \xrightarrow{(2)\ H_3O^+} CH_3CH_2C(CH_3)_2-OH$

(3) $CH_3CH_2C(=O)-OCH_3 + 2CH_3MgBr \xrightarrow{(2)\ H_3O^+} CH_3CH_2C(CH_3)_2-OH$

(b) 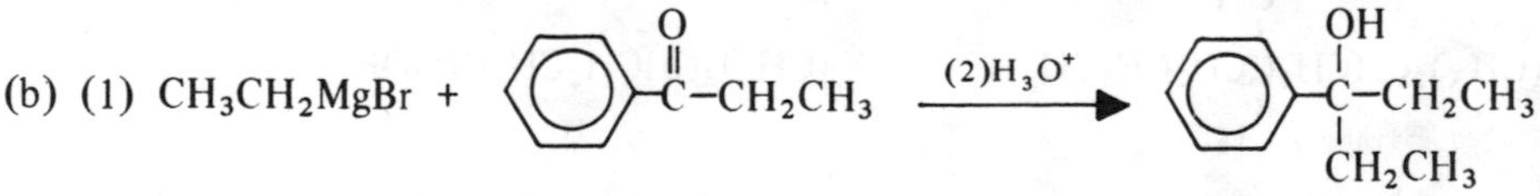 (1) CH_3CH_2MgBr + $C_6H_5C(=O)-CH_2CH_3 \xrightarrow{(2)H_3O^+} C_6H_5C(OH)(CH_2CH_3)-CH_2CH_3$

(2) 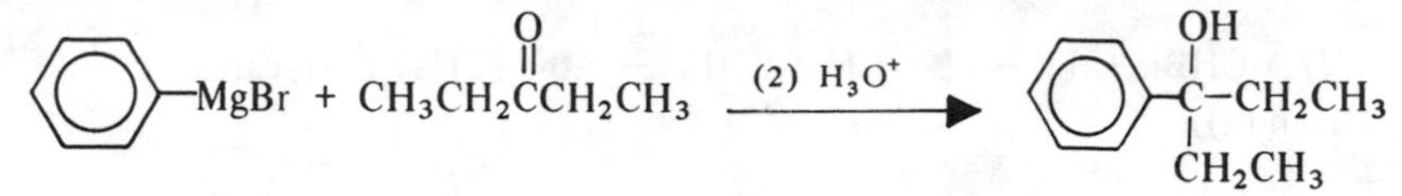 $C_6H_5MgBr + CH_3CH_2\overset{O}{\overset{\|}{C}}CH_2CH_3 \xrightarrow{(2)\ H_3O^+} C_6H_5\text{-}\underset{CH_2CH_3}{\overset{OH}{C}}\text{-}CH_2CH_3$

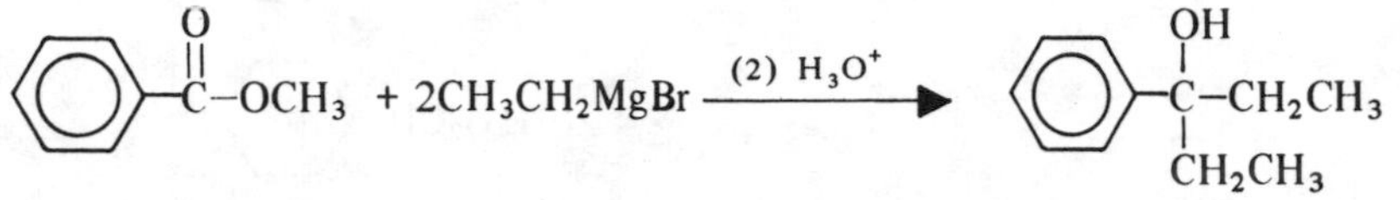

(3) $C_6H_5\overset{O}{\overset{\|}{C}}\text{-}OCH_3 + 2CH_3CH_2MgBr \xrightarrow{(2)\ H_3O^+} C_6H_5\text{-}\underset{CH_2CH_3}{\overset{OH}{C}}\text{-}CH_2CH_3$

(c) cyclohexanone $(C_6H_{10}{=}O)$ $+ C_6H_5\text{-}MgBr \xrightarrow{(2)\ H_3O^+}$ 1-phenylcyclohexanol (ring C bearing OH and C_6H_5)

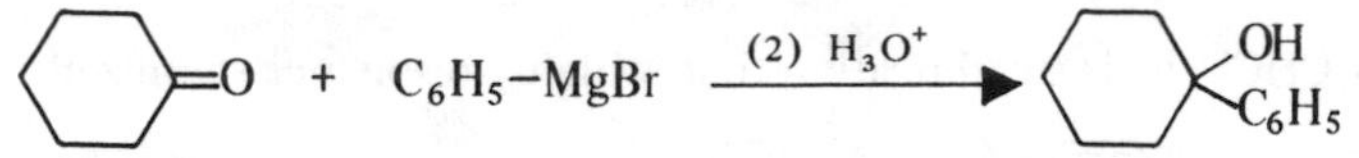

(d) 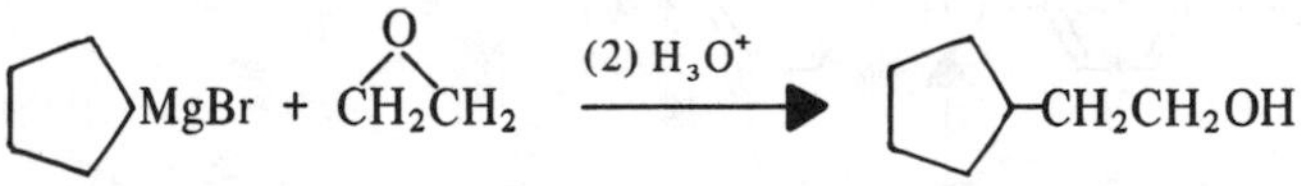 cyclopentyl-MgBr $+ \overset{O}{CH_2CH_2}$ (ethylene oxide) $\xrightarrow{(2)\ H_3O^+}$ cyclopentyl-CH_2CH_2OH

(e) (1) 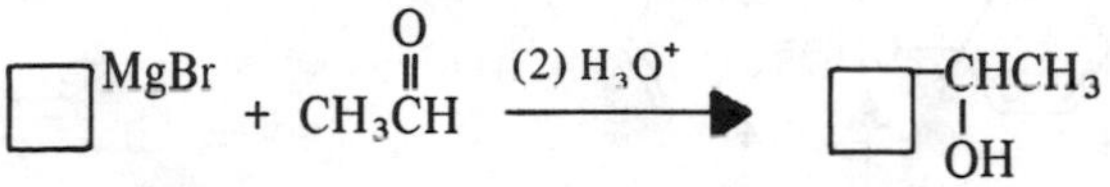 cyclobutyl-MgBr $+ CH_3\overset{O}{\overset{\|}{C}}H \xrightarrow{(2)\ H_3O^+}$ cyclobutyl-$\underset{OH}{CH}CH_3$

(2) cyclobutyl-$\overset{O}{\overset{\|}{C}}H + CH_3MgBr \xrightarrow{(2)\ H_3O^+}$ cyclobutyl-$\underset{OH}{CH}CH_3$

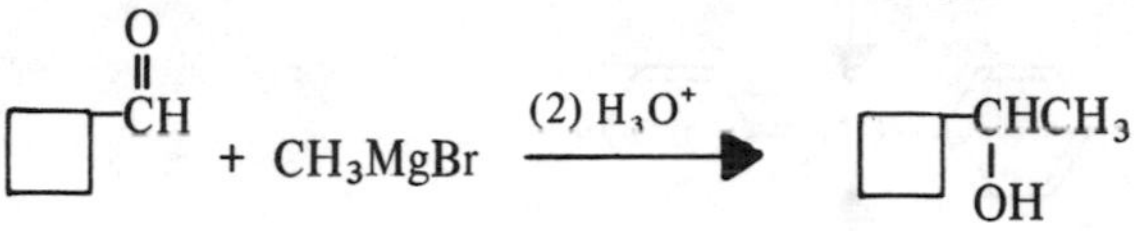

14.24

(a) $3(CH_3)_2CHOH + PBr_3 \longrightarrow (CH_3)_2CHBr + H_3PO_3$

$(CH_3)_2CHBr + Mg \xrightarrow{\text{ether}} (CH_3)_2CHMgBr$

$(CH_3)_2CHMgBr + CH_3\overset{O}{\overset{\|}{C}}H \xrightarrow{(2)\ H_3O^+} (CH_3)_2CH\overset{OH}{CH}CH_3$

(b) $(CH_3)_2CHMgBr + H\overset{O}{\overset{\|}{C}}H \xrightarrow{(2)\ H_3O^+} (CH_3)_2CHCH_2OH$
(from (a))

(c) $(CH_3)_2CHMgBr + \overset{O}{CH_2CH_2} \xrightarrow{(2)\ H_3O^+} (CH_3)_2CHCH_2CH_2OH \xrightarrow{SOCl_2} (CH_3)_2CHCH_2CH_2Cl$

(d) $(CH_3)_2CHMgBr + H\overset{O}{\overset{\|}{C}}CH(CH_3)_2 \xrightarrow{(2)\ H_3O^+} (CH_3)_2CH\overset{OH}{CH}CH(CH_3)_2$
(from (a))

(e) $(CH_3)_2CHMgBr + D_2O \longrightarrow (CH_3)_2CHD$

(f) $(CH_3)_2CHBr + Li \longrightarrow (CH_3)_2CHLi \xrightarrow{CuI} [(CH_3)_2CH)]CuLi$
(from (a))

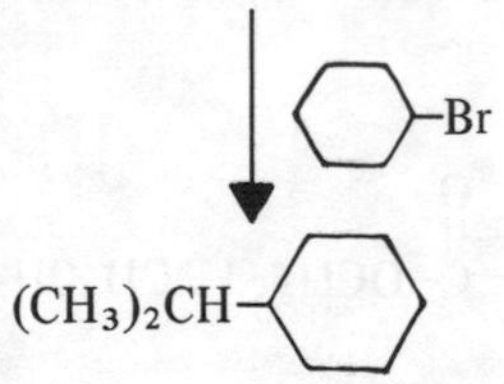

$\downarrow$ cyclohexyl–Br

$(CH_3)_2CH$–cyclohexyl

14.25

Ionization of $(C_6H_5)_2CHCl$ yields a carbocation that is highly resonance stabilized:

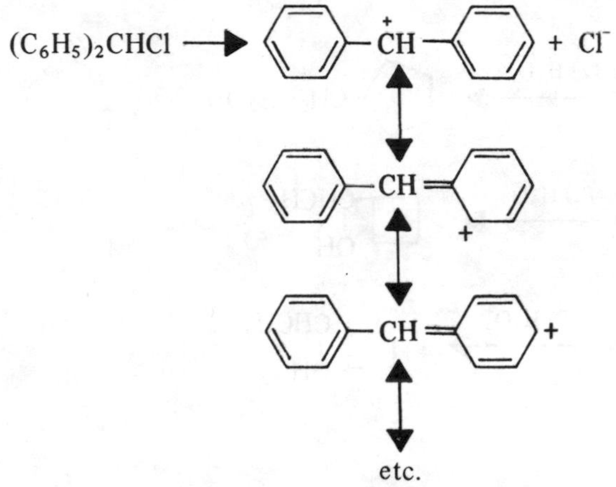

$(C_6H_5)_2CHCl \longrightarrow C_6H_5\overset{+}{C}HC_6H_5 + Cl^-$

$\updownarrow$

$C_6H_5CH=C_6H_5^+$ (positive charge on ortho carbon)

$\updownarrow$

$C_6H_5CH=C_6H_5^+$ (positive charge on para carbon)

$\updownarrow$

etc.

14.26

The Grignard reagent that forms early in the reaction reacts with the allyl halide in an S_N2 reaction:

$$RCH{=}CH\overset{\delta-}{C}H_2\!:\!\overset{\delta+}{Mg}X + RCH{=}CHCH_2{-}X \longrightarrow$$

$$RCH{=}CHCH_2{-}CH_2CH{=}CHR + MgX_2$$

14.27

(a) The carbocation stability is in the order

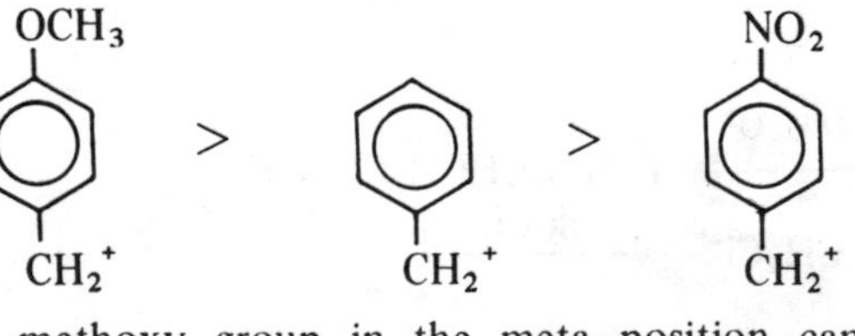

$p\text{-}CH_3OC_6H_4CH_2^+ > C_6H_5CH_2^+ > p\text{-}O_2NC_6H_4CH_2^+$

(b) The methoxy group in the meta position cannot stabilize the carbocation by resonance:

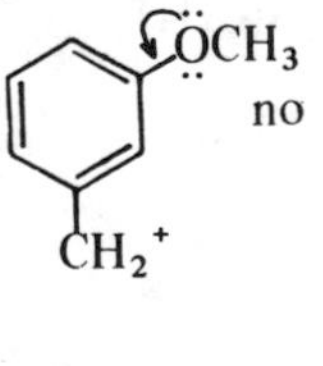

$m\text{-}CH_3\ddot{O}C_6H_4CH_2^+$ no especially stable structure is possible

The small stabilization resulting from the release of electrons into the ring through resonance is cancelled by oxygen's electron-withdrawing inductive effect.

14.28

(a) bromocyclopentane + $(CH_3)_2CuLi \xrightarrow[\text{ether}]{0°}$ methylcyclopentane + CH_3Cu + LiBr

(b) 3-bromocyclopentene + $(CH_3)_2CuLi \xrightarrow[\text{ether}]{0°}$ 3-methylcyclopentene + CH_3Cu + LiBr

(c) $CH_2{=}CH{-}CH_2Br + (CH_3CH_2)_2CuLi \xrightarrow[\text{ether}]{0°} CH_2{=}CH{-}CH_2{-}CH_2{-}CH_3 +$ CH_3CH_2Cu + LiBr

(d) $(CH_3)(H)C{=}C(CH_3)(I) + (CH_3CH_2CH_2CH_2)_2CuLi \xrightarrow[\text{ether}]{0°} (CH_3)(H)C{=}C(CH_3)(CH_2CH_2CH_2CH_3)$ + $CH_3CH_2CH_2CH_2Cu$ + LiI

14.29

(a) benzene + $CH_3COO^-\ Li^+$

(b) benzene + $CH_3O^-\ Li^+$

(c) CH_4 + $MgBrNH_2$

(d) $(CH_3)_4Si$ + 4MgBrCl

(e) $(C_6H_5)_3P$ + 3MgBrCl

(f) $(CH_3CH_2)_2Cd$ + 2MgBrCl

(g) $C_6H_5{-}CH_2OH$ + Mg^{++}

14.30

(a) Allyl bromide decolorizes Br_2/CCl_4 solution; *n*-propyl bromide does not.

(b) Benzyl bromide gives an AgBr precipitate with $AgNO_3$ in alcohol; *p*-bromotoluene does not.

(c) Benzyl chloride gives an AgCl precipitate with $AgNO_3$ in alcohol; or vinyl chloride decolorizes Br_2/CCl_4 solution.

(d) Phenyllithium (a small amount) reacts vigorously with water to give benzene and a strongly basic aqueous solution (LiOH). Diphenylmercury does not react in this way.

(e) Bromocyclohexane gives a AgBr precipitate with $AgNO_3$ in alcohol; bromobenzene does not.

14.31

A is 3-chloro-2-chloromethyl-1-propene.

$$\begin{array}{c}(b)\ H \qquad\qquad \overset{(a)}{CH_2{-}Cl} \\ C{=}C \quad (a) \\ (b)\ H \qquad\qquad CH_2{-}Cl\end{array}$$

(a) Singlet δ4.25 (4H)

(b) Singlet δ5.35 (2H)

B is 1,3-dichloro-2-butene

$$\overset{(a)}{CH_3}{-}\underset{Cl}{C}{=}\overset{(c)}{CH}{-}\overset{(b)}{CH_2Cl}$$

(a) Singlet δ2.2

(b) Doublet δ4.15

(c) Triplet δ5.7

14.32

Compound **C** is 1,3-dichlorobutane.

$$\overset{(a)}{CH_3}{-}\underset{Cl}{\overset{(b)}{CH}}{-}\overset{(c)}{CH_2}{-}\overset{(d)}{CH_2Cl}$$

(a) Doublet δ1.6

(b) Multiplet δ4.3

(c) Multiplet δ2.2 (resembles a quartet)

(d) Multiplet δ3.7 (resembles a triplet)

Note: the protons labeled (c) are actually diastereotopic, but their chemical shifts are approximately the same. Coupling constants J_{bc} and J_{cd} are also approximately the same, and thus the spectrum is fortuitously simple.

14.33

D is 3-chloro-2-methylpropene.

$$\begin{array}{c}(a)\ H \qquad\qquad \overset{(c)}{CH_2{-}Cl} \\ C{=}C \\ (b)\ H \qquad\qquad \underset{(d)}{CH_3}\end{array}$$

(a) Singlet, δ5.1

(b) Singlet, δ4.9

(c) Singlet, δ4.0

(d) Singlet, δ1.9

14.34

E is 1,3-butadiene. **F** and **G** are the products of the 1,2- and 1,4-addition of chlorine. **H** is 1,2,3,4-tetrachlorobutane.

$$CH_2{=}CH{-}CH{=}CH_2 \ (\mathbf{E}) \xrightarrow{Cl_2} CH_2{=}CH{-}\underset{\underset{\displaystyle Cl}{|}}{C}HCH_2Cl \ (\mathbf{F}) + ClCH_2CH{=}CHCH_2Cl \ (\mathbf{G})$$

(or vice versa)

$$\mathbf{F} \text{ or } \mathbf{G} \xrightarrow{Cl_2} \overset{(a)}{ClCH_2}{-}\overset{(b)}{\underset{\underset{\displaystyle Cl}{|}}{C}H}{-}\overset{(b)}{\underset{\underset{\displaystyle Cl}{|}}{C}H}{-}\overset{(a)}{CH_2Cl} \ (\mathbf{H})$$

(a) Doublet δ 3.8

(b) Triplet δ 4.6

Special Topic

I.1

$$\mathrm{Cl_3C{-}\overset{\overset{\displaystyle O}{\|}}{C}H} + \mathrm{H_2SO_4} \rightleftharpoons \left[\mathrm{Cl_3C{-}\overset{\overset{\displaystyle \overset{+}{O}H}{\|}}{C}H} \longleftrightarrow \mathrm{Cl_3C{-}\overset{\overset{\displaystyle OH}{|}}{\underset{+}{C}}H}\right] + \mathrm{HSO_4^-}$$

$$\mathrm{Cl_3C{-}\overset{\overset{\displaystyle OH}{|}}{\underset{+}{C}}H} + \mathrm{C_6H_5Cl} \longrightarrow \mathrm{Cl_3C{-}\overset{\overset{\displaystyle OH}{|}}{\underset{\underset{\displaystyle H}{|}}{C}}{-}(H)C_6H_4^{+}Cl} \xrightarrow{-H^+} \mathrm{Cl_3C{-}\overset{\overset{\displaystyle OH}{|}}{\underset{\underset{\displaystyle H}{|}}{C}}{-}C_6H_4Cl}$$

$$\mathrm{Cl_3C{-}\overset{\overset{\displaystyle OH}{|}}{C}H{-}C_6H_4Cl} + \mathrm{H_2SO_4} \rightleftharpoons \mathrm{Cl_3C{-}\overset{\overset{\displaystyle \overset{+}{O}H_2}{|}}{C}H{-}C_6H_4Cl} + \mathrm{HSO_4^-} \rightleftharpoons \mathrm{Cl_3C{-}\overset{+}{C}H{-}C_6H_4Cl} + \mathrm{H_2O}$$

$$\mathrm{C_6H_5Cl} + \mathrm{\underset{\underset{\displaystyle CCl_3}{|}}{\overset{+}{C}H}{-}C_6H_4{-}Cl} \longrightarrow \mathrm{Cl{-}\overset{+}{C_6H_4}(H){-}\underset{\underset{\displaystyle CCl_3}{|}}{C}H{-}C_6H_4{-}Cl} \xrightarrow{-H^+}$$

$$\mathrm{Cl{-}C_6H_4{-}\underset{\underset{\displaystyle CCl_3}{|}}{C}H{-}C_6H_4{-}Cl}$$

I.2

An elimination reaction.

I.3

(a)

$$\text{1,2,4,5-}\mathrm{C_6H_2Cl_4} + \mathrm{OH^-} \rightleftharpoons \left[\mathrm{\overset{-}{C_6H_2}Cl_3(Cl)(OH)} \longleftrightarrow \mathrm{\overset{-}{C_6H_2}Cl_3(Cl)(OH)} \longleftrightarrow \text{etc.}\right]$$

$$\text{2,4,5-trichloro intermediate} \longrightarrow \text{2,4,5-trichlorophenol} \xrightarrow{OH^-} \text{2,4,5-trichlorophenoxide} + H_2O$$

(b) 2,4,5-trichlorophenoxide + 2,4,5-trichlorophenoxide ⟶ intermediate ⟶ 2,3,7,8-tetrachlorodibenzo-p-dioxin

I.4

An S_N2 reaction:

$$\text{2,4,5-}Cl_3C_6H_2O^- + ClCH_2CO_2^- \longrightarrow \text{2,4,5-}Cl_3C_6H_2O{-}CH_2\overset{O}{\overset{\|}{C}}{-}O^- + Cl^-$$

CHAPTER FIFTEEN
Alcohols, Phenols, and Ethers

15.1

(a) Alcohols: $CH_2{=}CHCH_2OH$
2-Propen-l-ol
(Allyl alcohol)

$$\begin{array}{l} CH_2-CH-OH \\ \;\;\diagdown\;\diagup \\ \;\;\;CH_2 \end{array}$$

cyclopropanol

Ethers: $CH_2{=}CH{-}O{-}CH_3$
Methoxyethene
(Methyl vinyl ether)

(b) Alcohols: $CH_2{=}CHCH_2CH_2OH$
3-buten-l-ol

$$\begin{array}{l} CH_2{=}CH\underset{\displaystyle OH}{\underset{|}{C}}HCH_3 \end{array}$$

3-buten-2-ol

$CH_3CH{=}CHCH_2OH$
2-buten-1-ol (cis and trans)

$$CH_2{=}\overset{\displaystyle CH_3}{\overset{|}{C}}{-}CH_2OH$$

2-methyl-2-propen-1-ol

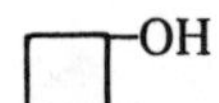

cyclobutanol

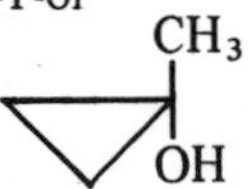

1-methylcyclopropanol

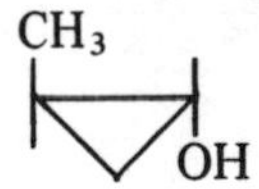

trans-2-methylcyclopropanol

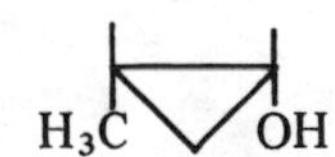

cis-2-methylcyclopropanol

Ethers: $CH_3CH{=}CH{-}OCH_3$ 1-methoxypropene

$CH_2{=}CHCH_2OCH_3$ 3-methoxypropene

$$\begin{array}{l} \quad\;\; CH_3 \\ \quad\;\;\; | \\ CH_2{=}C{-}OCH_3 \end{array}$$
2-methoxypropene

$CH_2{=}CHOCH_2CH_3$ ethoxyethene

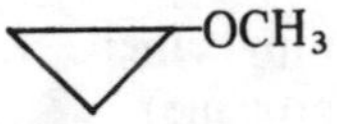

$$\bigtriangledown{-}OCH_3$$
methoxycyclopropane

(c) Alcohols:

$CH_3CH_2CH_2CH_2CH_2OH$ 1-pentanol

$$\begin{array}{l} CH_3CH_2CH_2CHCH_3 \\ \qquad\qquad\qquad\;\; | \\ \qquad\qquad\qquad OH \end{array}$$
2-pentanol

$$\begin{array}{l} CH_3CH_2CHCH_2CH_3 \\ \qquad\qquad\;\; | \\ \qquad\qquad OH \end{array}$$
3-pentanol

$$\begin{array}{l} \qquad\qquad CH_3 \\ \qquad\qquad\;\; | \\ CH_3CH_2CHCH_2OH \end{array}$$
2-methyl-1-butanol

$$\begin{array}{l} \qquad CH_3 \\ \qquad\;\; | \\ CH_3CHCH_2CH_2OH \end{array}$$
3-methyl-1-butanol

$$\begin{array}{l} \qquad\qquad CH_3 \\ \qquad\qquad\;\; | \\ CH_3CH_2CCH_3 \\ \qquad\qquad\;\; | \\ \qquad\qquad OH \end{array}$$
2-methyl-2-butanol

$$\begin{array}{l} \qquad CH_3 \\ \qquad\;\; | \\ CH_3CHCHCH_3 \\ \qquad\qquad\; | \\ \qquad\qquad OH \end{array}$$
3-methyl-2-butanol

$$\begin{array}{l} \qquad\;\; CH_3 \\ \qquad\quad | \\ CH_3{-}C{-}CH_2OH \\ \qquad\quad | \\ \qquad\;\; CH_3 \end{array}$$
2,2-dimethyl-1-propanol (neopentyl alcohol)

Ethers:

$CH_3CH_2CH_2CH_2{-}O{-}CH_3$ methyl *n*-butyl ether (1-methoxybutane)

$$\begin{array}{l} \qquad CH_3 \\ \qquad\;\; | \\ CH_3CHCH_2{-}O{-}CH_3 \end{array}$$
methyl isobutyl ether (2-methyl-1-methoxypropane)

$$\begin{array}{l} \qquad\;\; CH_3 \\ \qquad\quad | \\ CH_3{-}C{-}O{-}CH_3 \\ \qquad\quad | \\ \qquad\;\; CH_3 \end{array}$$
methyl *tert*-butyl ether (2-methyl-2-methoxypropane)

$$CH_3CH_2\underset{}{\overset{CH_3}{\overset{|}{C}H}}{-}O{-}CH_3$$ methyl *sec*-butyl ether (2-methoxybutane)

$$CH_3CH_2CH_2{-}O{-}CH_2CH_3$$ ethyl *n*-propyl ether (1-ethoxypropane)

$$CH_3\overset{CH_3}{\overset{|}{C}H}{-}O{-}CH_2CH_3$$ ethyl isopropyl ether (2-ethoxypropane)

15.2

The two hydroxyl groups in ethylene glycol allow the formation of more hydrogen bonds than in the monohydroxy alcohols. Thus a single diol molecule can be associated with many neighboring diol molecules.

15.3

(a) $$CH_3\overset{CH_3}{\overset{|}{C}}{=}CH_2 + H_2O \xrightarrow{H^+} CH_3\underset{OH}{\underset{|}{\overset{CH_3}{\overset{|}{C}}}}CH_3$$

(b) $$CH_3CH_2CH_2CH_2CH{=}CH_2 + H_2O \xrightarrow{H^+} CH_3CH_2CH_2CH_2\overset{OH}{\overset{|}{C}H}CH_3$$

(c) cyclopentene $+ H_2O \xrightarrow{H^+}$ cyclopentanol (cyclopentane ring with OH)

(d) 1-methylcyclohexene (CH_3 on ring) $+ H_2O \xrightarrow{H^+}$ 1-methylcyclohexanol (ring carbon bearing CH_3 and OH)

15.4

Rearrangement of the secondary carbocation to the more stable tertiary carbocation,

$$CH_3\underset{CH_3}{\underset{|}{\overset{CH_3}{\overset{|}{C}}}}{-}CH{=}CH_2 \underset{}{\overset{H^+}{\rightleftharpoons}} CH_3{-}\underset{CH_3}{\underset{|}{\overset{CH_3}{\overset{|}{C}}}}{-}\overset{+}{C}H{-}CH_3 \longrightarrow CH_3{-}\underset{CH_3}{\underset{|}{\overset{CH_3}{\overset{|}{\overset{+}{C}}}}}{-}CH{-}CH_3$$

followed by reaction of the resulting carbocation with water:

$$CH_3{-}\underset{CH_3}{\underset{|}{\overset{+}{C}}}{-}\overset{CH_3}{\overset{|}{C}H}CH_3 + H_2O \rightleftharpoons CH_3{-}\underset{CH_3}{\underset{|}{\overset{^+OH_2}{\overset{|}{C}}}}{-}\underset{CH_3}{\underset{|}{C}HCH_3} \rightleftharpoons CH_3{-}\underset{CH_3}{\underset{|}{\overset{OH}{\overset{|}{C}}}}{-}\underset{CH_3}{\underset{|}{C}HCH_3} + H^+$$

15.5

(a) $CH_3CH_2CH_2CH_2CH{=}CH_2 \xrightarrow[\text{THF-}H_2O]{Hg(OAc)_2} CH_3CH_2CH_2CH_2\underset{OH}{CH}CH_2{-}HgOAc \xrightarrow[OH^-]{NaBH_4} CH_3CH_2CH_2CH_2\overset{OH}{CH}CH_3$

(b) 1-methylcyclohexene $\xrightarrow[(2)\ NaBH_4,\ OH^-]{(1)\ Hg(OAc)_2,\ THF{-}H_2O}$ 1-methylcyclohexanol (CH_3, OH on the same ring carbon)

(c) $CH_3\underset{CH_3}{\overset{CH_3}{C}}CH_2\overset{CH_3}{C}{=}CH_2 \xrightarrow{\text{same as (b)}} CH_3\underset{CH_3}{\overset{CH_3}{C}}CH_2\underset{OH}{\overset{CH_3}{C}}CH_3$

(d) $CH_3\underset{C_6H_5}{C}{=}CH_2 \xrightarrow{\text{same as (b)}} CH_3\underset{C_6H_5}{\overset{OH}{C}}CH_3$

15.6

(a) $CH_3{-}\underset{CH_3}{\overset{CH_3}{C}}{-}CH{=}CH_2 + (BH_3)_2 \longrightarrow (CH_3{-}\underset{CH_3}{\overset{CH_3}{C}}{-}CH_2CH_2)_3B \xrightarrow[OH^-,H_2O]{H_2O_2} CH_3{-}\underset{CH_3}{\overset{CH_3}{C}}{-}CH_2CH_2OH$

(b) $CH_3CH_2CH_2CH_2CH{=}CH_2 \xrightarrow[(2) H_2O_2,\ OH^-,\ H_2O]{(1)(BH_3)_2} CH_3CH_2CH_2CH_2CH_2CH_2OH$

(c) $C_6H_5{-}CH{=}CH_2 \xrightarrow{\text{same as (b)}} C_6H_5{-}CH_2CH_2OH$

(d) 1-methylcyclohexene (CH_3) $\Longrightarrow$ same as (b) $\Longrightarrow$ trans-2-methylcyclohexanol (H, CH_3, OH, H) + enantiomer

15.7

(a) H–C(H)(H)–O–H

3H = 3(−1)
1O = +1
total = −2 = oxidation state of C

H–C(=O)–O–H

1H = −1
3O = +3
total = +2 = oxidation state of C

H–C(=O)–H

2H = −2
2O = +2
total = 0 = oxidation state of C

(b) CH_4, CH_3OH, HC(=O)H, HC(=O)OH, CO_2
−4, −2, 0, +2, +4

(c) a change from −2 to 0

(d) an oxidation, since the oxidation state increases

(e) a reduction from +6 to +3

(f) Methanol undergoes an increase in oxidation state of 2 and chromium undergoes a decrease of 3. Since the amount of oxidation must equal the amount of reduction, two moles of H_2CrO_4 will be required to oxidize three moles of CH_3OH, and thus two-thirds of a mole of H_2CrO_4 will be required for one mole of CH_3OH.

15.8

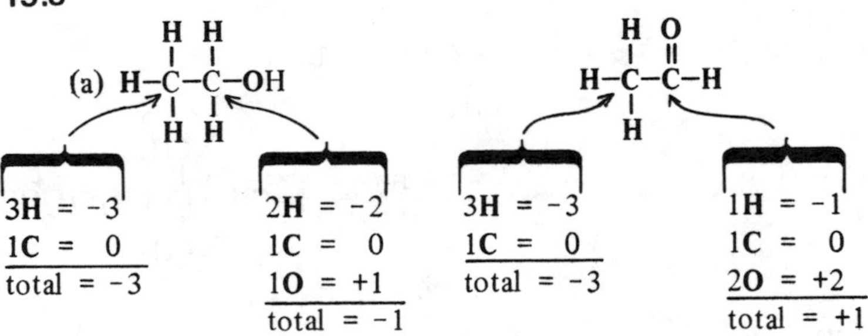

(b) Only the carbon of the $-CH_2OH$ group of ethanol undergoes a change in oxidation state. The oxidation state of the CH_3- group remains unchanged.

(c) H–C(H)(H)–C(=O)–OH

3H = −3	1C = 0
1C = 0	3O = +3
total = −3	total = +3

(d) Each atom of silver undergoes a decrease in oxidation state of 1 when $Ag_2O \longrightarrow 2\,Ag$ and each carbon undergoes an increase of 2 when $CH_3CHO \longrightarrow CH_3COOH$. Thus one mole of Ag_2O will be required to oxidize one mole of acetaldehyde.

15.9

(a) If we consider the hydrogenation of ethene as an example, we find that the oxidation state of carbon decreases. Thus, because the

$$H_2C=CH_2 + H_2 \xrightarrow{Ni} H_3C-CH_3$$

2H = −2	3H = −3
2C = 0	1C = 0
total = −2	total = −3

reaction involves the *addition* of *hydrogen*, it is both an *addition reaction* and a *reduction.*

(b) The hydrogenation of acetaldehyde is not only an addition reaction, it is also a *reduction* because the carbon atom of the $>C=O$ group goes from a + 1 to a − 1 oxidation state. The reverse reaction (the *dehydrogenation* of ethyl alcohol) is not only an *elimination* reaction, it is also an *oxidation.*

15.10

(a) $LiAlH_4$ (b) $NaBH_4$ (c) $LiAlH_4$

15.11

(a) $CH_3C\equiv\bar{C}: + CH_3CH_2OH \rightleftharpoons CH_3C\equiv CH + CH_3CH_2O^-$

stronger base	stronger acid	weaker acid	weaker base

(b) $CH_3CH_2CH_2\overset{\delta-}{CH_2}:\overset{\delta+}{Li} + CH_3CH_2OH \rightleftharpoons CH_3CH_2CH_2CH_3 + CH_3CH_2\bar{O}\,Li^+$

stronger base	stronger acid	weaker acid	weaker base

(c) $CH_3\overset{\delta-}{CH_2}:\overset{\delta+}{MgBr} + CH_3CH_2OH \rightleftharpoons CH_3CH_3 + CH_3CH_2\bar{O}\overset{+}{M}gBr$

stronger base	stronger acid	weaker acid	weaker base

15.12

(a) $C_6H_6 + SO_3 \xrightarrow{H_2SO_4} C_6H_5SO_3H \xrightarrow{PCl_5} C_6H_5SO_2Cl \xrightarrow{CH_3OH} C_6H_5SO_2-OCH_3$

(b) 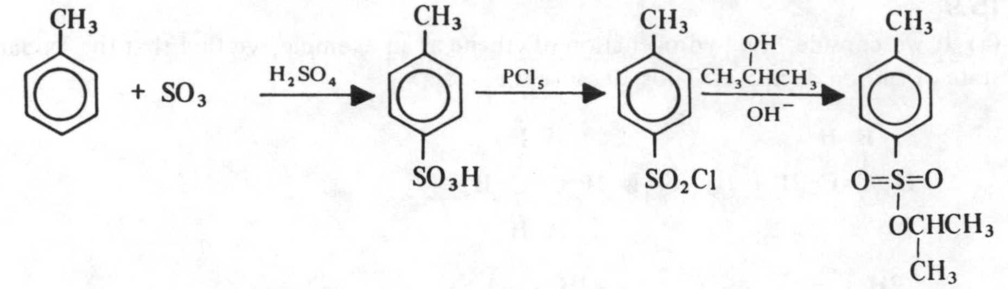 toluene + SO_3 $\xrightarrow{H_2SO_4}$ p-$CH_3C_6H_4SO_3H$ $\xrightarrow{PCl_5}$ p-$CH_3C_6H_4SO_2Cl$ $\xrightarrow[OH^-]{CH_3CH(OH)CH_3}$ p-$CH_3C_6H_4S(=O)_2OCH(CH_3)_2$

(c) 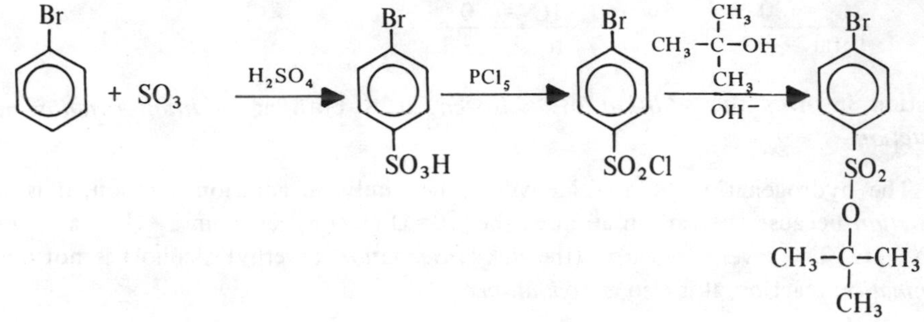 bromobenzene + SO_3 $\xrightarrow{H_2SO_4}$ p-$BrC_6H_4SO_3H$ $\xrightarrow{PCl_5}$ p-$BrC_6H_4SO_2Cl$ $\xrightarrow[OH^-]{(CH_3)_3C-OH}$ p-$BrC_6H_4SO_2-O-C(CH_3)_3$

15.13

(a) 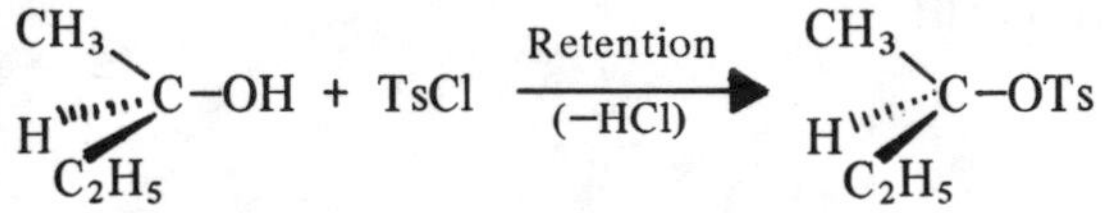 $CH_3(H)(C_2H_5)C-OH$ + TsCl $\xrightarrow[(-HCl)]{\text{Retention}}$ $CH_3(H)(C_2H_5)C-OTs$

R-2-butanol

(b) 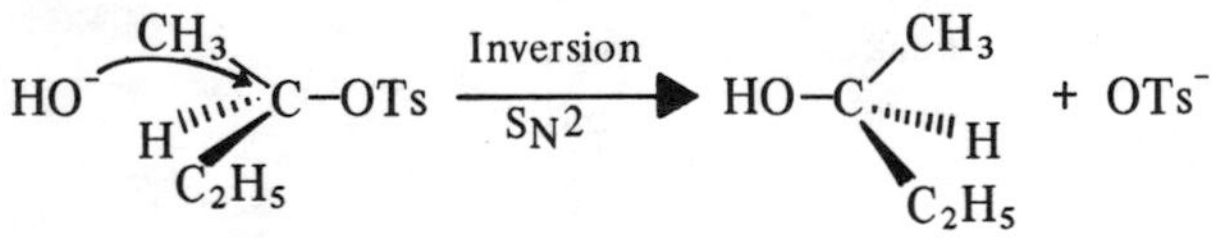 HO^- + $CH_3(H)(C_2H_5)C-OTs$ $\xrightarrow[S_N2]{\text{Inversion}}$ $HO-C(CH_3)(H)(C_2H_5)$ + OTs^-

(c) cyclohexane ring with H_3C and OH $\xrightarrow[\text{Retention}]{TsCl}$ ring with H_3C and OTs $\xrightarrow[\text{Inversion}]{Cl^-}$ ring with H_3C and Cl

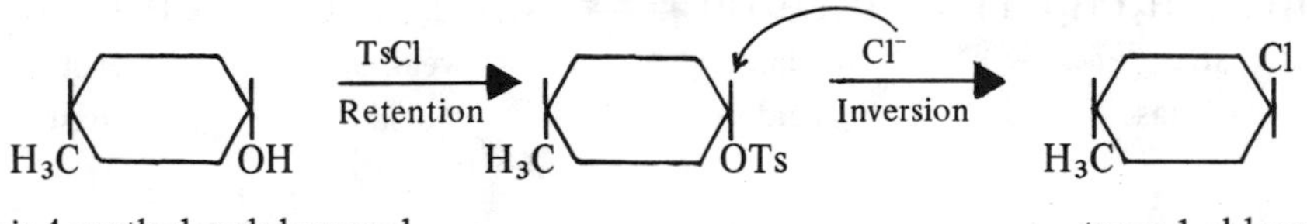

cis-4-methylcyclohexanol — *trans*-1-chloro-4-methylcyclohexane

15.14

(a) *Trans*-2-pentene because it is more stable.

(b) A 1-phenylpropene is conjugated and thus is more stable than the unconjugated 3-phenylpropene, and *trans*-1-phenylpropene is more stable than the *cis* isomer.

15.15

(a) 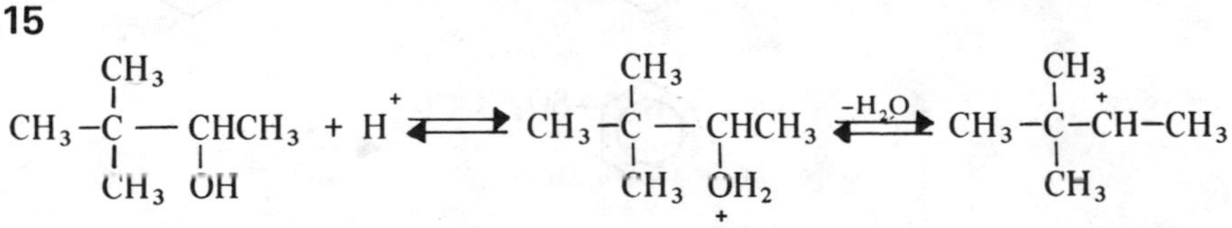 $CH_3-C(CH_3)_2-CH(OH)CH_3 + H^+ \rightleftarrows CH_3-C(CH_3)_2-CH(\overset{+}{O}H_2)CH_3 \overset{-H_2O}{\rightleftarrows} CH_3-C(CH_3)_2-\overset{+}{C}H-CH_3$

$$CH_3-\overset{\overset{CH_3}{|}}{\underset{\underset{CH_3}{|}}{\overset{+}{C}}}-CH-CH_2 \xrightleftharpoons{H_2O} CH_3-\overset{\overset{CH_3}{|}}{\underset{\underset{CH_3}{|}}{C}}-CH=CH_2 + H_3\overset{+}{O}$$

(H on CH; product I)

But this secondary carbocation can also rearrange to a tertiary carbocation before losing a proton:

$$CH_3-\overset{\overset{CH_3}{|}}{\underset{\underset{CH_3}{|}}{C}}-\overset{+}{C}H-CH_3 \longrightarrow \overset{H}{\overset{|}{C}H_2}-\overset{\overset{CH_3}{|}}{\underset{+}{C}}-\overset{\overset{H}{|}}{\underset{\underset{CH_3}{|}}{C}}-CH_3$$

(b) H_2O ↙ ; (a) H_2O ↘

$$CH_2=\overset{\overset{CH_3}{|}}{C}-\underset{\underset{CH_3}{|}}{C}H-CH_3 + H_3\overset{+}{O} \quad \text{II}$$

$$CH_3-\overset{\overset{CH_3}{|}}{C}=\underset{\underset{CH_3}{|}}{C}-CH_3 + H_3\overset{+}{O} \quad \text{III}$$

(b) 2,3-Dimethyl-2-butene (III) is the most substituted alkene, therefore it is most stable.

15.16

(a)
$$CH_3-\overset{\overset{CH_3}{|}}{\underset{\underset{CH_3}{|}}{C}}-OH \xrightarrow{H^+} CH_3-\overset{\overset{CH_3}{|}}{\underset{\underset{CH_3}{|}}{C}}-\overset{+}{O}H_2 \xrightarrow{-H_2O}$$

$$CH_3-\overset{\overset{CH_3}{|}}{\underset{\underset{CH_3}{|}}{C^+}} \xrightarrow[(1^\circ \text{ or } 2^\circ)]{R-OH} CH_3-\overset{\overset{CH_3}{|}}{\underset{\underset{CH_3}{|}}{C}}-\underset{\underset{H}{|}}{\overset{+}{O}}-R \xrightarrow{-H^+} CH_3-\overset{\overset{CH_3}{|}}{\underset{\underset{CH_3}{|}}{C}}-O-R$$

(b)
$$CH_2=\overset{\overset{CH_3}{|}}{\underset{\underset{CH_3}{|}}{C}} \xrightarrow{H^+} CH_3-\overset{\overset{CH_3}{|}}{\underset{\underset{CH_3}{|}}{C^+}} \xrightarrow[(1^\circ \text{ or } 2^\circ)]{R-OH} CH_3-\overset{\overset{CH_3}{|}}{\underset{\underset{CH_3}{|}}{C}}-\underset{\underset{H}{|}}{\overset{+}{O}}-R$$

$$\xrightarrow{-H^+} CH_3-\overset{\overset{CH_3}{|}}{\underset{\underset{CH_3}{|}}{C}}-O-R$$

15.17

(a) (1)
$$CH_3\overset{\overset{CH_3}{|}}{C}HO^-Na^+ + CH_3-L \longrightarrow CH_3\overset{\overset{CH_3}{|}}{C}HO-CH_3 + L^- + Na^+$$

(L = X, OSO_2R, or OSO_2OR)

(2)
$$CH_3O^- + CH_3-\overset{\overset{CH_3}{|}}{C}H-L \longrightarrow CH_3O-\overset{\overset{CH_3}{|}}{C}HCH_3 + L^-$$

(L = X, OSO_2R, or OSO_2OR)

(b) Both methods involve S_N2 reactions. Therefore, method (1) is better because substitution takes place at an unhindered methyl carbon. In method (2) where substitution must take place at a relatively hindered secondary carbon the reaction would be accompanied by considerable elimination.

15.18

Reaction of the alcohol with K and then of the resulting salt with C_2H_5Br does not break bonds to the chiral carbon, and these reactions therefore occur with retention.
Reaction of the tosylate, $C_6H_5CH_2\underset{\displaystyle OTs}{\underset{|}{C}}HCH_3$, with C_2H_5OH in K_2CO_3 solution, however, is an S_N2 reaction at the chiral carbon and thus it occurs with inversion.

15.19

$$Cl{-}CH_2CH_2CH_2CH_2OH \xrightleftharpoons{OH^-} Cl{-}CH_2CH_2CH_2CH_2{-}\ddot{O}{:}^- + H_2O \longrightarrow \text{(tetrahydrofuran ring: } CH_2{-}CH_2{-}CH_2{-}CH_2{-}O\text{)} + Cl^-$$

15.20

(a) $HO^- + HOCH_2{-}CH_2{-}Cl \rightleftharpoons H_2O + {}^-O{-}CH_2{-}CH_2{-}Cl \longrightarrow \overset{O}{CH_2{-}CH_2}$ (epoxide)

(b) The $-\ddot{O}{:}^-$ group must displace the $\overline{Cl}$ from the back side,

trans-2-chlorocyclohexanol $\xrightleftharpoons{OH^-}$ (alkoxide, O^- attacking C–Cl from back side) $\rightleftharpoons$ epoxide

trans-2-chloro-
cyclohexanol

Backside attack is not possible with the *cis*-isomer therefore it does not form an epoxide.

15.21

In strong acid, ethers are protonated:

$$R{-}O{-}R + HA \rightleftharpoons R{-}\overset{+}{\underset{H}{O}}{-}R + \bar{A}$$

In weak acid or neutral media, protonation does not occur to any appreciable extent. Thus, in strong acid, the leaving group is an alcohol molecule,

$$\bar{X} + R{-}\overset{+}{\underset{H}{O}}{-}R \longrightarrow X{-}R + HO{-}R$$

whereas in weak acid or neutral media, the leaving group would have to be an alkoxide ion,

$$X^- + R{-}O{-}R \longrightarrow X{-}R + {}^-O{-}R$$

R–OH is a good leaving group; $R{-}O^-$ is a very poor one.

15.22

(a) $CH_2{-}CH_2$ (epoxide, O bridging) $\xrightarrow{H^+}$ $CH_2{-}CH_2$ (protonated epoxide, $\overset{+}{O}{-}H$) $\xrightarrow{CH_3OH}$ $CH_2(\overset{+}{O}(H)CH_3){-}CH_2OH$ $\longrightarrow$ $HOCH_2CH_2OCH_3$ Methyl cellosolve

(b) An analogous reaction yields ethyl cellosolve, $HOCH_2CH_2OCH_2CH_3$

(c) $CH_2{-}CH_2$ (epoxide) $\xrightarrow{I^-}$ $ICH_2CH_2O^-$ $\xrightarrow{H_2O}$ $HOCH_2CH_2I + OH^-$

(d) $CH_2{-}CH_2$ (epoxide) $\xrightarrow{:NH_3}$ $\overset{+}{N}H_3CH_2CH_2O^-$ $\longrightarrow$ $HOCH_2CH_2NH_2$

(e) $CH_2{-}CH_2$ (epoxide) $\xrightarrow{CH_3O^-}$ $CH_3OCH_2CH_2O^-$ $\xrightarrow{CH_3OH}$ $HOCH_2CH_2OCH_3 + CH_3O^-$

15.23

The reaction is an S_N2 reaction and thus nucleophilic attack takes place much more rapidly at the primary carbon than at the more hindered secondary carbon.

$CH_3CH{-}CH_2$ (epoxide) $+ C_2H_5O^-$ $\xrightarrow[C_2H_5OH]{fast}$ $CH_3CH(OH)CH_2OC_2H_5$ Major product

$CH_3CH{-}CH_2$ (epoxide) $+ C_2H_5O^-$ $\xrightarrow[C_2H_5OH]{slow}$ $CH_3CH(OC_2H_5)CH_2OH$ Minor product

15.24

Ethoxide ion attacks the epoxide ring at the primary carbon because it is less hindered and the following reactions take place.

$Cl{-}CH_2{-}CH{-}\overset{*}{C}H_2$ (epoxide) $+ {}^-OC_2H_5 \longrightarrow Cl{-}CH_2{-}CH(O^-){-}\overset{*}{C}H_2OC_2H_5 \longrightarrow$

$CH_2{-}CH{-}\overset{*}{C}H_2OC_2H_5$ (epoxide, O bridging CH_2 and CH)

15.25

In structures **2 - 4** the carbon-oxygen bond is a double bond. Thus we would expect the carbon-oxygen bond of a phenol to be much stronger than that of an alcohol.

15.26

An electron-releasing group (i.e., $-CH_3$) destabilizes the phenoxide anion by intensifying its negative charge. This makes the substituted phenol less acidic than phenol itself.

$$\text{p-}CH_3C_6H_4OH + H_2O \rightleftharpoons \text{p-}CH_3C_6H_4O^- + H_3O^+$$

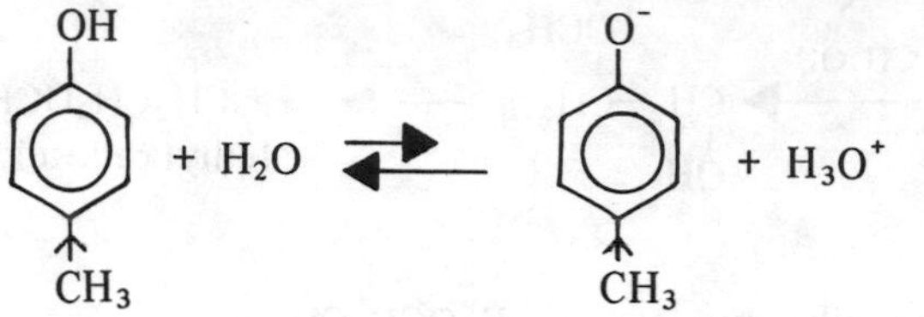

Electron-releasing $-CH_3$ destabilizes the anion more than the acid – Ka is smaller than for phenol.

An electron-withdrawing group such as chlorine can stabilize the phenoxide by dispersing its negative charge through an inductive effect. This makes the substituted phenol more acidic than phenol itself.

$$\text{p-}ClC_6H_4OH + H_2O \rightleftharpoons \text{p-}Cl^{\delta-}C_6H_4O^- + H_3O^+$$

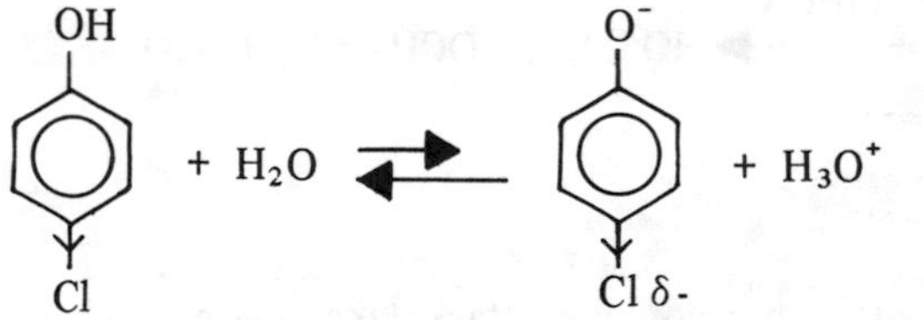

Electron-withdrawing chlorine stabilizes the anion by dispersing its negative charge. Ka is larger than for phenol.

Nitro groups are electron-withdrawing by their inductive and resonance effects. The resonance effect is especially important in stabilizing the phenoxide anion. In the 2,4,6-trinitrophenoxide anion, for example, structures, **B**, **C**, and **D** contribute to the resonance hybrid and stabilize it by dispersing the negative charge. This explains why 2,4,6-trinitrophenol (picric acid) is so exceptionally acidic.

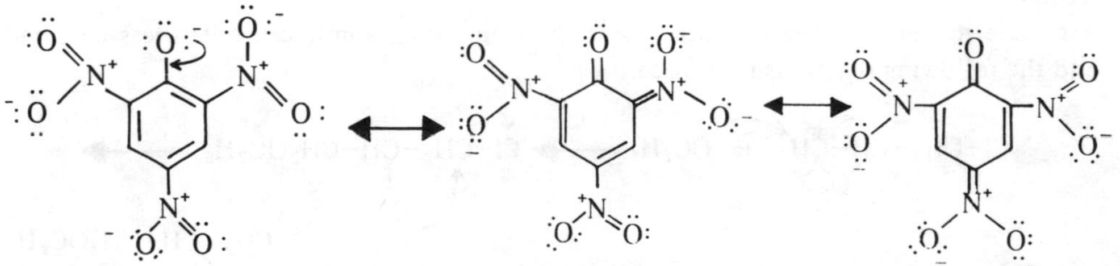

A ⟷ B ⟷ C

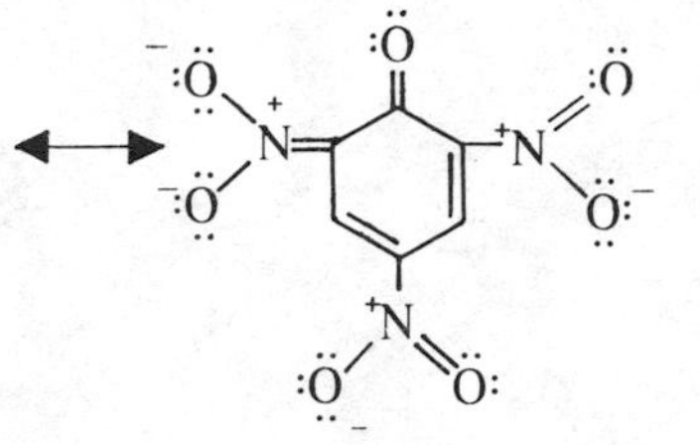

D

15.27

(d), (e), (f) All of these are stronger acids than H_2CO_3 (see Table 15.6), thus they would all be converted to their soluble sodium salts when treated with aqueous $NaHCO_3$. With 2,4-dinitrophenol, for example, the following reaction would take place.

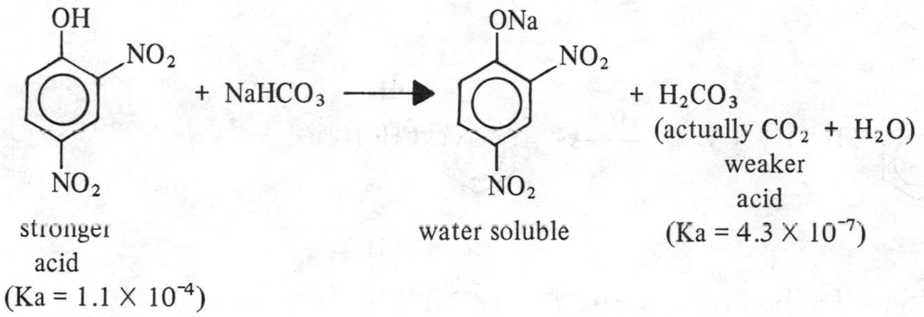

OH, NO_2, NO_2 + $NaHCO_3$ ⟶ ONa, NO_2, NO_2 + H_2CO_3 (actually CO_2 + H_2O)

stronger acid ($Ka = 1.1 \times 10^{-4}$) — water soluble — weaker acid ($Ka = 4.3 \times 10^{-7}$)

15.28

(a) The para-sulfonated phenol because it is the major product at the higher temperature–when the reaction is under equilibrium control.

(b) For ortho sulfonation, because it is the major reaction pathway at the lower temperature–when the reaction is under rate control.

15.29

(a)

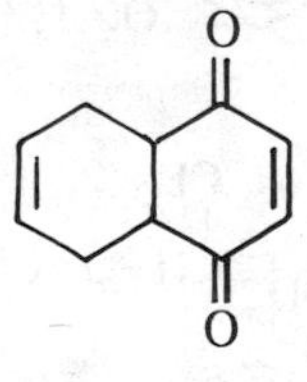

(b)

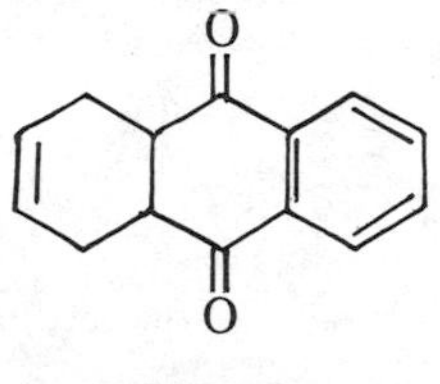

(c) 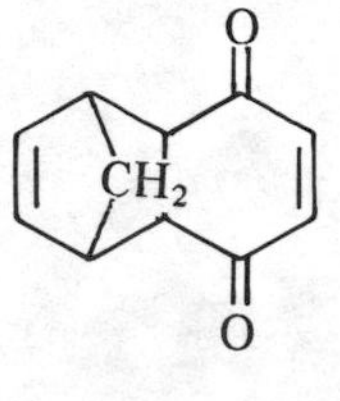 CH_2

15.30

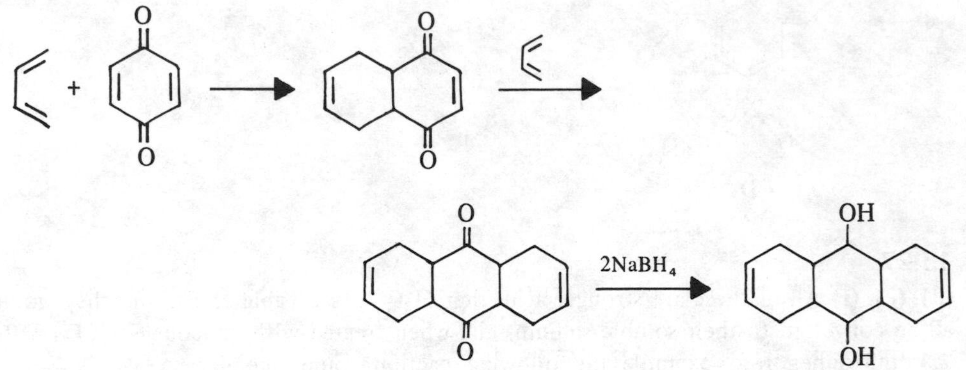

$$\xrightarrow{2NaBH_4}$$

OH / OH

15.31

(a) (1) C_6H_5–$CH{=}CH_2$ + H_2O $\xrightarrow{H^+}$ C_6H_5–$CH(OH)CH_3$

(2) C_6H_5–$CH{=}CH_2$ $\xrightarrow[(2)\ NaBH_4,OH^-]{(1)\ Hg(OAc)_2,THF,H_2O}$ C_6H_5–$CH(OH)CH_3$

(b) C_6H_5–$CH{=}CH_2$ $\xrightarrow{(BH_3)_2}$ $(C_6H_5$–$CH_2CH_2)_3B$ $\xrightarrow[OH^-,H_2O]{H_2O_2}$ C_6H_5–CH_2CH_2OH

(c) C_6H_5–$CH{=}CH_2$ $\xrightarrow[THF-CH_3OH]{Hg(OAc)_2}$ C_6H_5–$CH(OCH_3)CH_2HgOAc$ $\xrightarrow[OH^-]{NaBH_4}$ C_6H_5–$CH(OCH_3)CH_3$

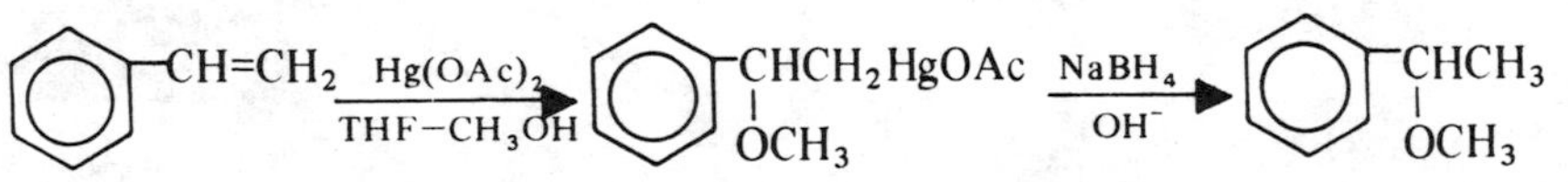

(d) C_6H_5–$CH(CH_3)$–OH $\xrightarrow{Na}$ C_6H_5–$CH(CH_3)$–O^- $\xrightarrow{CH_3CH_2Br}$ C_6H_5–$CH(CH_3)$–O–CH_2CH_3

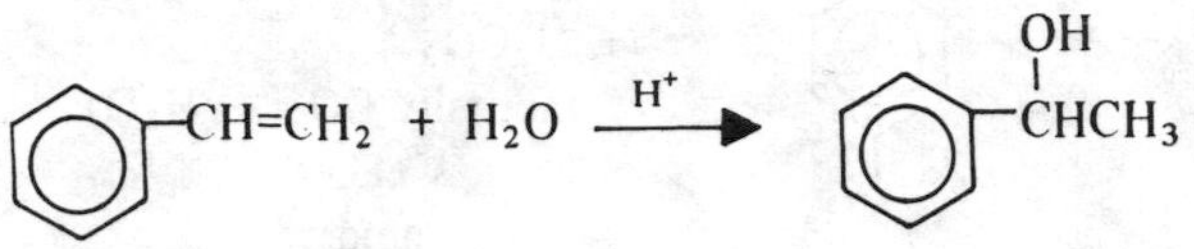

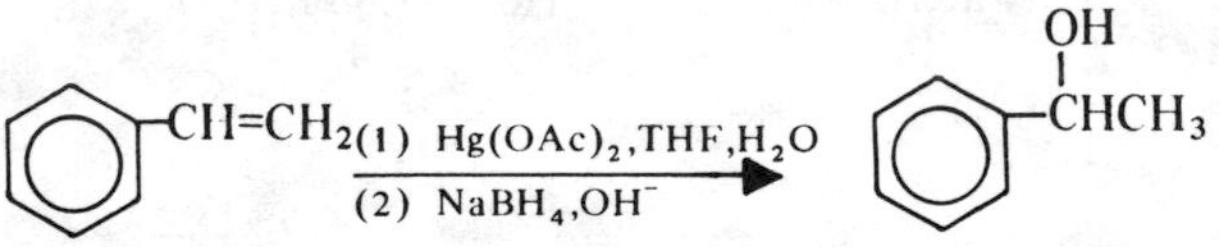

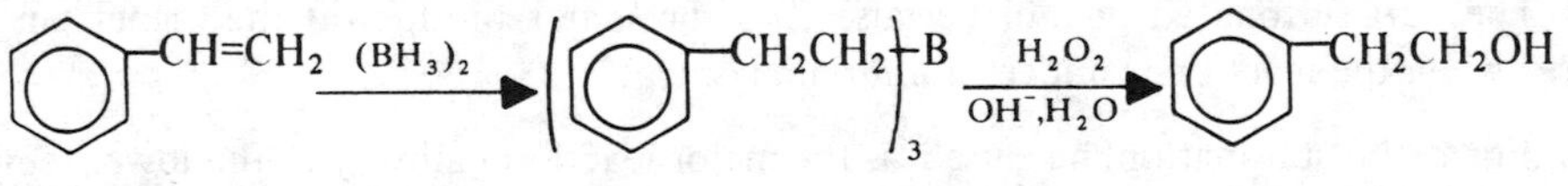

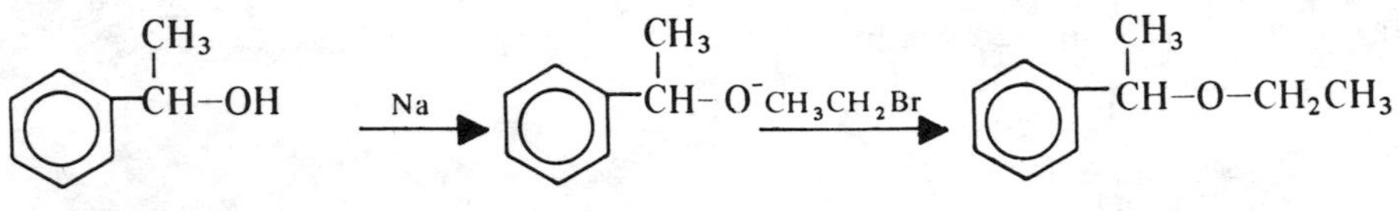

(e) C_6H_5–CH_2COOH $\xrightarrow[ether]{LiAlH_4}$ C_6H_5–CH_2CH_2OH

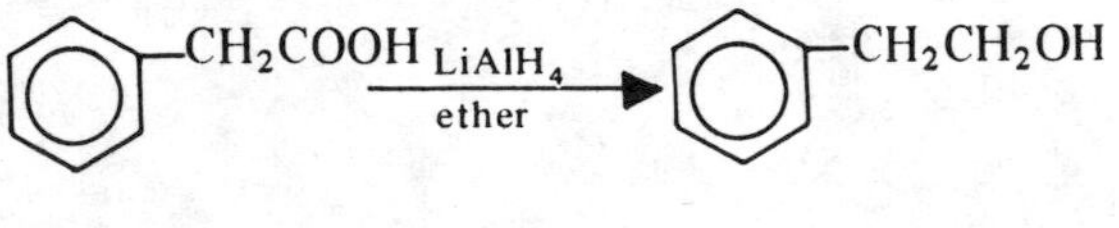

(f) C_6H_5–$C(=O)$–CH_3 $\xrightarrow[ether]{LiAlH_4\ or\ NaBH_4}$ C_6H_5–$CH(OH)CH_3$

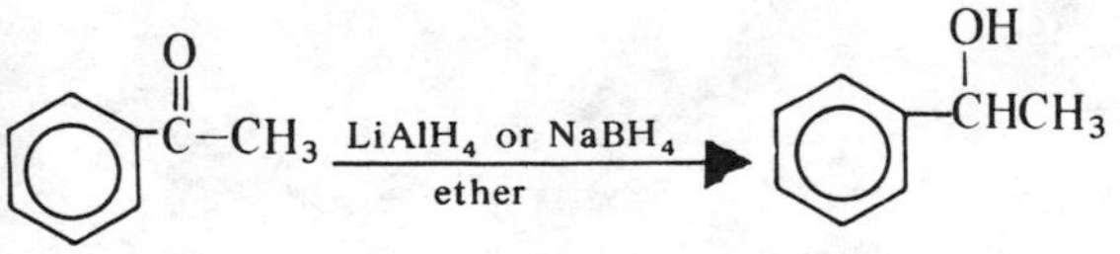

(g) $C_6H_5CH_3 \xrightarrow[CCl_4,\ light]{Br_2} C_6H_5CH_2Br \xrightarrow[ether]{Mg} C_6H_5CH_2MgBr$

$C_6H_5CH_2MgBr \xrightarrow[(2)\ H_3O^+]{(1)\ CH_2O} C_6H_5CH_2CH_2OH$

(h) $C_6H_6 \xrightarrow[FeBr_3]{Br_2} C_6H_5Br \xrightarrow[ether]{Mg} C_6H_5MgBr \xrightarrow[(2)\ H_3O^+]{(1)\ \overset{O}{CH_2-CH_2}} C_6H_5CH_2CH_2OH$

(i) $C_6H_5CH_2CO_2CH_3 \xrightarrow[ether]{LiAlH_4} C_6H_5CH_2CH_2OH$

15.32

(a) $CH_3CH_2CH_2CH_2OH \xrightarrow{PBr_3} CH_3CH_2CH_2CH_2Br$

$\xrightarrow{(CH_3)_3COK} CH_3CH_2CH{-}CH_2$

(b) $CH_3CH_2CH{=}CH_2 \xrightarrow[(2)\ NaBH_4,\ OH^-]{(1)\ Hg(OAc)_2,\ THF\text{-}H_2O} CH_3CH_2\overset{OH}{\overset{|}{C}}HCH_3$

(c) $CH_3CH_2\overset{OH}{\overset{|}{C}}HCH_3 \xrightarrow[H_2SO_4]{CrO_3} CH_3CH_2\overset{O}{\overset{\|}{C}}CH_3$

(d) $CH_3CH_2CH_2CH_2OH \xrightarrow{PBr_3} CH_3CH_2CH_2CH_2Br$

(e) $CH_3CH_2CH{=}CH_2 + HBr \xrightarrow[(no\ peroxides)]{} CH_3CH_2\overset{Br}{\overset{|}{C}}HCH_3$

(f) $CH_3CH_2CH_2CH_2Br \xrightarrow[ether]{Mg} CH_3CH_2CH_2CH_2MgBr$

$\xrightarrow[(2)\ H_3O^+]{(1)\ CH_2O} CH_3CH_2CH_2CH_2CH_2OH$

(g) $CH_3CH_2CH_2CH_2MgBr \xrightarrow[(2)\ H_3O^+]{(1)\ \overset{O}{CH_2-CH_2}} CH_3CH_2CH_2CH_2CH_2CH_2OH$

$\xrightarrow{PBr_3} CH_3CH_2CH_2CH_2CH_2CH_2Br \xrightarrow{(CH_3)_3COK} CH_3CH_2CH_2CH_2CH{=}CH_2$

(h) $CH_3CH_2CH_2CH_2MgBr + CH_3\overset{O}{\overset{\|}{C}}CH_2CH_3 \xrightarrow{(2)H_3O^+} CH_3CH_2CH_2CH_2\underset{CH_3}{\underset{|}{\overset{OH}{\overset{|}{C}}}}CH_2CH_3$

(i) $CH_3CH_2CH_2CH_2OH \xrightarrow{CrO_3 \cdot 2C_5H_5N} CH_3CH_2CH_2\overset{\overset{O}{\|}}{C}H$

(j) $CH_3CH_2CH_2CH_2MgBr + CH_3CH_2CH_2\overset{\overset{O}{\|}}{C}H \xrightarrow{(2)\ H_3O^+}$

$$CH_3CH_2CH_2CH_2\overset{\overset{OH}{|}}{C}HCH_2CH_2CH_3$$

(k) $CH_3CH_2\overset{\overset{Br}{|}}{C}HCH_3 \xrightarrow[ether]{Mg} CH_3CH_2\overset{\overset{CH_3}{|}}{C}HMgBr$

$$\xrightarrow[(2)H_3O^+]{(1)CH_3CH_2CH_2\overset{\overset{O}{\|}}{C}H} CH_3CH_2\overset{\overset{CH_3}{|}}{C}H-\overset{\overset{OH}{|}}{C}HCH_2CH_2CH_3$$

(l) $CH_3CH_2CH_2CH_2MgBr + CO_2 \xrightarrow{(2)H_3O^+} CH_3CH_2CH_2CH_2COOH$

(m) (1)$CH_3CH_2\overset{\overset{CH_3}{|}}{C}HOH \xrightarrow{Na} CH_3CH_2\overset{\overset{CH_3}{|}}{C}HONa$

$$\xrightarrow{CH_3CH_2CH_2CH_2Br} CH_3CH_2\overset{\overset{CH_3}{|}}{C}H-O-CH_2CH_2CH_2CH_3$$

(2)$CH_3CH_2CH=CH_2 + Hg(OAc)_2 \xrightarrow[CH_3CH_2CH_2CH_2OH]{THF} CH_3CH_2\underset{\underset{OCH_2CH_2CH_2CH_3}{|}}{C}H-CH_2-HgOAc$

$$\xrightarrow[OH^-]{NaBH_4} CH_3CH_2\overset{\overset{CH_3}{|}}{C}H-O-CH_2CH_2CH_2CH_3$$

(n) (1)$2CH_3CH_2CH_2CH_2OH \xrightarrow[140°]{H_2SO_4} (CH_3CH_2CH_2CH_2)_2O$

(2)$CH_3CH_2CH_2CH_2OH + Na \longrightarrow CH_3CH_2CH_2CH_2ONa \xrightarrow{CH_3CH_2CH_2CH_2Br}$

$$(CH_3CH_2CH_2CH_2)_2O$$

(o) $CH_3CH_2CH_2CH_2Br + 2Li \longrightarrow CH_3CH_2CH_2CH_2Li + LiBr$

(p) $CH_3CH_2CH_2CH_2Li \xrightarrow{CuI} (CH_3CH_2CH_2CH_2)_2CuLi \xrightarrow{CH_3CH_2CH_2CH_2Br}$

$$CH_3CH_2CH_2CH_2CH_2CH_2CH_2CH_3$$

15.33

(a) $CH_3CH_2CH_2\overset{-}{O}\ \overset{+}{Na}$ Sodium propoxide

(b) $CH_3CH_2CH_2-O-CH_2CH_2CH_2CH_3$ propyl butyl ether

(c) $CH_3-\underset{\underset{O}{\|}}{\overset{\overset{O}{\|}}{S}}-OCH_2CH_2CH_3$ propyl methanesulfonate

(d) 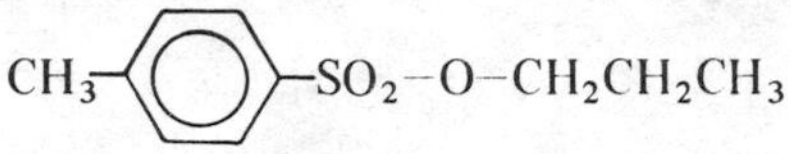 $CH_3-C_6H_4-SO_2-O-CH_2CH_2CH_3$ propyl *p*-toluenesulfonate (or propyl tosylate)

(e) $CH_3\overset{O}{\overset{\|}{C}}-O-CH_2CH_2CH_3$ propyl acetate

(f) $CH_3CH_2\overset{O}{\overset{\|}{C}}-\overset{-}{O}\ \overset{+}{K}$ potassium propanoate

(g) $CH_3CH_2CH_2Cl$ 1-chloropropane

(h) $CH_3CH_2CH_2Cl$ 1-chloropropane

(i) $CH_3CH_2CH_2-O-CH_2CH_2CH_3$ dipropyl ether

(j) $CH_3CH_2CH_2Br$ 1-bromopropane

(k) $CH_3-C(CH_3)_2-O-CH_2CH_2CH_3$ propyl *tert*-butyl ether

(l) 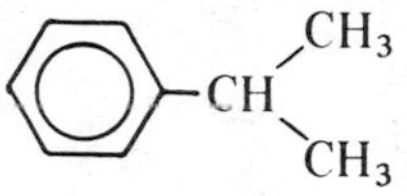 $C_6H_5-CH(CH_3)_2$ isopropylbenzene (major) + $C_6H_5-CH_2CH_2CH_3$ propylbenzene

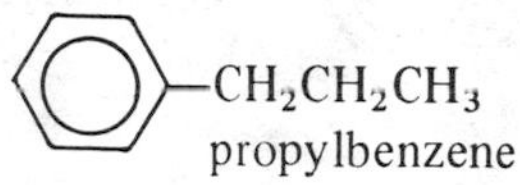

15.34

(a) $CH_3\underset{}{CH}(CH_3)\overset{-}{O}\ \overset{+}{Na}$ sodium isopropoxide

(b) $CH_3CH(CH_3)-O-CH_2CH_2CH_2CH_3$ isopropyl butyl ether

(c) $CH_3SO_2-O-CH(CH_3)CH_3$ isopropyl methanesulfonate

(d) 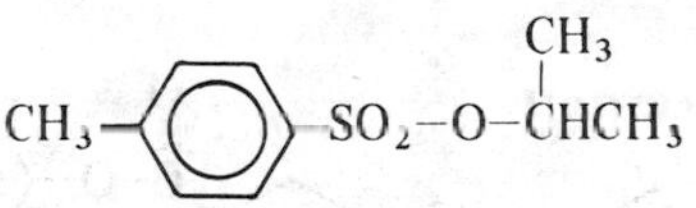 $CH_3-C_6H_4-SO_2-O-CH(CH_3)CH_3$ isopropyl *p*-toluenesulfonate

(e) $CH_3\overset{O}{\overset{\|}{C}}-O-CH(CH_3)CH_3$ isopropyl acetate

(f) $CH_3\overset{O}{\overset{\|}{C}}CH_3$ acetone (+ CH_3COOH and CO_2)

(g) $CH_3CH(CH_3)Cl$ 2-chloropropane

(h) Same as (g)

(i) $CH_3CH(CH_3)-O-CH(CH_3)CH_3$ diisopropyl ether

(j) $CH_3CH(CH_3)Br$ 2-bromopropane

(k) $CH_3C(CH_3)_2-O-CH(CH_3)CH_3$ isopropyl *tert*-butyl ether

(l) 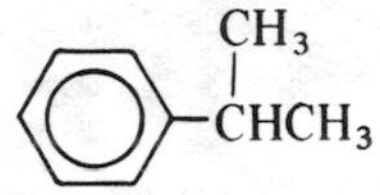 $C_6H_5-CH(CH_3)CH_3$ isopropyl benzene

15.35

(a) 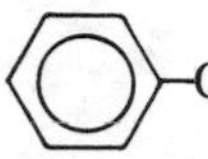 $C_6H_5-ONa + CH_3CH_2OH$

(b)  $C_6H_6 + CH_3CH_2OMgBr$

(c) 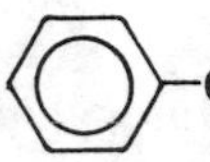 $C_6H_5-ONa + H_2O$,

(d) 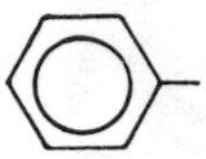 $C_6H_5-OH + NaCl$

(e) $CH_3CH_2OH + NaOH$

15.36

(a) $CH_3Br + CH_3CH_2Br$

(b) 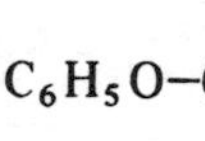 $C_6H_5-OH + CH_3CH_2Br$

(c) $Br-CH_2CH_2CH_2CH_2-Br$

(d) $Br-CH_2CH_2-Br$ (2 moles)

15.37

(a) $CH_3CH(OH)CH_2(OH)$

(b) $CH_3CH_2-O-CH_2CH_2OH$

(c) $C_6H_5O-CH_2CH_2OH$

(d) 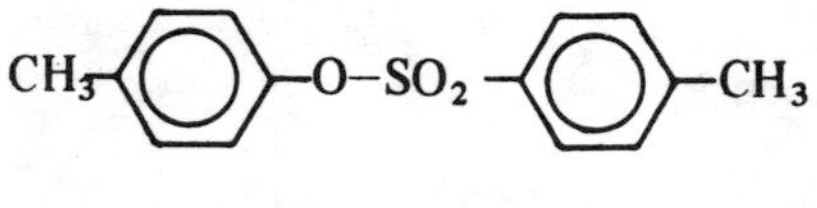 $CH_3-C_6H_4-O-SO_2-C_6H_4-CH_3$

(e) 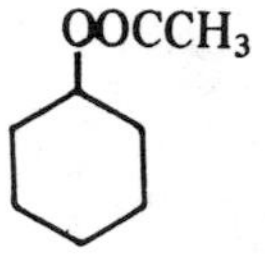 $C_6H_{11}-OOCCH_3$

(f) 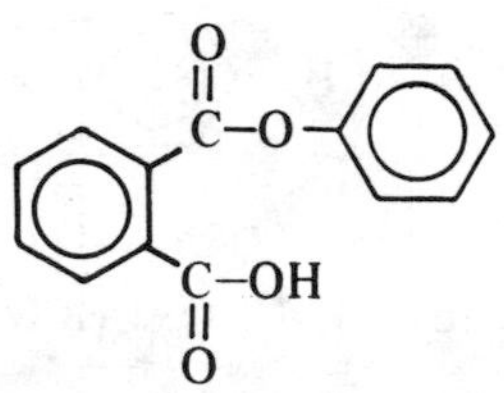 $C_6H_4(C(=O)-O-C_6H_5)(C(=O)-OH)$

(g) 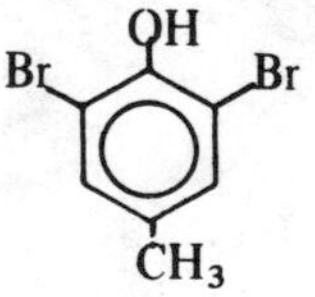 $C_6H_2(OH)(Br)_2(CH_3)$

(h) 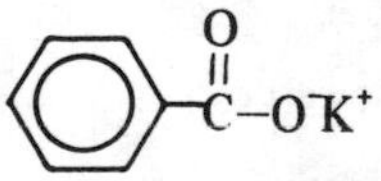 $C_6H_5-C(=O)-O^-K^+$

(i) $C_6H_5-C(=O)H$

(j) $C_6H_5-CH_2O^-Na^+$

(k) $C_6H_5-CH_2OCH_3$

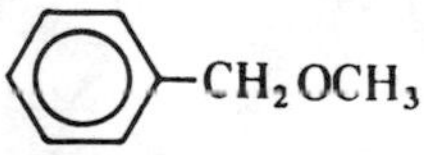

(l) 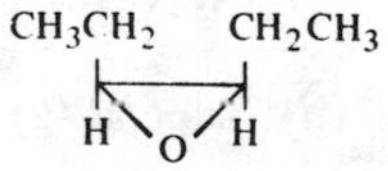 $CH_3CH_2(H)C-O-C(H)CH_2CH_3$

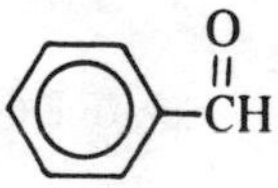

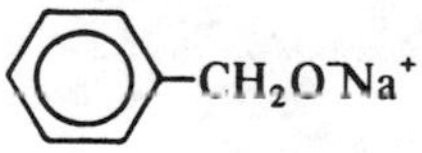

(m) [Newman-type projection: CH_3CH_2–C(H)(OH)–C(HO)(H)–CH_2CH_3] + enantiomer

15.38

(a) *p*-Cresol is soluble in aqueous NaOH; benzyl alcohol is not.

(b) Cyclohexanol is soluble in cold, concentrated H_2SO_4; cyclohexane is not. (Cyclohexanol also gives a positive test with CrO_3 in H_2SO_4, while cyclohexane does not.)

(c) Cyclohexene will decolorize Br_2/CCl_4 solution; cyclohexanol will not.

(d) Allyl propyl ether will decolorize Br_2/CCl_4 solution; dipropyl ether will not.

(e) *p*-Cresol is soluble in aqueous NaOH; anisole is not.

(f) Picric acid is soluble in aqueous $NaHCO_3$: 2,4,6-trimethylphenol is not (cf. Problem 15.27).

15.39

2,2-dimethylcyclohexanol $\xrightleftharpoons{H^+}$ protonated alcohol ($-OH_2^+$) $\rightleftharpoons$ secondary carbocation + H_2O $\longrightarrow$ (methyl shift) tertiary carbocation

tertiary carbocation $\xrightarrow{-H^+}$ 1,2-dimethylcyclohexene

15.40

(a) 3-methylphenol + $CH_2{=}CHCH_3$ $\xrightarrow{H_2SO_4}$ 5-methyl-2-isopropylphenol (ring bearing CH_3, OH, and $CH(CH_3)_2$)

(b) 5-methyl-2-isopropylphenol (ring bearing CH_3, OH, CH_3CHCH_3) $\xrightarrow{H_2/Ni}$ 5-methyl-2-isopropylcyclohexanol (ring bearing CH_3, OH, CH_3CHCH_3)

15.41

The position ortho to the isopropyl group is sterically more hindered than the position ortho to the methyl group:

15.42

$$CH_3CH_2CH_2CH_2CH_2\overset{O}{\overset{\|}{C}}-OH \xrightarrow{SOCl_2} CH_3CH_2CH_2CH_2CH_2\overset{O}{\overset{\|}{C}}-Cl \xrightarrow[AlCl_3]{\text{1,3-dihydroxybenzene}}$$

$$\text{2,4-dihydroxyphenyl}-C(=O)-CH_2CH_2CH_2CH_2CH_3 \xrightarrow[HCl]{Zn(Hg)} \text{2,4-dihydroxyphenyl}-CH_2CH_2CH_2CH_2CH_2CH_3$$

15.43

$$CH_3CH_2\overset{O}{\overset{\|}{C}}-OH \xrightarrow{SOCl_2} CH_3CH_2\overset{O}{\overset{\|}{C}}-Cl \xrightarrow[AlCl_3]{C_6H_5OCH_3} p\text{-}CH_3O-C_6H_4-\overset{O}{\overset{\|}{C}}CH_2CH_3$$

$$CH_3O-C_6H_4-\overset{O}{\overset{\|}{C}}CH_2CH_3 \xrightarrow[ether]{NaBH_4} CH_3O-C_6H_4-\overset{OH}{\overset{|}{C}}HCH_2CH_3 \xrightarrow[heat]{H_2SO_4} CH_3O-C_6H_4-CH{=}CH-CH_3$$

15.44

X is a phenol because it dissolves in aqueous NaOH but not in aqueous $NaHCO_3$. It gives a dibromo derivative, and must therefore be substituted in the ortho or para position. The broad infrared peak at 3250 cm^{-1} also suggests a phenol. The peak at 830 cm^{-1} indicates para substitution. The pmr singlet at δ1.3 (9H) suggests 9 methyl hydrogens which must be a *tert*-butyl group. The structure of X is:

$$HO-C_6H_4-C(CH_3)_3 \quad (\text{para; } CH_3-\overset{|}{\underset{CH_3}{C}}-CH_3)$$

15.45

The broad infrared peak at 3400 cm^{-1} indicates a hydroxy group and the two bands at 720 and 770 cm^{-1} suggest a monosubstituted benzene ring. The presence of these groups is also indicated by the peaks at $\delta 2.7$ and $\delta 7.2$ in the pmr spectrum. The pmr spectrum also shows a triplet at $\delta 0.7$ indicating a $-CH_3$ group coupled with an adjacent $-CH_2-$ group. What appears at first to be a quartet at $\delta 1.9$ actually shows further splitting. There is also a triplet at $\delta 4.35$ (1H). Putting these pieces together in the only way possible gives us the following structure for Y.

$C_6H_5-\underset{\displaystyle OH}{\underset{|}{C}}HCH_2CH_3$

Analysed spectra are as follows

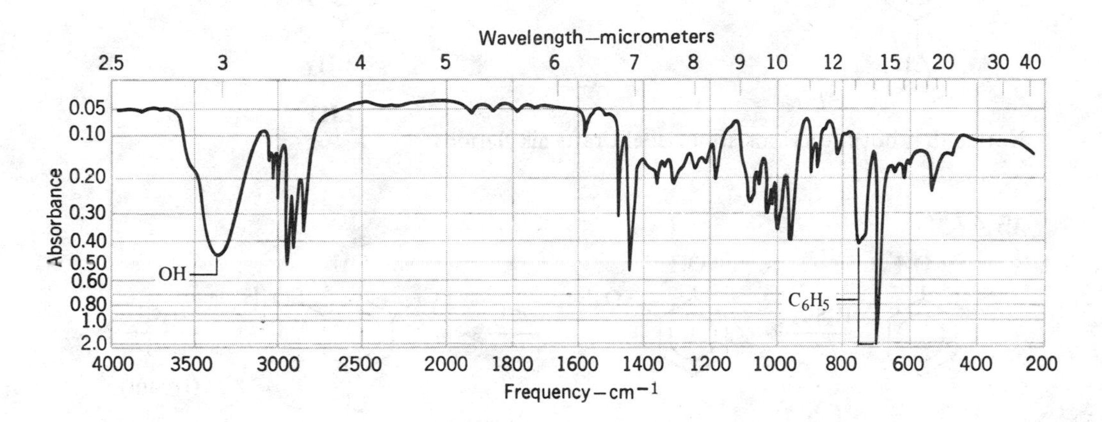

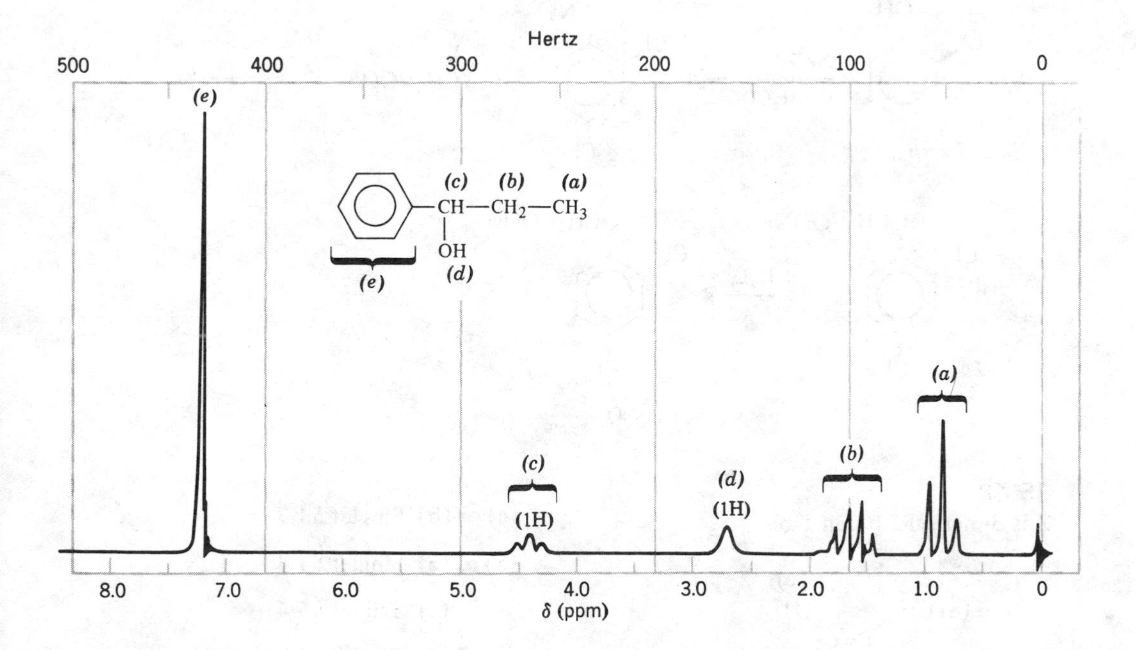

Fig. 15.2. The infrared and proton nmr spectra of compound Y, problem 15.45 (Spectra courtesy of Sadtler Research Laboratories Inc.)

15.46

(a) 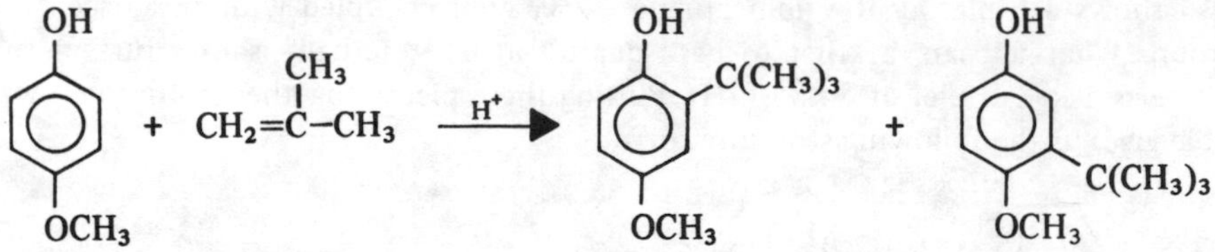 4-Methoxyphenol (OH / OCH_3) + $CH_2{=}C(CH_3){-}CH_3$ $\xrightarrow{H^+}$ 2-$C(CH_3)_3$-4-methoxyphenol (OH, $C(CH_3)_3$, OCH_3) + 3-$C(CH_3)_3$-4-methoxyphenol (OH, $C(CH_3)_3$, OCH_3)

BHA

(b) 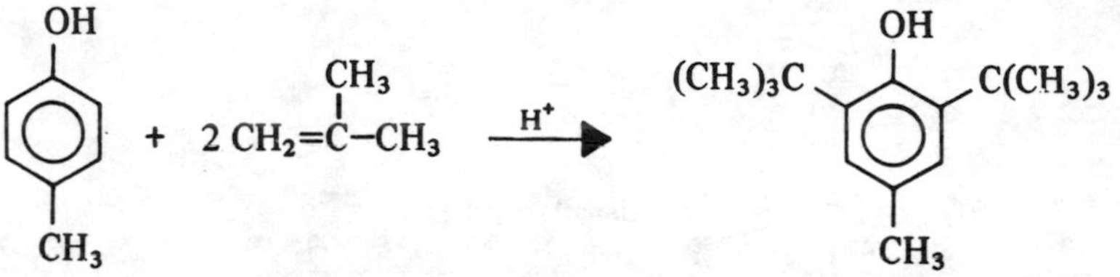 4-Methylphenol (OH / CH_3) + $2\ CH_2{=}C(CH_3){-}CH_3$ $\xrightarrow{H^+}$ $(CH_3)_3C$, $C(CH_3)_3$ (OH / CH_3)

BHT

Notice that both reactions are Friedel-Crafts alkylations.

15.47

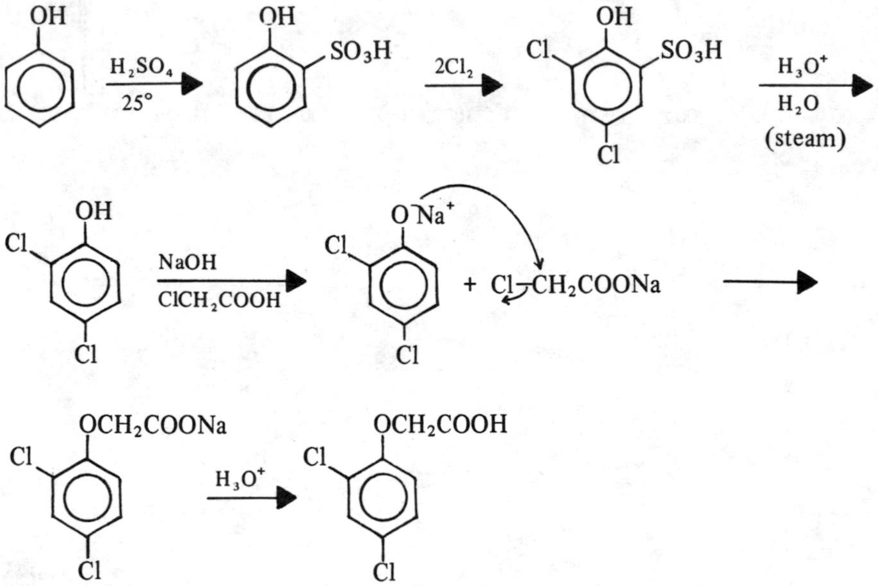

Phenol (OH) $\xrightarrow[25°]{H_2SO_4}$ OH, SO_3H $\xrightarrow{2Cl_2}$ Cl, OH, SO_3H, Cl $\xrightarrow[H_2O\ (steam)]{H_3O^+}$

OH, Cl, Cl $\xrightarrow[ClCH_2COOH]{NaOH}$ O^-Na^+, Cl, Cl + $Cl{-}CH_2COONa$ $\longrightarrow$

OCH_2COONa, Cl, Cl $\xrightarrow{H_3O^+}$ OCH_2COOH, Cl, Cl

2,4-D

15.48

Z is 3-methyl-2-buten-1-ol.

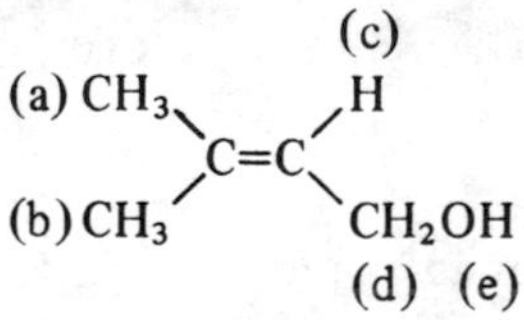

(a) CH_3, (b) CH_3 $\text{C}{=}\text{C}$ (c) H, CH_2OH (d) (e)

(a) or (b) Singlet δ 1.7

(b) or (a) Singlet δ 1.8

(c) Triplet δ 5.4

(d) Doublet δ 4.1

(e) Singlet δ 1.9

15.49

(a) A is $CH_2{=}CHC(CH_3)(OH)C{\equiv}CH$, C is $BrMgOCH_2CH{=}C(CH_3)C{\equiv}CMgBr$

(b) A is an allylic alcohol and thus forms a carbocation readily. B is a conjugated enyne and is therefore more stable than A.

$$\underset{\textbf{A}}{CH_2{=}CH{-}\overset{CH_3}{\underset{OH}{C}}{-}C{\equiv}CH} \xrightarrow[(-H_2O)]{H^+} CH_2{=}CH{-}\overset{CH_3}{\underset{+}{C}}{-}C{\equiv}CH \longleftrightarrow$$

$$\overset{+}{C}H_2{-}CH{=}\overset{CH_3}{C}{-}C{\equiv}CH \xrightarrow[(-H^+)]{H_2O} \underset{\textbf{B}}{HOCH_2{-}CH{=}\overset{CH_3}{C}{-}C{\equiv}CH}$$

15.50

$$\text{(trimethylcyclohexenyl)}{-}CH_2CH{=}C(CH_3){-}\underset{O}{\overset{\|}{C}}H + BrMgOCH_2CH{=}\overset{CH_3}{C}C{\equiv}CMgBr \xrightarrow[(2)\ H^+]{} $$

$$\underset{\textbf{D}}{\text{(ring)}{-}CH_2CH{=}C(CH_3){-}\underset{OH}{CH}{-}C{\equiv}C{-}\overset{CH_3}{C}{=}CHCH_2OH} \xrightarrow[\text{quinoline}]{H_2,\ Pd/CaCO_3}$$

$$\underset{\textbf{E}}{\text{(ring)}{-}CH_2CH{=}C(CH_3){-}\underset{OH}{CH}{-}CH{=}CH{-}\overset{CH_3}{C}{=}CHCH_2OH} \xrightarrow{(CH_3\overset{O}{\overset{\|}{C}})_2O}$$

$$\underset{\textbf{F}}{\text{(ring)}{-}CH_2CH{=}C(CH_3){-}\underset{OH}{CH}{-}CH{=}CH{-}\overset{CH_3}{C}{=}CHCH_2O\overset{O}{\overset{\|}{C}}CH_3} \xrightarrow{-H_2O}$$

$$\text{(ring)}{-}CH{=}CH{-}C(CH_3){=}CH{-}CH{=}CH{-}C(CH_3){=}CH{-}CH_2O\overset{O}{\overset{\|}{C}}CH_3$$

Vitamin A acetate

15.51

$$CH_3\overset{O}{\overset{\|}{C}}CH_3 + H^+ \rightleftharpoons H\overset{+}{O}{=}\overset{CH_3}{\underset{CH_3}{C}} \longleftrightarrow HO{-}\overset{CH_3}{\underset{CH_3}{C^+}} \xrightarrow{C_6H_5OH}$$

$$HO\text{-}C(CH_3)_2\text{-}(\overset{+}{C}_6H_5(H))\text{-}OH \xrightarrow{-H^+} HO\text{-}C(CH_3)_2\text{-}C_6H_4\text{-}OH$$

$$\xrightarrow{H^+} H\text{-}\overset{+}{O}(H)\text{-}C(CH_3)_2\text{-}C_6H_4\text{-}OH \xrightarrow{-H_2O}$$

$$^{+}C(CH_3)_2\text{-}C_6H_4\text{-}OH \xrightarrow{C_6H_5OH}$$

$$HO\text{-}(\overset{+}{C}_6H_5(H))\text{-}C(CH_3)_2\text{-}C_6H_4\text{-}OH \xrightarrow{-H^+}$$

$$HO\text{-}C_6H_4\text{-}C(CH_3)_2\text{-}C_6H_4\text{-}OH$$

"Bisphenol A"

Special Topic

J.1

(a) 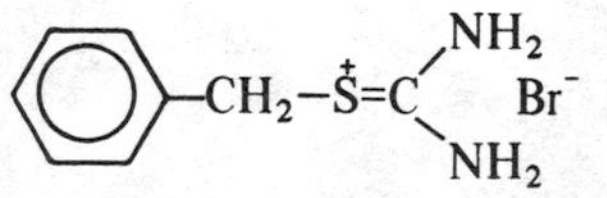 $C_6H_5-CH_2-\overset{+}{S}=C(NH_2)_2\ Br^-$ (b) $C_6H_5-CH_2SH$

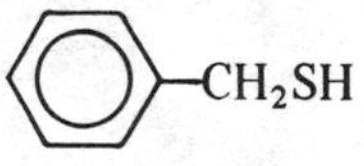

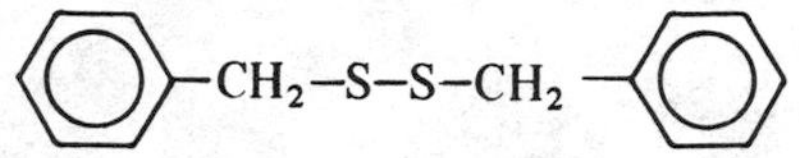

(c) $C_6H_5-CH_2-S-S-CH_2-C_6H_5$ (d) 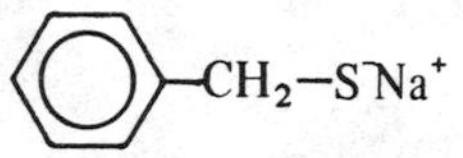 $C_6H_5-CH_2-S^-Na^+$

(e) 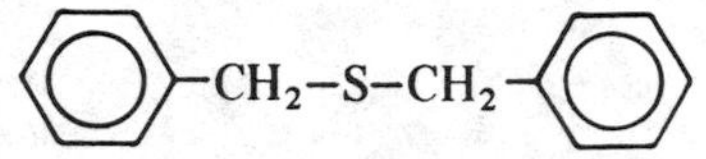 $C_6H_5-CH_2-S-CH_2-C_6H_5$

J.2

$$CH_2{=}CHCH_2Br + S{=}C(NH_2)_2 \xrightarrow[(2)\ OH^-,\ H_2O]{(1)\ CH_3CH_2OH} CH_2{=}CHCH_2SH$$

$$\xrightarrow{H_2O_2} CH_2{=}CHCH_2-S-S-CH_2CH{=}CH_2$$

J.3

$$CH_2{=}CHCH_2OH \xrightarrow{Br_2} CH_2BrCHBrCH_2OH \xrightarrow{NaSH} \underset{SH}{CH_2}-\underset{SH}{CH}-CH_2OH$$

J.4

(a) $ClCH_2CH_2\overset{O}{\overset{\|}{C}}(CH_2)_4CO_2C_2H_5$ (this step is the Friedel-Crafts acylation of an alkene)

(b) $SOCl_2$

(c) 2 $C_6H_5CH_2SH$ and KOH

(d) H_3O^+

(e) $CH_2(CH_2SH)CH_2$–$CH(SH)(CH_2)_4COOH$

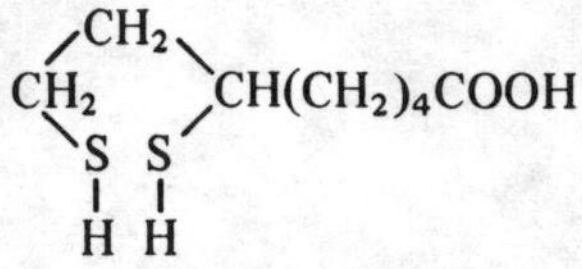

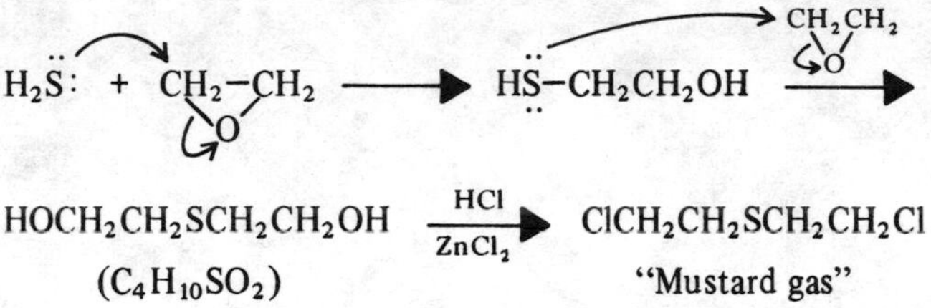

$$\begin{array}{ccc} & CH_2 & \\ CH_2 & & CH(CH_2)_4COOH \\ | & & | \\ S & & S \\ | & & | \\ H & & H \end{array}$$

J.5

$$H_2\ddot{S}: + \underset{\diagdown O \diagup}{CH_2-CH_2} \longrightarrow H\ddot{\underset{..}{S}}-CH_2CH_2OH \xrightarrow{\underset{\diagdown O \diagup}{CH_2CH_2}}$$

$$\underset{(C_4H_{10}SO_2)}{HOCH_2CH_2SCH_2CH_2OH} \xrightarrow[ZnCl_2]{HCl} \underset{\text{"Mustard gas"}}{ClCH_2CH_2SCH_2CH_2Cl}$$

CHAPTER SIXTEEN
Aldehydes and Ketones

16.1

(a) $CH_3CH_2CH_2CH_2\overset{O}{\overset{\|}{C}}H$ Pentanal

$CH_3CH_2\underset{CH_3}{\underset{|}{C}}H\overset{O}{\overset{\|}{C}}H$ 2-Methylbutanal

$CH_3\underset{CH_3}{\underset{|}{C}}HCH_2\overset{O}{\overset{\|}{C}}H$ 3-Methylbutanal

$CH_3\overset{CH_3}{\overset{|}{\underset{CH_3}{\underset{|}{C}}}}-CHO$ 2,2-Dimethylpropanal

$CH_3CH_2CH_2\underset{O}{\underset{\|}{C}}CH_3$ 2-Pentanone

$CH_3CH_2\underset{O}{\underset{\|}{C}}CH_2CH_3$ 3-Pentanone

$CH_3\underset{CH_3}{\underset{|}{C}}H\overset{O}{\overset{\|}{C}}CH_3$ 3-Methyl-2-butanone

(b) and (c) $C_6H_5-\overset{O}{\overset{\|}{C}}CH_3$ Acetophenone or phenyl methyl ketone

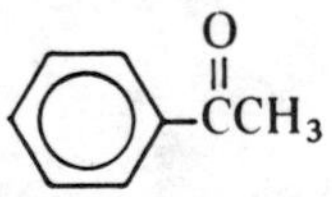

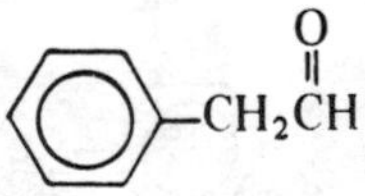

$C_6H_5-CH_2\overset{O}{\overset{\|}{C}}H$ Phenylethanal or phenylacetaldehyde

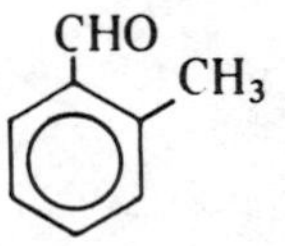

o-Tolualdehyde (CHO and CH_3 on adjacent ring positions)

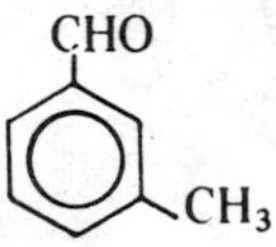

m-Tolualdehyde (CHO and CH_3 at 1,3-positions)

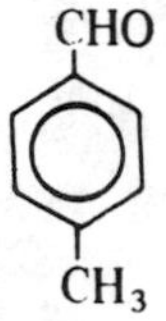

p-Tolualdehyde (CHO and CH_3 at 1,4-positions)

16.2

(a) 1-Pentanol, because its molecules form hydrogen bonds to each other.

(b) 2-Pentanol, because its molecules form hydrogen bonds to each other.

(c) Pentanal, because its molecules are more polar.

(d) 2-Phenylethanol, because its molecules form hydrogen bonds to each other.

(e) Benzyl alcohol because its molecules form hydrogen bonds to each other.

16.3

(a) $C_6H_6 \xrightarrow{Br_2,\ Fe} C_6H_5Br \xrightarrow[ether]{Mg} C_6H_5MgBr \xrightarrow[(2)\ H^+]{(1)\ HCHO}$

$C_6H_5CH_2OH \xrightarrow[CH_2Cl_2]{CrO_3.C_5H_5N} C_6H_5CHO$

(b) $C_6H_5CH_3 \xrightarrow[(2)\ H^+]{(1)\ KMnO_4,\ OH^-,\ heat} C_6H_5COOH \xrightarrow{SOCl_2}$

$C_6H_5COCl \xrightarrow[ether]{LiAlH[OC(CH_3)_3]_3} C_6H_5CHO$

$C_6H_5COCl \xrightarrow[Pd(S)]{H_2} C_6H_5CHO$

(c) $CH_3CH_2Br \xrightarrow{HC{\equiv}CNa} CH_3CH_2C{\equiv}CH \xrightarrow[(2)\ H_2O_2,\ OH^-]{(1)\ Sia_2BH} CH_3CH_2CH_2\overset{O}{\overset{\|}{C}}H$

(d) $CH_3C{\equiv}CCH_3 \xrightarrow[H_2O]{H_3O^+,\ Hg^{++}} CH_3\overset{O}{\overset{\|}{C}}CH_2CH_3$

(e) $C_6H_5\underset{OH}{\underset{|}{C}}HCH_3 \xrightarrow[H_2SO_4]{CrO_3} C_6H_5\underset{O}{\underset{\|}{C}}CH_3$

(f) $C_6H_6 \xrightarrow[AlCl_3]{CH_3COCl} C_6H_5\underset{O}{\underset{\|}{C}}CH_3$

(g) $C_6H_5\underset{O}{\underset{\|}{C}}Cl \xrightarrow[\text{or } (CH_3)_2Cd]{(CH_3)_2CuLi} C_6H_5\underset{O}{\underset{\|}{C}}CH_3$

16.4

(a) The nucleophile is the negatively charged carbon of the Grignard reagent *acting as a carbanion.*

(b) The magnesium portion of the Grignard reagent acts as a Lewis acid and accepts an electron pair of the carbonyl oxygen. This makes the carbonyl carbon even more positive and, therefore, even more susceptible to nucleophilic attack.

$$\overset{\delta+}{-\overset{|}{C}}=\overset{\delta-}{\ddot{O}} \quad \overset{\delta+}{Mg}-X \quad \underset{\delta-}{R} \longrightarrow -\underset{R}{\overset{|}{C}}-\ddot{\underset{..}{O}}-MgX$$

(c) The product that forms initially (above) is a magnesium halide salt of an alcohol.

(d) On addition of water, the organic product that forms is an alcohol.

16.5

The nucleophile is a hydride ion.

16.6

(a) $\underset{\text{I}}{RCH=\overset{+}{\underset{..}{O}}-R} \longleftrightarrow \underset{\text{II}}{R\overset{+}{C}H-\ddot{\underset{..}{O}}-R}$

(b) and (c) Structure I should make a greater contribution because in it both the carbon atom and the oxygen atom have an octet of electrons and because it has one more bond.

16.7

$$(CH_3)_2C=O \underset{-HOCH_2CH_2OH}{\overset{+HOCH_2CH_2OH}{\rightleftharpoons}} (CH_3)_2C(O{-}H)(OCH_2CH_2OH)$$

$$\underset{-H^+}{\overset{+H^+}{\rightleftharpoons}} (CH_3)_2C(\overset{+}{O}H_2)(\ddot{O}CH_2CH_2OH) \underset{+H_2O}{\overset{-H_2O}{\rightleftharpoons}} (CH_3)_2C=\overset{+}{O}CH_2CH_2\ddot{O}H$$

$$\rightleftharpoons (CH_3)_2C\langle\overset{+}{O}(H){-}CH_2{-}CH_2{-}O\rangle \underset{+H^+}{\overset{-H^+}{\rightleftharpoons}} (CH_3)_2C\langle O{-}CH_2{-}CH_2{-}O\rangle$$

16.8

(a) A (3-oxocyclopentane with $CO_2C_2H_5$) $\xrightarrow[H^+]{HOCH_2CH_2OH}$ ethylene ketal of A (with $CO_2C_2H_5$)

$\xrightarrow{2\ CH_3MgI}$ ethylene ketal bearing $-\underset{CH_3}{\overset{OMgI}{C}}-CH_3$ $\xrightarrow[H_2O]{H_3O^+}$ cyclopentanone bearing $-\underset{CH_3}{\overset{OH}{C}}-CH_3$

C

(b) Addition would take place at the ketone group as well as at the ester group. The product (after hydrolysis) would be,

$$\text{1-methylcyclopentan-1-ol bearing at C-3 a } -C(OH)(CH_3)-CH_3 \text{ group (ring C: HO, } CH_3\text{)}$$

16.9

(a)
$$\text{cyclohexanone} + HSCH_2CH_2SH \xrightarrow{BF_3} \text{cyclohexane spiro-1,3-dithiolane } (\langle^{S-CH_2}_{S-CH_2})$$

$$\xrightarrow[Ni]{Raney} \text{cyclohexane} + CH_3CH_3 + NiS$$

(b)
$$CH_3\overset{O}{\overset{\|}{C}}CH_2CH_2CO_2C_2H_5 + HSCH_2CH_2SH \xrightarrow{BF_3}$$

$$CH_3-\underset{}{\overset{S\quad S\ (CH_2-CH_2)}{C}}-CH_2CH_2CO_2C_2H_5 \xrightarrow[Ni]{Raney} CH_3CH_2CH_2CH_2CO_2C_2H_5 + CH_3CH_3 + NiS$$

16.10

(a)
$$CH_3\overset{O}{\overset{\|}{C}}H \xrightarrow[H_2O]{NaHSO_3} CH_3\overset{OH}{\overset{|}{C}}HSO_3Na \xrightarrow[H_2O]{NaCN} CH_3\overset{OH}{\overset{|}{C}}HCN$$

$$\xrightarrow[reflux]{HCl} CH_3\overset{OH}{\overset{|}{C}}HCOOH$$

Lactic acid

(b) A racemic modification.

16.11

(a)
$$CH_3I \xrightarrow{(C_6H_5)_3P} CH_3-P(C_6H_5)_3^+\ I^- \xrightarrow{RLi} CH_2{=}P(C_6H_5)_3$$

(b)
$$\text{cyclopentyl}-Br \xrightarrow{P(C_6H_5)_3} \text{cyclopentyl}-P(C_6H_5)_3^+\ Br^-$$

$$\xrightarrow{RLi} \text{cyclopentylidene}{=}P(C_6H_5)_3$$

(c)
$$CH_3(CH_2)_4CH_2Br \xrightarrow{(C_6H_5)_3P} CH_3(CH_2)_4CH_2-P(C_6H_5)_3^+\ Br^-$$

$$\xrightarrow{RLi} CH_3(CH_2)_4CH{=}P(C_6H_5)_3$$

16.12

(a) $C_6H_5CH_2Br \xrightarrow[(2)\ RLi]{(1)\ (C_6H_5)_3P} C_6H_5CH{=}P(C_6H_5)_3 \xrightarrow{CH_3\overset{O}{\overset{\|}{C}}CH_3} C_6H_5CH{=}\underset{\displaystyle CH_3}{\underset{|}{C}}CH_3$

(b) $CH_3I \xrightarrow[(2)\ RLi]{(1)\ (C_6H_5)_3P} CH_2{=}P(C_6H_5)_3 \xrightarrow{C_6H_5\overset{O}{\overset{\|}{C}}CH_3} C_6H_5\underset{\displaystyle CH_3}{\underset{|}{C}}{=}CH_2$

(c) $CH_3CH_2Br \xrightarrow[(2)\ RLi]{(1)\ (C_6H_5)_3P} CH_3CH{=}P(C_6H_5)_3 \xrightarrow{C_6H_5\overset{O}{\overset{\|}{C}}CH_3} C_6H_5\underset{\displaystyle CH_3}{\underset{|}{C}}{=}CHCH_3$

(d) $CH_2{=}P(C_6H_5)_3 \xrightarrow{CH_3\overset{O}{\overset{\|}{C}}CH_3} (CH_3)_2C{=}CH_2$
(from part b)

(e) $CH_2{=}P(C_6H_5)_3 \xrightarrow{\text{cyclopentanone}}$ methylenecyclopentane ($C_5H_8{=}CH_2$)
(from part b)

(f) $CH_3CH_2CH_2Br \xrightarrow[(2)\ RLi]{(1)\ (C_6H_5)_3P} CH_3CH_2CH{=}P(C_6H_5)_3$

$$\xrightarrow{CH_3\overset{O}{\overset{\|}{C}}CH_2CH_3} CH_3CH_2CH{=}\overset{\displaystyle CH_3}{\overset{|}{C}}CH_2CH_3$$

(g) $CH_2{=}CHCH_2Br \xrightarrow[(2)\ RLi]{(1)\ (C_6H_5)_3P} CH_2{=}CHCH{=}P(C_6H_5)_3$

$$\xrightarrow{C_6H_5\overset{O}{\overset{\|}{C}}H} C_6H_5CH{=}CHCH{=}CH_2$$

(h) $C_6H_5CH{=}P(C_6H_5)_3 \xrightarrow{C_6H_5\overset{O}{\overset{\|}{C}}H} C_6H_5CH{=}CHC_6H_5$
(from part (a))

16.13

$$(C_6H_5)_3P: + C_6H_5CH\overset{O}{-}CHC_6H_5 \longrightarrow$$

$$C_6H_5\underset{(C_6H_5)_3\overset{+}{P}}{\underset{|}{C}H}-\overset{O^-}{\overset{|}{C}}HC_6H_5 \longrightarrow \begin{array}{c} C_6H_5CH-CHC_6H_5 \\ | \qquad\quad | \\ (C_6H_5)_3P - O \end{array}$$

$$\longrightarrow C_6H_5CH{=}CHC_6H_5 + (C_6H_5)_3P{=}O$$

16.14

(a) $CH_3OCH_2Br + (C_6H_5)_3P \xrightarrow{(2)\ base} CH_3OCH{=}P(C_6H_5)_3$

(b) Hydrolysis of the ether yields a hemiacetal (see the example in part (c) below) that then goes on to form an aldehyde.

(c) cyclohexanone $+ CH_3OCH{=}P(C_6H_5)_3 \longrightarrow$

cyclohexylidene$=CHOCH_3$ $\underset{-H^+}{\overset{+H^+}{\rightleftharpoons}}$ (H)cyclohexyl$-CH{=}\overset{+}{O}CH_3$ $\underset{-H_2O}{\overset{+H_2O}{\rightleftharpoons}}$

cyclohexyl$-CH(\overset{+}{O}H_2)-OCH_3$ $\underset{+H^+}{\overset{-H^+}{\rightleftharpoons}}$ cyclohexyl$-CH(OH)-OCH_3$ $\underset{+CH_3OH}{\overset{-CH_3OH}{\rightleftharpoons}}$ cyclohexyl$-\overset{O}{\overset{\|}{C}}H$

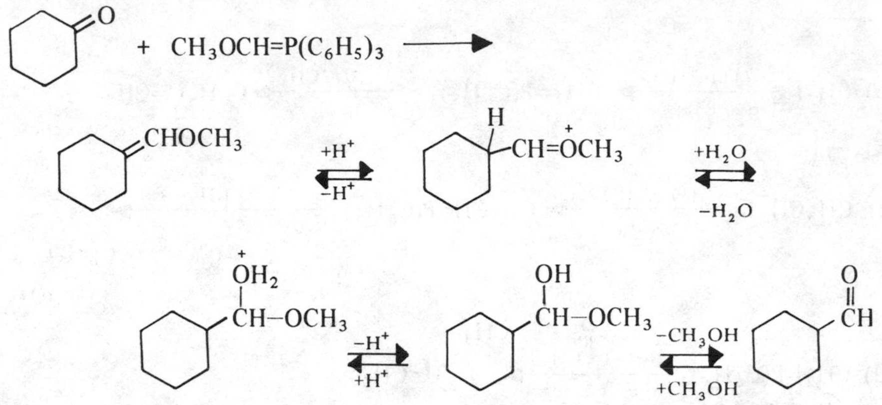

16.15

(a) cyclohexanone$=O + CH_2{=}S(CH_3)_2 \longrightarrow$ spiro-epoxide (cyclohexane spiro to CH_2 epoxide, with H, H) $+ CH_3SCH_3$

(b) $(CH_3)_2C{=}O + CH_2{=}S(CH_3)_2 \longrightarrow (CH_3)_2C\overset{O}{—}CH_2 + CH_3SCH_3$

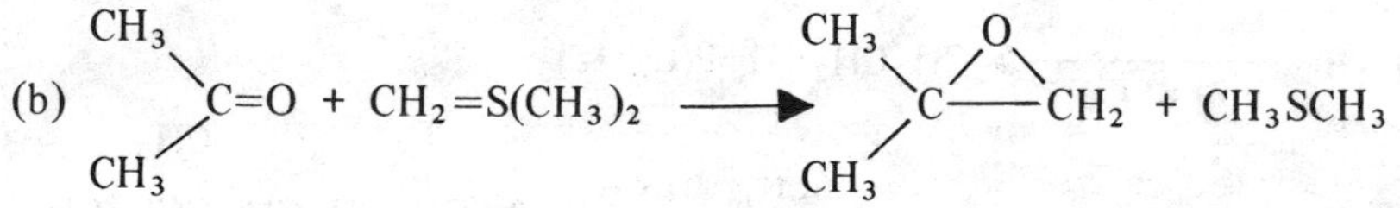

16.16

No. 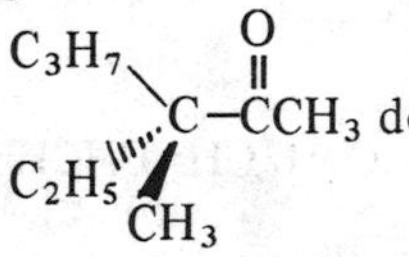 $(C_3H_7)(C_2H_5)(CH_3)C-\overset{O}{\overset{\|}{C}}CH_3$ does not have a hydrogen attached to its chiral α-carbon and thus enol formation involving the chiral carbon is not possible. With $(C_2H_5)(CH_3)(H)CCH_2\overset{O}{\overset{\|}{C}}CH_3$ the chiral carbon is a β-carbon and thus enol formation does not affect it.

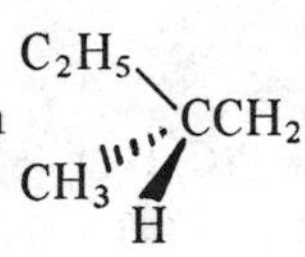

16.17

In OD^- / D_2O

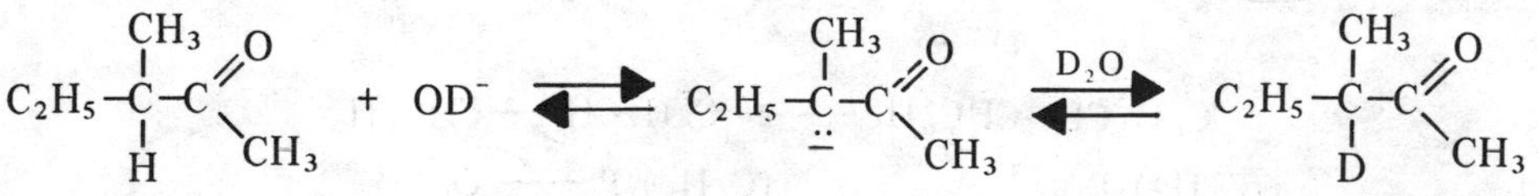

$$C_2H_5-\underset{H}{\overset{CH_3}{C}}-\overset{O}{\overset{\|}{C}}-CH_3 + OD^- \rightleftharpoons C_2H_5-\underset{\ddot{}}{\overset{CH_3}{C^-}}-\overset{O}{\overset{\|}{C}}-CH_3 \overset{D_2O}{\rightleftharpoons} C_2H_5-\underset{D}{\overset{CH_3}{C}}-\overset{O}{\overset{\|}{C}}-CH_3$$

In D_3O^+ / D_2O

$$C_2H_5-\underset{H}{\overset{CH_3}{C}}-C(=O)CH_3 + D_3O^+ \xrightleftharpoons{(+D^+)} C_2H_5-\underset{H}{\overset{CH_3}{C}}-C(=\overset{+}{O}-D)CH_3$$

$$\xrightleftharpoons{-H^+} (CH_3)(C_2H_5)C{=}C(OD)(CH_3) \xrightleftharpoons{+D^+} C_2H_5-\underset{D}{\overset{CH_3}{C}}-C(=\overset{+}{O}-D)CH_3$$

$$\xrightleftharpoons{-D^+} C_2H_5-\underset{D}{\overset{CH_3}{C}}-C(=O)CH_3$$

16.18

The reaction is said to be "base promoted" because base is consumed as the reaction takes place. A catalyst is, by definition, not consumed.

16.19

(a) The slow step in base-catalyzed racemization is the same as that in base-promoted halogenation–*the formation of an enolate ion*. (Formation of an enolate ion from phenyl *sec*-butyl ketone leads to racemization because the enolate ion is achiral. When it accepts a proton it yields a racemic modification.) The slow step in acid-catalyzed racemization is also the same as that in acid-catalyzed halogenation–*the formation of an enol*. (The enol, like the enolate ion, is achiral and tautomerizes to yield a racemic modification of the ketone.)

(b) According to the mechanism given, the slow step for acid-catalyzed iodination (formation of the enol) is the same as that for acid-catalyzed bromination. Thus we would expect both reactions to occur at the same rate.

(c) Again, the slow step for both reactions (formation of the enolate ion) is the same, and consequently, both reactions take place at the same rate.

16.20

(a)

$$\overset{\beta}{CH_3}\,\overset{\alpha}{CH_2}\overset{O}{\overset{\|}{C}}H + \overset{-}{O}H \rightleftharpoons CH_3\overset{\cdot\cdot -}{C}H\overset{O}{\overset{\|}{C}}H + H_2O$$

$$CH_3CH_2\overset{O}{\overset{\|}{C}}H + \overset{-}{:}\underset{CH_3}{\underset{|}{C}H}\overset{O}{\overset{\|}{C}}H \rightleftharpoons CH_3CH_2\overset{O^-}{\overset{|}{C}H}\underset{CH_3}{\underset{|}{C}H}\overset{O}{\overset{\|}{C}}H$$

$$CH_3CH_2\overset{O^-}{\overset{|}{C}H}\underset{CH_3}{\underset{|}{C}H}\overset{O}{\overset{\|}{C}}H + HOH \rightleftharpoons CH_3CH_2\overset{OH}{\overset{|}{C}H}\underset{CH_3}{\underset{|}{C}H}\overset{O}{\overset{\|}{C}}H$$

(b) For $CH_3CH_2\underset{}{\overset{OH}{C}}HCH_2CH_2\overset{O}{\overset{\|}{C}}H$ to form, a hydroxide ion would have to remove a β-proton in the first step. This does not happen because the anion that would be produced, i.e., $^{-}{:}CH_2CH_2CHO$, cannot be stabilized by resonance.

(c) $CH_3CH_2CH{=}\underset{CH_3}{C}\overset{O}{\overset{\|}{C}}H$

16.21

(a) $2CH_3CH_2CH_2CHO \xrightarrow[H_2O]{OH^-} CH_3CH_2CH_2\overset{OH}{C}H\underset{CH_2CH_3}{C}HCHO$

(b) Product of (a) $\xrightarrow[(-H_2O)]{H^+} CH_3CH_2CH_2CH{=}\underset{CH_2CH_3}{C}CHO \xrightarrow{NaBH_4} CH_3CH_2CH_2CH{=}\underset{CH_2CH_3}{C}CH_2OH$

(c) Product of (b) $\xrightarrow[Ni]{H_2} CH_3CH_2CH_2CH_2\underset{CH_2CH_3}{C}HCH_2OH$

(d) Product of (a) $\xrightarrow{NaBH_4} CH_3CH_2CH_2\overset{OH}{C}H\underset{CH_2CH_3}{C}HCH_2OH$

16.22

(a) $CH_3\overset{O}{\overset{\|}{C}}CH_3 + OH^- \rightleftharpoons CH_3\overset{O}{\overset{\|}{C}}CH_2{:}^- + H_2O$

$CH_3\overset{O}{\overset{\|}{C}}CH_2{:}^- + CH_3\overset{O}{\overset{\|}{C}}CH_3 \rightleftharpoons CH_3\overset{O}{\overset{\|}{C}}CH_2\underset{CH_3}{\overset{O^-}{C}}CH_3$

$CH_3\overset{O}{\overset{\|}{C}}CH_2\underset{CH_3}{\overset{O^-}{C}}CH_3 + HOH \rightleftharpoons CH_3\overset{O}{\overset{\|}{C}}CH_2\underset{CH_3}{\overset{OH}{C}}CH_3 + OH^-$

(b) $CH_3\overset{O}{\overset{\|}{C}}CH{=}\underset{CH_3}{\underset{|}{C}}CH_3$

16.23

$$CH_3CH_2\overset{O}{\overset{\|}{C}}H + OH^- \rightleftharpoons CH_3\overset{..}{\overset{-}{C}}H\overset{O}{\overset{\|}{C}}H + H_2O$$

$$CH_3\overset{O}{\overset{\|}{C}}H + CH_3\overset{..}{\overset{-}{C}}H\overset{O}{\overset{\|}{C}}H \rightleftharpoons CH_3\overset{O^-}{\overset{|}{C}}H\underset{CH_3}{\underset{|}{C}}H\overset{O}{\overset{\|}{C}}H$$

$$CH_3\overset{O^-}{\overset{|}{C}}H\underset{CH_3}{\underset{|}{C}}H\overset{O}{\overset{\|}{C}}H + H_2O \rightleftharpoons CH_3\overset{OH}{\overset{|}{C}}H\underset{CH_3}{\underset{|}{C}}H\overset{O}{\overset{\|}{C}}H$$

2-Methyl-3-hydroxybutanal

$$CH_3\overset{O}{\overset{\|}{C}}H + OH^- \rightleftharpoons {}^-{:}CH_2\overset{O}{\overset{\|}{C}}H + H_2O$$

$$CH_3\,CH_2\overset{O}{\overset{\|}{C}}H + {}^-{:}CH_2\overset{O}{\overset{\|}{C}}H \rightleftharpoons CH_3CH_2\overset{O^-}{\overset{|}{C}}HCH_2\overset{O}{\overset{\|}{C}}H$$

$$CH_3CH_2\overset{O^-}{\overset{|}{C}}HCH_2\overset{O}{\overset{\|}{C}}H + H_2O \rightleftharpoons CH_3CH_2\overset{OH}{\overset{|}{C}}HCH_2\overset{O}{\overset{\|}{C}}H$$

3-Hydroxypentanal

16.24

Three successive aldol additions occur.

First Aldol Addition:

$$CH_3\overset{O}{\overset{\|}{C}}H + OH^- \rightleftharpoons \overset{-}{:}CH_2\overset{O}{\overset{\|}{C}}H + H_2O$$

$$H\overset{O}{\overset{\|}{C}}H + \overset{-}{:}CH_2\overset{O}{\overset{\|}{C}}H \rightleftharpoons {}^-OCH_2CH_2\overset{O}{\overset{\|}{C}}H$$

$${}^-OCH_2CH\overset{O}{\overset{\|}{C}}H + H_2O \rightleftharpoons HOCH_2CH_2\overset{O}{\overset{\|}{C}}H$$

Second Aldol Addition

$$HOCH_2CH_2\overset{O}{\overset{\|}{C}}H + OH^- \rightleftharpoons HOCH_2\ddot{C}^-H\overset{O}{\overset{\|}{C}}H + H_2O$$

$$H\overset{O}{\overset{\|}{C}}H + HOCH_2\ddot{C}^-H\overset{O}{\overset{\|}{C}}H \rightleftharpoons HOCH_2CH(CH_2O^-)CHO$$

$$HOCH_2CH(CH_2O^-)CHO + H_2O \rightleftharpoons HOCH_2CH(CH_2OH)CHO + OH^-$$

Third Aldol Addition

$$HOCH_2CH(CH_2OH){-}CHO + OH^- \rightleftharpoons HOCH_2\ddot{C}^-(CH_2OH){-}CHO$$

$$H\overset{O}{\overset{\|}{C}}H + HOCH_2\ddot{C}^-(CH_2OH){-}CHO \rightleftharpoons HOCH_2{-}C(CH_2OH)(CH_2O^-){-}CHO$$

$$HOCH_2{-}C(CH_2OH)(CH_2O^-){-}CHO + H_2O \rightleftharpoons HOCH_2{-}C(CH_2OH)_2{-}CHO + OH^-$$

16.25

(a) $CH_3COOH + BF_3 \rightleftharpoons CH_3COOBF_3^- + H^+$

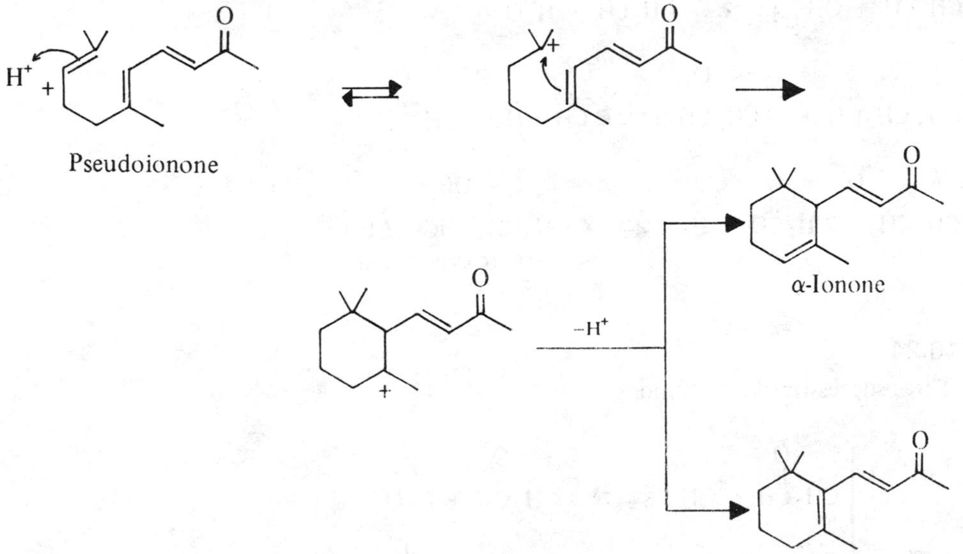

H^+ + Pseudoionone ⇌ (cyclized cation) → (ring cation) $\xrightarrow{-H^+}$ α-Ionone + β-Ionone

Pseudoionone

α-Ionone

β-Ionone

(b) In β-ionone the double bonds and the carbonyl group are conjugated, thus it is more stable.

(c) β-Ionone, because it is a conjugated unsaturated system.

16.26

$$HCH(=O) + OH^- \rightleftharpoons H-C(O^-)(H)-O-H$$

$$H-C(O^-)(H)-O-H + HC(=O)-C(CH_2OH)_2-CH_2OH \longrightarrow$$

(from Prob. 16.24)

$$H_2O + H-C(=O)-O^- + H-C(O^-)(H)-C(CH_2OH)_2-CH_2OH \rightleftharpoons$$

$$HOCH_2-C(CH_2OH)_2-CH_2OH + OH^- + HC(=O)-O^-$$

Pentaerythritol

16.27

(b) 2-Methyl-1,3-cyclohexanedione is more acidic because its enolate ion is stabilized by an additional resonance structure.

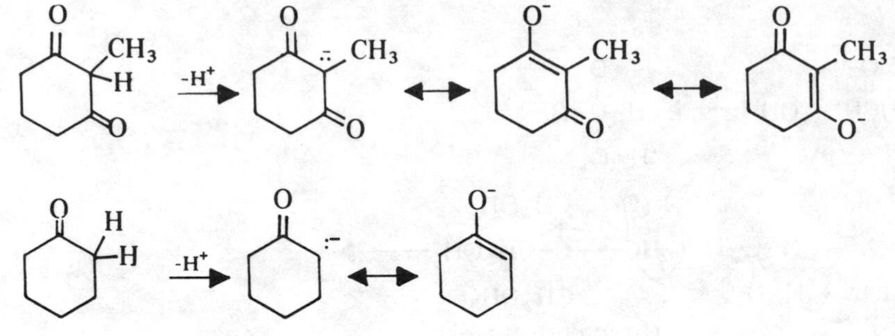

16.28

(a) $C_6H_5\overset{O}{\overset{\|}{C}}CH_3 \underset{+H^+}{\overset{-H^+}{\rightleftharpoons}} C_6H_5\overset{O}{\overset{\|}{C}}CH_2{:}^-$

$C_6H_5\overset{O}{\overset{\|}{C}}CH_2{:}^- + C_6H_5CH{=}CH\overset{O}{\overset{\|}{C}}C_6H_5 \rightleftharpoons$

$C_6H_5CH(CH_2COC_6H_5){-}CH{=}C(O^-)C_6H_5 \underset{-H^+}{\overset{+H^+}{\rightleftharpoons}} C_6H_5CH(CH_2COC_6H_5)CH_2\overset{O}{\overset{\|}{C}}C_6H_5$

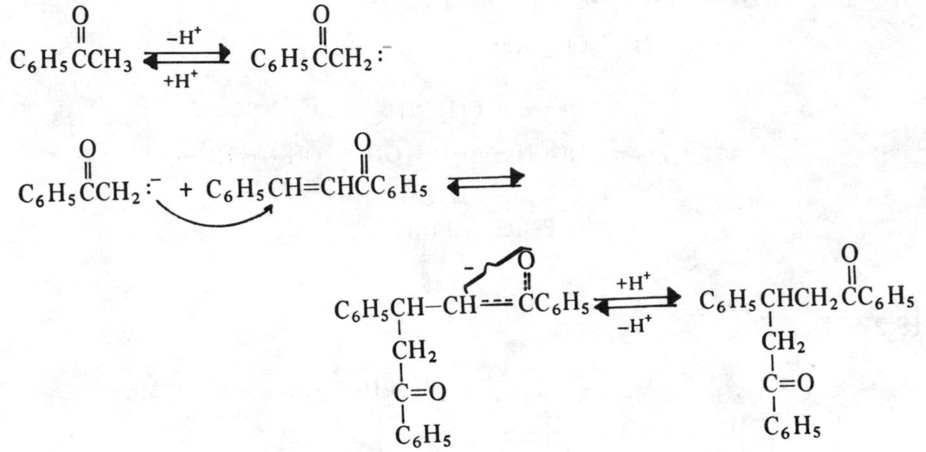

(b) cyclopentadiene $\underset{+H^+}{\overset{-H^+}{\rightleftharpoons}}$ cyclopentadienyl anion

cyclopentadienyl anion $+ C_6H_5CH{=}CH\overset{O}{\overset{\|}{C}}C_6H_5 \rightleftharpoons$

$C_6H_5CH(C_5H_5){-}CH{=}C(O^-)C_6H_5 \underset{-H^+}{\overset{+H^+}{\rightleftharpoons}} C_6H_5CH(C_5H_5)CH_2\overset{O}{\overset{\|}{C}}C_6H_5$

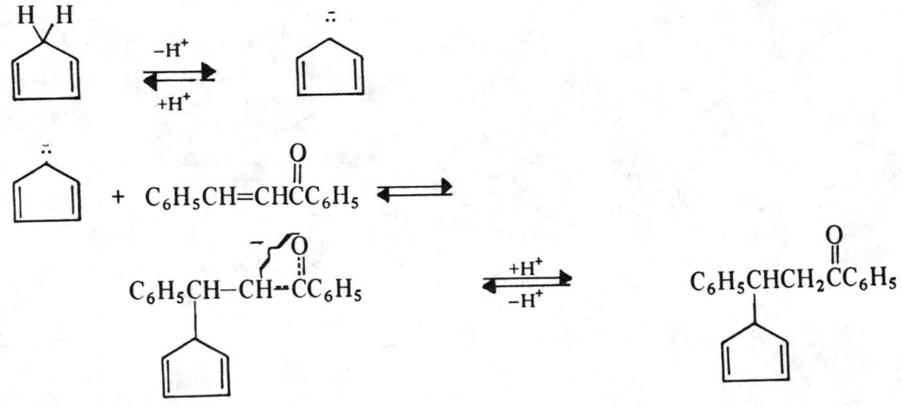

16.29

$H_2\ddot{N}{-}\ddot{N}H_2 + CH_2{=}CH{-}\overset{O}{\overset{\|}{C}}H \xrightarrow[\text{addition}]{\text{conjugate}}$

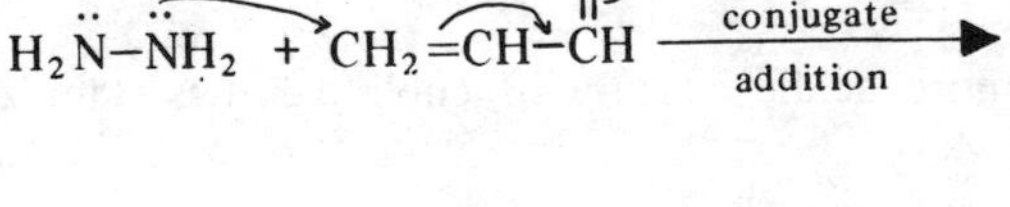

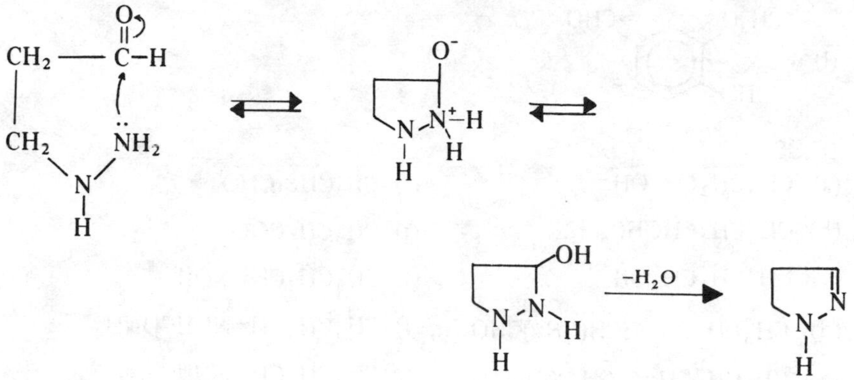

CH_2—C(=O)—H ; CH_2 ; $\ddot{N}H_2$; N—H $\rightleftarrows$ O$^-$; N—N^+—H ; H ; H $\rightleftarrows$

OH ; N—N—H ; H $\xrightarrow{-H_2O}$ N=N ; N—H

16.30

(a) Acetone, $CH_3\overset{O}{\overset{\|}{C}}CH_3$

(b) Acetophenone, $C_6H_5\overset{O}{\overset{\|}{C}}CH_3$

(d) 2-Pentanone, $CH_3CH_2CH_2\overset{O}{\overset{\|}{C}}CH_3$

(f) 1-Phenylethanol, $C_6H_5\overset{OH}{\overset{|}{C}}HCH_3$

(h) 2-Butanol, $CH_3CH_2\overset{OH}{\overset{|}{C}}HCH_3$

(i) 1-Acetylnaphthalene, $\overset{O}{\overset{\|}{C}}CH_3$

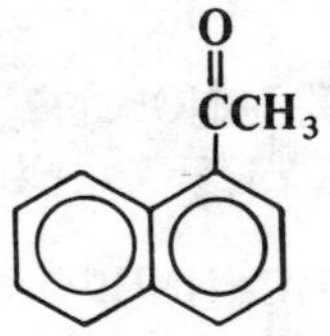

16.31

(a) HCHO

(b) CH_3CHO

(c) $C_6H_5CH_2CHO$

(d) CH_3COCH_3

(e) $CH_3COCH_2CH_3$

(f) $CH_3COC_6H_5$

(g) $C_6H_5CH{=}CHCOCH_3$

(h) $C_6H_5CH{=}CHCOC_6H_5$

(i) $C_6H_5COC_6H_5$

(j) 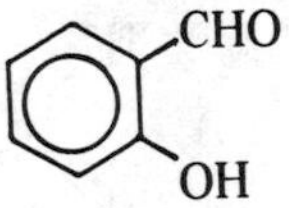 CHO ; OH

(k) 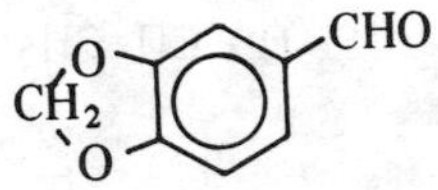 CHO ; O ; CH_2 ; O

(l) 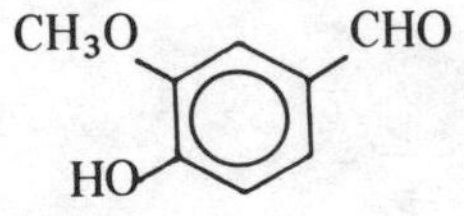 CH_3O, HO on benzene ring with CHO (4-hydroxy-3-methoxybenzaldehyde)

16.32

(a) $CH_3CH_2CH_2OH$

(b) $CH_3CH_2CHOHC_6H_5$

(c) $CH_3CH_2CH_2OH$

(d) $CH_3CH_2CHOHCH(CH_3)CHO$

(e) $CH_3CH_2CH{=}C(CH_3)CHO$

(f) $CH_3CH_2CH_2OH$

(g) 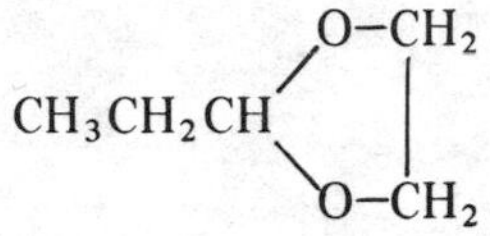 $CH_3CH_2CH\langle O{-}CH_2 / O{-}CH_2 \rangle$ (ring)

(h) $CH_3CH_2CH{=}CHCH_3$

(i) $CH_3CHBrCHO$

(j) $CH_3CH_2COO^- + Ag\downarrow$

(k) $CH_3CH_2CH{=}NOH$

(l) $CH_3CH_2CH{=}NNHCONH_2$

(m) $CH_3CH_2CH{=}NNHC_6H_5$

(n) CH_3CH_2COOH

(o) 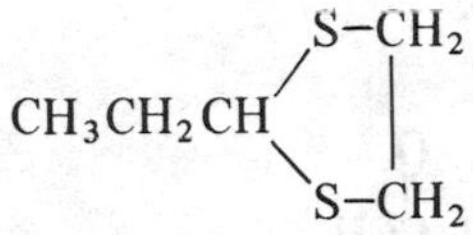 $CH_3CH_2CH\langle S{-}CH_2 / S{-}CH_2 \rangle$ (ring)

(p) $CH_3CH_2CH_3 + CH_3CH_3 + NiS$

16.33

(a) $CH_3CHOHCH_3$

(b) $C_6H_5COHCH_3$ with CH_3 on C, i.e. $C_6H_5C(OH)(CH_3)CH_3$

(c) $CH_3CHOHCH_3$

(d) $CH_3COCH_2C(OH)(CH_3)CH_3$

(e) $CH_3COCH{=}C(CH_3)CH_3$

(f) $CH_3CHOHCH_3$

(g) 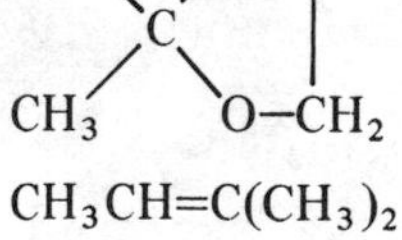 $(CH_3)_2C\langle O{-}CH_2 / O{-}CH_2 \rangle$ (ring)

(h) $CH_3CH{=}C(CH_3)_2$

(i) CH_3COCH_2Br

(j) No reaction

(k) $CH_3C(CH_3){=}NOH$

(l) $CH_3C(CH_3){=}NNHCONH_2$

(m) $CH_3C(CH_3){=}NNHC_6H_5$

(n) No reaction

(o) 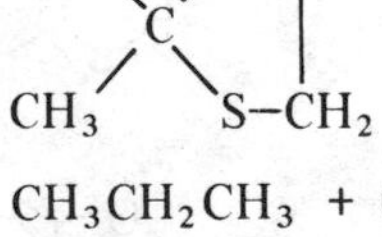 $(CH_3)_2C\langle S{-}CH_2 / S{-}CH_2 \rangle$ (ring)

(p) $CH_3CH_2CH_3 + CH_3CH_3 + NiS$

16.34

(a) 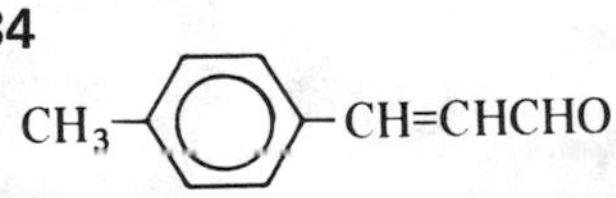 $CH_3{-}C_6H_4{-}CH{=}CHCHO$ (para)

(b) 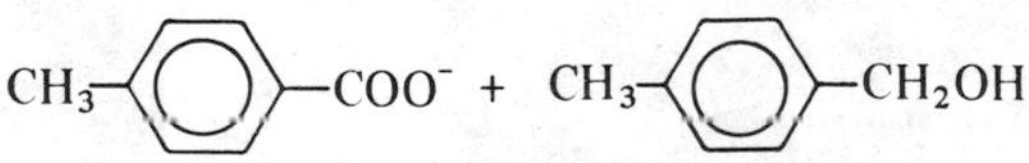 $CH_3{-}C_6H_4{-}COO^- + CH_3{-}C_6H_4{-}CH_2OH$ (para)

(c) 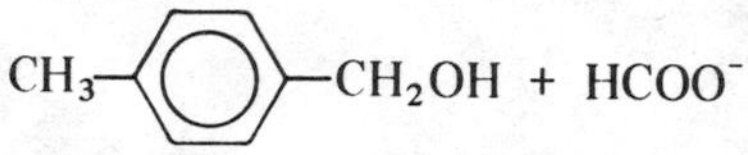 $CH_3-C_6H_4-CH_2OH + HCOO^-$

(e) 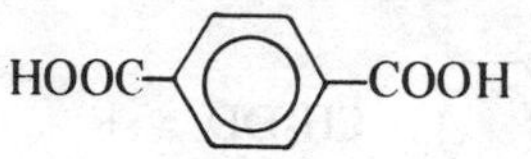 $HOOC-C_6H_4-COOH$

(d) 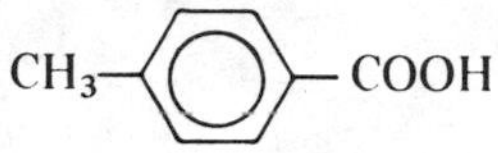 $CH_3-C_6H_4-COOH$

(f) 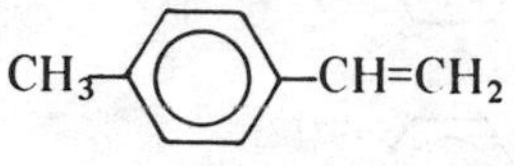 $CH_3-C_6H_4-CH{=}CH_2$

16.35

(a) 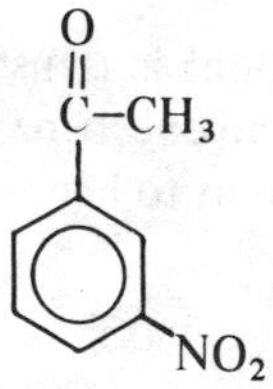 $C_6H_4(NO_2)-C(=O)-CH_3$ (3-nitro)

(d) 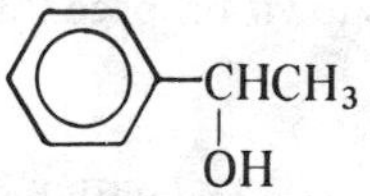 $C_6H_5-CH(OH)CH_3$

(b) 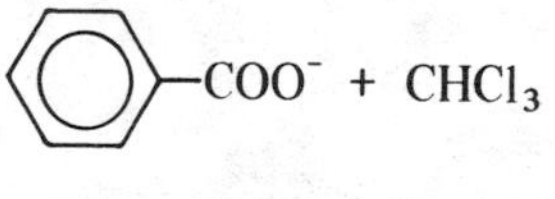 $C_6H_5-COO^- + CHCl_3$

(e) $C_6H_5-C(CH_3)(OH)-C_6H_5$

(c) $C_6H_5-C(=CH_2)-CH_3$

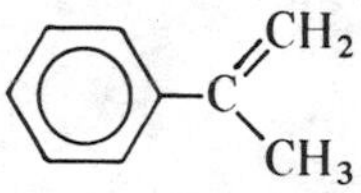

16.36

(a) 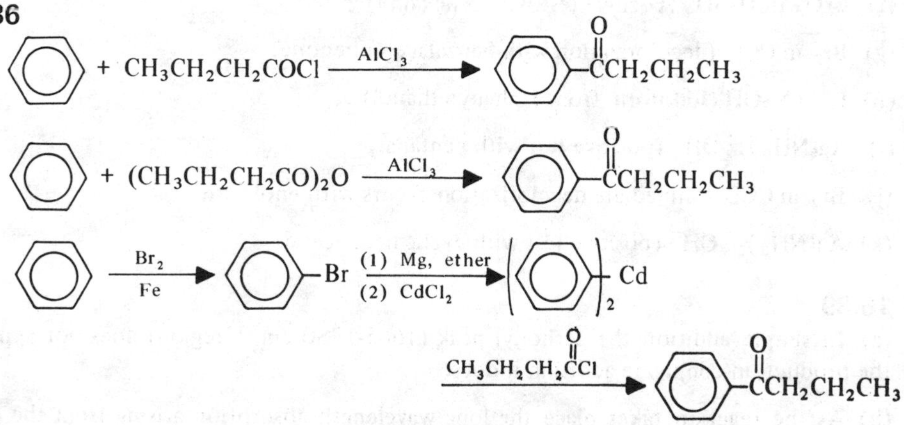 $C_6H_6 + CH_3CH_2CH_2COCl \xrightarrow{AlCl_3} C_6H_5-\overset{O}{\overset{\|}{C}}CH_2CH_2CH_3$

$C_6H_6 + (CH_3CH_2CH_2CO)_2O \xrightarrow{AlCl_3} C_6H_5-\overset{O}{\overset{\|}{C}}CH_2CH_2CH_3$

$C_6H_6 \xrightarrow[Fe]{Br_2} C_6H_5-Br \xrightarrow[(2)\ CdCl_2]{(1)\ Mg,\ ether} (C_6H_5)_2Cd \xrightarrow{CH_3CH_2CH_2\overset{O}{\overset{\|}{C}}Cl} C_6H_5-\overset{O}{\overset{\|}{C}}CH_2CH_2CH_3$

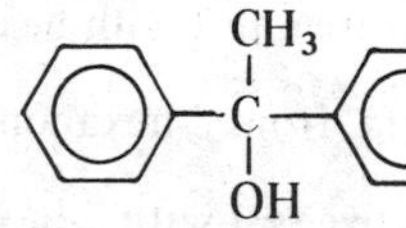

(b) 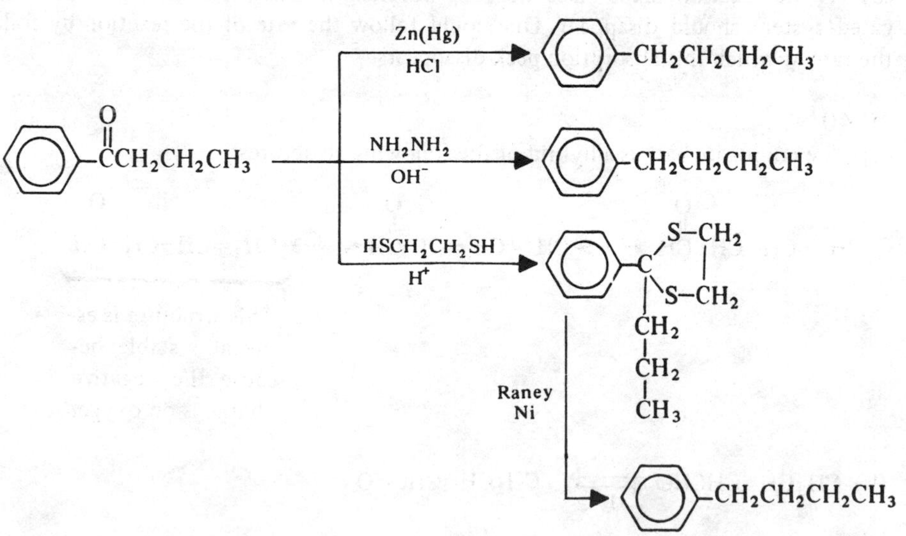 $C_6H_5-\overset{O}{\overset{\|}{C}}CH_2CH_2CH_3$

$\xrightarrow[HCl]{Zn(Hg)} C_6H_5-CH_2CH_2CH_2CH_3$

$\xrightarrow[OH^-]{NH_2NH_2} C_6H_5-CH_2CH_2CH_2CH_3$

$\xrightarrow[H^+]{HSCH_2CH_2SH} C_6H_5-C(S-CH_2)(S-CH_2)-CH_2CH_2CH_3 \xrightarrow{Raney\ Ni} C_6H_5-CH_2CH_2CH_2CH_3$

16.37

(a) $C_6H_5-CH_2OD$ + $C_6H_5-COO^-$

(b) $C_6H_5-CH_2OH$ + $C_6H_5-COO^-$

(c) Yes. In both reactions a hydride ion (rather than a deuteride ion) is transferred to benzaldehyde. This shows that the hydride ion is transferred from one benzaldehyde molecule to another (as shown on page 734) and not from the solvent to benzaldehyde.

16.38

(a) $Ag(NH_3)_2{}^+OH^-$ (positive test with benzaldehyde)

(b) $Ag(NH_3)_2{}^+OH^-$ (positive test with hexanal)

(c) I_2 in NaOH (Iodoform, from 2-hexanone)

(d) CrO_3 in H_2SO_4 (positive test with 2-hexanol)

(e) I_2 in NaOH (Iodoform from 2-hexanol)

(f) CrO_3 in H_2SO_4 (positive test with 3-hexanol)

(g) Br_2 in CCl_4 (decolorization with benzalacetophenone)

(h) I_2 in NaOH (Iodoform from 1-phenylethanol)

(i) $Ag(NH_3)_2{}^+OH^-$ (positive test with pentanal)

(j) Br_2 in CCl_4 (immediate decolorization occurs with enol form)

(k) $Ag(NH_3)_2{}^+OH^-$ (positive test with cyclic hemiacetal)

16.39

(a) In simple addition the carbonyl peak (1665-1780 cm^{-1} region) does not appear in the product; in conjugate addition it does.

(b) As the reaction takes place the long wavelength absorption arising from the conjugated system should disappear. One could follow the rate of the reaction by following the rate at which this absorption peak disappears.

16.40

(a) The conjugate base is a hybrid of the following structures:

$$^{-}\!:CH_2-CH=CH-\overset{\overset{\displaystyle O}{\|}}{C}H \longleftrightarrow CH_2=CH-\overset{..}{\bar{C}}H-\overset{\overset{\displaystyle O}{\|}}{C}H \longleftrightarrow CH_2=CH-CH=\overset{\overset{\displaystyle O^-}{|}}{C}H$$

This structure is especially stable because the negative charge is on oxygen

(b) $CH_3CH=CHCHO \underset{+H^+}{\overset{-H^+}{\rightleftharpoons}} \bar{:}CH_2CH-CHCHO$

$$C_6H_5CH{=}CH\overset{O}{\overset{\|}{C}}H + \bar{:}CH_2CH{=}CHCHO \rightleftharpoons$$

$$C_6H_5CH{=}CH\overset{O^-}{\overset{|}{C}}H{-}CH_2CH{=}CHCHO \underset{-H^+}{\overset{+H^+}{\rightleftharpoons}} C_6H_5CH{=}CH\overset{OH}{\overset{|}{C}}H{-}CH_2CH{=}CHCHO$$

$$\xrightarrow{-H_2O} C_6H_5CH{=}CHCH{=}CHCH{=}CHCHO$$

16.41

(a) 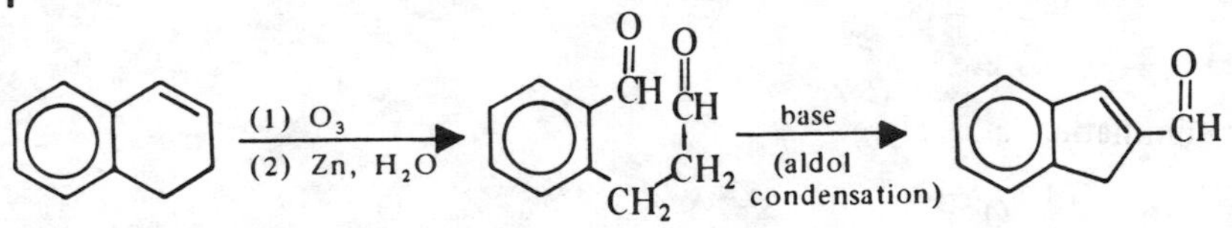 [dihydronaphthalene] $\xrightarrow[(2)\ Zn,\ H_2O]{(1)\ O_3}$ [dialdehyde: $\overset{O}{\overset{\|}{C}}H$, $\overset{O}{\overset{\|}{C}}H$, CH_2, CH_2] $\xrightarrow[\text{(aldol condensation)}]{\text{base}}$ [indene-$\overset{O}{\overset{\|}{C}}H$]

(b) [octalin] $\xrightarrow[(2)\ Zn,\ H_2O]{(1)\ O_3}$ [cyclodecanedione] $\xrightarrow[\text{(aldol condensation)}]{\text{base}}$ [bicyclic enone]

(c) [CH_3-substituted octalin] $\xrightarrow[(2)\ Na_2SO_3]{(1)\ OsO_4}$ [CH_3, diol: OH, OH] $\xrightarrow[\text{(sect. 19.4D)}]{HIO_4}$

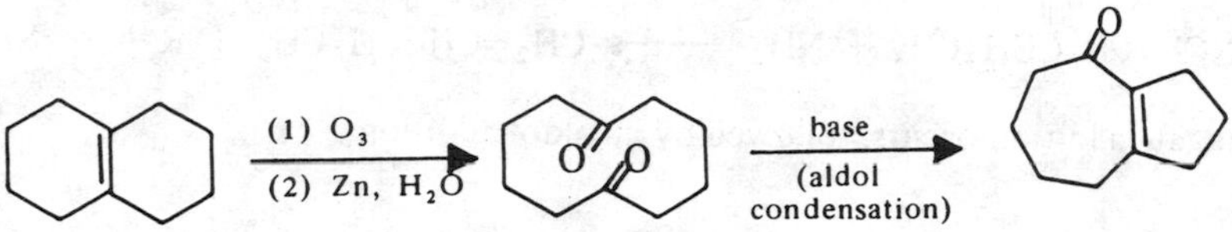

[CH_3, CHO, CHO] $\xrightarrow[\text{(aldol condensation)}]{\text{base}}$ [CH_3, CHO] + [CH_3, CHO]

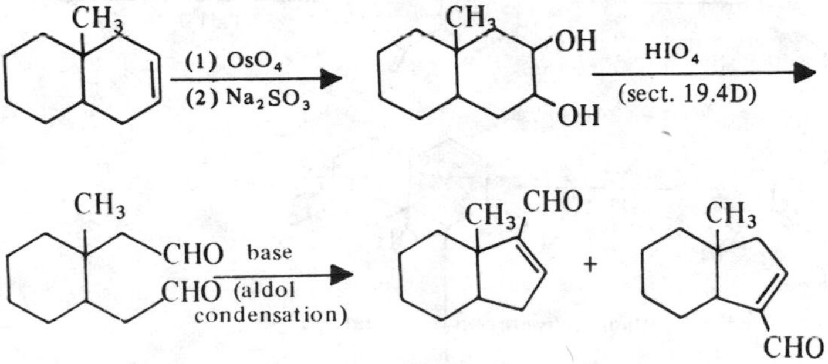

(d) 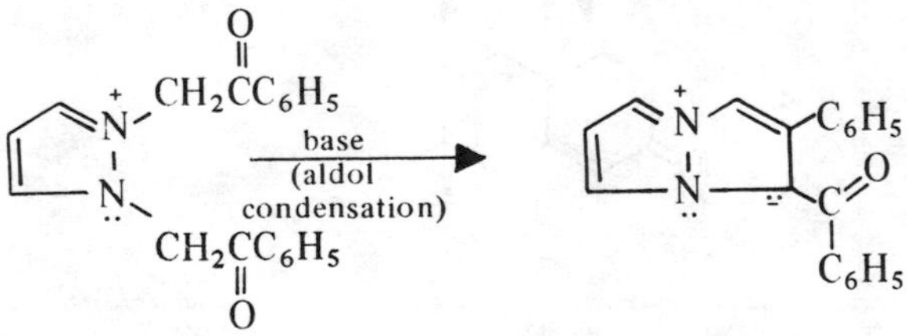 [pyrazolium with $\overset{+}{N}$–$CH_2\overset{O}{\overset{\|}{C}}C_6H_5$ and N–$CH_2\overset{\|}{\underset{O}{C}}C_6H_5$] $\xrightarrow[\text{(aldol condensation)}]{\text{base}}$ [fused bicyclic product: $\overset{+}{N}$, N, C_6H_5, $\bar{C}$–C=O, C_6H_5]

16.42

(a) 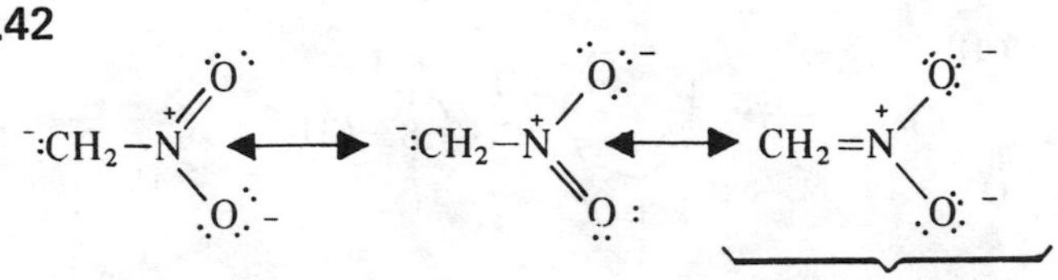 $\bar{:}CH_2{-}\overset{+}{N}(=\ddot{O})(-\ddot{O}:^-) \longleftrightarrow \bar{:}CH_2{-}\overset{+}{N}(-\ddot{O}:^-)(=\ddot{O}:) \longleftrightarrow CH_2{=}\overset{+}{N}(-\ddot{O}:^-)(-\ddot{O}:^-)$

This structure is especially stable because both negative charges are on oxygen

(b) All of these syntheses are variations of a crossed aldol addition or condensation.

$$HCH(=O) + CH_3NO_2 \xrightarrow{\text{base}} HOCH_2CH_2NO_2$$

$$C_6H_5CHO + CH_3NO_2 \xrightarrow[(-H_2O)]{\text{base}} C_6H_5CH{=}CHNO_2$$

$$C_6H_5CHO + CH_3CH_2NO_2 \xrightarrow[(-H_2O)]{\text{base}} C_6H_5CH{=}C(CH_3)NO_2$$

16.43

First an elimination takes place,

$$R_3\overset{+}{N}CH_2CH_2CCH_2CH_3\,(C{=}O) + NH_2^- \longrightarrow CH_2{=}CHCCH_2CH_3\,(C{=}O) + R_3N + NH_3$$

then a conjugate addition occurs, followed by an aldol addition:

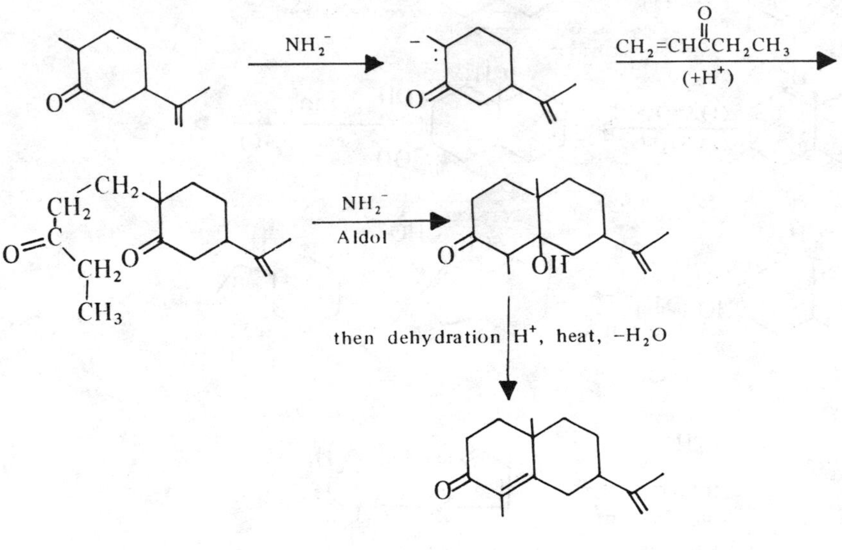

NH_2^-

$CH_2{=}CHCCH_2CH_3$ (C=O)

$(+H^+)$

CH_2, CH_2, $O{=}C$, CH_2, CH_3, O

NH_2^-

Aldol

O, OH

then dehydration H^+, heat, $-H_2O$

16.44

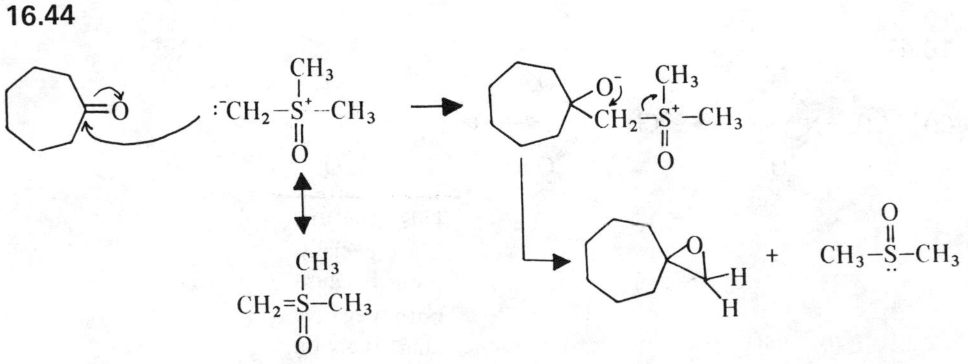

CH_3

$:^-CH_2-S^+-CH_3$ (S=O)

$CH_2{=}S-CH_3$ (CH_3, S=O)

O^-, $CH_2-S^+-CH_3$ (CH_3, S=O)

O, H, H

$+ \quad CH_3-\ddot{S}-CH_3$ (S=O)

16.45

(a) Compound U is phenyl ethyl ketone:

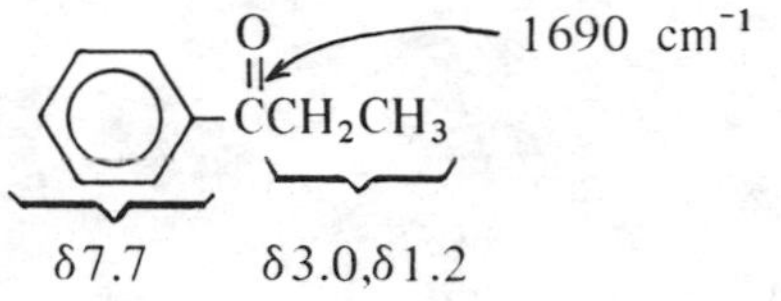

C_6H_5–$\overset{O}{\overset{\|}{C}}CH_2CH_3$ (1690 cm^{-1})

δ7.7 δ3.0,δ1.2

(b) Compound V is benzyl methyl ketone:

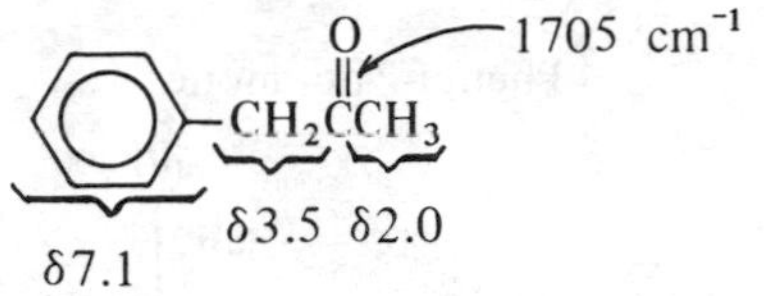

C_6H_5–$CH_2\overset{O}{\overset{\|}{C}}CH_3$ (1705 cm^{-1})

δ7.1 δ3.5 δ2.0

16.46

Compound W is:

multiplet, δ 7.3 (aromatic ring with CH_2–C=O–CH_2) singlet δ 3.4; infrared peak near 1715 cm^{-1}

heat (1) $KMnO_4$, OH^-, (2) H_3O^+ → Phthalic aicd ($C_6H_4(COOH)_2$)

Compound X is:

multiplet, δ 7.5 (aromatic ring with C=O, CH_2, CH_2) CH_2 triplet, δ 2.5; CH_2 triplet δ 3.1

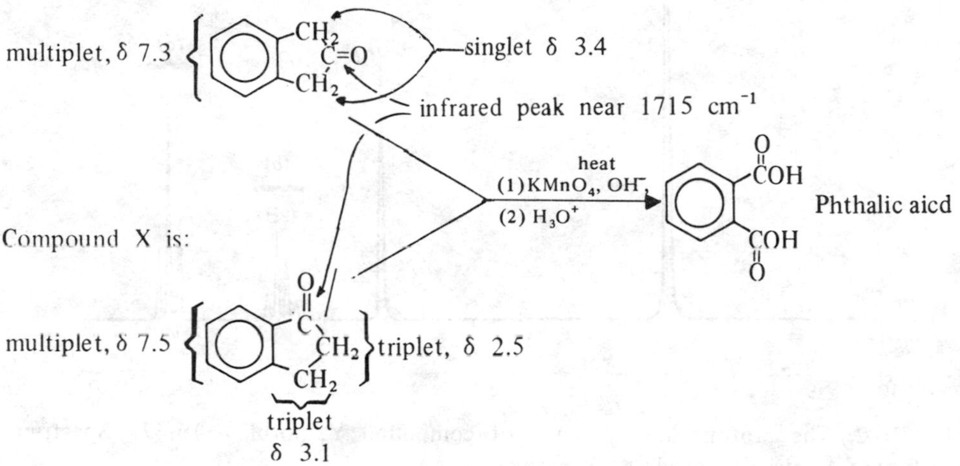

16.47

The pmr spectra (Figures 16.2 and 16.3) each have a five hydrogen peak near δ7.1, suggesting that Y and Z each have a C_6H_5– group. The infrared spectrum of each compound shows a strong peak near 1705 cm^{-1}. This indicates that each compound has a C=O group. We have, therefore, the following pieces,

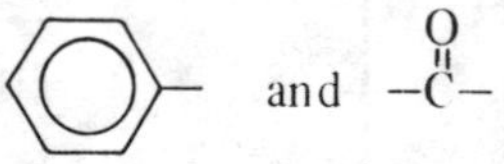

C_6H_5– and $-\overset{O}{\overset{\|}{C}}-$

If we subtract the atoms of these pieces from the molecular formula,

$$\begin{array}{l} C_{10}H_{12}O \\ -\underline{C_7\ H_5\ O}\ (C_6H_5 + C{=}O) \\ C_3\ H_7 \end{array}$$

We are left with, C_3 H_7

In the pmr spectrum of Y we see an ethyl group [triplet, δ1.0 (3H) and quartet, δ2.3 (2H)] and an unsplit $-CH_2-$ group [singlet, δ3.7 (2H)]. This means that Y must be,

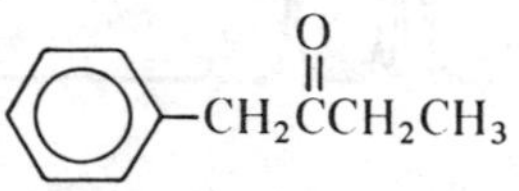

C_6H_5–$CH_2\overset{O}{\overset{\|}{C}}CH_2CH_3$

1-Phenyl-2-butanone

In the pmr spectrum of Z, we see an unsplit $-CH_3$ group [singlet, δ2.0 (3H)] and a multiplet (actually two superimposed triplets) at δ2.8. This means Z must be,

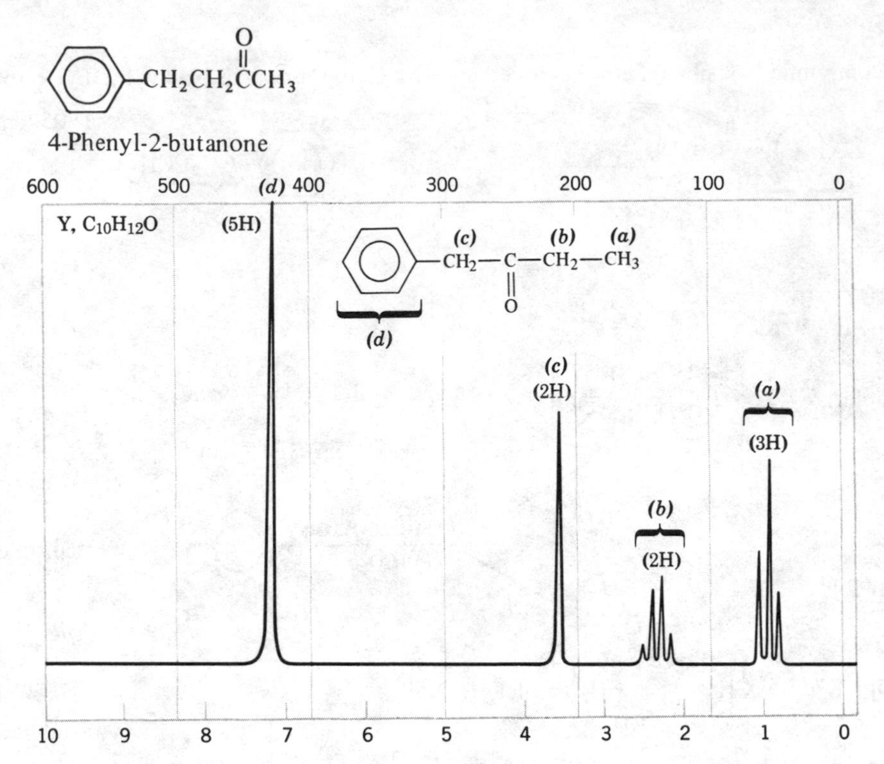

FIG. 16.2. The proton nmr spectrum of compound Y, porblem 16.47. (Spectrum courtesy of Aldrich Chemical Co., Milwaukee, Wis.)

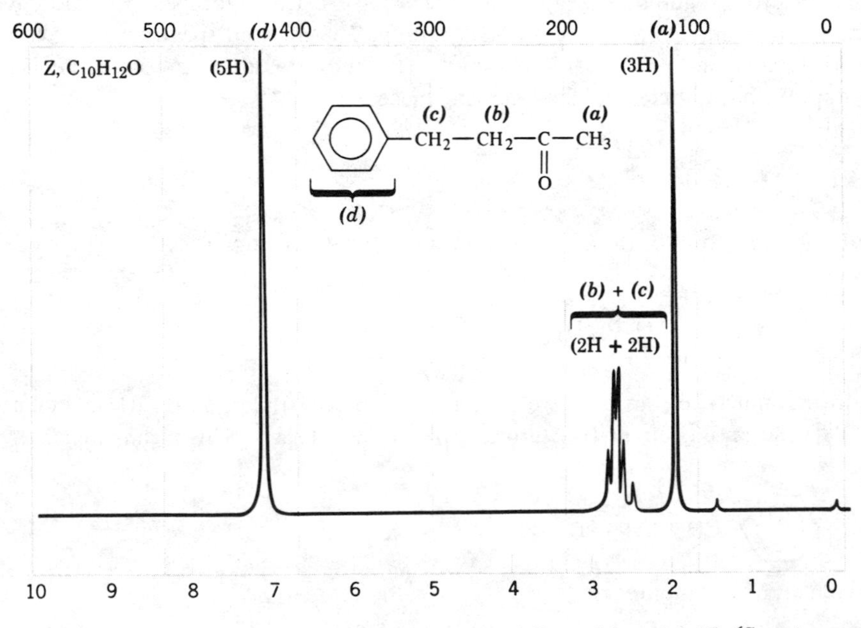

FIG. 16.3. The proton nmr spectrum of compound Z, problem 16.47. (Spectrum courtesy of Aldrich Chemical Co., Milwaukee, Wis.)

16.48

A is $CH_3\overset{O}{\overset{\|}{C}}CH_2CH(OCH_3)_2$

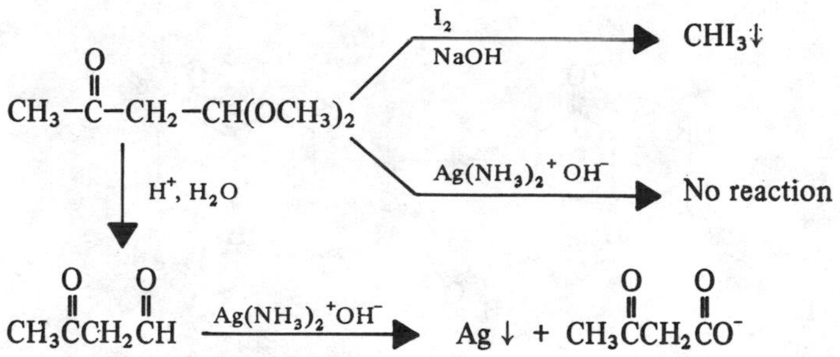

$$CH_3-\overset{O}{\overset{\|}{C}}-CH_2-CH(OCH_3)_2 \xrightarrow[NaOH]{I_2} CHI_3\downarrow$$

$$CH_3-\overset{O}{\overset{\|}{C}}-CH_2-CH(OCH_3)_2 \xrightarrow{Ag(NH_3)_2^+\,OH^-} \text{No reaction}$$

$$CH_3-\overset{O}{\overset{\|}{C}}-CH_2-CH(OCH_3)_2 \xrightarrow{H^+,\ H_2O} CH_3\overset{O}{\overset{\|}{C}}CH_2\overset{O}{\overset{\|}{C}}H$$

$$CH_3\overset{O}{\overset{\|}{C}}CH_2\overset{O}{\overset{\|}{C}}H \xrightarrow{Ag(NH_3)_2^+OH^-} Ag\downarrow + CH_3\overset{O}{\overset{\|}{C}}CH_2\overset{O}{\overset{\|}{C}}O^-$$

$$\underset{(a)}{CH_3}-\overset{O}{\overset{\|}{C}}-\underset{(b)}{CH_2}-\underset{(d)}{CH}(\underset{(c)}{OCH_3})_2$$

(a) Singlet $\delta 2.1$

(b) Doublet $\delta 2.6$

(c) Singlet $\delta 3.2$

(d) Triplet $\delta 4.7$

16.49

Reasonance contributions such as those shown below render the carbonyl groups of 2-hydroxybenzaldehyde and 4-hydroxybenzaldehyde highly unreactive toward nucleophilic attack. Hence no hydride transfer takes place.

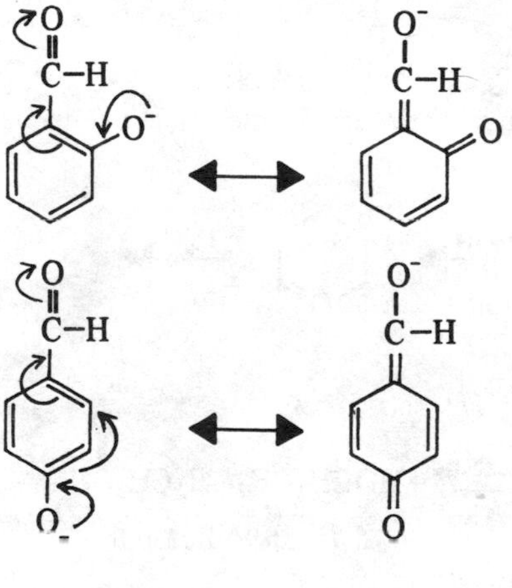

$$\text{2-}(O^-)C_6H_4CH{=}O \longleftrightarrow \text{(cyclohexadienone)}{=}CH{-}O^-$$

$$\text{4-}(O^-)C_6H_4CH{=}O \longleftrightarrow O{=}C_6H_4{=}CH{-}O^-$$

16.50

The two nitrogens of semicarbazide that are adjacent to the C=O group bear partial positive charges because of resonance contributions made by the second and third structures below,

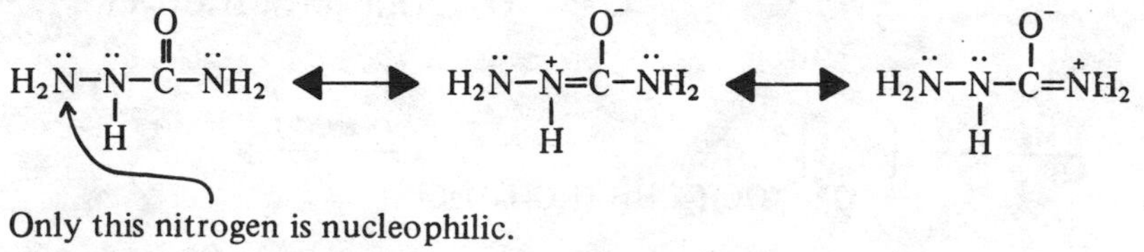

$$H_2\ddot{N}-\underset{H}{\underset{|}{\ddot{N}}}-\overset{O}{\overset{\|}{C}}-\ddot{N}H_2 \longleftrightarrow H_2\ddot{N}-\underset{H}{\underset{|}{\overset{+}{N}}}{=}\overset{O^-}{\overset{|}{C}}-\ddot{N}H_2 \longleftrightarrow H_2\ddot{N}-\underset{H}{\underset{|}{\ddot{N}}}-\overset{O^-}{\overset{|}{C}}{=}\overset{+}{N}H_2$$

Only this nitrogen is nucleophilic.

16.51

Abstraction of an α-hydrogen at the ring junction yields an enolate ion that can then accept a proton to form either *trans*-1-decalone or *cis*-1-decalone. Since *trans*-1-decalone is more stable, it predominates at equilibrium.

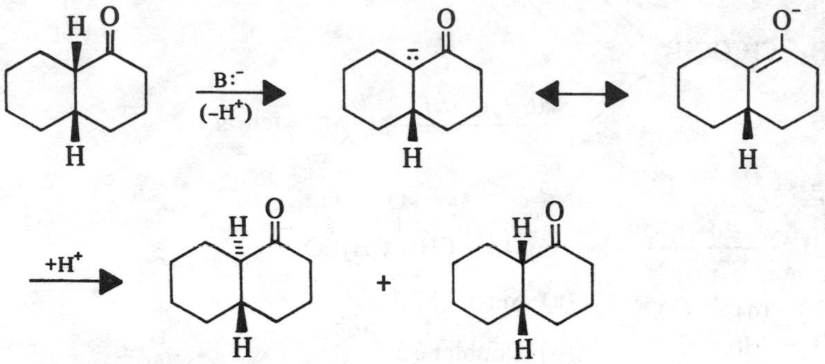

B:⁻ (–H⁺)

+H⁺

(95%) *trans*-1-Decalone (more stable)

(5%) *cis*-1-Decalone (less stable)

16.52

(a) +H⁺ → +ROH → –H⁺ → O–R

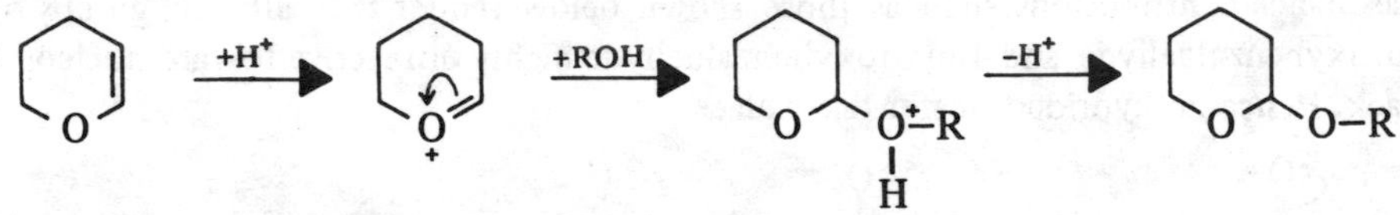

(b) Tetrahydropyranyl ethers are acetals; thus they are stable in aqueous base and hydrolyze readily in aqueous acid.

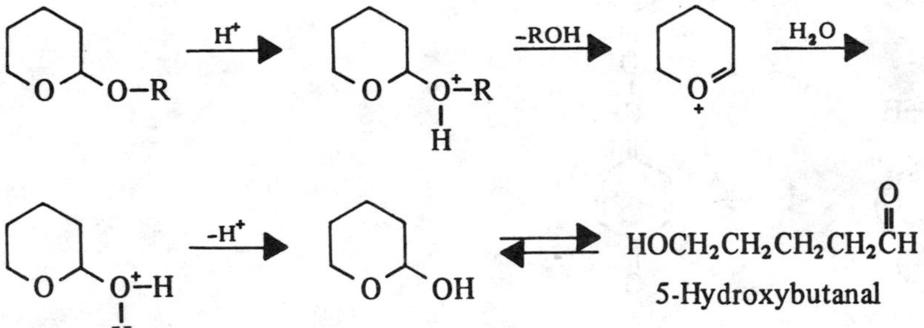

O–R →(H^+) O⁺–R (H) →(–ROH) →(H_2O)

O⁺–H (H) →(–H⁺) OH ⇌ $HOCH_2CH_2CH_2CH_2\overset{O}{\overset{\|}{C}}H$

5-Hydroxybutanal

(c) $HOCH_2CH_2CH_2CH_2Cl$ →(dihydropyran, H^+) $OCH_2CH_2CH_2CH_2Cl$

→(Mg, ether) $OCH_2CH_2CH_2CH_2MgCl$ →($CH_3\overset{O}{\overset{\|}{C}}CH_3$)

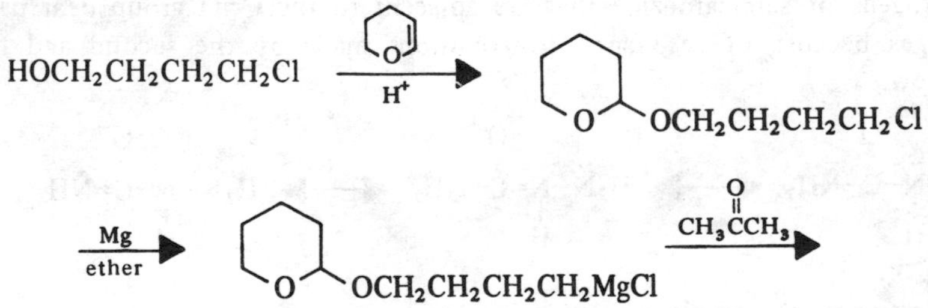

$$\text{(tetrahydropyranyl)}-OCH_2CH_2CH_2CH_2\overset{CH_3}{\underset{CH_3}{|}}COMgCl \xrightarrow[H_2O]{H^+} HOCH_2CH_2CH_2CH_2\overset{CH_3}{\underset{CH_3}{|}}COH$$

$$(+\ HOCH_2CH_2CH_2CH_2\overset{O}{\overset{\|}{C}}H)$$

CHAPTER SEVENTEEN

Carboxylic Acids and their Derivatives: Nucleophilic Substitution at Acyl Carbon

17.1

(a) Carbon dioxide is an acid; it converts an aqueous solution of the strong base, NaOH, into an aqueous solution of the weaker base, $NaHCO_3$.

$$NaOH_{(aq)} + CO_2 \longrightarrow NaHCO_{3\,(aq)}$$

In this new solution, the more strongly basic *p*-cresoxide ion accepts a proton and becomes *p*-cresol,

$$p\text{-}CH_3C_6H_4O^- + HCO_3^- \rightleftharpoons \underset{\text{water-insoluble}}{p\text{-}CH_3C_6H_4OH} + CO_3^{=}$$

The more weakly basic benzoate ion remains in solution.

(b) Dissolve all three compounds in an organic solvent such as CH_2Cl_2, then extract with aqueous NaOH. The organic layer will contain cyclohexanol, which can be separated by distillation. The aqueous layer will contain the benzoic acid, as sodium benzoate, and the *p*-cresol, as sodium *p*-cresoxide.

Now pass CO_2 into the aqueous layer; this will cause *p*-cresol to separate (it can then be extracted into an organic solvent and purified by distillation). After separation of the *p*-cresol, the aqueous phase can be acidified with aqueous HCl to yield benzoic acid as a precipitate.

17.2

An electron-withdrawing group stabilizes the carboxylate ion more than the carboxylic acid. It does this by assisting in delocalization of the negative charge of the carboxylate ion through an inductive effect.

$$ClCH_2\text{—}C(=O)\text{—}O\text{—}H + H_2O \rightleftharpoons \left\{ClCH_2\text{—}C(\cdots O)_2\right\}^- + H_2O$$

Negative charge is delocalized by the electron-withdrawing chlorine.

Of course, the greater the number of electron-withdrawing groups the greater will be the acid-strengthening effect. Thus dichloroacetic acid is stronger than chloroacetic acid and trichloroacetic acid is stronger yet.

17.3

(a) CH_2FCOOH (F– is more electronegative than H–)

(b) CH_2FCOOH (F– is more electronegative than Cl–)

(c) $CH_2ClCOOH$ (Cl– is more electronegative than Br–)

(d) $CH_3CHClCH_2COOH$ (Cl– is closer to –COOH)

(e) $CH_3CH_2CHClCOOH$ (Cl– is closer to –COOH)

(f) 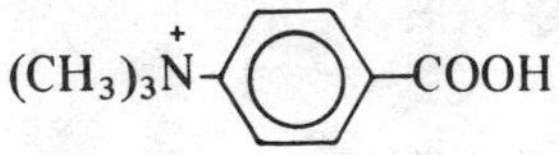 $(CH_3)_3\overset{+}{N}-C_6H_4-COOH$ [$(CH_3)_3\overset{+}{N}-$ is more electronegative than H–]

(g) 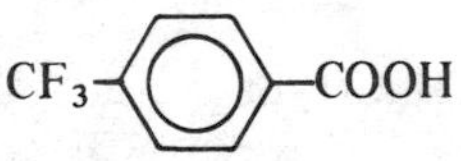 $CF_3-C_6H_4-COOH$ (CF_3- is more electronegative than CH_3-)

17.4

(a) The carboxyl group is an electron-withdrawing group; thus in a dicarboxylic acid such as those in Table 17.3, one carboxyl group increases the acidity of the other.

(b) As the distance between the carboxyl groups increases the acid-strengthening, inductive effect decreases.

17.5

These syntheses are easy to see if we work backward.

(a) $C_6H_5CH_2COOH \xleftarrow[(2)\ H^+]{(1)\ CO_2} C_6H_5CH_2MgBr$

$\uparrow$ Mg, ether

$C_6H_5CH_2Br$

(b) $CH_3CH_2CH_2C(CH_3)_2COOH \xleftarrow[(2)\ H^+]{(1)\ CO_2} CH_3CH_2CH_2C(CH_3)_2MgBr$

$\uparrow$ Mg, ether

$CH_3CH_2CH_2C(CH_3)_2Br$

(c) $CH_2{=}CHCH_2COOH \xleftarrow[(2)\ H^+]{(1)\ CO_2} CH_2{=}CHCH_2MgBr$

$\uparrow$ Mg, ether

$CH_2{=}CHCH_2Br$

(d) $CH_3-C_6H_4-COOH \xleftarrow[(2)\ H^+]{(1)\ CO_2} CH_3-C_6H_4-MgBr$

$\uparrow$ Mg, ether

$CH_3-C_6H_4-Br$

(e) $CH_3CH_2CH_2CH_2CH_2COOH \xleftarrow[(2)\ H_2O]{(1)\ CO_2} CH_3CH_2CH_2CH_2CH_2MgBr$

$\uparrow$ Mg, ether

$CH_3CH_2CH_2CH_2CH_2Br$

17.6

(a) $C_6H_5CH_2COOH \xleftarrow[(2)\ H^+,\ H_2O,\ heat]{(1)\ CN^-} C_6H_5CH_2Br$

$CH_2{=}CHCH_2COOH \xleftarrow[(2)\ H^+,\ H_2O,\ heat]{(1)\ CN^-} CH_2{=}CHCH_2Br$

$CH_3CH_2CH_2CH_2COOH \xleftarrow[(2)\ H^+,\ H_2O,\ heat]{(1)\ CN^-} CH_3CH_2CH_2CH_2Br$

(b) A nitrile synthesis. Preparation of a Grignard reagent from $HOCH_2CH_2CH_2CH_2Br$ would not be possible because of the presence of the acidic hydroxyl group.

17.7

(a) $CH_3COOH + C_6H_5COCl \xrightarrow{pyridine} CH_3\overset{O}{\overset{\|}{C}}O\overset{O}{\overset{\|}{C}}C_6H_5$

(b) $CH_3(CH_2)_4COOH + (CH_3CO)_2O \xrightarrow{heat} CH_3(CH_2)_4\overset{O}{\overset{\|}{C}}O\overset{O}{\overset{\|}{C}}(CH_2)_4CH_3 + 2CH_3COOH$ (remove by distillation)

(c) $HO\overset{O}{\overset{\|}{C}}CH_2CH_2CH_2\overset{O}{\overset{\|}{C}}OH \xrightarrow[(-H_2O)]{heat}$ cyclic anhydride (glutaric anhydride: ring of C(=O)–O–C(=O)–CH₂–CH₂–CH₂)

17.8

Since maleic acid is a *cis*-dicarboxylic acid, dehydration occurs readily:

$$\underset{\text{Maleic acid}}{cis\text{-}HC(COOH){=}CH(COOH)} \xrightarrow{200^\circ} \underset{\text{Maleic anhydride}}{HC{=}CH\ (\text{ring: } C(=O)\text{–}O\text{–}C(=O))} + H_2O$$

Maleic acid

Maleic anhydride

Being a *trans*-dicarboxylic acid, fumaric acid must undergo isomerization to maleic acid first. This requires a higher temperature.

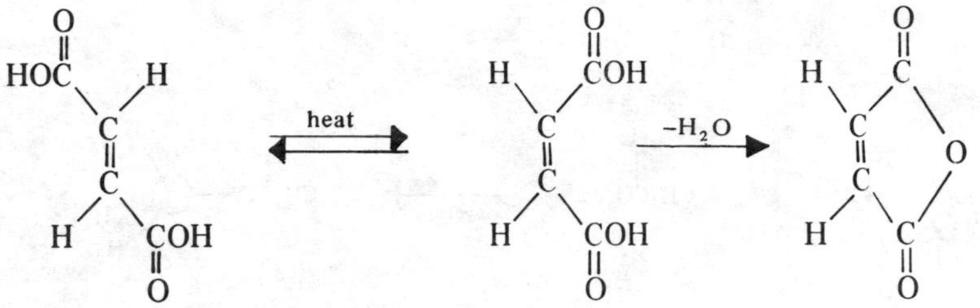

$$\text{Fumaric acid (trans-HO}_2\text{C–CH=CH–CO}_2\text{H)} \xrightarrow{\text{heat}} \text{cis-HO}_2\text{C–CH=CH–CO}_2\text{H} \xrightarrow{-H_2O} \text{maleic anhydride}$$

Fumaric acid

17.9

The labeled oxygen should appear in the carboxyl group of the acid. (Follow the reverse steps of the mechanism on page 769 of the text using $H_2{}^{18}O$.)

17.10

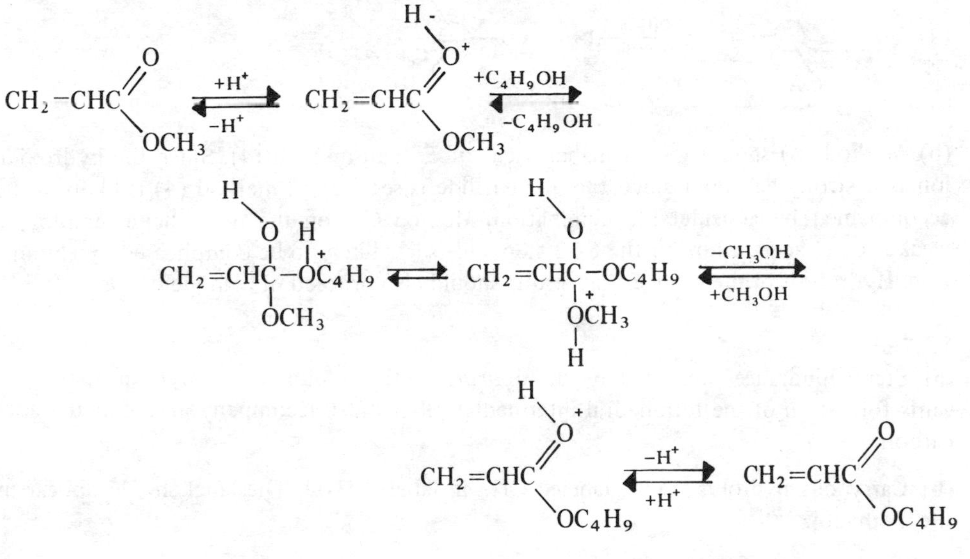

$$CH_2{=}CHC(=O)OCH_3 \underset{-H^+}{\overset{+H^+}{\rightleftharpoons}} CH_2{=}CHC(={}^+OH)OCH_3 \underset{-C_4H_9OH}{\overset{+C_4H_9OH}{\rightleftharpoons}}$$

$$CH_2{=}CHC(OH)(OCH_3)\overset{+}{O}(H)C_4H_9 \rightleftharpoons CH_2{=}CHC(OH)(OC_4H_9)\overset{+}{O}(H)CH_3 \underset{+CH_3OH}{\overset{-CH_3OH}{\rightleftharpoons}}$$

$$CH_2{=}CHC(={}^+OH)OC_4H_9 \underset{+H^+}{\overset{-H^+}{\rightleftharpoons}} CH_2{=}CHC(=O)OC_4H_9$$

17.11

(a)

(1) trans-2-methylcyclopentanol (H_3C, OH) + $C_6H_5SO_2Cl$ ⟶ A (H_3C, $OSO_2C_6H_5$) $\xrightarrow[\text{(inversion)}]{OH^-,\ \text{heat}}$

B (H_3C, OH) + $C_6H_5SO_3^-$

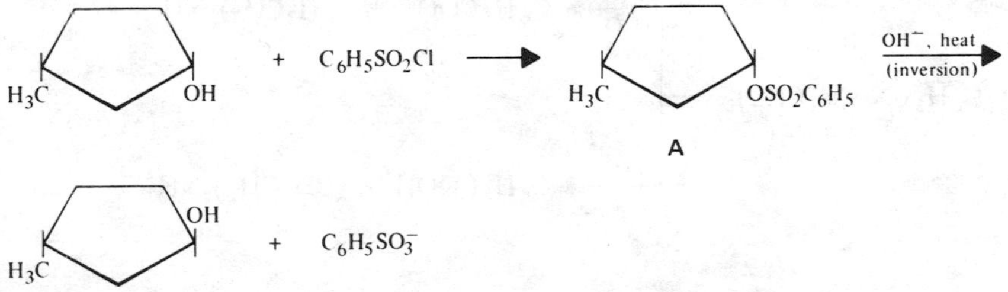

A

B

(2) 2-methylcyclopentanol (H_3C, OH) + $C_6H_5C(=O){-}Cl$ ⟶ C (H_3C, $O{-}C(=O)C_6H_5$)

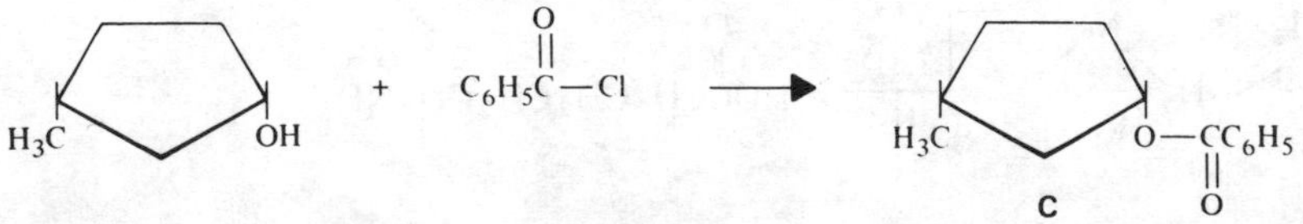

C

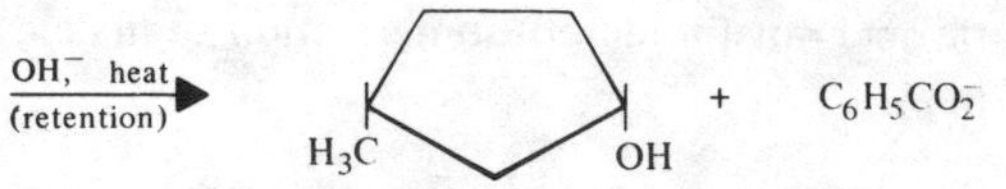

$\xrightarrow[\text{(retention)}]{OH^-,\ \text{heat}}$ (cyclopentane with H_3C and OH) + $C_6H_5CO_2^-$

D

(3) (H, C_6H_{13}, CH_3)C–Br + $CH_3\overset{O}{\overset{\|}{C}}O^-Na^+$ $\xrightarrow{\text{(inversion)}}$ $CH_3\overset{O}{\overset{\|}{C}}O$–C(H, C_6H_{13}, CH_3)

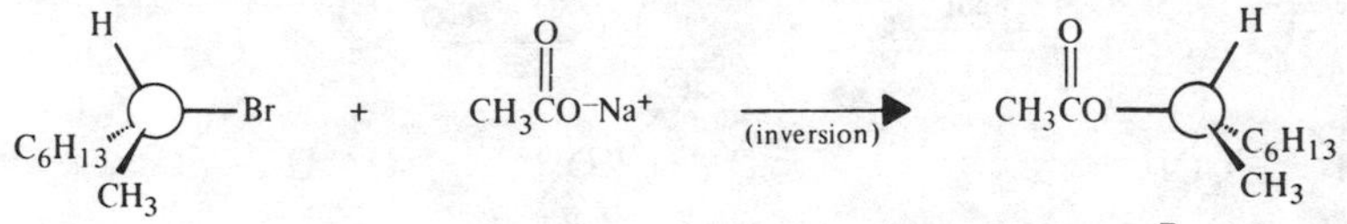

E

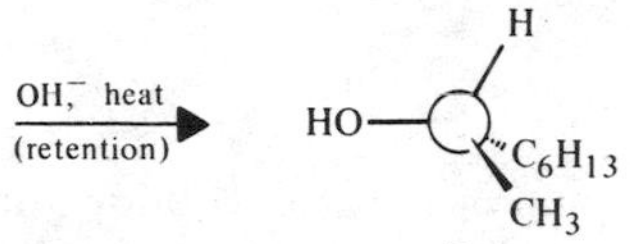

$\xrightarrow[\text{(retention)}]{OH^-,\ \text{heat}}$ HO–C(H, C_6H_{13}, CH_3)

F

(4) (H, C_6H_{13}, CH_3)C–Br $\xrightarrow[\text{(inversion)}]{OH^-,\ \text{heat}}$ HO–C(H, C_6H_{13}, CH_3)

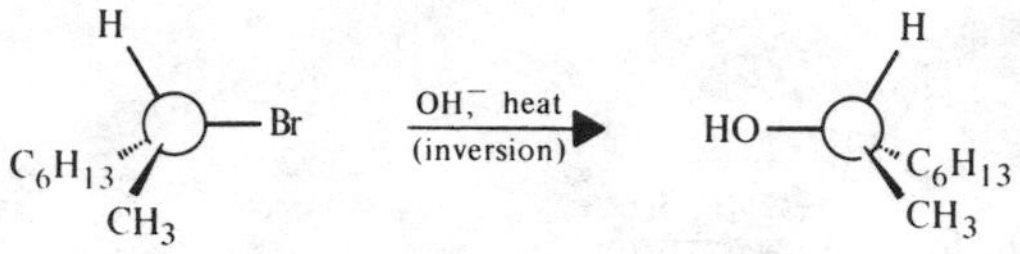

(b) Method (3) should give a higher yield of **F** than method (4). Since the hydroxide ion is a strong base and since the alkyl halide is secondary, method (4) is likely to be accompanied by considerable elimination. Method (3), on the other hand, employs a weaker base, acetate ion, in the S_N2 step and is less likely to be complicated by elimination. Hydrolysis of the ester **E** that results should also proceed in high yield.

17.12

(a) Steric hindrance presented by the di-*ortho* methyl groups of methyl mesitoate prevents formation of the tetrahedral intermediate that must accompany attack at the acyl carbon.

(b) Carry out hydrolysis with labeled OH^- in labeled H_2O. The label should appear in the methanol.

17.13

(a) $C_6H_5\overset{O}{\overset{\|}{C}}N(CH_2CH_3)_2$

$\xrightarrow[H_2O]{OH^-}$ $C_6H_5COO^-$ + $(CH_3CH_2)_2NH$

$\xrightarrow[H_2O]{H^+}$ C_6H_5COOH + $(CH_3CH_2)_2\overset{+}{N}H_2$

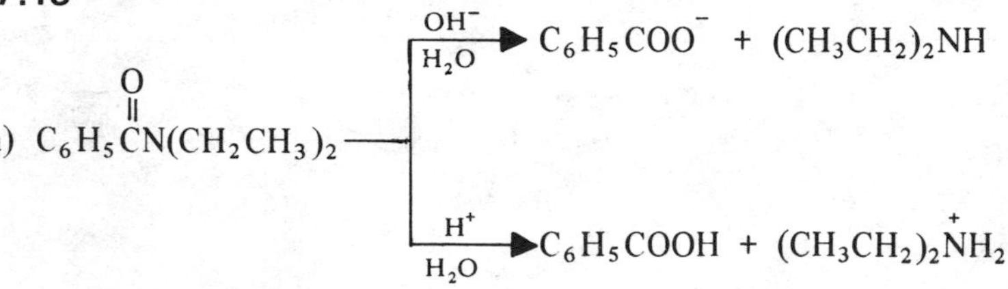

(b) (cyclic lactam: CH_2–CH_2–CH_2–CH_2–C(=O)–N–H)

$\xrightarrow[H_2O]{OH^-}$ $^-O\overset{O}{\overset{\|}{C}}CH_2CH_2CH_2CH_2NH_2$

$\xrightarrow[H_2O]{H^+}$ $HO\overset{O}{\overset{\|}{C}}CH_2CH_2CH_2CH_2\overset{+}{N}H_3$

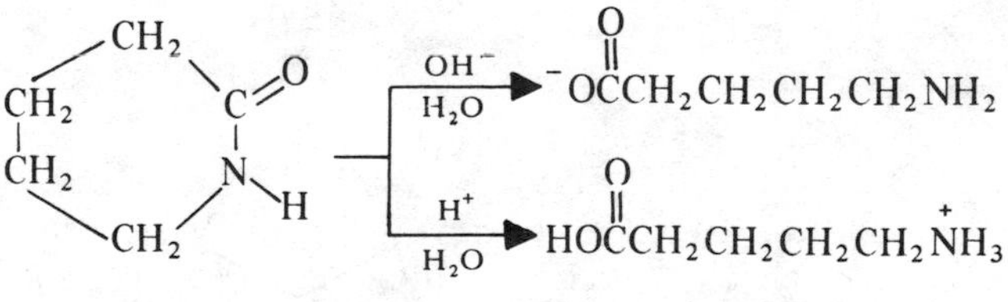

(c) $HOOCCH(CH_3)-NHC(=O)CHNH_2(CH_2C_6H_5)$

$$\xrightarrow[H_2O]{OH^-} {}^-OOCCH(CH_3)NH_2 + {}^-OOCCH(CH_2C_6H_5)NH_2$$

$$\xrightarrow[H_2O]{H^+} HOOCCH(CH_3)\overset{+}{N}H_3 + HOOCCH(CH_2C_6H_5)\overset{+}{N}H_3$$

17.14

(a) $(CH_3)_3CCOOH \xrightarrow{SOCl_2} (CH_3)_3CCOCl \xrightarrow{NH_3} (CH_3)_3CCONH_2 \xrightarrow[heat]{P_2O_5} (CH_3)_3CC{\equiv}N$

(b) An elimination reaction would take place.

$$CN^- + H-CH_2-C(CH_3)_2-Br \longrightarrow HCN + CH_2{=}C(CH_3)_2 + Br^-$$

17.15

(a) by a Kolbe electrolysis of hexanoic acid:

(1) $CH_3(CH_2)_4\overset{O}{\overset{\|}{C}}-O^- \xrightarrow[(-e^-)]{anode} CH_3(CH_2)_4\overset{O}{\overset{\|}{C}}-O\cdot$

(2) $CH_3(CH_2)_4\overset{O}{\overset{\|}{C}}-O\cdot \longrightarrow CH_3(CH_2)_3CH_2\cdot + CO_2$

(3) $2CH_3(CH_2)_3CH_2\cdot \longrightarrow CH_3(CH_2)_8CH_3$

(b) by decarboxylation of a β-keto acid:

$$CH_3(CH_2)_3\overset{O}{\overset{\|}{C}}CH_2\overset{O}{\overset{\|}{C}}OH \xrightarrow{100\text{-}150°} CH_3(CH_2)_3\overset{O}{\overset{\|}{C}}CH_3 + CO_2$$

(c) by decarboxylation of a substituted malonic acid

$$CH_3CH_2C(COOH)(CH_3)-COOH \xrightarrow{100\text{-}150°} CH_3CH_2CH(CH_3)COOH + CO_2$$

(d) by a Hunsdiecker reaction

$$C_6H_5CH_2COOAg + Br_2 \xrightarrow[heat]{CCl_4} C_6H_5CH_2Br + CO_2 + AgBr$$

17.16

(a) Because of the weakness of the oxygen-oxygen bond of the diacyl peroxide.

$$R{-}\overset{O}{\overset{\|}{C}}{-}O{-}O{-}\overset{O}{\overset{\|}{C}}R \longrightarrow 2R{-}\overset{O}{\overset{\|}{C}}{-}O\cdot \qquad \Delta H \simeq 35 \text{ kcal/mole}$$

(b) By decarboxylation of the carboxylate radical produced in part (a).

$$R{-}\overset{O}{\overset{\|}{C}}{-}O\cdot \longrightarrow R\cdot + CO_2$$

(c) (1) $R{-}\overset{O}{\overset{\|}{C}}{-}O{-}O{-}\overset{O}{\overset{\|}{C}}{-}R \xrightarrow{\text{heat}} 2R{-}\overset{O}{\overset{\|}{C}}{-}O\cdot$

(2) $R{-}\overset{O}{\overset{\|}{C}}{-}O\cdot \longrightarrow R\cdot + CO_2$

} Chain-Initiating Steps

(3) $R\cdot + CH_2{=}CH_2 \longrightarrow RCH_2CH_2\cdot$

(4) $RCH_2CH_2\cdot + CH_2{=}CH_2 \longrightarrow RCH_2CH_2CH_2CH_2\cdot$

} Chain-Propagating Steps

(5) etc.

17.17

(a) $CH_3(CH_2)_4COOH$

(b) $CH_3(CH_2)_4CONH_2$

(c) $CH_3(CH_2)_4CONHC_2H_5$

(d) $CH_3(CH_2)_4CON(C_2H_5)_2$

(e) $CH_3CH_2CH{=}CHCH_2COOH$

(f) $CH_3CH{=}CHCH_2\underset{\displaystyle CH_3}{\underset{|}{C}H}COOH$

(g) $HOOCCH_2CH_2CH_2CH_2COOH$

(h) benzene ring with COOH groups at adjacent (ortho) positions

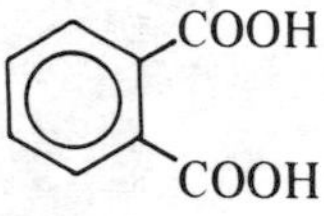

(i) benzene ring with COOH groups at meta positions

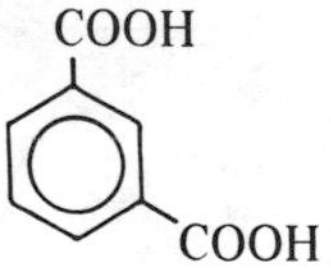

(l) $C_2H_5OOC(CH_2)_4COOC_2H_5$

(m) $CH_3CH_2COOCH_2CH(CH_3)_2$

(n) naphthalene ring with COOH

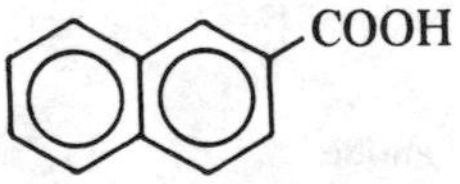

(o) $\overset{HOOC}{\underset{H}{}}\!\!>\!C{=}C\!<\!\overset{COOH}{\underset{H}{}}$

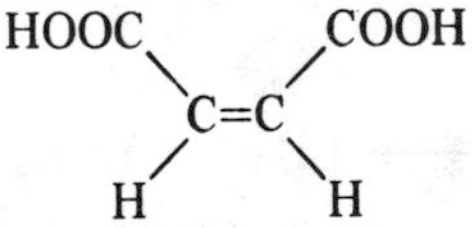

(p) $HOOCCHOHCH_2COOH$

(q) $\overset{HOOC}{\underset{H}{}}\!\!>\!C{=}C\!<\!\overset{H}{\underset{COOH}{}}$

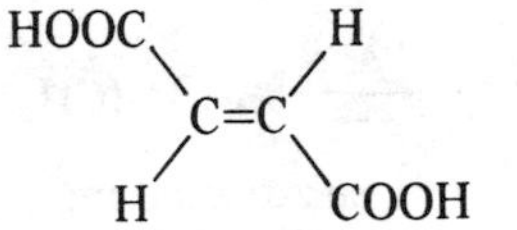

(r) $HOOCCH_2CH_2COOH$

(s) five-membered ring: $CH_2{-}CH_2$ connected to two $C{=}O$ groups joined through NH

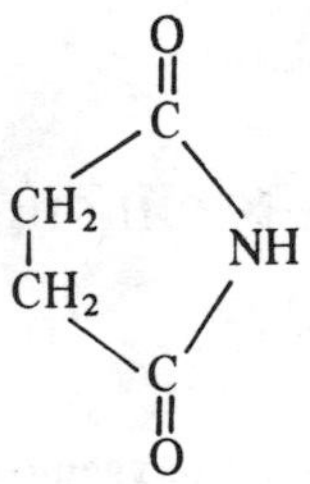

(j) 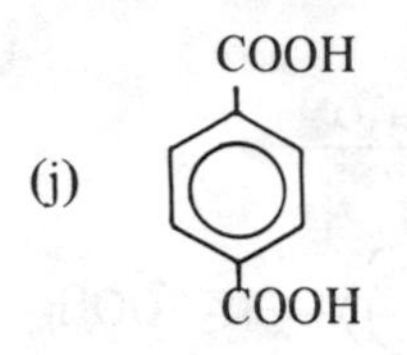

(t) $HOOCCH_2COOH$

(k) $C_2H_5OOC{-}COOC_2H_5$

(u) $C_2H_5OOCCH_2COOC_2H_5$

17.18

(a) Benzoic acid

(b) Benzoyl chloride

(c) Benzamide

(d) Benzoic anhydride

(e) Benzyl benzoate

(f) Phenyl benzoate

(g) Isopropyl acetate

(h) N,N-Dimethylacetamide

(i) Acetonitrile

(j) Maleic anhydride

(k) Phthalic anhydride

(l) Phthalimide

(m) Glyceryl tripalmitate

(n) α-Ketosuccinic acid (oxaloacetic acid)

(o) Methyl salicylate

17.19

(a) $C_6H_5{-}Br \xrightarrow{\text{Mg, ether}} C_6H_5{-}MgBr \xrightarrow[\text{(2) } H_3O^+]{\text{(1) } CO_2} C_6H_5{-}COOH$

(b) $C_6H_5{-}CH_3 \xrightarrow[\text{(2) } H_3O^+]{\text{(1) } KMnO_4,\ OH^-,\ \text{heat}} C_6H_5{-}COOH$

(c) $C_6H_5{-}CN \xrightarrow[\text{heat}]{H_3O^+,\ H_2O} C_6H_5{-}COOH + NH_4^+$

(d) $C_6H_5{-}\overset{O}{\overset{\|}{C}}CH_3 \xrightarrow[(-CHCl_3)]{Cl_2,\ OH^-} C_6H_5{-}COO^- \xrightarrow{H_3O^+} C_6H_5{-}COOH$

(e) $C_6H_5{-}CHO \xrightarrow{Ag(NH_3)_2^+\ OH^-} C_6H_5{-}COO^- \xrightarrow{H_3O^+} C_6H_5{-}COOH$

(f) $C_6H_5{-}CH{=}CH_2 \xrightarrow[\text{(2) } H_3O^+]{\text{(1) } KMnO_4,\ OH^-,\ \text{heat}} C_6H_5{-}COOH$

(g) $C_6H_5{-}CH_2OH \xrightarrow[\text{(2) } H_3O^+]{\text{(1) } KMnO_4,\ OH^-,\ \text{heat}} C_6H_5{-}COOH$

17.20

(a) $C_6H_5CH_2CHO \xrightarrow{Ag(NH_3)_2^+\ OH^-} C_6H_5CH_2COO^- \xrightarrow{H_3O^+} C_6H_5CH_2COOH$

(b) $C_6H_5CH_2Br \xrightarrow[(2)\ CO_2]{(1)\ Mg,\ ether} C_6H_5CH_2COOMgBr \xrightarrow{H_3O^+} C_6H_5CH_2COOH$

$C_6H_5CH_2Br \xrightarrow{CN^-} C_6H_5CH_2CN \xrightarrow[heat]{H_3O^+,\ H_2O} C_6H_5CH_2COOH$

17.21

(a) $CH_3CH_2CH_2CH_2CH_2OH \xrightarrow[(2)\ H_3O^+]{(1)\ KMnO_4,\ OH^-,\ heat} CH_3CH_2CH_2CH_2COOH$

(b) $CH_3CH_2CH_2CH_2Br \xrightarrow[(2)\ CO_2]{(1)\ Mg,\ ether} CH_3CH_2CH_2CH_2COOMgBr \xrightarrow{H_3O^+} CH_3CH_2CH_2CH_2COOH$

$CH_3CH_2CH_2CH_2Br \xrightarrow{CN^-} CH_3CH_2CH_2CH_2CN \xrightarrow[heat]{H_3O^+,\ H_2O} CH_3CH_2CH_2CH_2COOH$

(c) $CH_3CH_2CH_2CH_2C(=O)CH_3 \xrightarrow[(-CHCl_3)]{Cl_2,\ OH^-} CH_3CH_2CH_2CH_2COO^- \xrightarrow{H_3O^+} CH_3CH_2CH_2CH_2COOH$

(d) $CH_3(CH_2)_3CH{=}CH(CH_2)_3CH_3 \xrightarrow[(2)\ H_3O^+]{(1)\ KMnO_4,\ OH^-,\ heat} 2CH_3(CH_2)_3COOH$

(e) $CH_3CH_2CH_2CH_2CHO \xrightarrow[(2)\ H_3O^+]{(1)\ Ag(NH_3)_2^+OH^-} CH_3CH_2CH_2CH_2COOH$

17.22

(a) $CH_3COOH + HCl$

(b) $CH_3COOH + AgCl$

(c) $CH_3COOCH_2(CH_2)_2CH_3$

(d) CH_3CONH_2

(e) $o\text{-}CH_3C_6H_4C(=O)CH_3 + p\text{-}CH_3C_6H_4C(=O)CH_3$

(f) CH_3CHO

(g) CH_3COCH_3

(h) $CH_3COCH_2CH_3$

(i) $CH_3CONHCH_3$

(j) $CH_3CONHC_6H_5$

(k) $CH_3CON(CH_3)_2$

(l) $CH_3COSCH_2CH_3$

(m) $(CH_3CO)_2O$

(n) $(CH_3CO)_2O$

(o) $CH_3COOC_6H_5$

17.23

(a) $CH_3CONH_2 + CH_3COONH_4$

(b) $2CH_3COOH$

(c) $CH_3COOCH_2CH_2CH_3 + CH_3COOH$

(d) $C_6H_5COCH_3 + CH_3COOH$

(e) $CH_3CONHCH_2CH_3 + CH_3COO^- CH_3CH_2NH_3^+$

(f) $CH_3CON(CH_2CH_3)_2 + CH_3COO^- (CH_3CH_2)_2NH_2^+$

17.24

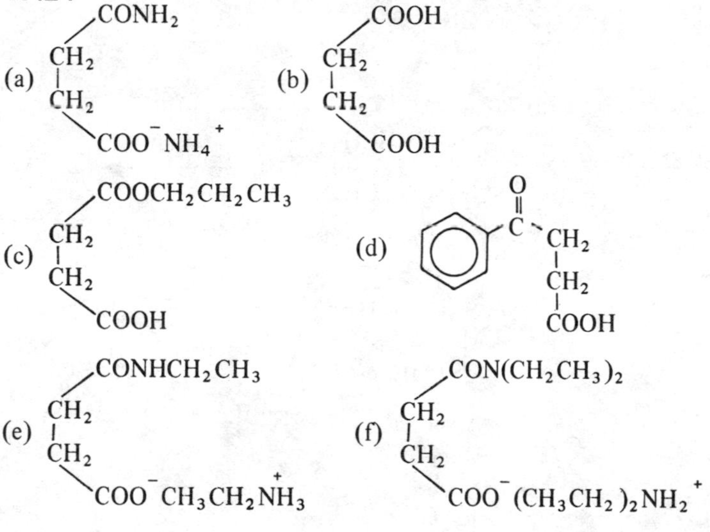

(a)
$CONH_2$
CH_2
CH_2
$COO^- NH_4^+$

(b)
$COOH$
CH_2
CH_2
$COOH$

(c)
$COOCH_2CH_2CH_3$
CH_2
CH_2
$COOH$

(d)
$C_6H_5-C(=O)-CH_2$
CH_2
$COOH$

(e)
$CONHCH_2CH_3$
CH_2
CH_2
$COO^- CH_3CH_2\overset{+}{N}H_3$

(f)
$CON(CH_2CH_3)_2$
CH_2
CH_2
$COO^- (CH_3CH_2)_2NH_2^+$

17.25

(a)
O
C
CH_2 O + C_6H_6 $\xrightarrow{AlCl_3}$ $C_6H_5-C(=O)-CH_2$ $\xrightarrow[HCl,\ heat]{Zn(Hg)}$
CH_2
C
O

$C_6H_5-CH_2-CH_2$
CH_2
$COOH$
$\xrightarrow{SOCl_2}$
$C_6H_5-CH_2-CH_2$
CH_2
$C-Cl$
O
$\xrightarrow{AlCl_3}$
O

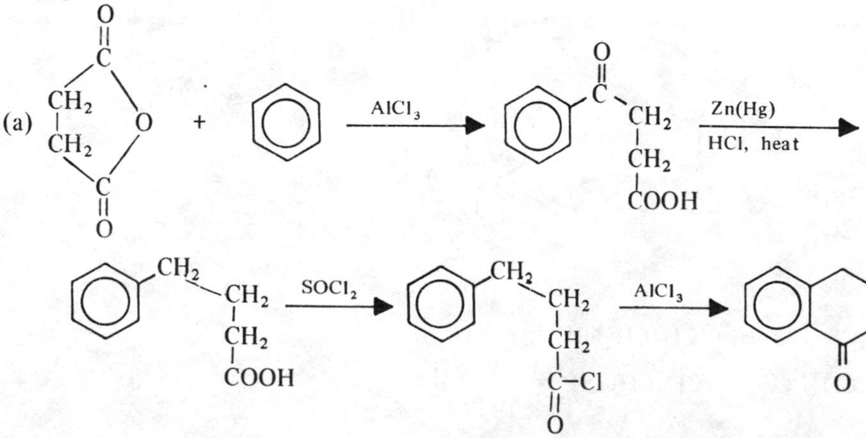

(b) [1-tetralone] $\xrightarrow[(2)\ H_3O^+]{(1)\ CH_3MgBr}$ [1-methyl-1-tetralol: OH, CH_3] $\xrightarrow[(-H_2O)]{H^+,\ heat}$ [4-methyl-1,2-dihydronaphthalene: CH_3]

(c) [4-methyl-1,2-dihydronaphthalene: CH_3] $\xrightarrow[(-H_2)]{Pt,\ Heat}$ [1-methylnaphthalene: CH_3]

(d) [phthalic anhydride] + [benzene] $\xrightarrow{AlCl_3}$ [2-benzoylbenzoic acid: C=O, COH, O] $\xrightarrow[heat]{Zn(Hg),\ HCl}$

[2-benzylbenzoic acid: CH_2, COH, O] $\xrightarrow[(2)\ AlCl_3]{(1)\ SOCl_2}$ [anthrone: O]

(e) [phthalic anhydride] $\xrightarrow{CH_3NH_2\ (excess)}$ [$C(=O)NHCH_3$, $CO^-\ CH_3NH_3^+$] $\xrightarrow{H_3O^+}$

[$C(=O)NHCH_3$, COH] $\xrightarrow[(-H_2O)]{heat}$ [N-methylphthalimide: NCH_3]

(f) [cyclopentadiene] + [maleic anhydride] $\longrightarrow$ [bicyclic Diels–Alder adduct anhydride]

(g) $CH_3CH{=}CH{-}CH{=}CHCH_3$ + [maleic anhydride] $\longrightarrow$ [3,6-dimethylcyclohex-4-ene-1,2-dicarboxylic anhydride: CH_3, CH_3]

17.26

(a) $CH_3CH_2COOH + CH_3CH_2OH$

(b) $CH_3CH_2COO^- + CH_3CH_2OH$

(c) $CH_3CH_2COO(CH_2)_7CH_3$ + CH_3CH_2OH

(d) $CH_3CH_2CONHCH_3$ + CH_3CH_2OH

(e) $CH_3CH_2CH_2OH$ + CH_3CH_2OH

(f) $CH_3CH_2C(C_6H_5)(OH)-C_6H_5$ + CH_3CH_2OH

17.27

(a) CH_3CH_2COOH + NH_4^+

(b) $CH_3CH_2COO^-$ + NH_3

(c) CH_3CH_2CN

17.28

(a) Benzoic acid dissolves in aqueous $NaHCO_3$. Methyl benzoate does not.

(b) Benzoyl chloride gives a precipitate (AgCl) when treated with alcoholic $AgNO_3$. Benzoic acid does not.

(c) Benzoic acid dissolves in aqueous $NaHCO_3$. Benzamide does not.

(d) Benzoic acid dissolves in aqueous $NaHCO_3$. *p*-Cresol does not.

(e) Refluxing benzamide with aqueous NaOH liberates NH_3 which can be detected in the vapors with moist red litmus paper. Ethyl benzoate does not liberate NH_3.

(f) Cinnamic acid, because it has a double bond, decolorizes Br_2 in CCl_4. Benzoic acid does not.

(g) Benzoyl chloride gives a precipitate (AgCl) when treated with alcoholic $AgNO_3$. Ethyl benzoate does not.

(h) 2-Chlorobutanoic acid gives a precipitate (AgCl) when treated with alcoholic silver nitrate. Butanoic acid does not.

17.29

(a) 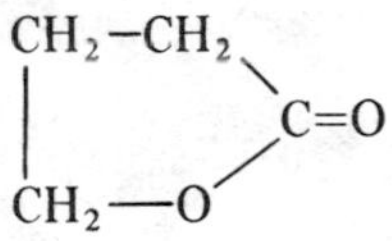 Ring: $CH_2-CH_2-C(=O)-O-CH_2$ (five-membered lactone: CH_2-CH_2, $C=O$, CH_2-O)

(d) 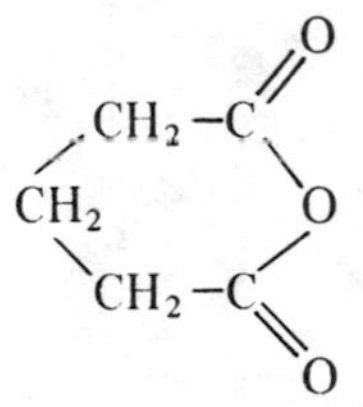 Ring: $CH_2-C(=O)-O-C(=O)-CH_2-CH_2$ (cyclic anhydride: $CH_2-C=O$, CH_2, O, $CH_2-C=O$)

(b) $CH_3CH=CHCOOH$

(e) 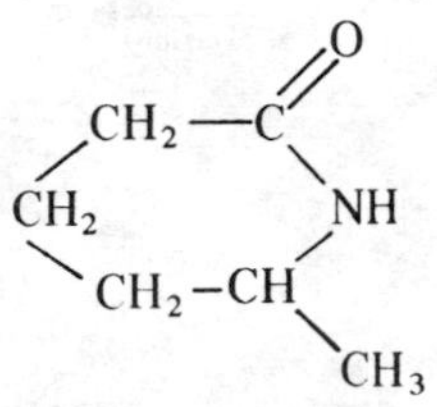 Ring: $CH_2-C(=O)-NH-CH(CH_3)-CH_2-CH_2$ (lactam: $CH_2-C=O$, CH_2, NH, CH_2-CH, CH_3)

(c) CH_3CH_2CH [cyclic diester ring: C(=O)—O / O—C(=O)] $CHCH_2CH_3$

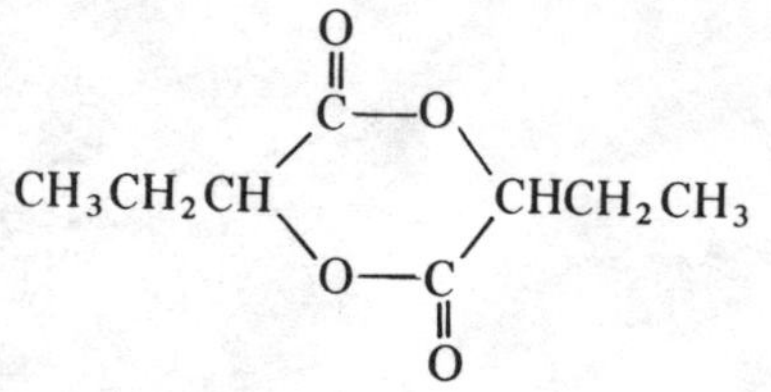

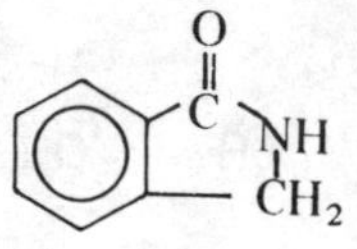

(f) [benzene ring fused: C(=O)—NH—CH_2]

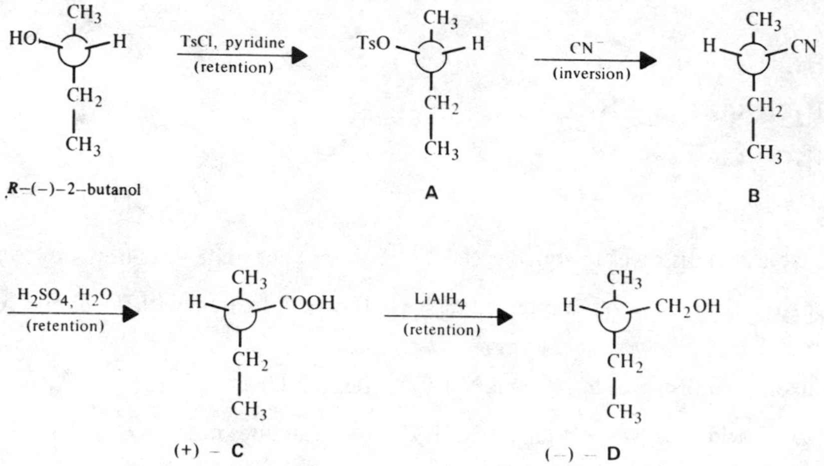

17.30

(a) R–(–)–2-butanol [HO, H, CH_3, CH_2CH_3] $\xrightarrow[\text{(retention)}]{\text{TsCl, pyridine}}$ A [TsO, H, CH_3, CH_2CH_3] $\xrightarrow[\text{(inversion)}]{CN^-}$ B [H, CN, CH_3, CH_2CH_3]

$\xrightarrow[\text{(retention)}]{H_2SO_4,\ H_2O}$ (+) – C [H, COOH, CH_3, CH_2CH_3] $\xrightarrow[\text{(retention)}]{LiAlH_4}$ (–) – D [H, CH_2OH, CH_3, CH_2CH_3]

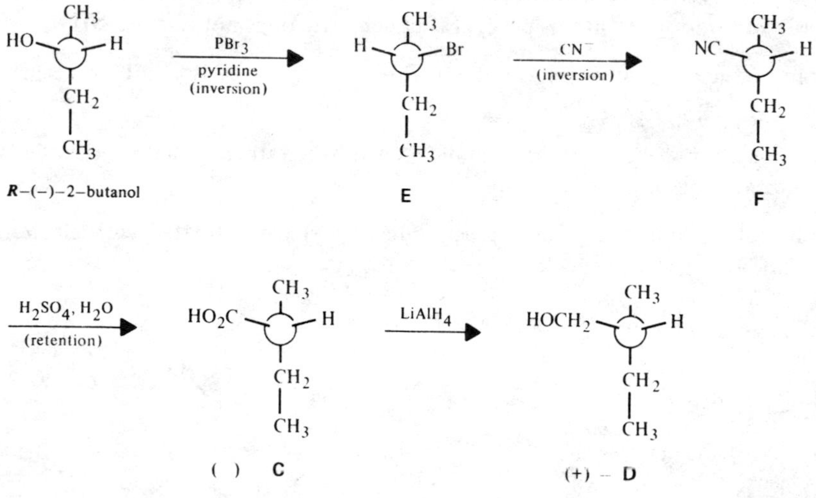

(b) R–(–)–2-butanol [HO, H, CH_3, CH_2CH_3] $\xrightarrow[\text{pyridine (inversion)}]{PBr_3}$ E [H, Br, CH_3, CH_2CH_3] $\xrightarrow[\text{(inversion)}]{CN^-}$ F [NC, H, CH_3, CH_2CH_3]

$\xrightarrow[\text{(retention)}]{H_2SO_4,\ H_2O}$ ( ) C [HO_2C, H, CH_3, CH_2CH_3] $\xrightarrow{LiAlH_4}$ (+) – D [$HOCH_2$, H, CH_3, CH_2CH_3]

(c) A [TsO, H, CH_3, CH_2CH_3] $\xrightarrow[\text{(inversion)}]{CH_3COO^-}$ G [H, $OCCH_3$ (C=O), CH_3, CH_2CH_3] $\xrightarrow[\text{(retention)}]{OH^-}$ (+) – H [H, OH, CH_3, CH_2CH_3]

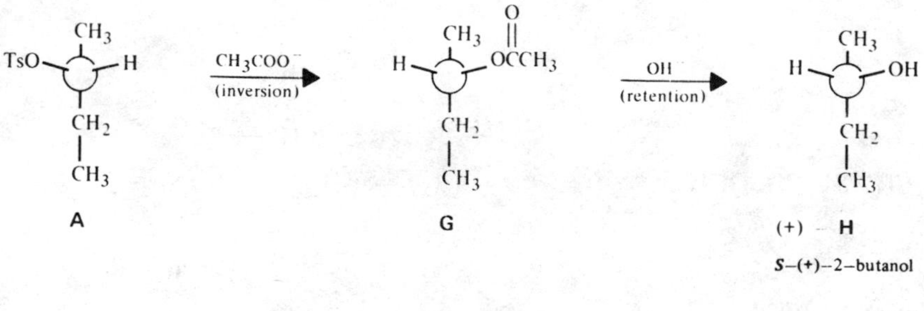

S–(+)–2-butanol

(d)

CH_3 / H / CH_2OH / CH_2 / CH_3 $\xrightarrow[\text{(retention)}]{PBr_3}$ CH_3 / H / CH_2Br / CH_2 / CH_3 $\xrightarrow[\text{ether (retention)}]{Mg}$ CH_3 / H / CH_2MgBr / CH_2 / CH_3

(−) − **D**   **J**   **K**

$\xrightarrow[\text{(2) } H^+ \text{ (retention)}]{\text{(1) } CO_2}$ CH_3 / H / CH_2COOH / CH_2 / CH_3

L

(e)

CHO / H / OH / CH_2OH $\xrightarrow{HCN}$ CN / H / OH / H / OH / CH_2OH + CN / HO / H / H / OH / CH_2OH

R–(+)–glyceraldehyde   **M**   **N**

(f) **M** $\xrightarrow[\text{heat}]{H_2SO_4,\ H_2O}$ COOH / H / OH / H / OH / CH_2OH $\xrightarrow[HNO_3]{(O)}$ COOH / H / OH / H / OH / COOH

P   *meso*–tartaric acid

(g) **N** $\xrightarrow[\text{heat}]{H_2SO_4,\ H_2O}$ COOH / HO / H / H / OH / CH_2OH $\xrightarrow[HNO_3]{(O)}$ COOH / HO / H / H / OH / COOH

(−)–tartaric acid

17.31

(a) $CH_3\overset{CH_3}{C}HCHO + H\overset{O}{\overset{\|}{C}}H \xrightarrow[H_2O]{K_2CO_3} CH_3\overset{CH_3}{\underset{CH_2OH}{C}}CHO \xrightarrow[\text{(2) KCN}]{\text{(1) } NaHSO_3}$

A

$CH_3\overset{CH_3}{\underset{CH_2OH}{C}}—\overset{OH}{C}HCN \xrightarrow{H_3O^+} \left[CH_3\overset{CH_3}{\underset{CH_2OH}{C}}—\overset{OH}{C}HCOOH \right] \xrightarrow{-H_2O}$

(±)-**B**   (±)-**C**

$$\underset{(\pm)\text{-}\mathbf{D}}{\text{lactone: } CH_3-C(CH_3)(CH_2-O-)-CH(OH)-C(=O)} \xrightarrow{H_2NCH_2CH_2\overset{O}{\overset{\|}{C}}OH} (\pm)\text{-Pantothenic acid}$$

$$(\pm)\text{-}\mathbf{D} \xrightarrow{H_2NCH_2CH_2\overset{O}{\overset{\|}{C}}NHCH_2CH_2SH} (\pm)\text{-Pantetheine}$$

(b)

$$(CH_3)_2C(CH_2OH)-CH(OH)(H)-\overset{O}{\overset{\|}{C}}-NHCH_2CH_2\overset{O}{\overset{\|}{C}}-NHCH_2CH_2SH$$

$$\xrightarrow[\text{heat}]{OH^-,\ H_2O} (CH_3)_2C(CH_2OH)-CH(OH)(H)-COO^- + H_2NCH_2CH_2COO^- + H_2NCH_2CH_2S^-$$

17.32

$$CH_3CH_2O-C_6H_4-NH-\overset{O}{\overset{\|}{C}}-CH_3 \xrightarrow[\substack{H_2O \\ \text{reflux}}]{OH^-} CH_3CH_2O-C_6H_4-NH_2$$

Phenacetin → Phenetidine + CH_3COO^-

An interpretation of the spectral data for phenacetin is given in Fig. 17.1.

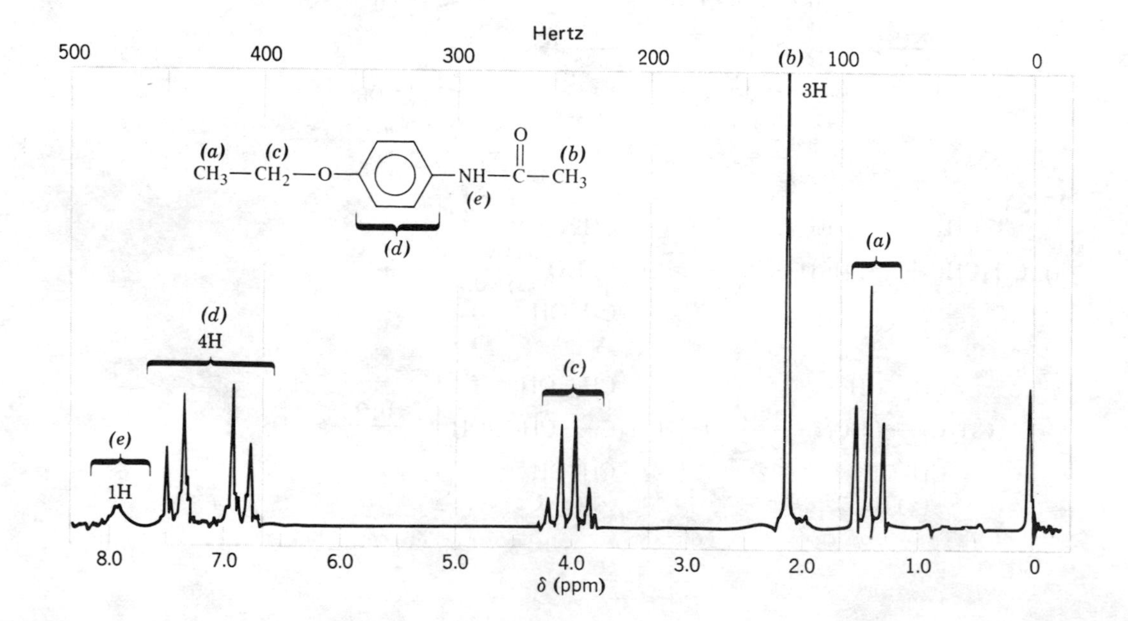

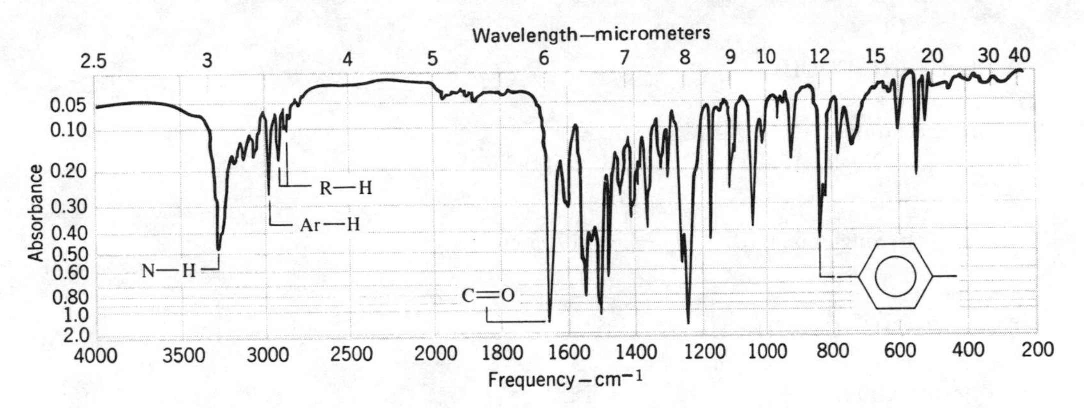

FIG. 17.1. The proton nmr and infrared spectra of phenacetin. (proton nmr spectrum courtesy of Varian Associates. IR spectrum courtesy of Sadtler Inc.)

17.33

(a) $\underset{a}{CH_3}\underset{c}{CH_2}-O-\overset{O}{\overset{\|}{C}}-\underset{b}{CH_2}\underset{b}{CH_2}-\overset{O}{\overset{\|}{C}}-O-\underset{c}{CH_2}\underset{a}{CH_3}$

Interpretation:

a Triplet δ1.2 (6H) $\quad 2-\overset{O}{\overset{\|}{C}}-$, 1740 cm^{-1}

b Singlet δ2.5 (4H)

c Quartet δ4.1 (4H)

(b) $\underset{d}{C_6H_5}-\overset{O}{\overset{\|}{C}}-O-\underset{c}{CH_2}-\underset{b}{\overset{\overset{CH_3\ a}{|}}{CH}}-\underset{a}{CH_3}$

Interpretation:

a Doublet δ1.0 (6H) $\quad -\overset{O}{\overset{\|}{C}}-$, 1720 cm^{-1}

b Multiplet δ2.1 (1H)

c Doublet δ4.1 (2H)

d Multiplet δ7.8 (5H)

(c) $\underset{d}{C_6H_5}-\underset{b}{CH_2}-\overset{O}{\overset{\|}{C}}-O-\underset{c}{CH_2}\underset{a}{CH_3}$

Interpretation:

a Triplet δ1.2 (3H) $\quad -\overset{O}{\overset{\|}{C}}-$, 1740 cm^{-1}

b Singlet δ3.5 (2H)

c Quartet δ4.1 (2H)

d Multiplet δ7.3 (5H)

(d) $Cl-CH(Cl)-COOH$ (a = CH, b = COOH)

Interpretation:

a Singlet δ6.0 — $-OH$, 2500-2700 cm^{-1}

b Singlet δ11.70 — $-C(=O)-$, 1705 cm^{-1}

(e) $Cl-CH_2-C(=O)-OCH_2CH_3$ (b = CH_2Cl, c = OCH_2, a = CH_3)

Interpretation:

a Triplet δ1.3 — $-C(=O)-$, 1745 cm^{-1}

b Singlet δ4.0

c Quartet δ4.2

17.34

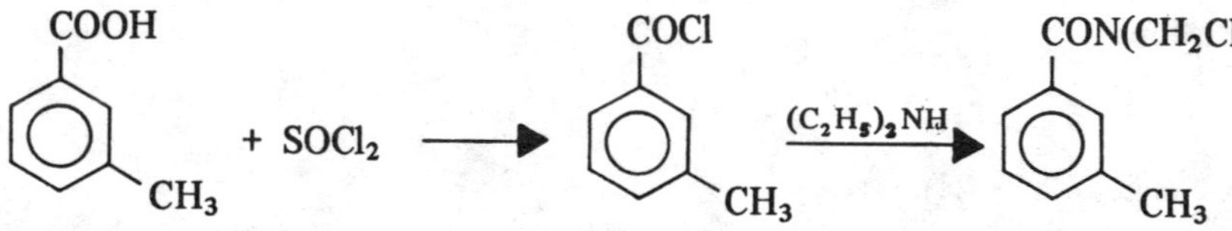

m-Toluic acid (COOH, CH_3) + $SOCl_2$ ⟶ m-toluoyl chloride (COCl, CH_3) $\xrightarrow{(C_2H_5)_2NH}$ $CON(CH_2CH_3)_2$ (with CH_3 meta)

17.35

Alkyl groups are electron releasing; they help disperse the positive charge of an alkylammonium salt and thereby help to stabilize it.

$$R\ddot{N}H_2 + H_3O^+ \longrightarrow R{\rightarrow}NH_3^+ + H_2O$$

Stabilized by electron-releasing alkyl group.

Alkylamines, consequently, are somewhat stronger bases than ammonia.

Amides, on the other hand, have acyl groups, $R-C(=O)-$, attached to nitrogen, and acyl groups are electron withdrawing. They are especially electron withdrawing because of resonance contributions of the kind shown below,

$$R-C(=\ddot{O}:)-\ddot{N}H_2 \longleftrightarrow R-C(-\ddot{O}:^-)=\overset{+}{N}H_2$$

This kind of resonance also *stabilizes* the amide. The tendency of the acyl group to be electron withdrawing, however, *destabilizes* the conjugate acid of an amide and reactions such as the following do not take place to an appreciable extent.

$$RC(=O)-\ddot{N}H_2 + H_3O^+ \not\longrightarrow RC(=O)-NH_3^+ + H_2O$$

Stabilized by resonance. — *Destabilized by electron-withdrawing acyl group.*

17.36

(a) The conjugate base of an amide is stabilized by resonance.

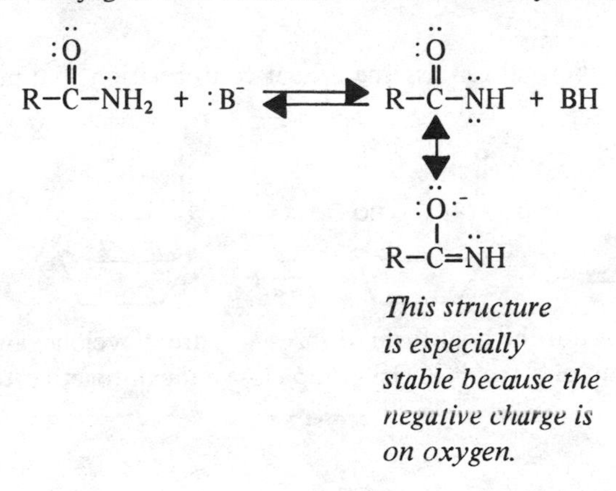

(b) The conjugate base of an imide is stabilized by an additional resonance structure,

$$RC(=O)-NH-CR(=O) + OH^- \longrightarrow RC(=O)-\ddot{N}^--CR(=O) + H_2O$$

An imide

$$\longleftrightarrow RC(-O^-)=N-CR(=O) \longleftrightarrow RC(=O)-N=CR(-O^-)$$

17.37

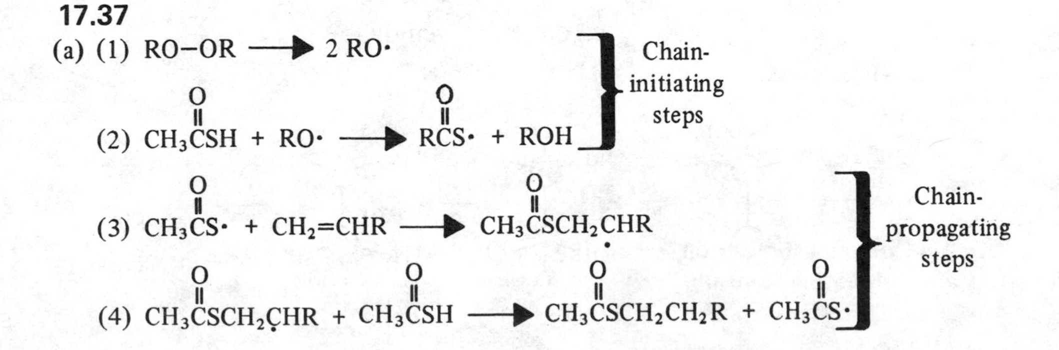

(b) $CH_3C(CH_3){=}CHCH_3 + CH_3C(=O)SH \xrightarrow{ROOR} CH_3CH(CH_3)CH(SC(=O)CH_3)CH_3$

$\xrightarrow[(2)\ H_3O^+]{(1)\ OH^-,\ heat} CH_3CH(CH_3)CH(SH)CH_3 + CH_3C(=O)OH$

17.38

cis-4-Hydroxycyclohexane carboxylic acid can assume a boat conformation that permits lactone formation.

HOC(=O)–(chair)–OH ⇄ (boat: COH(=O), HO) $\xrightarrow{-H_2O}$ (lactone: C(=O)–O)

Neither of the chair conformations nor the boat form of *trans*-4-hydroxycyclohexane carboxylic acid places the –OH group and the –COOH group close enough together to permit lactonization.

17.39

(a)

CHO / H–C–OH / CH_2OH $\xrightarrow[\text{oxidation (see p. 884)}]{Br_2,\ H_2O}$ COOH / H–C–OH / CH_2OH $\xrightarrow{PBr_3}$ COOH / H–C–OH / CH_2Br $\xrightarrow{NaCN}$ COOH / H–C–OH / CH_2 / CN $\xrightarrow[heat]{H_3O^+}$ COOH / H–C–OH / CH_2 / COOH

(–)–glyceric acid; (–)–3–bromo–2–hydroxypropanoic acid; ($C_4H_5NO_3$); R–(+)–malic acid

17.40

CHO / H–C–OH / CH_2OH $\xrightarrow{HCN}$ CN / H–C–OH / H–C–OH / CH_2OH + CN / HO–C–H / H–C–OH / CH_2OH

R–(+)–glyceraldehyde; **M**; **N**

N $\xrightarrow[H_2O]{H_2SO_4}$ COOH / HO–C–H / H–C–OH / CH_2OH $\xrightarrow[HNO_3]{(O)}$ COOH / HO–C–H / H–C–OH / COOH $\xrightarrow{PBr_3}$ COOH / HO–C–H / Br–C–H / COOH $\xrightarrow[H^+]{Zn}$ COOH / HO–C–H / CH_2 / COOH

(–)–tartaric acid

(b) Replacement of either alcoholic –OH by a reaction that proceeds with inversion produces the same stereoisomer.

COOH | HO–C1–H | H–C2–OH | COOH $\xrightarrow[\text{(inversion at C-2)}]{PBr_3}$ COOH | HO–C–H | Br–C–H | COOH ≡ COOH | H–C–Br | H–C–OH | COOH $\xleftarrow[\text{(inversion at C-1)}]{PBr_3}$ COOH | HO–C1–H | H–C2–OH | COOH

(c) Two. The stereoisomer given in (b) above and the one given below.

COOH | HO–C1–H | H–C2–OH | COOH $\xrightarrow[\text{(retention at C-2)}]{PBr_3}$ COOH | HO–C–H | H–C–Br | COOH ≡ COOH | Br–C–H | H–C–OH | COOH $\xleftarrow[\text{(retention at C-1)}]{PBr_3}$ COOH | HO–C1–H | H–C2–OH | COOH

(d) It would have made no difference because treating either isomer (or both together) with zinc and acid produces (–)-malic acid.

COOH | HO–C–H | Br–C–H | COOH $\xrightarrow[H^+]{Zn}$ COOH | HO–C–H | CH_2 | COOH ≡ COOH | HO–C–H | CH_2 | COOH $\xleftarrow[H^+]{Zn}$ COOH | HO–C–H | H–C–Br | COOH

(–)-malic acid

17.41

(a) $CH_3O_2C{-}C{\equiv}C{-}CO_2CH_3$. This is a Diels-Alder reaction.

(b) H_2, Pd. The disubstituted double bond is less hindered than the tetrasubstituted double bond and hence is more reactive.

(c) $CH_2{=}CH{-}CH{=}CH_2$. Another Diels-Alder reaction.

(d) $LiAlH_4$

(e) $CH_3S(=O)_2{-}Cl$ and pyridine

(f) $CH_3CH_2S^-$

(g) OsO_4

(h) Raney Ni

(i) Base. This is an aldol condensation.

(j) C_6H_5Li (or C_6H_5MgBr) followed by H_3O^+

(k) H_3O^+. This is an acid-catalyzed rearrangement of an allylic alcohol.

(l) $CH_3\overset{\overset{O}{\|}}{C}Cl$, pyridine

(m) Heat. This is an acetate pyrolysis.

(n) O_3 followed by oxidation.

(o) Heat

17.42

(a)

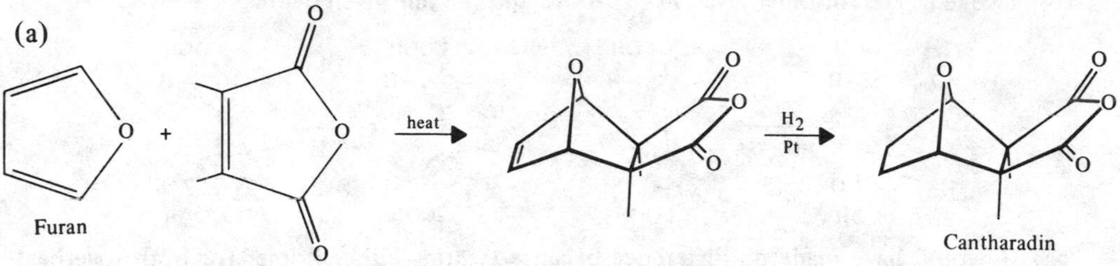

Furan + Dimethylmaleic anhydride $\xrightarrow{\text{heat}}$ $\xrightarrow[\text{Pt}]{H_2}$ Cantharadin

(b) Cantharidin apparently undergoes dehydrogenation to the Diels-Alder adduct shown above and then the adduct spontaneously decomposes through a reverse Diels-Alder reaction to furan and dimethylmaleic anhydride. These results suggest that the attempted Diels-Alder synthesis fails because the position of equilibrium favors reactants rather than products.

K

Special Topic

K.1

(a) cyclohexanone $\xrightarrow[H_2Cr_2O_7]{(O)}$ $HOOC(CH_2)_4COOH$

(b) $HOOC(CH_2)_4COOH + 2NH_3 \longrightarrow NH_4OOC(CH_2)_4COONH_4$

$$\xrightarrow{heat} H_2N\overset{O}{\overset{\|}{C}}(CH_2)_4\overset{O}{\overset{\|}{C}}NH_2 \xrightarrow[catalyst]{350^\circ} N{\equiv}C(CH_2)_4C{\equiv}N$$

$$\xrightarrow[catalyst]{4H_2} H_2NCH_2(CH_2)_4CH_2NH_2$$

(c) $CH_2{=}CH{-}CH{=}CH_2 \xrightarrow{Cl_2} ClCH_2CH{=}CHCH_2Cl \xrightarrow{2NaCN}$

$$N{\equiv}CCH_2CH{=}CHCH_2C{\equiv}N \xrightarrow[Ni]{H_2} N{\equiv}C(CH_2)_4C{\equiv}N$$

$$\xrightarrow[catalyst]{4H_2} H_2NCH_2(CH_2)_4CH_2NH_2$$

(d) tetrahydrofuran $\xrightarrow{2HCl}$ $ClCH_2CH_2CH_2CH_2Cl \xrightarrow{2NaCN}$

$$N{\equiv}C(CH_2)_4C{\equiv}N \xrightarrow[catalyst]{4H_2} H_2NCH_2(CH_2)_4CH_2NH_2$$

K.2

(a) $HOCH_2CH_2OH + :\bar{B} \rightleftharpoons HOCH_2CH_2O^- + HB$

$$RO\overset{O}{\overset{\|}{C}}{-}C_6H_4{-}\overset{O}{\overset{\|}{C}}OCH_3 + {}^-OCH_2CH_2OH \rightleftharpoons$$

$$RO\overset{O}{\overset{\|}{C}}{-}C_6H_4{-}\underset{OCH_3}{\overset{O^-}{C}}{-}OCH_2CH_2OH \rightleftharpoons RO\overset{O}{\overset{\|}{C}}{-}C_6H_4{-}\overset{O}{\overset{\|}{C}}OCH_2CH_2OH + CH_3O^-$$

$$CH_3O^- + HB \rightleftharpoons CH_3OH + :B^-$$

$R = CH_3{-}$ or $HOCH_2CH_2{-}$

(b) $ROC(=O)-C_6H_4-C(=O)-OCH_3 \underset{-H^+}{\overset{+H^+}{\rightleftharpoons}} ROC(=O)-C_6H_4-C(=\overset{+}{O}H)-OCH_3$

$\underset{-HOCH_2CH_2OH}{\overset{+HOCH_2CH_2OH}{\rightleftharpoons}} ROC(=O)-C_6H_4-C(OH)(OCH_3)-\overset{+}{O}(H)-CH_2CH_2OH \rightleftharpoons$

$ROC(=O)-C_6H_4-C(OH)(\overset{+}{O}(H)CH_3)-OCH_2CH_2OH \underset{+CH_3OH}{\overset{-CH_3OH}{\rightleftharpoons}} ROC(=O)-C_6H_4-C(=\overset{+}{O}H)OCH_2CH_2OH$

$\underset{+H^+}{\overset{-H^+}{\rightleftharpoons}} ROC(=O)-C_6H_4-C(=O)-OCH_2CH_2OH$

R = CH_3- or $HOCH_2CH_2-$

K.3

(a) $CH_3OC(=O)-C_6H_4-C(=O)OCH_3 + HOCH_2-C_6H_{10}-CH_2OH$

(b) By high-pressure catalytic hydrogenation

K.4

etc.$-O-CH_2CH(O-)CH_2OC(=O)-C_6H_4-C(=O)OCH_2CH(OH)CH_2OC(=O)-C_6H_4-C(=O)OCH_2CH(O-)CH_2O-$etc.

(each pendant O of the upper chain attached through $O=C-C_6H_4-C=O$ phthalate bridges to the lower chain)

etc.$-O-CH_2CH(O-)CH_2OC(=O)-C_6H_4-C(=O)OCH_2CH(O-)CH_2OC(=O)-C_6H_4-C(=O)OCH_2CH(O-)CH_2O-$etc.

(central pendant: $-O-C(=O)-C_6H_4-C(=O)-O-$etc.)

K.5

$$HO-C_6H_4-C(CH_3)_2-C_6H_4-OH + Cl-\overset{O}{\overset{\|}{C}}-Cl \xrightarrow[(-HCl)]{pyridine}$$

$$-C_6H_4-C(CH_3)_2-C_6H_4-O-\overset{O}{\overset{\|}{C}}-\left[O-C_6H_4-C(CH_3)_2-C_6H_4-O-\overset{O}{\overset{\|}{C}}-O\right]_n \text{etc.}$$

Lexan

K.6

(a) The resin is probably formed in the following way. Base converts the bisphenol A to a phenoxide ion that attacks a carbon of the epoxide ring of epichlorohydrin:

$$ClCH_2\underset{\backslash O/}{CH-CH_2} + {}^{-}O-C_6H_4-C(CH_3)_2-C_6H_4-O^{-} + \underset{\backslash O/}{CH_2-CH}CH_2Cl \longrightarrow$$

$$Cl-CH_2-\underset{O^-}{\underset{|}{CH}}-CH_2-O-C_6H_4-C(CH_3)_2-C_6H_4-OCH_2-\underset{O^-}{\underset{|}{CH}}-CH_2-Cl$$

$$\xrightarrow{-2Cl^-} \underset{\backslash O/}{CH_2-CH}CH_2O-C_6H_4-C(CH_3)_2-C_6H_4-OCH_2\underset{\backslash O/}{CH-CH_2}$$

$$\xrightarrow{{}^{-}O-C_6H_4-C(CH_3)_2-C_6H_4-O^{-}} \text{then} \xrightarrow{\underset{\backslash O/}{CH_2-CH}CH_2Cl}$$

$$\underset{\backslash O/}{CH_2-CH}CH_2-\left[O-C_6H_4-C(CH_3)_2-C_6H_4-OCH_2\underset{OH}{\underset{|}{CH}}CH_2\right]_n-O-C_6H_4-C(CH_3)_2-C_6H_4-OCH_2\underset{\backslash O/}{CH-CH_2}$$

(b) The excess of epichlorohydrin limits the molecular weight and insures that the resin has epoxy ends.

(c) Adding the hardener brings about cross linking by reacting at the terminal epoxide groups of the resin:

$$H_2NCH_2CH_2NHCH_2CH_2\ddot{N}H_2 + \underset{\backslash O/}{CH_2-CH}CH_2-[\text{polymer}]-CH_2\underset{\backslash O/}{CH-CH_2} \longrightarrow$$

$$-CH_2-\underset{OH}{CH}CH_2-\underset{H}{N}CH_2CH_2-\underset{\underset{\underset{\underset{[polymer]}{CH_2}}{CHOH}}{CH_2}}{N}-CH_2CH_2-\underset{H}{N}-CH_2\underset{OH}{CH}CH_2\,[polymer]\,CH_2\underset{OH}{CH}CH_2-etc.$$

$$\underset{\underset{H}{N}}{\overset{}{}}\;\; CH_2-CHOH-CH_2-\underset{H}{N}-CH_2CH_2\underset{H}{N}-CH_2CH_2-\underset{H}{N}-CH_2\underset{OH}{CH}CH_2\,[polymer]\,CH_2\underset{OH}{\overset{etc.}{C}}HCH_2$$

K.7

(a) $$\left[C_6H_4 \,NH\overset{O}{\overset{\|}{C}}OCH_2CH_2O\overset{O}{\overset{\|}{C}}(CH_2)_6\overset{O}{\overset{\|}{C}}OCH_2CH_2O\overset{O}{\overset{\|}{C}}NH \right]_n$$

(b) To ensure that the polyester chain has $-CH_2OH$ end groups.

K.8

Because the para-position is occupied by a methyl group, cross-linking does not occur and the resulting polymer remains thermoplastic.

K.9

$$H-\overset{O}{\overset{\|}{C}}-H \xrightarrow{H^+} H-\overset{O^+-H}{\overset{\|}{C}}-H \xrightarrow{\text{phenol (OH)}} [\text{cyclohexadienone intermediate with } \overset{+}{O}H,\ H,\ H-\overset{O-H}{C}-H] \xrightarrow{-H^+}$$

$$\text{OH, }CH_2OH\text{ (o-hydroxybenzyl alcohol)} \xrightarrow[\text{as before}]{H-\overset{O^+-H}{\overset{\|}{C}}-H} HOCH_2,\ OH,\ CH_2OH \xrightarrow[\text{as before}]{H-\overset{O^+-H}{\overset{\|}{C}}-H}$$

$$HOCH_2,\ OH,\ CH_2OH,\ CH_2OH \xrightarrow{H^+} HOCH_2,\ OH,\ CH_2OH_2^+,\ CH_2OH \xrightarrow{-H_2O}$$

HOCH$_2$–C$_6$H$_2$(OH)(CH$_2^+$)(CH$_2$OH) $\xrightarrow{C_6H_5OH}$ HOCH$_2$–C$_6$H$_2$(OH)(CH$_2$OH)–CH$_2$–C$_6$H$_4$OH $\xrightarrow{H_2C{=}OH^+}$ etc.

$\xrightarrow{C_6H_5OH}$ Bakelite

CHAPTER EIGHTEEN
Amines

18.1

Dissolve both compounds in ether and extract with aqueous HCl. This gives an ether layer that contains cyclohexane and an aqueous layer that contains cyclohexylammonium chloride. Cyclohexane may then be recovered from the ether layer by distillation. Cyclohexylamine may be recovered from the aqueous layer by adding aqueous NaOH (to convert cyclohexylammonium chloride to cyclohexylamine) and then by ether extraction and distillation

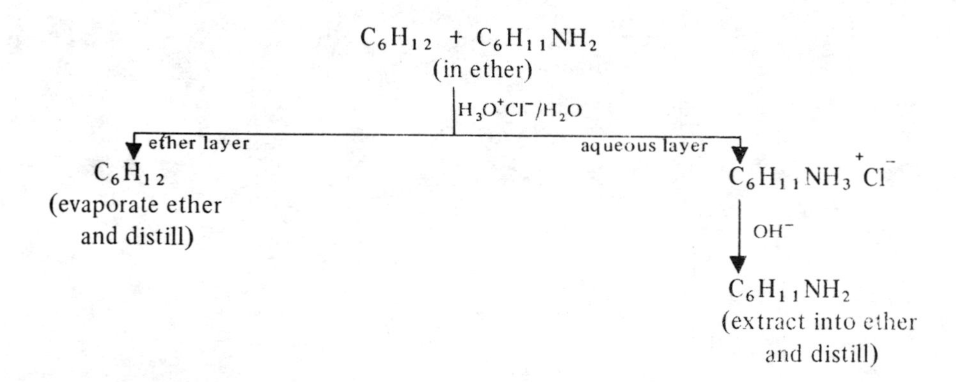

18.2

We begin by dissolving the mixture in a water-immiscible organic solvent such as CH_2Cl_2 or ether. Then, extractions with aqueous acids and bases allow us to separate the components. (We separate *p*-cresol from benzoic acid by taking advantage of benzoic acid's solubility in the more weakly basic aqueous $NaHCO_3$, whereas, *p*-cresol requires the more strongly basic, aqueous NaOH.

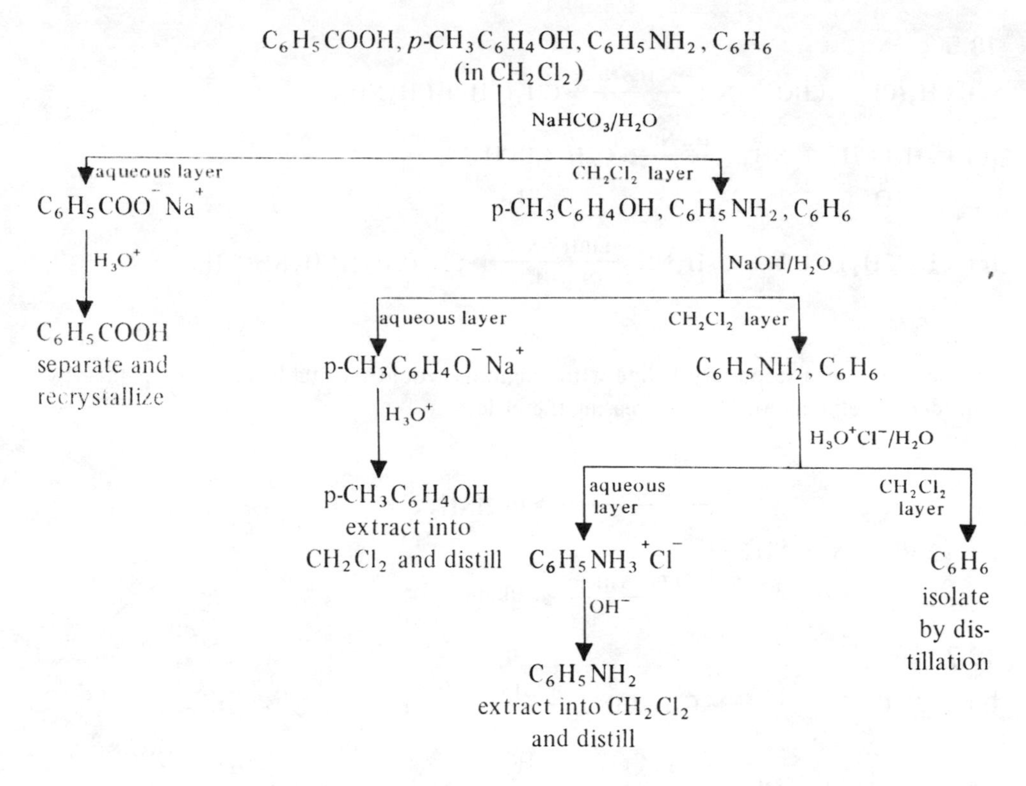

18.3

(a) Neglecting Kekule´ forms of the ring, we can write the following resonance structures for the phthalimide anion.

(b) Phthalimide is more acidic than benzamide because its anion is stabilized by resonance to a greater extent than the anion of benzamide. (Benzamide has only one carbonyl group attached to the nitrogen and thus fewer resonance contributors are possible.)

18.4

KOH; $C_6H_5CH_2Br$ (−KBr)

NH_2NH_2, ethanol, reflux → $C_6H_5CH_2NH_2$ +

Benzylamine

18.5

(a) $CH_3(CH_2)_3CHO + NH_3 \xrightarrow{H_2,\ Ni} CH_3(CH_2)_3CH_2NH_2$

(b) $C_6H_5\overset{O}{\overset{\|}{C}}CH_3 + NH_3 \xrightarrow{H_2,\ Ni} C_6H_5\underset{NH_2}{\underset{|}{C}}HCH_3$

(c) $CH_3(CH_2)_4CHO + C_6H_5NH_2 \xrightarrow[CH_3OH]{LiBH_3CN} CH_3(CH_2)_4CH_2NHC_6H_5$

18.6

The reaction of a secondary halide with ammonia would inevitably be accompanied by considerable elimination thus decreasing the yield.

$$R\overset{R'}{\overset{|}{C}}H{-}X + \underset{excess}{NH_3} \begin{cases} \xrightarrow{\text{Substitution}} R\overset{R'}{\overset{|}{C}}HNH_2 \\ \xrightarrow{\text{Elimination}} \text{alkene} \end{cases}$$

18.7

(a) $C_6H_5COOH \xrightarrow{SOCl_2} C_6H_5COCl \xrightarrow{CH_3CH_2NH_2}$

$C_6H_5CONHCH_2CH_3 \xrightarrow{LiAlH_4} C_6H_5CH_2NHCH_2CH_3$

(b) $CH_3CH_2CH_2CH_2CH_2Br \xrightarrow{NaCN} CH_3CH_2CH_2CH_2CH_2CN$

$\xrightarrow{LiAlH_4} CH_3CH_2CH_2CH_2CH_2CH_2NH_2$

(c) $CH_3CH_2COOH \xrightarrow{SOCl_2} CH_3CH_2COCl \xrightarrow{(CH_3CH_2CH_2)_2NH}$

$CH_3CH_2CON(CH_2CH_2CH_3)_2 \xrightarrow{LiAlH_4} (CH_3CH_2CH_2)_3N$

(d) $CH_3\overset{O}{\overset{\|}{C}}CH_2CH_3 \xrightarrow{NH_2OH} CH_3\overset{NOH}{\overset{\|}{C}}CH_2CH_3 \xrightarrow{Na/C_2H_5OH} CH_3\overset{NH_2}{\overset{|}{C}}HCH_2CH_3$

18.8

(a) $CH_3O{-}C_6H_5 \xrightarrow[H_2SO_4]{HNO_3} CH_3O{-}C_6H_4{-}NO_2 \xrightarrow[HCl]{Fe}$

$CH_3O{-}C_6H_4{-}NH_2$

(b) $CH_3O{-}C_6H_5 \xrightarrow[AlCl_3]{CH_3COCl} CH_3O{-}C_6H_4{-}\overset{O}{\overset{\|}{C}}CH_3 \xrightarrow[H_2,\ Ni]{NH_3}$

$$CH_3O-C_6H_4-CHCH_3(NH_2)$$

(c)

$$C_6H_5CH_3 \xrightarrow{Cl_2,\ h\nu} C_6H_5CH_2Cl$$

$$\xrightarrow{(CH_3)_3N} C_6H_5CH_2\overset{+}{N}(CH_3)_3\ Cl^-$$

(d)

$$NO_2-C_6H_4-CH_3 \xrightarrow[(2)\ H_3O^+]{(1)\ KMnO_4,\ OH^-} NO_2-C_6H_4-COOH \xrightarrow{SOCl_2}$$

$$NO_2-C_6H_4-\overset{O}{\overset{\|}{C}}-Cl \xrightarrow{NH_3} NO_2-C_6H_4-\overset{O}{\overset{\|}{C}}NH_2 \xrightarrow{Br_2,\ OH^-} NO_2-C_6H_4-NH_2$$

(e)

$$CH_3-C_6H_5 + Tl(OOCCF_3)_3 \xrightarrow{CF_3COOH} CH_3-C_6H_4-Tl(OOCCF_3)_2$$

$$\xrightarrow{KCN,\ H_2O,\ h\nu} CH_3-C_6H_4-CN \xrightarrow{LiAlH_4} CH_3-C_6H_4-CH_2NH_2$$

18.9

An amine acting as a base.

$$CH_3CH_2\ddot{N}H_2 + H_3O^+ \rightleftharpoons CH_3CH_2NH_3^+ + H_2O$$

An amine acting as a nucleophile in an alkylation reaction.

$$(CH_3CH_2)_3N: + CH_3-I \longrightarrow (CH_3CH_2)_3\overset{+}{N}-CH_3\ I^-$$

An amine acting as a nucleophile in an acylation reaction.

$$\underset{(excess)}{(CH_3)_2\ddot{N}H} + CH_3C(=O)Cl \longrightarrow (CH_3)_2N\overset{O}{\overset{\|}{C}}CH_3 + (CH_3)_2NH_2Cl$$

An amino group acting as an activating group and as an ortho-para director in electrophilic aromatic substituion.

$$C_6H_5NH_2 \xrightarrow[H_2O,\ room\ temp.]{Br_2} 2,4,6\text{-}Br_3C_6H_2NH_2$$

18.10

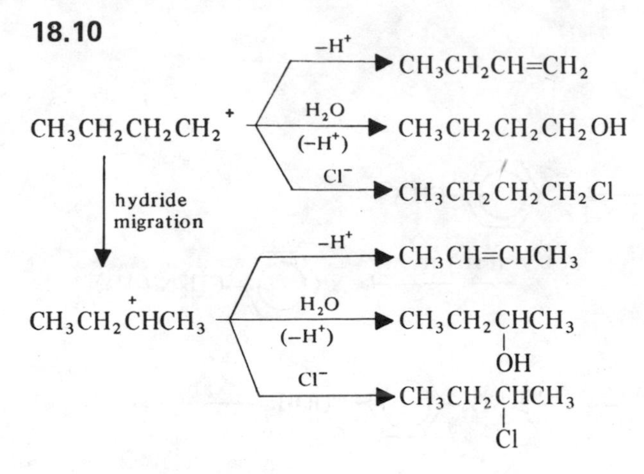

18.11

(a) $^{-}O{-}N{=}O + H_3O^+ \rightleftharpoons HO{-}N{=}O$

$HO{-}N{=}O + H_3O^+ \rightleftharpoons HO^+(H){-}N{=}O$

$HO^+(H){-}N{=}O \rightleftharpoons H_2O + \overset{+}{N}{=}O$

(b) $:N(CH_3)_2$-benzene + $\overset{+}{N}O$ → $\overset{+}{N}(CH_3)_2$ (ring with H, NO) $\xrightarrow{-H^+}$ $:N(CH_3)_2$-benzene with NO (para)

(c) The $\overset{+}{N}O$ ion is a weak electrophile. For it to react with an aromatic ring, the ring must have a powerful activating group such as $-OH$ or $-NR_2$.

18.12

(a) benzene $\xrightarrow[H_2SO_4,\ heat]{fuming\ HNO_3}$ m-dinitrobenzene (NO_2, NO_2) $\xrightarrow[(2)\ OH^-]{(1)\ Fe,\ HCl,\ heat}$ m-phenylenediamine (NH_2, NH_2)

(b) m-dinitrobenzene (NO_2, NO_2) (from part a) $\xrightarrow[NH_3,\ C_2H_5OH]{H_2S}$ m-nitroaniline (NO_2, NH_2)

(c) benzene $\xrightarrow[H_2SO_4]{HNO_3}$ nitrobenzene (NO_2) $\xrightarrow[Fe]{Cl_2}$

NO_2 (m-chloronitrobenzene) $\xrightarrow[(2)\ OH^-]{(1)\ Fe,\ HCl,\ heat}$ NH_2 (m-chloroaniline), Cl

(d) NO_2 (nitrobenzene) (as in c) $\xrightarrow[Fe]{Br_2}$ m-bromonitrobenzene (NO_2, Br) $\xrightarrow[(2)\ OH^-]{(1)\ Fe,\ HCl,\ heat}$ m-bromoaniline (NH_2, Br)

(e) NO_2 (nitrobenzene) (as in c) $\xrightarrow[(2)\ OH^-]{(1)\ Fe,\ HCl,\ heat}$ NH_2 (aniline) $\xrightarrow{(CH_3CO)_2O}$

$NHCOCH_3$ (acetanilide) $\xrightarrow[H_2SO_4]{conc.}$ $NHCOCH_3$, SO_3H (para) $\xrightarrow{HNO_3}$ $NHCOCH_3$, NO_2 (ortho), SO_3H (para)

$\xrightarrow[heat]{57\%\ H_2SO_4}$ NH_2, NO_2 (o-nitroaniline) (56% yield from acetanilide cf., page 508 of text)

(f) $NHCOCH_3$ (from part e) $\xrightarrow[H_2SO_4]{HNO_3}$ $NHCOCH_3$, NO_2 (para) (90%, cf. page 508 of text) $\xrightarrow[H_2O]{H^+}$ NH_2, NO_2 (p-nitroaniline)

18.13

CH_3, NH_2 (*p*-Toluidine) $\xrightarrow[H_2O]{Br_2}$ CH_3, Br, NH_2, Br (2,6-dibromo-4-methylaniline) $\xrightarrow[H_2O,\ 0\text{-}5^\circ]{H_2SO_4,\ NaNO_2}$

CH_3, Br, N_2^+, Br $\xrightarrow[H_2O,\ 130^\circ]{H_3PO_2}$ CH_3, Br, Br (3,5-Dibromotoluene) $+ N_2$

18.14

(a)

$$CH_3C_6H_4NH_2 \xrightarrow{(CH_3CO)_2O} CH_3C_6H_4NHCOCH_3 \xrightarrow[H_2SO_4]{HNO_3} CH_3C_6H_3(NO_2)NHCOCH_3 \xrightarrow[H_2O]{H^+}$$

$$CH_3C_6H_3(NO_2)NH_2 \xrightarrow[0\text{-}5^\circ]{HONO} CH_3C_6H_3(NO_2)N_2^+ \xrightarrow[(-N_2)]{H_3PO_2} CH_3C_6H_4NO_2$$

$$\xrightarrow[(2)\ OH^-]{(1)\ Fe,\ HCl} CH_3C_6H_4NH_2$$

$$CH_3C_6H_4NH_2 \xrightarrow{HONO} CH_3C_6H_4N_2^+$$

(b) $$\xrightarrow{CuCl} CH_3C_6H_4Cl$$

(c) $$\xrightarrow{CuBr} CH_3C_6H_4Br$$

(d) $$\xrightarrow{KI} CH_3C_6H_4I$$

(e) $$\xrightarrow{CuCN} CH_3C_6H_4CN$$

(f) $$CH_3C_6H_4CN \xrightarrow[H_2O]{H^+} CH_3C_6H_4COOH$$

18.15

$$NH_2C_6H_4NO_2 \xrightarrow[Fe]{Br_2} NH_2C_6H_2Br_2NO_2 \xrightarrow[0\text{-}5^\circ]{HONO} N_2^+C_6H_2Br_2NO_2$$

$\xrightarrow{CuBr}$ 3,4,5-tribromonitrobenzene (Br, Br, Br, NO_2) $\xrightarrow{H_2,\ Pt}$ 3,4,5-tribromoaniline (Br, Br, Br, NH_2) $\xrightarrow[(0\text{-}5^\circ)]{HONO}$

$\longrightarrow$ 3,4,5-tribromobenzenediazonium (Br, Br, Br, N_2^+) $\xrightarrow{H_3PO_2}$ 1,2,3-tribromobenzene (Br, Br, Br)

18.16

$HO_3S\text{-}C_6H_4\text{-}NH_2 \xrightarrow[0\text{-}5^\circ]{HONO} HO_3S\text{-}C_6H_4\text{-}N_2^+$

$\xrightarrow[\text{NaOH, pH 8-10}]{\text{2-naphthol (OH)}}$ 1-($N{=}N$-C_6H_4-SO_3Na)-2-naphthol (OH)

Orange II

18.17

$C_6H_6 \xrightarrow[H_2SO_4]{HNO_3} C_6H_5NO_2 \xrightarrow[HCl]{Fe} C_6H_5NH_2 \xrightarrow[O\text{-}S^\circ]{HONO}$

$C_6H_5N_2^+ \xrightarrow[\text{pH 5-7}]{C_6H_5N(CH_3)_2} C_6H_5\text{-}N{=}N\text{-}C_6H_4\text{-}N(CH_3)_2$

18.18

$C_6H_5NH_2 \xrightarrow{\text{(cf: Problem 18.12f)}}$ p-$NO_2C_6H_4NH_2$ (NH_2, NO_2) $\xrightarrow[0\text{-}5^\circ]{HONO}$

$$\text{O}_2\text{N-C}_6\text{H}_4\text{-N}_2^+ \xrightarrow[\text{pH 5-7}]{\text{NH}_2,\text{OH-naphthalene-(SO}_3\text{H)}_2} \text{O}_2\text{N-C}_6\text{H}_4\text{-N=N-C}_{10}\text{H}_3(\text{NH}_2)(\text{OH})(\text{SO}_3\text{H})_2$$

$$\xrightarrow[\text{pH 8-10 (NaOH)}]{\text{N}_2^+\text{-C}_6\text{H}_4\text{-C}_6\text{H}_4\text{-N}_2^+ \text{ from benzidine + HONO}} \text{O}_2\text{N-C}_6\text{H}_4\text{-N=N-C}_{10}\text{H}_2(\text{NH}_2)(\text{OH})(\text{SO}_3\text{Na})_2\text{-N=N-C}_6\text{H}_4\text{-C}_6\text{H}_4\text{-N}_2^+$$

$$\xrightarrow{\text{C}_6\text{H}_5\text{-OH}} \text{O}_2\text{N-C}_6\text{H}_4\text{-N=N-C}_{10}\text{H}_2(\text{NH}_2)(\text{OH})(\text{SO}_3\text{Na})_2\text{-N=N-C}_6\text{H}_4\text{-C}_6\text{H}_4\text{-N=N-C}_6\text{H}_4\text{-OH}$$

Diamine Green B

18.19

(1) That A reacts with benzenesulfonyl chloride in aqueous KOH to give a clear solution which on acidification yields a precipitate shows that A is a primary amine.

(2) That diazotization of A followed by treatment with 2-naphthol gives an intensely colored precipitate shows that A is a primary aromatic amine, that is, A is a substituted aniline.

(3) Consideration of the molecular formula of A leads us to conclude that A is a toluidine

$$\begin{array}{l} C_7H_9N \\ -\ \underline{C_6H_6N} \\ CH_3 \end{array} = \text{—C}_6\text{H}_4\text{—NH}_2$$

But is A *o*-toluidine, *m*-toluidine, or *p*-toluidine?

(4) This question is answered by the infrared data. A single absorption peak in the 680-840 cm^{-1} region at 815 cm^{-1} is indicative of a *para* substituted benzene. Thus A is *p*-toluidine.

$$p\text{-}CH_3\text{-}C_6H_4\text{-}NH_2$$

A

18.20

First convert the sulfonamide to its anion, then alkylate the anion with an alkyl halide, then remove the $-SO_2C_6H_5$ group by hydrolysis. For example

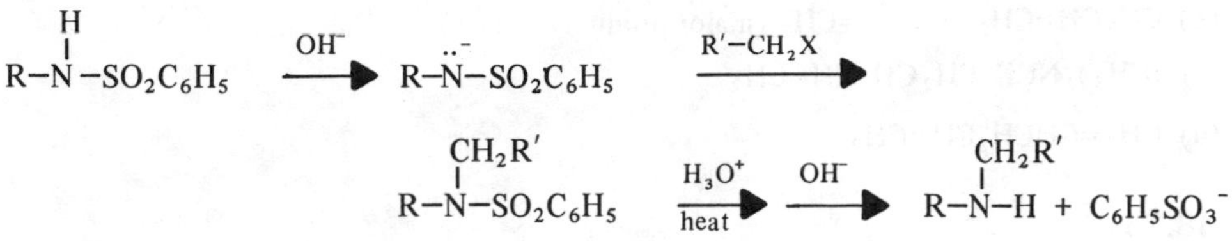

$$R-\overset{H}{\underset{|}{N}}-SO_2C_6H_5 \xrightarrow{OH^-} R-\ddot{N}^{-}-SO_2C_6H_5 \xrightarrow{R'-CH_2X}$$

$$R-\overset{CH_2R'}{\overset{|}{N}}-SO_2C_6H_5 \xrightarrow[heat]{H_3O^+} \xrightarrow{OH^-} R-\overset{CH_2R'}{\overset{|}{N}}-H + C_6H_5SO_3^-$$

18.21

(a)

Aniline ($C_6H_5NH_2$) $\xrightarrow{(CH_3CO)_2O}$ $C_6H_5NHCOCH_3$ $\xrightarrow{HOSO_2Cl}$

p-$CH_3CONH-C_6H_4-SO_2Cl$ $\xrightarrow{H_2N\text{-thiazole}}$ p-$CH_3CONH-C_6H_4-SO_2NH$-(2-thiazolyl)

$\xrightarrow[heat]{dil\ HCl}$ p-$H_2N-C_6H_4-SO_2NH$-(2-thiazolyl)

Sulfathiazole

(b) Sulfathiazole + succinic anhydride →

p-$(NHCOCH_2CH_2COOH)-C_6H_4-SO_2NH$-(2-thiazolyl)

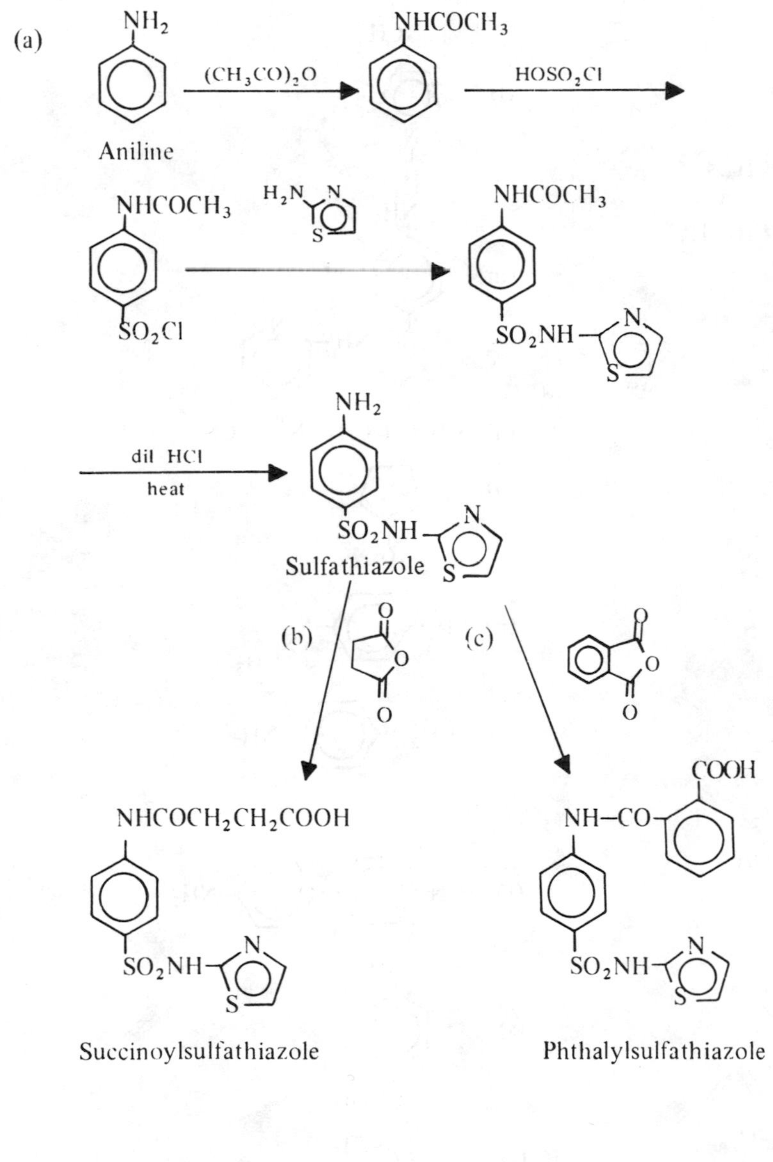

Succinoylsulfathiazole

(c) Sulfathiazole + phthalic anhydride →

p-$(NH-CO-C_6H_4\text{-}o\text{-}COOH)-C_6H_4-SO_2NH$-(2-thiazolyl)

Phthalylsulfathiazole

18.22

(a) $CH_3C(CH_3){=}CH_2$ + $CH_2{=}CH_2$ (major product)

(b) $CH_3CH{=}CH_2$ + $CH_2{=}CH_2$ (major product)

(c) $(CH_3)_2NCH_2CH_2CH_2CH{=}CH_2$

(d) $CH_2{=}CHCH_2CH{=}CH_2$

18.23

(a) $C_6H_5CH_2NHCH_3$

(b) $(CH_3CH(CH_3))_3N$

(c) 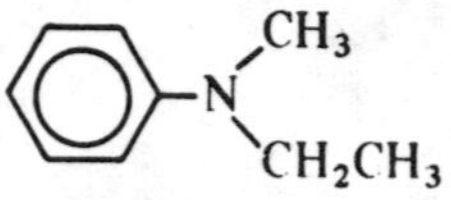 $C_6H_5N(CH_3)(CH_2CH_3)$

(d) 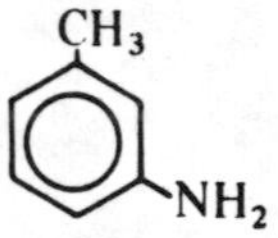 3-methylaniline (ring with CH_3 and NH_2 meta)

(e) 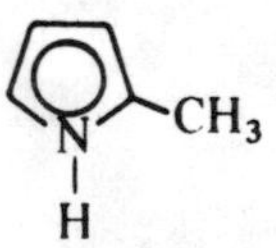 2-methylpyrrole (N–H, CH_3)

(f) N-ethylpiperidine ($N{-}CH_2CH_3$)

(g) N-ethylpyridinium bromide ($N^+{-}CH_2CH_3$ Br^-)

(h) pyridine-3-COOH

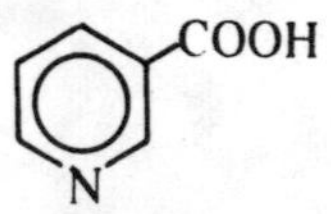

(i) 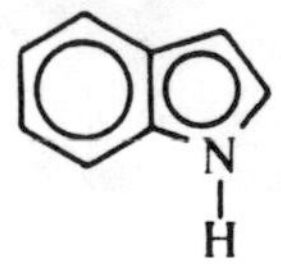 indole (N–H)

(j) 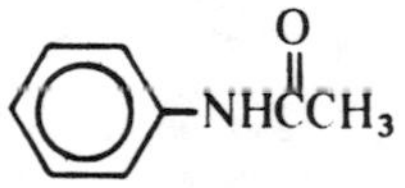 $C_6H_5NHCCH_3$ (C=O)

(k) $CH_3{-}N^+H_2{-}CH_3$ Cl^-

(l) 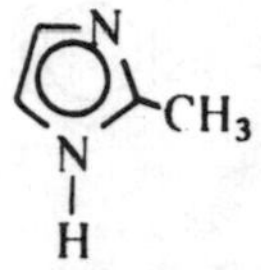 2-methylimidazole (N–H, CH_3)

(m) 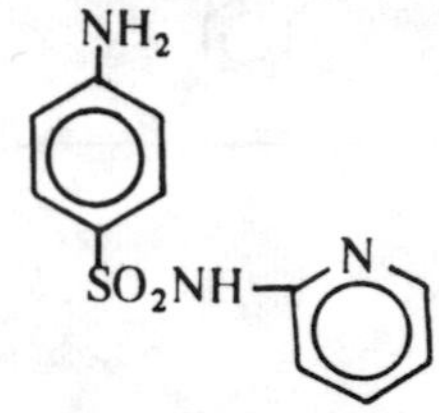 p-$NH_2C_6H_4SO_2NH$–(2-pyridyl)

(n) $(CH_3CH_2CH_2)_4N^+$ Cl^-

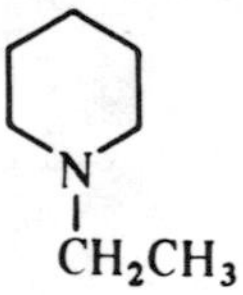

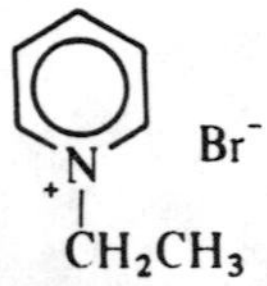

(o) 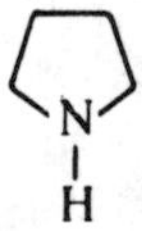 pyrrolidine (N–H)

(p) 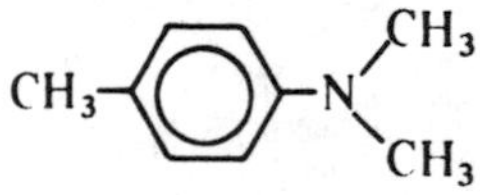 $CH_3{-}C_6H_4{-}N(CH_3)_2$

(q) 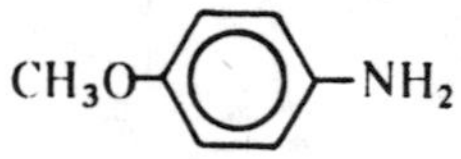 $CH_3O{-}C_6H_4{-}NH_2$

(r) 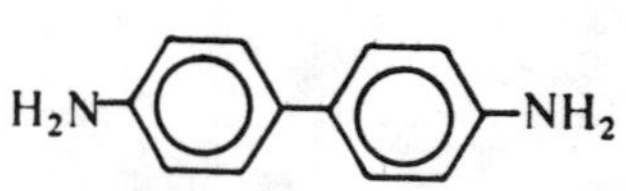 $H_2N{-}C_6H_4{-}C_6H_4{-}NH_2$

(s) 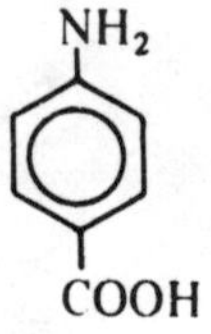 p-$NH_2C_6H_4COOH$

(t) 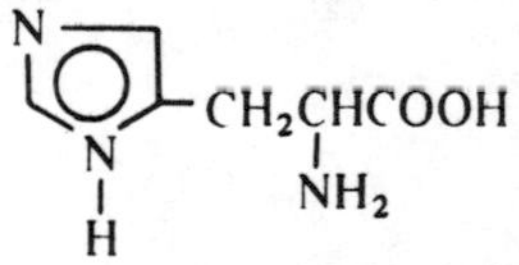 imidazole (N–H)–$CH_2CH(NH_2)COOH$

18.24

(a) Propylamine

(b) N-methylaniline

(c) Trimethylisopropylammonium iodide

(d) *o*-toluidine

(e) *o*-Anisidine (or *o*-methoxyaniline)

(f) Pyrazole

(g) 2-Aminopyrimidine

(h) Benzylammonium chloride

(i) N,N-Dipropylaniline

(j) Benzenesulfonamide

(k) Methylammonium acetate

(l) 3-Aminopropanol

(m) Purine

(n) N-Methylpyrrole

18.25

(a) $C_6H_5{-}C{\equiv}N + LiAlH_4 \longrightarrow C_6H_5{-}CH_2NH_2$

(b) $C_6H_5{-}C(=O){-}NH_2 + LiAlH_4 \longrightarrow C_6H_5{-}CH_2NH_2$

(c) $C_6H_5{-}CH_2Br + NH_{3\,(excess)} \longrightarrow C_6H_5{-}CH_2NH_2$

$C_6H_5{-}CH_2Br$ + potassium phthalimide (phthalimide NK) $\longrightarrow C_6H_5{-}CH_2{-}N$(phthalimide)

$\xrightarrow{NH_2NH_2} C_6H_5{-}CH_2NH_2$ + phthalhydrazide (cyclic C=O, N–H, N–H)

(d) $C_6H_5{-}CH_2OTs + NH_{3\,(excess)} \longrightarrow C_6H_5{-}CH_2NH_2$

(e) $C_6H_5{-}CHO + NH_3 \xrightarrow{H_2,\ Ni} C_6H_5{-}CH_2NH_2$

(f) $C_6H_5{-}CH_2NO_2 + 3H_2 \xrightarrow{Pt} C_6H_5{-}CH_2NH_2$

(g) $C_6H_5{-}CH_2C(=O)NH_2 \xrightarrow{Br_2,\ OH^-} C_6H_5{-}CH_2NH_2 + CO_3^{=}$

18.26

(a) Benzene $\xrightarrow[H_2SO_4]{HNO_3}$ nitrobenzene ($C_6H_5NO_2$) $\xrightarrow[(2)\ OH^-]{(1)\ Fe,\ HCl,\ heat}$ aniline ($C_6H_5NH_2$)

(b) Bromobenzene (C_6H_5Br) $\xrightarrow[liq\ NH_3]{NaNH_2}$ aniline ($C_6H_5NH_2$)

(c) Benzamide ($C_6H_5CONH_2$) $\xrightarrow{Br_2,\ OH^-}$ aniline ($C_6H_5NH_2$)

18.27

(a) $CH_3(CH_2)_2CH_2OH \xrightarrow{PBr_3} CH_3(CH_2)_2CH_2Br \xrightarrow{\text{potassium phthalimide (phthalimide NK)}}$

N-butylphthalimide ($C_6H_4(CO)_2NCH_2(CH_2)_2CH_3$) $\xrightarrow{NH_2NH_2} CH_3(CH_2)_2CH_2NH_2$ + phthalhydrazide ($C_6H_4(CO)_2N_2H_2$)

(b) $CH_3(CH_2)_2CH_2Br \xrightarrow{NaCN} CH_3(CH_2)_3CN \xrightarrow{LiAlH_4} CH_3(CH_2)_3CH_2NH_2$
(from part a)

(c) $CH_3(CH_2)_2CH_2OH \xrightarrow[(2)\ H_3O^+]{(1)\ KMnO_4,\ OH^-} CH_3CH_2CH_2COOH$

$\xrightarrow[(2)\ NH_3]{(1)\ SOCl_2} CH_3CH_2CH_2CONH_2 \xrightarrow{Br_2,\ OH^-} CH_3CH_2CH_2NH_2$

(d) $CH_3CH_2CH_2CH_2OH \xrightarrow[C_5H_5N]{CrO_3} CH_3CH_2CH_2CHO \xrightarrow[H_2,\ Ni]{CH_3NH_2}$

$CH_3CH_2CH_2CH_2NHCH_3$

18.28

N-methylpiperidine + CH_3I $\longrightarrow$ N,N-dimethylpiperidinium iodide ($\overset{+}{N}(CH_3)_2$ I^-) **A** $\xrightarrow[H_2O]{Ag_2O}$

B (N,N-dimethylpiperidinium hydroxide, CH_3, CH_3, OH^-) $\xrightarrow[(-H_2O)]{heat}$ C (ring-opened alkene, $N(CH_3)_2$) $\xrightarrow{CH_3I}$ D ($(CH_3)_3\overset{+}{N}$– alkene, I^-)

$$\xrightarrow[H_2O]{Ag_2O} \underset{\mathbf{E}}{CH_2{=}CHCH_2CH_2CH_2\overset{+}{N}(CH_3)_3\ \overset{-}{OH}} \xrightarrow{heat}$$

$$\underset{\mathbf{F}}{CH_2{=}CHCH_2CH{=}CH_2} + H_2O + (CH_3)_3N$$

18.29

(a) aniline (NH_2) $\xrightarrow{(CH_3CO)_2O}$ acetanilide ($NHCOCH_3$)

(b) aniline (NH_2) + phthalic anhydride $\xrightarrow{heat}$ N-phenylphthalimide

(c) acetanilide ($NHCOCH_3$) (from part a) $\xrightarrow[H_2SO_4]{HNO_3}$ p-nitroacetanilide ($NHCOCH_3$, NO_2) $\xrightarrow[(2)\ OH^-]{(1)\ H^+,\ H_2O}$ p-nitroaniline (NH_2, NO_2)

(d) acetanilide ($NHCOCH_3$) (from part a) $\xrightarrow{HOSO_2Cl}$ ($NHCOCH_3$, SO_2Cl) $\xrightarrow[(2)\ H_3O^+,\ heat]{(1)\ NH_3}$ (NH_2, SO_2NH_2)

(e) aniline (NH_2) $\xrightarrow[base]{2CH_3I}$ N,N-dimethylaniline ($N(CH_3)_2$)

(f) aniline (NH_2) $\xrightarrow{HONO}$ ($N_2^+\ X^-$) $\xrightarrow{HBF_4}$ ($N_2^+\ BF_4^-$) $\xrightarrow{heat}$ fluorobenzene (F)

(g) ($N_2^+\ X^-$) (from part f) $\xrightarrow{CuCl}$ chlorobenzene (Cl)

(h) $C_6H_5N_2^+ X^-$ (from part f) $\xrightarrow{CuBr}$ C_6H_5Br

(i) $C_6H_5N_2^+ X^-$ (from part f) $\xrightarrow{KI}$ C_6H_5I

(j) $C_6H_5N_2^+ X^-$ (from part f) $\xrightarrow{CuCN}$ C_6H_5CN

(k) C_6H_5CN (from part j) $\xrightarrow[\text{heat}]{H_3O^+,\ H_2O}$ C_6H_5COOH

(l) $C_6H_5N_2^+ X^-$ (from part f) $\xrightarrow[\text{heat}]{H_3O^+,\ H_2O}$ C_6H_5OH

(m) $C_6H_5N_2^+ X^-$ (from part f) $\xrightarrow{H_3PO_2}$ C_6H_6

(n) $C_6H_5N_2^+ X^-$ (from part f) + C_6H_5OH (from part l) $\xrightarrow[(pH\ 8\text{-}10)]{OH^-}$ $C_6H_5\text{-}N{=}N\text{-}C_6H_4\text{-}OH$

(o) $C_6H_5N_2^+ X^-$ (from part f) + $C_6H_5N(CH_3)_2$ (from part e) $\xrightarrow[(pH\ 5\text{-}7)]{H_3O^+}$ $C_6H_5\text{-}N{=}N\text{-}C_6H_4\text{-}N(CH_3)_2$

18.30

(a) $CH_3CH_2CH_2NH_2 \xrightarrow[(NaNO_2/HCl)]{HONO} [CH_3CH_2CH_2N_2^+] \xrightarrow{-N_2}$

$$[CH_3CH_2CH_2^+] \xrightarrow{\text{hydride shift}} [CH_3\overset{+}{C}HCH_3]$$

$[CH_3CH_2CH_2^+] \xrightarrow{Cl^-} CH_3CH_2CH_2Cl$; $\xrightarrow{H_2O} CH_3CH_2CH_2OH$; $\xrightarrow{-H^+} CH_3CH{=}CH_2$

$[CH_3\overset{+}{C}HCH_3] \xrightarrow{-H^+} CH_3CH{=}CH_2$; $\xrightarrow{H_2O} CH_3CH(OH)CH_3$; $\xrightarrow{Cl^-} CH_3CH(Cl)CH_3$

(b) $(CH_3CH_2)_2NH \xrightarrow[(NaNO_2/HCl)]{HONO} (CH_3CH_2)_2N{-}N{=}O$

(c) $C_6H_5{-}N(H)CH_2CH_2CH_3 \xrightarrow[(NaNO_2/HCl)]{HONO} C_6H_5{-}N(N{=}O)CH_2CH_2CH_3$

(d) $C_6H_5{-}N(CH_2CH_2CH_3)_2 \xrightarrow[(NaNO_2/HCl)]{HONO} O{=}N{-}C_6H_4{-}N(CH_2CH_2CH_3)_2$

(e) $CH_3CH_2CH_2{-}C_6H_4{-}NH_2 \xrightarrow[(NaNO_2/HCl)]{HONO,\ 0\text{-}5^\circ} CH_3CH_2CH_2{-}C_6H_4{-}N_2^+\ Cl^-$

18.31

(a) $CH_3CH_2CH_2NH_2 + C_6H_5SO_2Cl \xrightarrow[H_2O]{KOH} CH_3CH_2CH_2\bar{N}SO_2C_6H_5\ K^+$

clear solution

$\xrightarrow{H_3O^+} CH_3CH_2CH_2NHSO_2C_6H_5$

precipitate

(b) $(CH_3CH_2CH_2)_2NH + C_6H_5SO_2Cl \xrightarrow[H_2O]{KOH} (CH_3CH_2CH_2)_2NSO_2C_6H_5$

precipitate

$\xrightarrow{H_3O^+}$ No reaction (precipitate remains)

(c) $C_6H_5{-}N(H)CH_2CH_2CH_3 + C_6H_5SO_2Cl \xrightarrow[H_2O]{KOH} C_6H_5{-}N(SO_2C_6H_5)CH_2CH_2CH_3$

precipitate

$\xrightarrow{H_3O^+}$ No reaction (precipitate remains)

(d) $C_6H_5{-}N(CH_2CH_2CH_3)_2 + C_6H_5SO_2Cl \xrightarrow[H_2O]{KOH}$ No reaction (3° amine is insoluble)

$\xrightarrow{H_3O^+} C_6H_5{-}\overset{+}{N}H(CH_2CH_2CH_3)_2$

3° amine dissolves

(e) $C_3H_7-C_6H_4-NH_2 + C_6H_5SO_2Cl \xrightarrow[H_2O]{KOH} C_3H_7-C_6H_4-\bar{N}SO_2C_6H_5\ K^+$ (clear solution)

$\xrightarrow{H_3O^+} C_3H_7-C_6H_4-NHSO_2C_6H_5$ (precipitate)

18.32

(a) $(CH_2)_5N-H \xrightarrow[(NaNO_2/HCl)]{HONO} (CH_2)_5N-N{=}O$

(b) $(CH_2)_5N-H + C_6H_5SO_2Cl \xrightarrow[H_2O]{KOH} (CH_2)_5N-SO_2C_6H_5$

18.33

(a) $2CH_3CH_2NH_2 + C_6H_5COCl \longrightarrow CH_3CH_2NHCOC_6H_5 + CH_3CH_2\overset{+}{N}H_3\ Cl^-$

(b) $2CH_3NH_2 + (CH_3\overset{O}{\overset{\|}{C}})_2O \longrightarrow CH_3NH\overset{O}{\overset{\|}{C}}CH_3 + CH_3\overset{+}{N}H_3\ CH_3\overset{O}{\overset{\|}{C}}O^-$

(c) Succinic anhydride (cyclic $CH_2-C(=O)-O-C(=O)-CH_2$) $+ 2CH_3NH_2 \longrightarrow$ $CH_3NH-\overset{O}{\overset{\|}{C}}-CH_2-CH_2-\overset{O}{\overset{\|}{C}}O^-\ CH_3NH_3^+$

(d) (product of c) $\xrightarrow{heat}$ N-methylsuccinimide (cyclic $CH_2-C(=O)-N(CH_3)-C(=O)-CH_2$) $+ H_2O$

(e) Pyrrolidine (N–H) + phthalic anhydride $\longrightarrow$ C_6H_4 with $-\overset{O}{\overset{\|}{C}}-N$(pyrrolidinyl) and $-\overset{O}{\overset{\|}{C}}OH$ (ortho)

(f) Pyrrole (N–H) + $(CH_3CO)_2O$ ⟶ N-acetylpyrrole (N–C(=O)CH_3) + CH_3COOH

(g) $2\,C_6H_5NH_2 + CH_3CH_2\overset{O}{\overset{\|}{C}}Cl \longrightarrow C_6H_5NH\overset{O}{\overset{\|}{C}}CH_2CH_3 + C_6H_5NH_3^+\ Cl^-$

(h) $CH_3CH_2-\overset{+}{N}(CH_2CH_3)_2-CH_2CH_3\ \ ^-OH \longrightarrow CH_2{=}CH_2 + (CH_3CH_2)_3N + H_2O$

(i) 1,3-dinitrobenzene (NO_2, NO_2) + H_2S $\xrightarrow[C_2H_5OH]{NH_3}$ 3-nitroaniline (NO_2, NH_2)

(j) p-toluidine (CH_3, NH_2) + $Br_{2\,(excess)}$ $\xrightarrow{H_2O}$ 2,6-dibromo-4-methylaniline (CH_3, Br, Br, NH_2)

18.34

(a) Toluene (CH_3) $\xrightarrow[H_2SO_4]{HNO_3}$ o-nitrotoluene (CH_3, NO_2) + p-nitrotoluene (CH_3, NO_2)

Separate isomers

o-nitrotoluene (CH_3, NO_2) $\xrightarrow[(2)\ OH]{(1)\ Fe,\ HCl,\ heat}$ o-toluidine (CH_3, NH_2) $\xrightarrow{HONO}$

o-methylbenzenediazonium (CH_3, N_2^+ X^-) $\xrightarrow[heat]{H_3O^+}$ o-cresol (CH_3, OH)

(b) m-toluidine (CH_3, NH_2) $\xrightarrow[(2)\ H_3O^+,\ heat]{(1)\ HONO}$ m-cresol (CH_3, OH)

(from Problem 18.14a)

(c) CH_3–C_6H_4–NO_2 (from part a) $\xrightarrow[(2)\ OH^-]{(1)\ Fe,\ HCl,\ heat}$ CH_3–C_6H_4–NH_2 $\xrightarrow[(2)\ H_3O^+,\ heat]{(1)\ HONO}$ CH_3–C_6H_4–OH

(d) $C_6H_4(NH_2)_2$ (from Problem 18.12a) $\xrightarrow{HONO}$ $C_6H_4(N_2^+\ X^-)_2$ $\xrightarrow{CuCl}$ $C_6H_4Cl_2$

(e) $C_6H_4(N_2^+\ X^-)_2$ (cf. part d) $\xrightarrow{CuCN}$ $C_6H_4(CN)_2$

(f) NO_2–C_6H_4–NH_2 (from Problem 18.12a) $\xrightarrow[(2)\ KI]{(1)\ HONO}$ NO_2–C_6H_4–I $\xrightarrow[(2)\ OH^-]{(1)\ Fe,\ HCl,\ heat}$ NH_2–C_6H_4–I $\xrightarrow[(2)\ H_3O^+,\ heat]{(1)\ HONO}$ OH–C_6H_4–I

(g) NO_2–C_6H_4–NH_2 (from Problem 18.12a) $\xrightarrow[(2)\ CuBr]{(1)\ HONO}$ NO_2–C_6H_4–Br $\xrightarrow[(2)\ OH^-]{(1)\ Fe,\ HCl,\ heat}$ NH_2–C_6H_4–Br $\xrightarrow[(2)\ CuCN]{(1)\ HONO}$ CN–C_6H_4–Br

(h) NO_2–C_6H_4–NH_2 (from Problem 18.12f) $\xrightarrow[Fe]{Br_2}$ NO_2–$C_6H_2Br_2$–NH_2 (Br, Br ortho to NH_2) $\xrightarrow{HONO}$

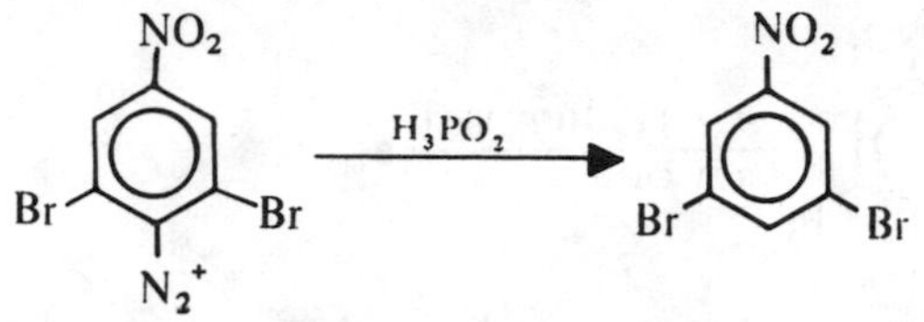

NO_2 Br Br N_2^+ $\xrightarrow{H_3PO_2}$ NO_2 Br Br

(i) NO_2 Br Br $\xrightarrow[(2)\ OH^-]{(1)\ Fe,\ HCl,\ heat}$ NH_2 Br Br

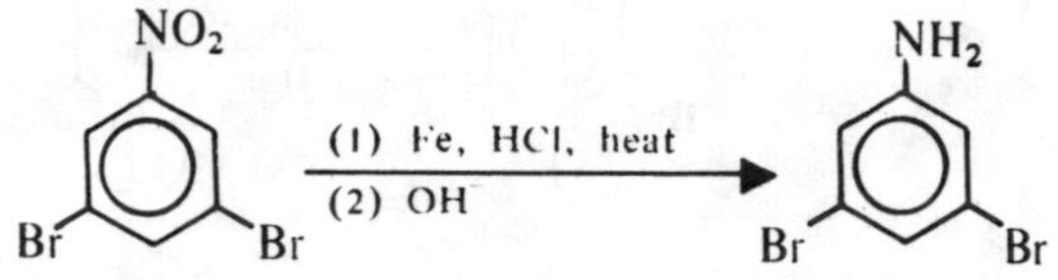

(from part h)

(j) NO_2 Br Br NH_2 $\xrightarrow[(2)\ CuBr]{(1)\ HONO}$ NO_2 Br Br Br $\xrightarrow[(2)\ OH^-]{(1)\ Fe,\ HCl,\ heat}$

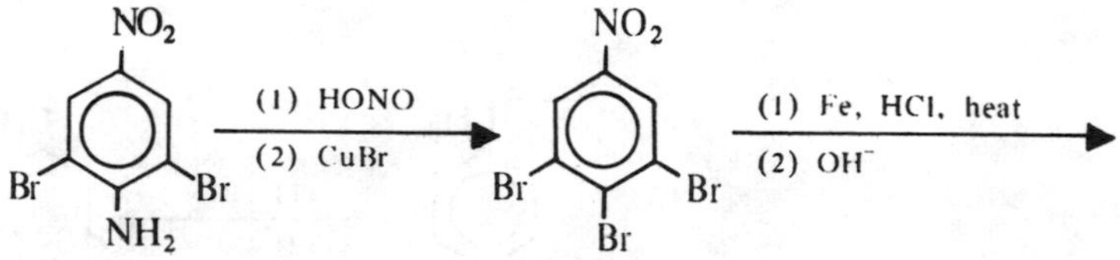

(from part h)

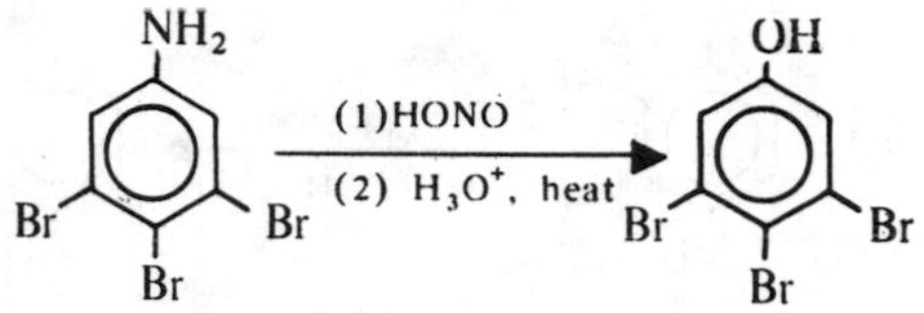

NH_2 Br Br Br $\xrightarrow[(2)\ H_3O^+,\ heat]{(1) HONO}$ OH Br Br Br

(k) NH_2 Br Br Br $\xrightarrow[(2)\ CuCN]{(1)\ HONO}$ CN Br Br Br

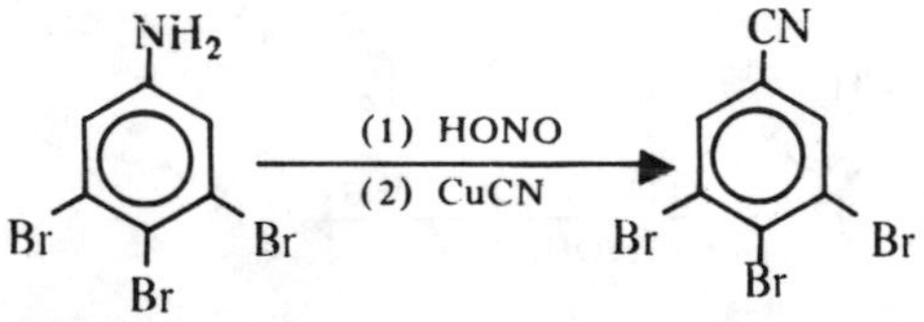

(from part j)

(l) NO_2 Br Br NH_2 $\xrightarrow[(2)\ CuCN]{(1)\ HONO}$ NO_2 Br Br CN $\xrightarrow[heat]{H_3O^+}$

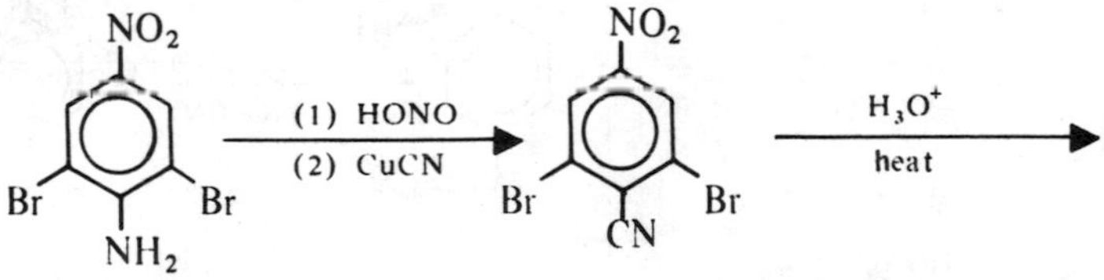

(from part h)

NO_2 Br Br COOH $\xrightarrow{H_2,\ Pt}$ NH_2 Br Br COOH $\xrightarrow[(2)\ H_3PO_2]{(1)\ HONO}$ Br Br COOH

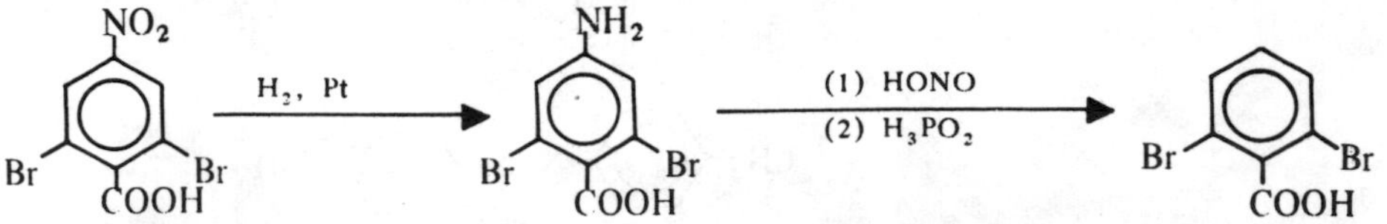

(m) 2,6-dibromo-4-nitroaniline (NO_2; Br, Br; NH_2) (from part h) $\xrightarrow[\text{(2) KI}]{\text{(1) HONO}}$ 1,3-dibromo-2-iodo-5-nitrobenzene (NO_2; Br, I, Br) $\xrightarrow[\text{(2) } OH^-]{\text{(1) Fe, HCl, heat}}$ 3,5-dibromo-4-iodoaniline (NH_2; Br, I, Br) $\xrightarrow[\text{(2) } H_3PO_2]{\text{(1) HONO}}$ 1,3-dibromo-2-iodobenzene (Br, I, Br)

(n) toluene (CH_3) $\xrightarrow[H_2SO_4]{\text{fuming } HNO_3}$ 2,4-dinitrotoluene (CH_3; NO_2; NO_2) $\xrightarrow{H_2S,\ NH_3}$ 4-amino-2-nitrotoluene (CH_3; NO_2; NH_2) $\xrightarrow[\text{(2) CuBr}]{\text{(1) HONO}}$ 4-bromo-2-nitrotoluene (CH_3; NO_2; Br)

(o) 4-amino-2-nitrotoluene (CH_3; NO_2; NH_2) (from part n) $\xrightarrow[\text{(2) } H_3O^+\text{, heat}]{\text{(1) HONO}}$ 4-methyl-3-nitrophenol (CH_3; NO_2; OH)

(p) 4-amino-2-nitrotoluene (CH_3; NO_2; NH_2) (from part n) $\xrightarrow[\text{(2) CuBr}]{\text{(1) HONO}}$ 4-bromo-2-nitrotoluene (CH_3; NO_2; Br) $\xrightarrow[\text{(2) } OH^-]{\text{(1) Fe, HCl, heat}}$ 5-bromo-2-methylaniline (CH_3; NH_2; Br) $\xrightarrow[\text{(2) CuCN}]{\text{(1) HONO}}$ 5-bromo-2-methylbenzonitrile (CH_3; CN; Br)

(q) $CH_3-C_6H_4-NH_2$ (from part c) $\xrightarrow{\text{HONO}}$ $CH_3-C_6H_4-N_2^+$ X^-

$\downarrow$ C_6H_5-OH, pH 8-10

$CH_3-C_6H_4-N{=}N-C_6H_4-OH$

(r) $CH_3-C_6H_4-N_2^+ X^-$ (from part q) $\xrightarrow[\text{pH 8-10}]{CH_3-C_6H_4-OH \text{ (from part c)}}$ $CH_3-C_6H_4-N{=}N-C_6H_3(OH)(CH_3)$

18.35

(a) Benzylamine dissolves in dilute HCl at room temperature,

$$C_6H_5CH_2NH_2 + H_3O^+ + Cl^- \xrightarrow{25^\circ} C_6H_5CH_2\overset{+}{N}H_3\overset{-}{Cl}$$

benzamide does not dissolve:

$$C_6H_5CONH_2 + H_3O^+ + Cl^- \xrightarrow{25^\circ} \text{No reaction}$$

(b) Allylamine reacts with (and decolorizes) bromine in carbon tetrachloride instantly,

$$CH_2{=}CHCH_2NH_2 + Br_2 \xrightarrow{CCl_4} \underset{Br}{CH_2}\underset{Br}{CH}CH_2NH_2$$

propylamine does not:

$$CH_3CH_2CH_2NH_2 + Br_2 \xrightarrow{CCl_4}$$ No reaction if the mixture is not heated or irradiated.

(c) The Hinsberg test:

$$CH_3-C_6H_4-NH_2 + C_6H_5SO_2Cl \xrightarrow[H_2O]{KOH} CH_3-C_6H_4-\overset{-}{N}SO_2C_6H_5\ K^+ \text{ (soluble)} \xrightarrow{H_3O^+} CH_3-C_6H_4-NHSO_2C_6H_5 \text{ (precipitate)}$$

$$C_6H_5-NHCH_3 + C_6H_5SO_2Cl \xrightarrow[H_2O]{KOH} C_6H_5-N(CH_3)SO_2C_6H_5 \text{ (precipitate)} \xrightarrow{H_3O^+} \text{Precipitate remains}$$

(d) The Hinsberg test:

$$C_6H_{11}-NH_2 + C_6H_5SO_2Cl \xrightarrow[H_2O]{KOH} C_6H_{11}-\overset{-}{N}SO_2C_6H_5\ K^+ \text{ (soluble)} \xrightarrow{H_3O^+} C_6H_{11}-NHSO_2C_6H_5 \text{ (precipitate)}$$

$$\text{(cyclohexyl)N–H} + C_6H_5SO_2Cl \xrightarrow[H_2O]{KOH} \underset{\text{precipitate}}{\text{(cyclohexyl)N–}SO_2C_6H_5} \xrightarrow{H_3O^+} \text{Precipitate remains}$$

(e) Pyridine dissolves in dilute HCl,

$$\text{C}_5\text{H}_5\text{N} + H_3O^+ + Cl^- \longrightarrow \text{C}_5\text{H}_5\overset{+}{\text{N}}\text{–H} \quad Cl^-$$

benzene does not:

$$C_6H_6 + H_3O^+ + Cl^- \longrightarrow \text{No reaction}$$

(f) Aniline reacts with nitrous acid at 0-5° to give a stable diazonium salt that couples with 2-naphthol yielding an intensely colored azo compound.

$$C_6H_5\text{–}NH_2 \xrightarrow[0\text{-}5^\circ]{HONO} C_6H_5\text{–}N_2^+ \xrightarrow{\text{2-naphthol}} C_6H_5\text{–N=N–}C_{10}H_6\text{OH}$$

Cyclohexylamine reacts with nitrous acid at 0-5° to yield a highly unstable diazonium salt—one that decomposes so rapidly that the addition of 2-naphthol gives no azo compound.

$$C_6H_{11}\text{–}NH_2 \xrightarrow[0\text{-}5^\circ]{HONO} \left[C_6H_{11}\text{–}N_2^+\right] \xrightarrow{-N_2} \left[C_6H_{11}^+\right] \longrightarrow$$

$$\text{alkenes, alcohols, etc.} \xrightarrow{\text{2-naphthol}} \text{No reaction}$$

(g) The Hinsberg test:

$$(C_2H_5)_3N + C_6H_5SO_2Cl \xrightarrow[H_2O]{KOH} \text{No reaction} \xrightarrow{H_3O^+} \underset{\text{soluble}}{(C_2H_5)_3\overset{+}{N}H}$$

$$(C_2H_5)_2NH + C_6H_5SO_2Cl \xrightarrow[H_2O]{KOH} \underset{\text{precipitate}}{(C_2H_5)_2NSO_2C_6H_5} \xrightarrow{H_3O^+} \text{Precipitate remains}$$

(h) Tripropylammonium chloride reacts with aqueous NaOH to give a water insoluble tertiary amine.

$$\underset{\text{water soluble}}{(CH_3CH_2CH_2)_3\overset{+}{N}H\ Cl^-} \xrightarrow[H_2O]{NaOH} \underset{\text{water insoluble}}{(CH_3CH_2CH_2)_3N}$$

Tetrapropylammonium chloride does not react with aqueous NaOH (at room temperature) and the tetrapropylammonium ion remains in solution.

$$\underset{\text{water soluble}}{(CH_3CH_2CH_2)_4\overset{+}{N}Cl^-} \xrightarrow[H_2O]{NaOH} \underset{\text{water soluble}}{(CH_3CH_2CH_2)_4\overset{+}{N}\ [Cl^- \text{ or } OH^-]}$$

(i) Tetrapropylammonium chloride dissolves in water to give a neutral solution. Tetrapropylammonium hydroxide dissolves in water to give a strongly basic solution.

18.36

Follow the procedure outlined in the answer to Problem 18.2. Toluene will show the same solubility behavior as benzene.

18.37

$$\text{succinic anhydride } (CH_2CH_2(CO)_2O) + NH_3 \longrightarrow H_2N\overset{O}{\overset{\|}{C}}CH_2CH_2\overset{O}{\overset{\|}{C}}OH$$

$$\xrightarrow[(-CO_3^{=})]{Br_2,\ OH^-} H_2NCH_2CH_2COO^- \xrightarrow{H^+} H_2NCH_2CH_2COOH$$

18.38

(a) $HOCH_2(CH_2)_8CH_2OH \xrightarrow{PBr_3} BrCH_2(CH_2)_8CH_2Br \xrightarrow{2\ (CH_3)_3N}$

$$(CH_3)_3\overset{+}{N}CH_2(CH_2)_8CH_2\overset{+}{N}(CH_3)_3\ 2Br^-$$

(b) $HOOCCH_2CH_2COOH + 2\ BrCH_2CH_2OH \xrightarrow{H^+}$

$$BrCH_2CH_2OOCCH_2CH_2COOCH_2CH_2Br \xrightarrow{2\ (CH_3)_3N}$$

$$(CH_3)_3\overset{+}{N}CH_2CH_2OOCCH_2CH_2COOCH_2CH_2\overset{+}{N}(CH_3)_3\ 2\ Br^-$$

(c) $(CH_3)_3N + CH_2\text{—}CH_2\ (\text{epoxide, bridged by O}) \longrightarrow (CH_3)_3\overset{+}{N}CH_2CH_2O^- \xrightarrow{CH_3\overset{O}{\overset{\|}{C}}Cl}$

$$(CH_3)_3\overset{+}{N}CH_2CH_2O\overset{O}{\overset{\|}{C}}CH_3\ Cl^-$$

18.39

Compound (b) would probably be inactive because of the meta orientation of the groups. Compound (d) would probably be inactive because the distance that separates the $-NH_2$ from the $-SO_2NH-$ group is too large.

18.40

$$\text{2-amino-4,5-diamino-6-hydroxypyrimidine } (H_2N,\ NH_2,\ NH_2,\ OH) + \overset{CHBr_2}{|}{C}(=O)\text{—}CH_2Br \xrightarrow{(-H_2O)} \text{pyrimidine with } \ddot{N}H_2\text{ and } N{=}C(\text{—}CH_2Br)\text{—}CH\text{—}Br \text{ (arrows: } NH_2 \text{ attacks CH, Br leaves)}$$

$\xrightarrow{(-HBr)}$ H_2N, N, N, N, H, Br, N, CH_2Br, OH $\xrightarrow[(-HBr)]{HCO_3^-}$

H_2N, N, N, N, N, CH_2 + $H_2\ddot{N}$—C₆H₄—CONHCH(CH_2CH_2COOH)COOH, OH, Br $\xrightarrow{-HBr}$ Folic acid

18.41

The results of the Hinsberg test indicate that compound W is a tertiary amine. The pmr spectrum provides evidence for the following:

(1) Two different C_6H_5— groups (one absorbing at δ7.2 and one at δ6.7.)
(2) A CH_3CH_2— group (the quartet at δ3.3 and the triplet at δ1.2).
(3) An unsplit $-CH_2-$ group (the singlet at δ4.4).

There is only one reasonable way to put all of this together.

C_6H_5—CH_2—N(CH_2CH_3)—C_6H_5

Thus W is N-benzyl-N-ethylaniline.

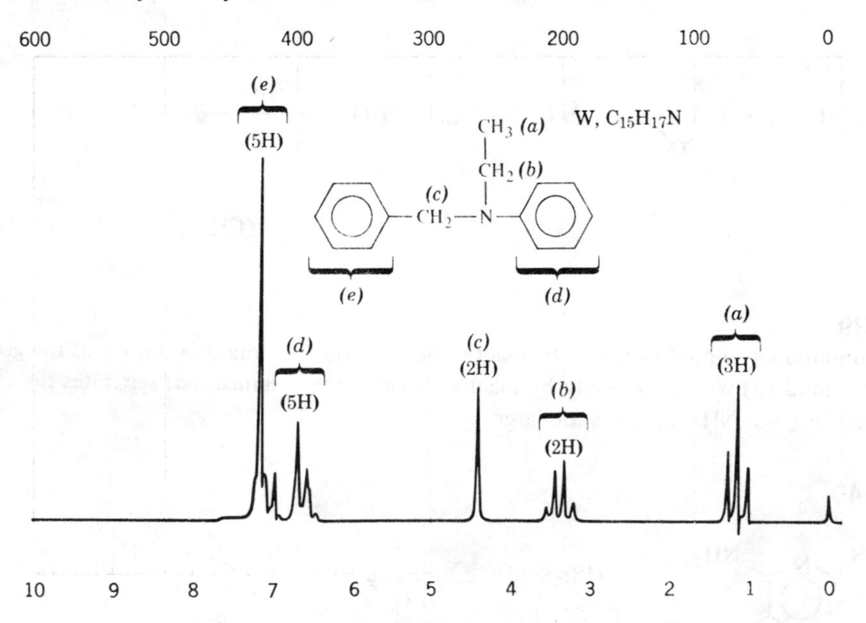

FIG. 18.3. The proton nmr spectrum of W (problem 18.41). (Spectrum courtesy of Aldrich Chemical Co.)

18.42

Compound X is benzyl bromide, $C_6H_5CH_2Br$. This is the only structure consistent with the pmr and infrared data. (The mono-substituted benzene ring is strongly indicated by the (5H), δ7.3 pmr absorption and is confirmed by the peaks at 690 cm^{-1} and 770 cm^{-1} in the infrared spectrum.)

Compound Y, therefore must be phenylacetonitrile, $C_6H_5CH_2CN$, and Z must be 2-phenylethylamine, $C_6H_5CH_2CH_2NH_2$.

$$C_6H_5CH_2Br \xrightarrow{NaCN} C_6H_5CH_2CN \xrightarrow{LiAlH_4} C_6H_5CH_2CH_2NH_2$$

X (C_7H_7Br) Y (C_8H_7N) Z ($C_8H_{11}N$)

Interpretations of the infrared and pmr spectra of Z are given in Fig. 18.4.

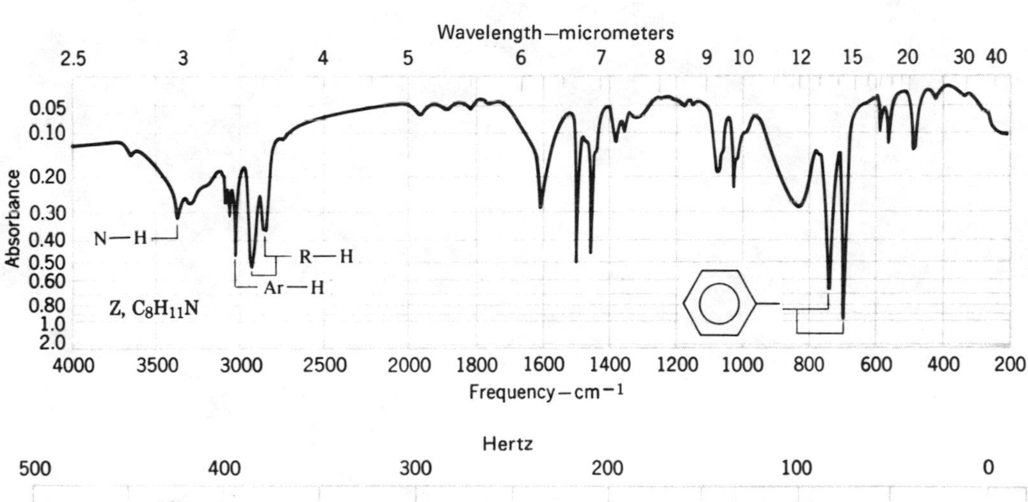

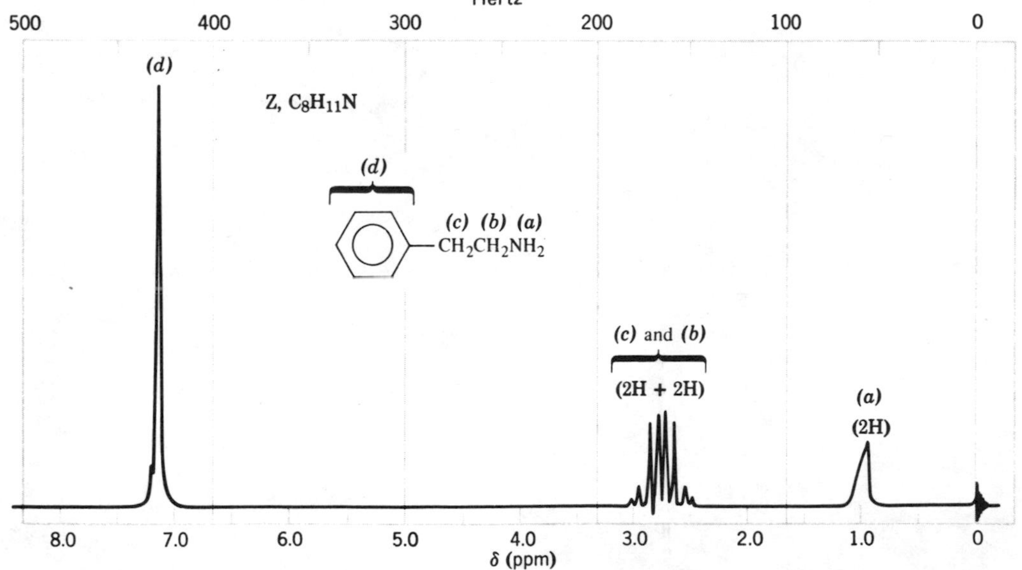

FIG. 18.4. Infrared and proton nmr spectra for compound Z, problem 18.42. (Spectra courtesy of Sadtler Inc.)

18.43

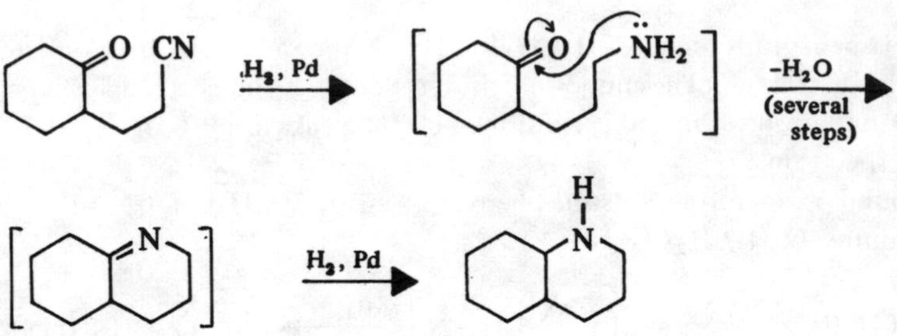

$\xrightarrow{H_2,\ Pd}$ $\xrightarrow[\text{(several steps)}]{-H_2O}$

$\xrightarrow{H_2,\ Pd}$

L

Special Topic

L.1

(a) and (b)

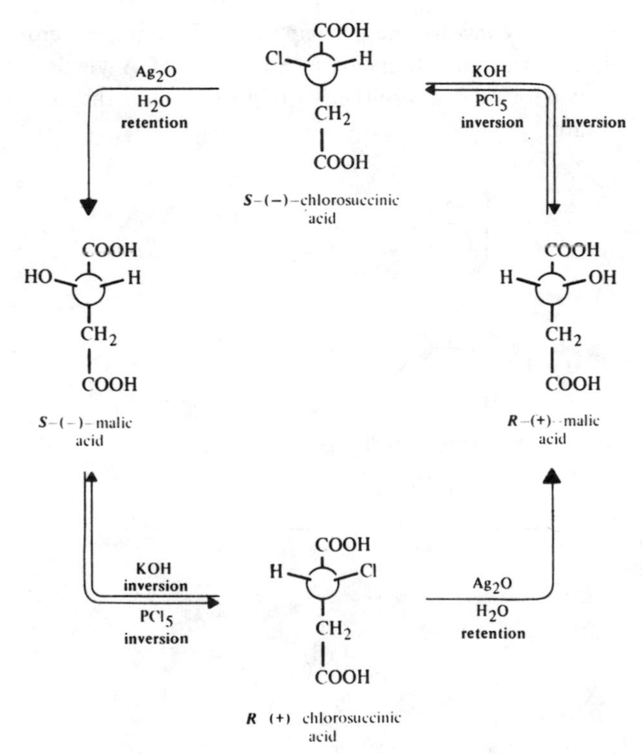

(c) The reaction takes place with retention of configuration.

(d)

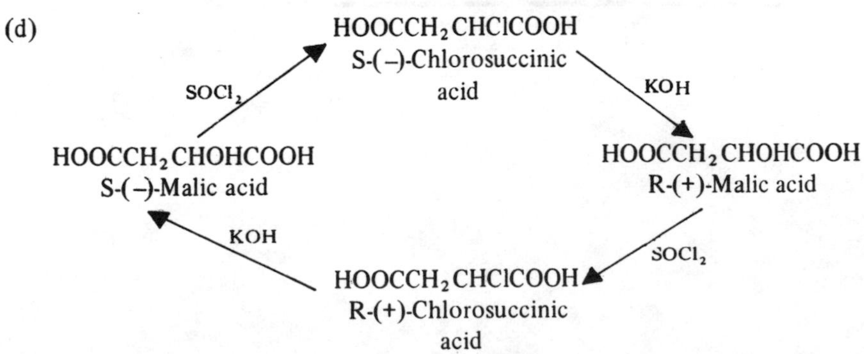

L.2

(a) Reaction of an alkene with halogen in water solution.

(b) $CH_3CH{=}CH_2 + Cl_2 \xrightarrow{H_2O} CH_3CH(OH)CH_2Cl \xrightarrow{NaOH} CH_3CH{-}CH_2$ (epoxide, O bridging)

L.3

2-(p-hydroxyphenyl)-1-chloropropane > 2-(p-tolyl)-1-chloropropane > 2-phenyl-1-chloropropane > 2-(p-nitrophenyl)-1-chloropropane

L.4

In each case, the reactions apparently involve the participation of the phenyl group and the formation of a phenonium ion as an intermediate. Solvolysis of **A** yields a chiral phenonium ion—one that reacts with solvent at either carbon to produce the same chiral (and thus optically active) acetate.

C_6H_5 CH_3 H H 2 1 CH_3 OTs

A

$(-TsO^-)$

H 2 1 CH_3 CH_3 H

Chiral phenonium ion

CH_3COOH $(-H^+)$

Reaction at C-1 Reaction at C-2

C_6H_5 CH_3 H H CH_3 $OCCH_3$ (C=O) ≡ H H_3C C_6H_5 CH_3 CH_3CO (C=O) H

B **B**

Optically active

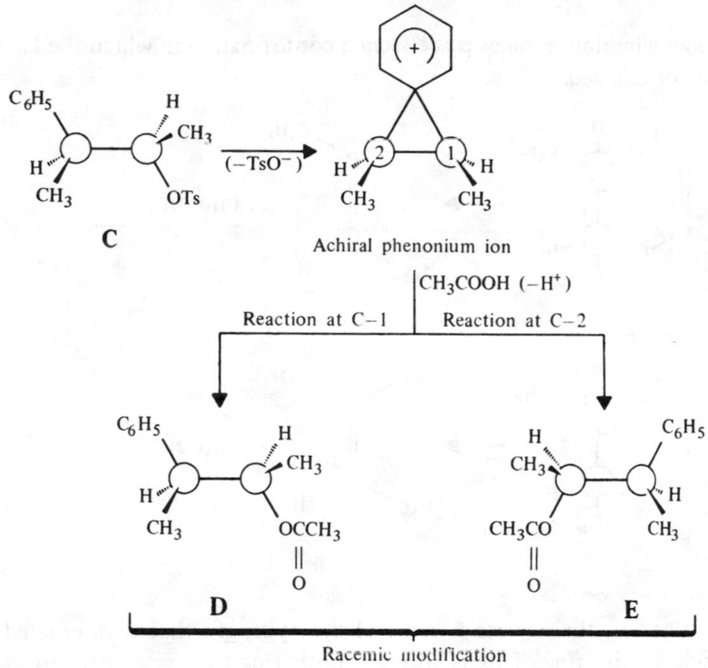

C

Achiral phenonium ion

CH_3COOH $(-H^+)$

Reaction at C–1 Reaction at C–2

D E

Racemic modification

(For an extensive discussion of this rearrangement, see J. M. Harris and C. C. Wamser, *Organic Reaction Mechanisms*. Wiley, New York, 1976, pp. 166-171.

L.5

Only a proton or deuteron *anti* to the bromine can be eliminated. The two conformations of *erythro*-2-bromo-butane-3-*d* in which a proton or deuteron is *anti* to the bromine are I and II below.

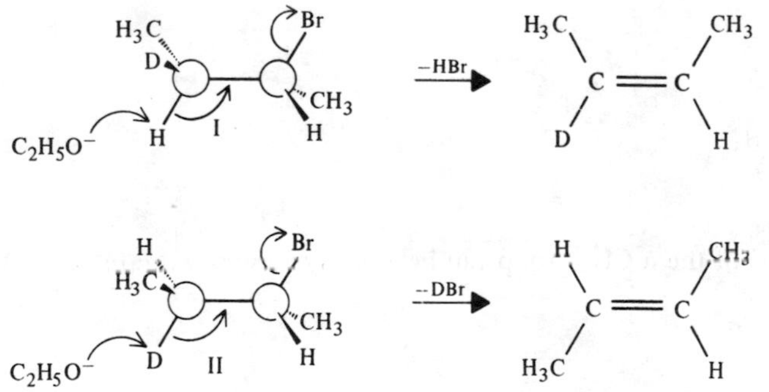

$\xrightarrow{-HBr}$

$\xrightarrow{-DBr}$

Conformation I can undergo loss of HBr to yield *cis*-2-butene-2-*d*. Conformation II can undergo loss of DBr to yield *trans*-2-butene.

L.6

In each case a *syn* elimination takes place from a conformation in which the large phenyl substituents *are not eclipsed.*

A

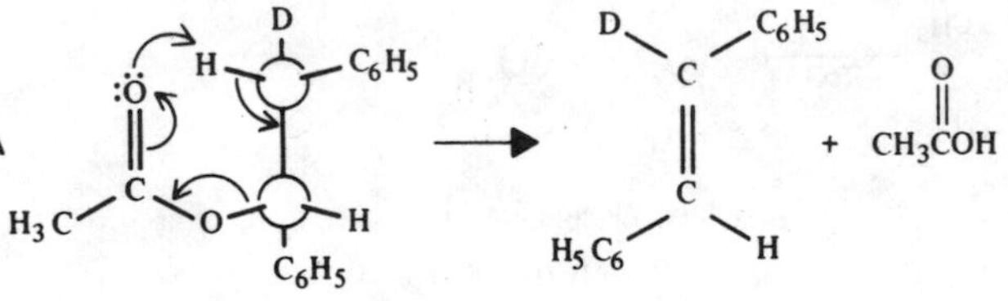

$$H_3C-C(=\ddot{O})-O \ldots \xrightarrow{} (D)C(C_6H_5)=C(H_5C_6)(H) + CH_3COOH$$

B

$$H_3C-C(=\ddot{O})-O \ldots \xrightarrow{} (H)C(C_6H_5)=C(H_5C_6)(H) + CH_3COOD$$

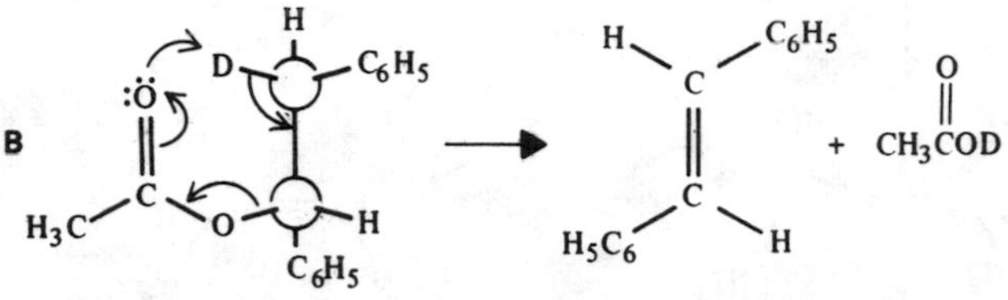

L.7

Elimination is syn; i.e., the acetate group and the hydrogen that is eliminated must be on the same side of the ring. This is true for both ring hydrogens beta to the acetate group in

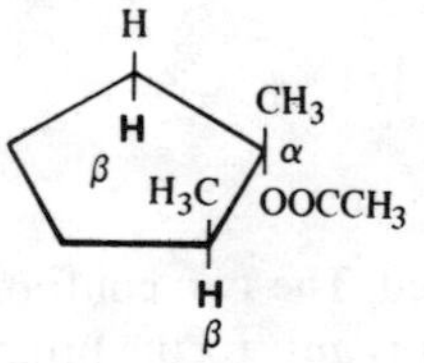

H, H β, CH_3, α, H_3C, $OOCCH_3$, H β

but for only one ring hydrogen in

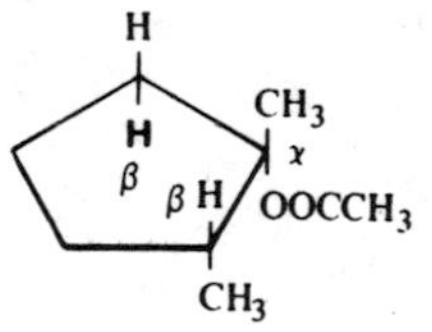

H, H β, CH_3, α, β H, $OOCCH_3$, CH_3

(In either case hydrogens of the α-CH_3 group can become syn to the acetate group.)

M

Special Topic

M.1

(a) (b)

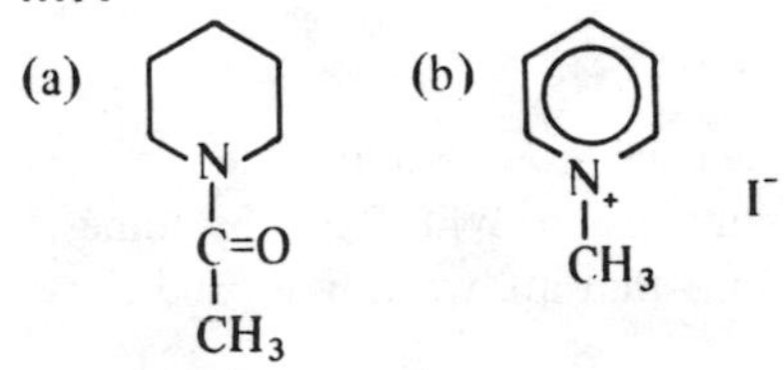

(c) (d) I^-

(e) $CH_2=CHCH_2CH_2N(CH_3)-CH_3$

M.2

(a) The cyclopentadienide ion.

(b) The pyrrole anion is a resonance hybrid of the following structures:

The imidazole anion is a hybrid of these:

M.3

A mechanism involving a "pyridyne" intermediate would involve a net loss (of 50%) of the deuterium label.

$+ \ddot{N}H_3 \xrightarrow{-HD}$

$$\text{C}_5\text{H}_3\text{N (2-Pyridyne)} + :\ddot{\text{N}}\text{H}_2^- \longrightarrow \text{(3-pyridyl anion, 2-NH}_2\text{)} \longrightarrow \text{(2-}\ddot{\text{N}}\text{H}^-\text{ pyridine)}$$

2-Pyridyne

Since in the actual experiment there was no loss of deuterium this mechanism was disallowed.

The mechanism given on page 870 would not be expected to result in a loss of deuterium, thus it is consistent with the labeling experiment.

$$\text{3-D-pyridine} + :\ddot{\text{N}}\text{H}_2^- \longrightarrow \text{(anionic adduct, 2-H, 2-}\ddot{\text{N}}\text{H}_2\text{)} \xrightarrow[-\text{H}_2]{\ddot{\text{N}}\text{H}_3} \text{2-NH}_2\text{-3-D-pyridine} + :\ddot{\text{N}}\text{H}_2^-$$

M.4

When pyridine undergoes nucleophilic substitution, the leaving group is a hydride ion—an ion that is a strong base and, consequently, a poor leaving group. With 2-halopyridines, on the other hand, the leaving groups are halide ions—ions that are weak bases and thus good leaving groups.

CHAPTER NINETEEN
Carbohydrates

19.1

(a) Two,

$$\begin{array}{c} CHO \\ | \\ {}^{*}CHOH \\ | \\ {}^{*}CHOH \\ | \\ CH_2OH \end{array}$$

(b) Two,

$$\begin{array}{c} CH_2OH \\ | \\ C{=}O \\ | \\ {}^{*}CHOH \\ | \\ {}^{*}CHOH \\ | \\ CH_2OH \end{array}$$

(c) There would be four stereoisomers (two sets of enantiomers) with each general structure: $2^2 = 4$.

19.2

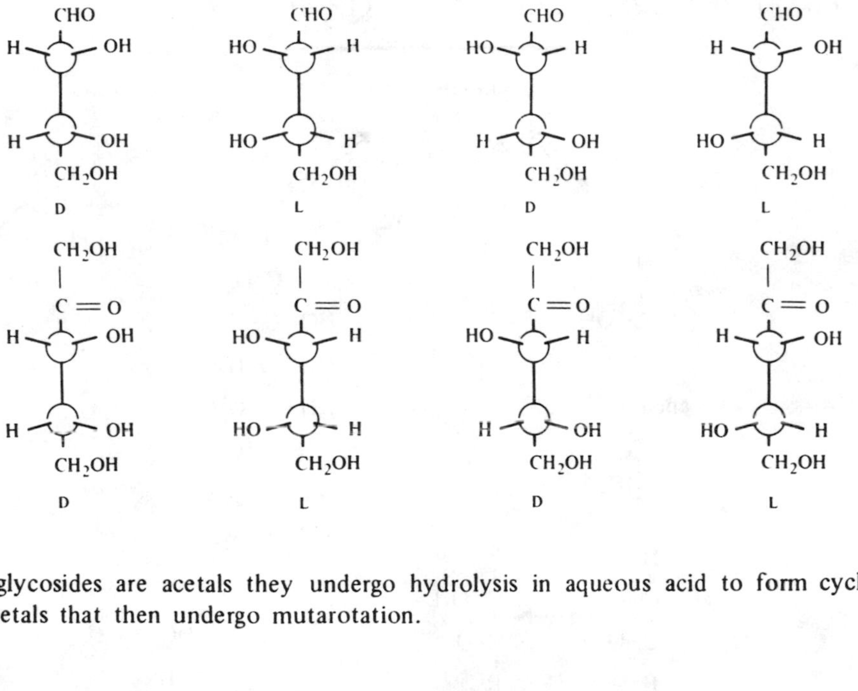

CHO; H–C–OH; H–C–OH; CH_2OH — D

CHO; HO–C–H; HO–C–H; CH_2OH — L

CHO; HO–C–H; H–C–OH; CH_2OH — D

CHO; H–C–OH; HO–C–H; CH_2OH — L

CH_2OH; C=O; H–C–OH; H–C–OH; CH_2OH — D

CH_2OH; C=O; HO–C–H; HO–C–H; CH_2OH — L

CH_2OH; C=O; HO–C–H; H–C–OH; CH_2OH — D

CH_2OH; C=O; H–C–OH; HO–C–H; CH_2OH — L

19.3

Since glycosides are acetals they undergo hydrolysis in aqueous acid to form cyclic hemiacetals that then undergo mutarotation.

19.4

CH_2OH, O, HO, HO, HO, OCH_3

methyl-α-D-glucopyranoside

CH_2OH, O, HO, OCH_3, HO, HO

methyl-β-D-glucopyranoside

19.5

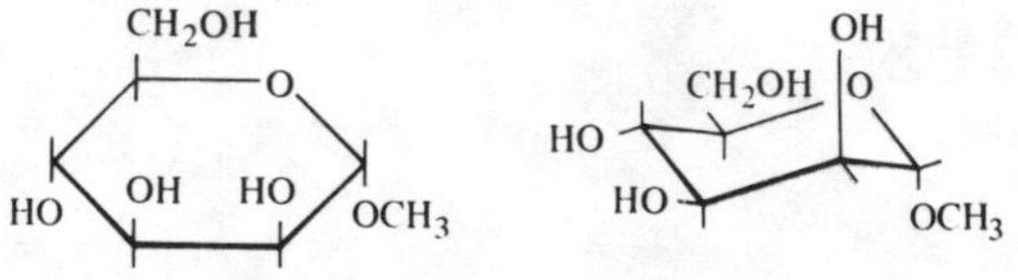

Haworth formula   conformational formula

methyl-α-D mannopyranoside

19.6

α-D-glucopyranose will give a positive test with Benedict's or Tollens' solution because it is a cyclic hemiacetal. Methyl-α-D-glucopyranoside, because it is a cyclic acetal, will not.

19.7

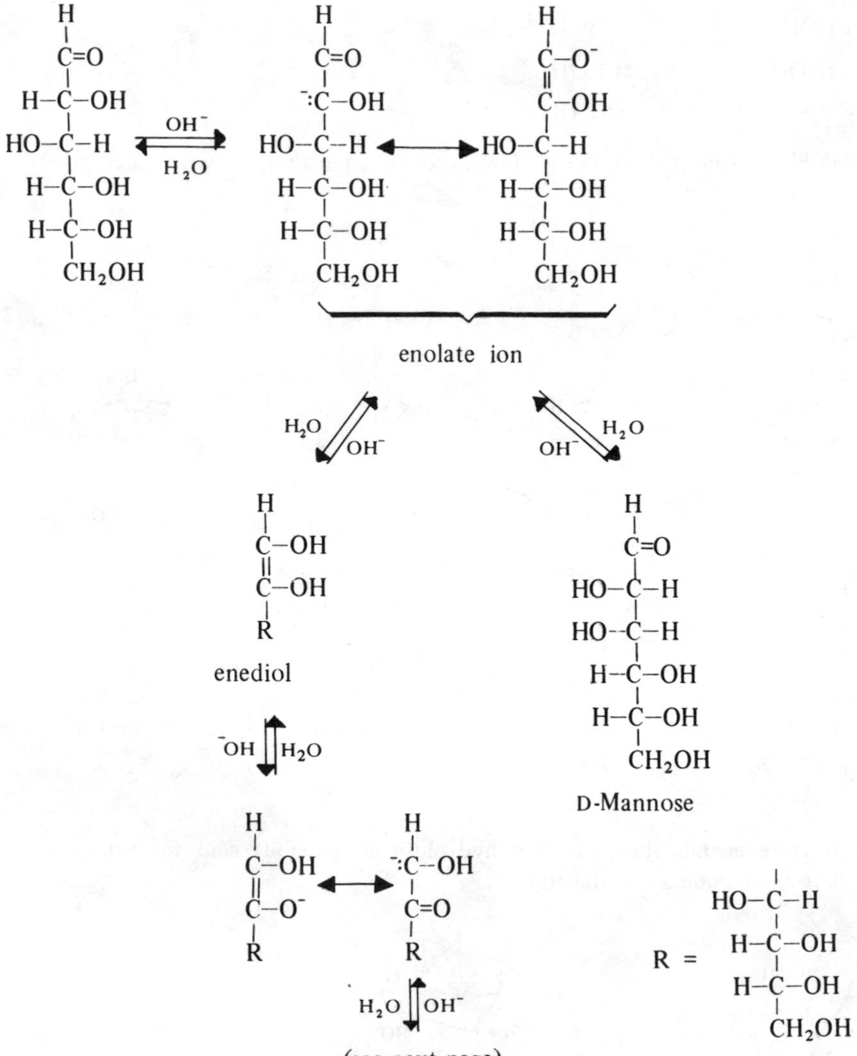

enolate ion

enediol

D-Mannose

R =

(see next page)

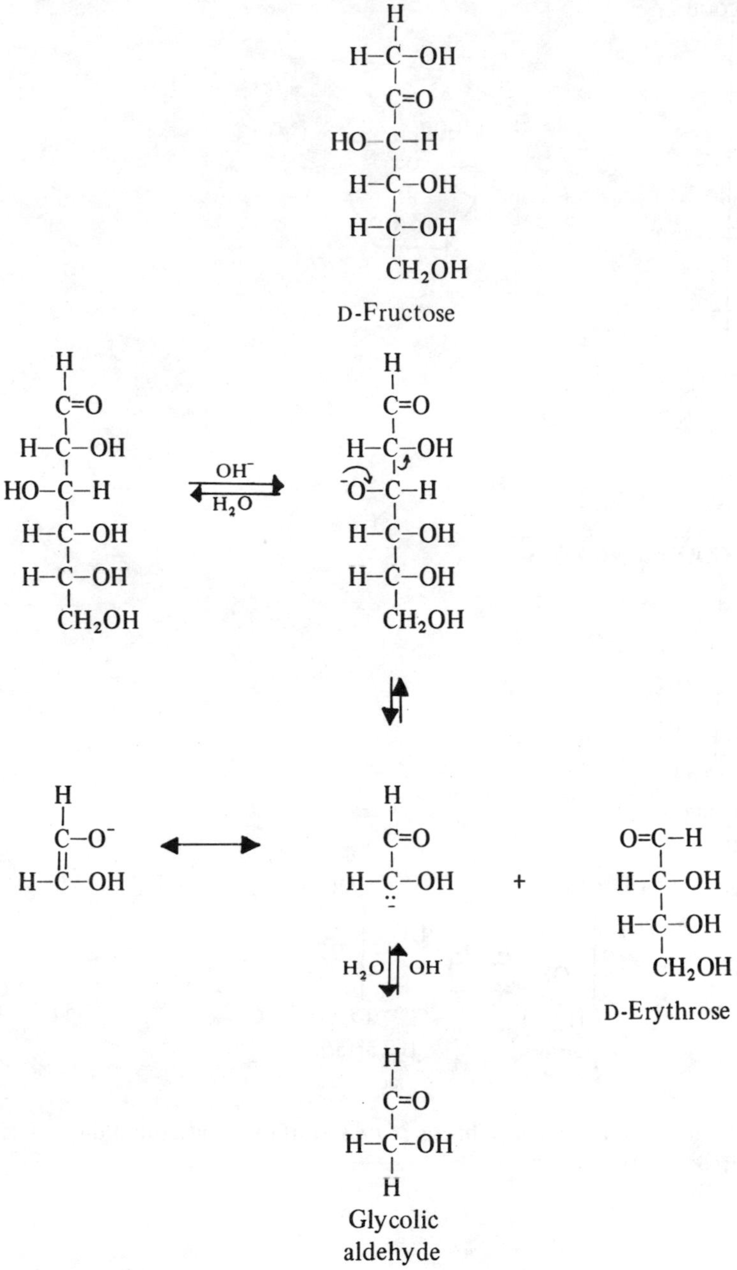

```
      H
      |
    H-C-OH
      |
      C=O
      |
   HO-C-H
      |
    H-C-OH
      |
    H-C-OH
      |
      CH2OH
   D-Fructose
```

```
      H                          H
      |                          |
      C=O                        C=O
      |                          |
    H-C-OH         OH-         H-C-OH
      |          <=====>         |
   HO-C-H          H2O        -O-C-H
      |                          |
    H-C-OH                     H-C-OH
      |                          |
    H-C-OH                     H-C-OH
      |                          |
      CH2OH                      CH2OH

                                 ⇅

      H                          H
      |                          |
      C-O-       <------>        C=O                  O=C-H
      ||                         |                      |
    H-C-OH                     H-C-OH        +        H-C-OH
                                 ..                     |
                                                      H-C-OH
                            H2O  ⇅  OH-                 |
                                                        CH2OH
                                 H                  D-Erythrose
                                 |
                                 C=O
                                 |
                               H-C-OH
                                 |
                                 H
                              Glycolic
                              aldehyde
```

19.8

$\xrightarrow[H_2O]{Br_2}$

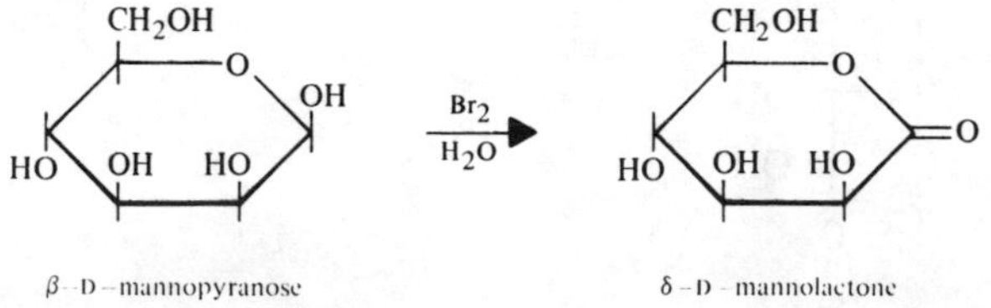

β-D-mannopyranose → δ-D-mannolactone

(Ring structures labelled CH_2OH, O, OH, HO, OH, HO; product ring with =O.)

COOH, HO, HO, OH, OH, CH_2OH

D-mannonic acid

CH_2OH, HO, O, =O, OH, HO

γ-D-mannolactone

19.9

(a) Yes (b) COOH, HO, HO, OH, OH, COOH (c) Yes

D-mannaric acid

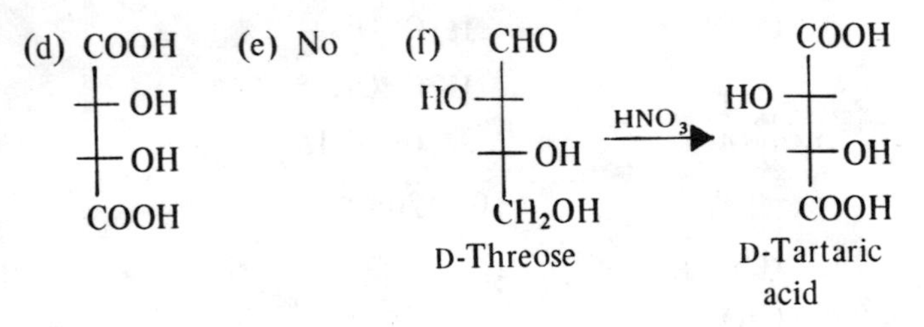

(g) The aldaric acid obtained from D-erythrose is *meso*-tartaric acid; the aldaric acid obtained from D-Threose is D-tartaric acid.

19.10

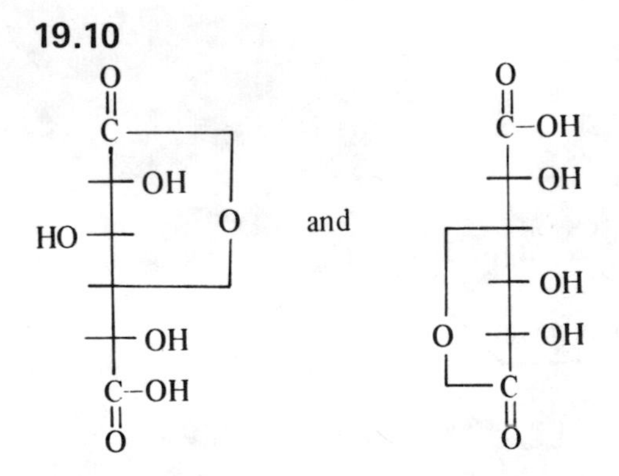

19.11

One way of predicting the products from a periodate oxidation is to place an $-OH$ group on each carbon at the point where C–C bond cleavage has occurred:

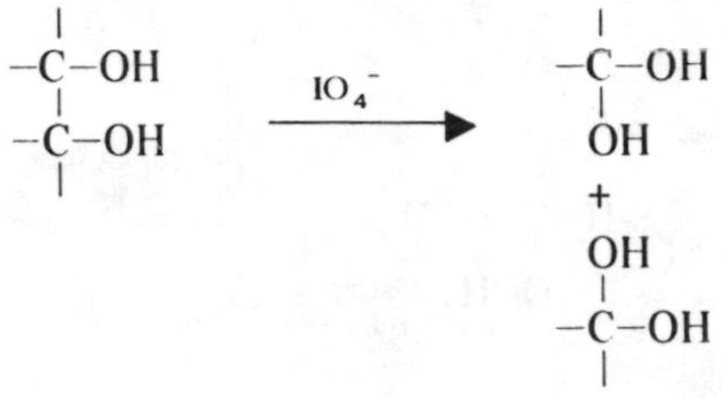

$$\begin{array}{c} | \\ -C-OH \\ | \\ -C-OH \\ | \end{array} \xrightarrow{IO_4^-} \begin{array}{c} | \\ -C-OH \\ | \\ OH \\ + \\ OH \\ | \\ -C-OH \\ | \end{array}$$

Then if we recall (p. 666) that gem-diols are usually unstable and lose water to produce carbonyl compounds, we get the following results:

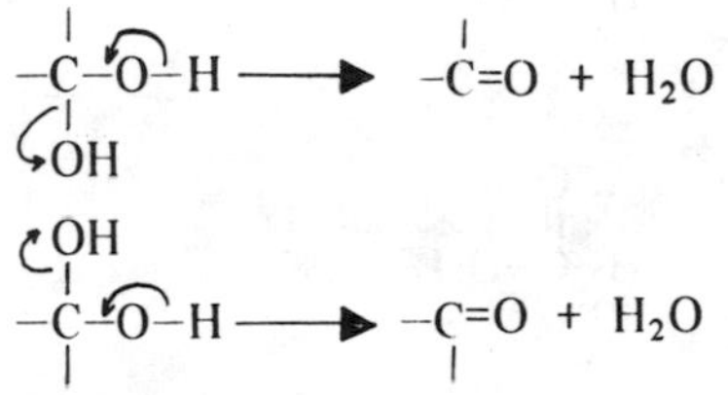

$$\begin{array}{c} | \\ -C-O-H \\ | \\ OH \end{array} \longrightarrow \begin{array}{c} | \\ -C=O \end{array} + H_2O$$

$$\begin{array}{c} OH \\ | \\ -C-O-H \\ | \end{array} \longrightarrow \begin{array}{c} -C=O \\ | \end{array} + H_2O$$

Let us apply this procedure to several examples here while we remember that for every C–C bond that is broken one mole of HIO_4 is consumed.

(a)

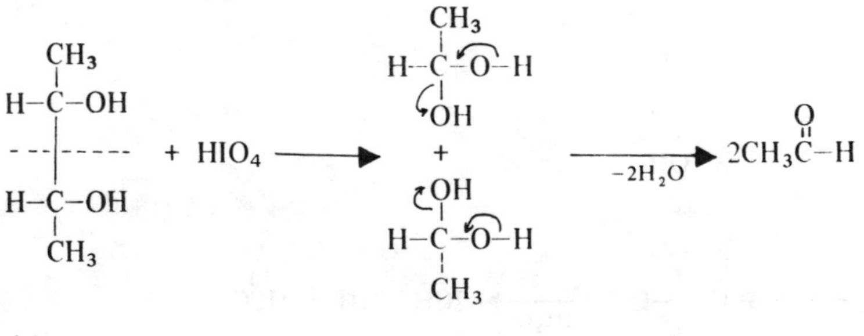

$$\begin{array}{c} CH_3 \\ | \\ H-C-OH \\ \text{- - -}|\text{- - - - -} \\ H-C-OH \\ | \\ CH_3 \end{array} + HIO_4 \longrightarrow \begin{array}{c} CH_3 \\ | \\ H-C-O-H \\ | \\ OH \\ + \\ OH \\ | \\ H-C-O-H \\ | \\ CH_3 \end{array} \xrightarrow[-2H_2O]{} 2CH_3\overset{O}{\overset{\|}{C}}-H$$

(b)

$$\begin{array}{c} H \\ | \\ H-C-OH \\ \text{- - -}|\text{- - -} \\ H-C-OH \\ \text{- -}|\text{- - -} \\ H-C-OH \\ | \\ CH_3 \end{array} + 2HIO_4 \longrightarrow \begin{array}{c} H \\ | \\ H-C-O-H \\ | \\ OH \\ + \\ O-H \\ | \\ H-C-OH \\ | \\ OH \\ + \\ OH \\ | \\ H-C-O-H \\ | \\ CH_3 \end{array} \xrightarrow[-3H_2O]{} \begin{array}{c} H \\ | \\ H-C=O \\ + \\ O \\ \| \\ H-C-OH \\ + \\ H-C=O \\ | \\ CH_3 \end{array}$$

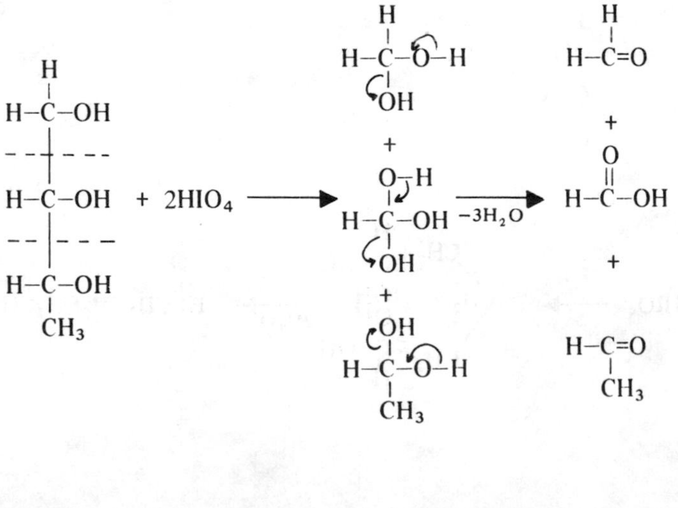

(c)

$$\begin{array}{c} H \\ | \\ H{-}C{-}OH \\ - - | - - - \\ H{-}C{-}OH \\ | \\ H{-}C{-}OCH_3 \\ | \\ OCH_3 \end{array} + HIO_4 \longrightarrow \begin{array}{c} H \\ | \\ H{-}C{-}OH \\ | \\ OH \\ + \\ OH \\ | \\ H{-}C{-}OH \\ | \\ H{-}C{-}OCH_3 \\ | \\ OCH_3 \end{array} \xrightarrow[-2H_2O]{} \begin{array}{c} H \\ | \\ H{-}C{=}O \\ + \\ O \\ \| \\ H{-}C \\ | \\ H{-}C{-}OCH_3 \\ | \\ OCH_3 \end{array}$$

(d)

$$\begin{array}{c} H \\ | \\ H{-}C{-}OH \\ - - | - - - \\ H{-}C{-}OH \\ - - | - - - \\ C{=}O \\ | \\ CH_3 \end{array} + 2HIO_4 \longrightarrow \begin{array}{c} H \\ | \\ H{-}C{-}OH \\ | \\ OH \\ + \\ OH \\ | \\ H{-}C{-}OH \\ | \\ OH \\ + \\ OH \\ | \\ C{=}O \\ | \\ CH_3 \end{array} \xrightarrow[-2H_2O]{} \begin{array}{c} H \\ | \\ H{-}C{=}O \\ + \\ O \\ \| \\ H{-}C{-}OH \\ + \\ O \\ \| \\ CH_3COH \end{array}$$

(e)

$$\begin{array}{c} CH_3 \\ | \\ C{=}O \\ - - | - - \\ H{-}C{-}OH \\ - - | - - \\ C{=}O \\ | \\ CH_3 \end{array} + 2HIO_4 \longrightarrow \begin{array}{c} CH_3 \\ | \\ C{=}O \\ | \\ OH \\ + \\ OH \\ | \\ H{-}C{-}OH \\ | \\ OH \\ + \\ OH \\ | \\ C{=}O \\ | \\ CH_3 \end{array} \xrightarrow[-2H_2O]{} 2CH_3\overset{O}{\overset{\|}{C}}OH + H\overset{O}{\overset{\|}{C}}OH$$

(f)

$$\begin{array}{l} CH_2{-}\underset{}{\overset{H}{C}}{-}OH \\ CH_2 \quad - - | - - \\ CH_2{-}\underset{H}{C}{-}OH \end{array} + HIO_4 \longrightarrow \begin{array}{l} CH_2{-}\overset{H}{C}{-}OH \\ \quad\quad\ \ OH \\ CH_2 \\ \quad\quad\ \ OH \\ CH_2{-}\underset{H}{C}{-}OH \end{array} \xrightarrow[-2H_2O]{} H\overset{O}{\overset{\|}{C}}CH_2CH_2CH_2\overset{O}{\overset{\|}{C}}H$$

(g)

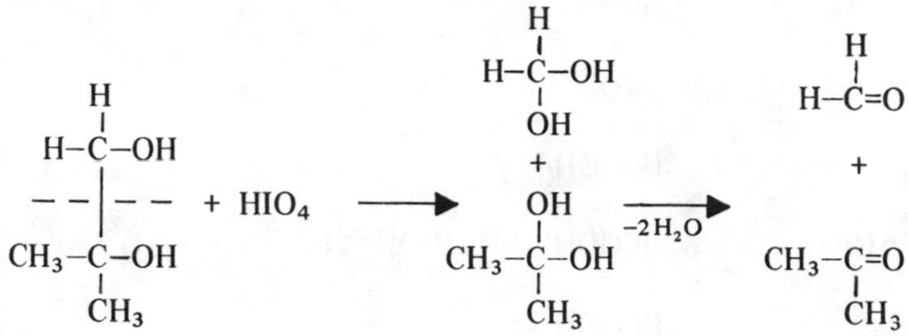

$$\begin{array}{c} H \\ | \\ H{-}C{-}OH \\ --|-- \\ CH_3{-}C{-}OH \\ | \\ CH_3 \end{array} + HIO_4 \longrightarrow \begin{array}{c} H \\ | \\ H{-}C{-}OH \\ | \\ OH \\ + \\ OH \\ | \\ CH_3{-}C{-}OH \\ | \\ CH_3 \end{array} \xrightarrow{-2H_2O} \begin{array}{c} H \\ | \\ H{-}C{=}O \\ + \\ CH_3{-}C{=}O \\ | \\ CH_3 \end{array}$$

(h)

$$\begin{array}{c} O \\ \| \\ H{-}C \\ --|-- \\ H{-}C{-}OH \\ --|-- \\ H{-}C{-}OH \\ --|-- \\ H{-}C{-}OH \\ | \\ H \end{array} + 3HIO_4 \longrightarrow \begin{array}{c} O \\ \| \\ H{-}C{-}OH \\ + \\ OH \\ | \\ H{-}C{-}OH \\ | \\ OH \\ + \\ OH \\ | \\ H{-}C{-}OH \\ | \\ OH \\ + \\ OH \\ | \\ H{-}C{-}OH \\ | \\ H \end{array} \xrightarrow{-3H_2O} 3H\overset{O}{\overset{\|}{C}}OH + H\overset{O}{\overset{\|}{C}}H$$

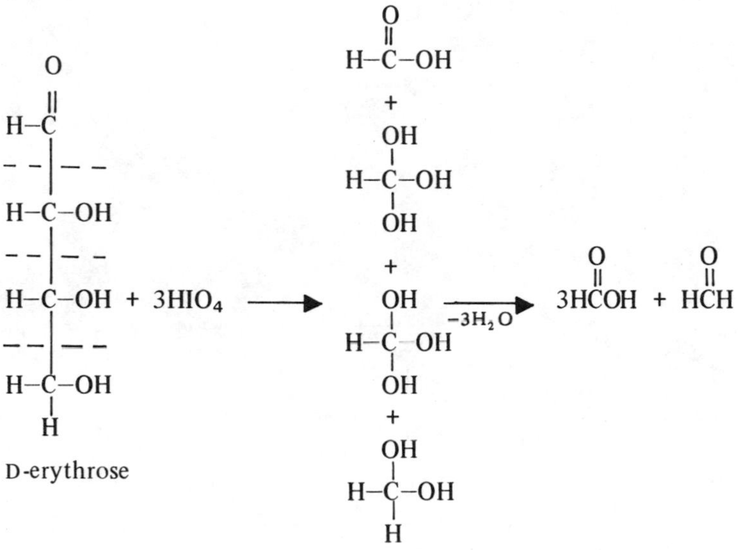

D-erythrose

19.12

Oxidation of an aldohexose and a ketohexose would each require five moles of HIO_4 but would give different results

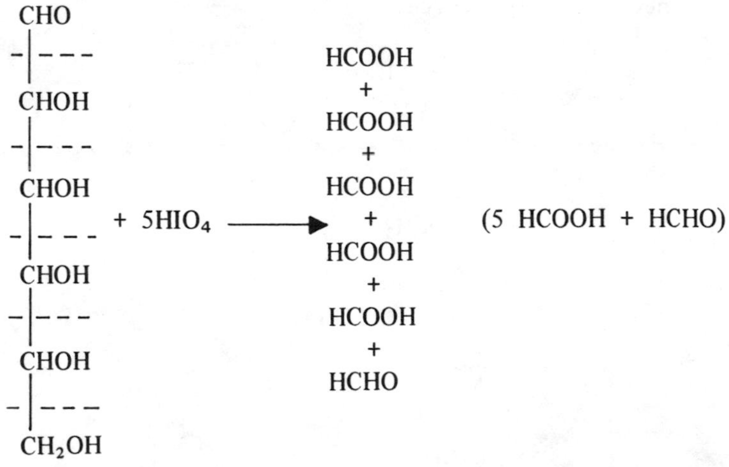

$$\begin{array}{c} CHO \\ -|--- \\ CHOH \\ -|--- \\ CHOH \\ -|--- \\ CHOH \\ -|--- \\ CHOH \\ -|--- \\ CH_2OH \end{array} + 5HIO_4 \longrightarrow \begin{array}{c} HCOOH \\ + \\ HCOOH \\ + \\ HCOOH \\ + \\ HCOOH \\ + \\ HCOOH \\ + \\ HCHO \end{array} \quad (5\ HCOOH + HCHO)$$

Aldohexose

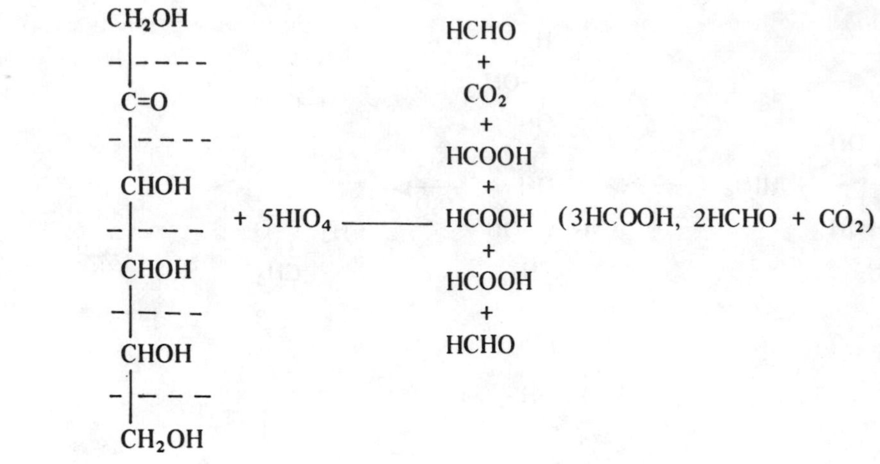

$$
\begin{array}{l}
CH_2OH \\
-|- - - - \\
C{=}O \\
-|- - - - \\
CHOH \\
-|- - - - \\
CHOH \\
-|- - - - \\
CHOH \\
-|- - - - \\
CH_2OH
\end{array}
+ 5HIO_4 \longrightarrow
\begin{array}{c}
HCHO \\ + \\ CO_2 \\ + \\ HCOOH \\ + \\ HCOOH \\ + \\ HCOOH \\ + \\ HCHO
\end{array}
\quad (3HCOOH,\ 2HCHO + CO_2)
$$

Ketohexose

19.13

(a)

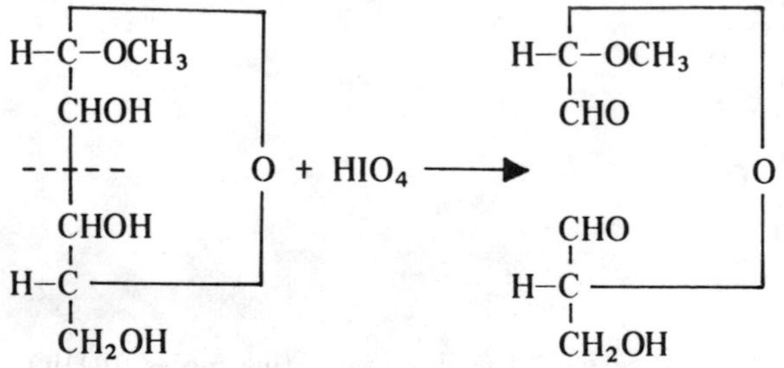

$$
\begin{array}{l}
H{-}C{-}OCH_3 \\
\quad CHOH \\
- - + - - \quad O \\
\quad CHOH \\
H{-}C \\
\quad CH_2OH
\end{array}
+ HIO_4 \longrightarrow
\begin{array}{l}
H{-}C{-}OCH_3 \\
\quad CHO \\
\qquad\qquad O \\
\quad CHO \\
H{-}C \\
\quad CH_2OH
\end{array}
$$

Any methylfuranoside of the α-D-pentose series

2

Dialdehyde

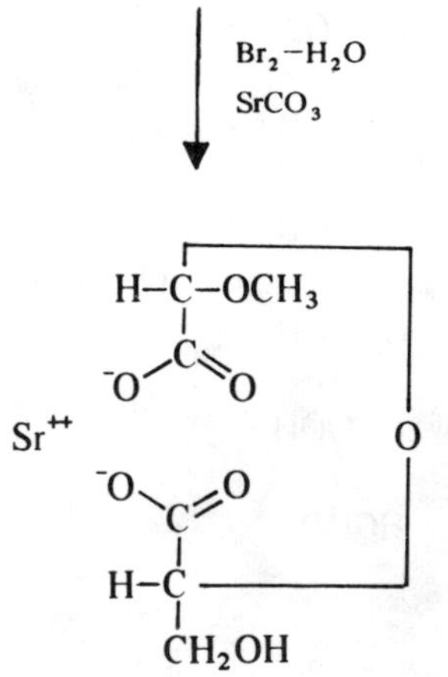

$\downarrow$ $Br_2{-}H_2O$, $SrCO_3$

$$
\begin{array}{l}
H{-}C{-}OCH_3 \\
\quad {}^{-}O{-}C{=}O \\
Sr^{++} \qquad\qquad O \\
\quad {}^{-}O{-}C{=}O \\
H{-}C \\
\quad CH_2OH
\end{array}
$$

3

Same strontium salt

(b) Although both compounds yield the same dialdehyde 2 (and Strontium salt 3), periodate oxidation of a methyl-α-D-pentofuranoside consumes only one mole of HIO_4 and produces no formic acid.

19.14

(a)

4	5
H	COOH
|	|
C=O	H—C—OH
|	|
COOH	CH_2OH
Glyoxylic acid	D-(−)-Glyceric acid

(b) This relates the configuration of the highest numbered carbon of the aldose to that of D-(+)-glyceraldehyde, and thus allows us to place the aldose in the D-family.

19.15

(a) Yes, D-glucitol would be optically active; only those alditols (below) whose molecules possess a plane of symmetry would be optically inactive.

(b)

```
CHO                        CH2OH
 ├─OH                       ├─OH
 ├─OH        NaBH4          ├─OH
 ├─OH     ─────────▶  - - - ┼ - - -   plane of symmetry
 ├─OH                       ├─OH
CH2OH                       ├─OH
                           CH2OH
                          optically
                          inactive
```

```
    CHO                         CH2OH
     ├─OH                        ├─OH
HO ─┤                       HO ─┤
             NaBH4          - - ┼ - -      plane of symmetry
HO ─┤     ─────────▶        HO ─┤
     ├─OH                        ├─OH
    CH2OH                       CH2OH
                          optically  inactive
```

19.16

(a)

```
      CH2OH                                 CH=NNHC6H5
       |                                    |
       C=O                                  C=NNHC6H5
            C6H5NHNH2
HO ───┤    ───────────▶              HO ───┤
       ├─OH                                 ├─OH
       ├─OH                                 ├─OH
      CH2OH                                CH2OH
```

(b) This experiment shows that D-glucose and D-fructose have the same configurations at C-3, C-4, and C-5.

19.17

(a)

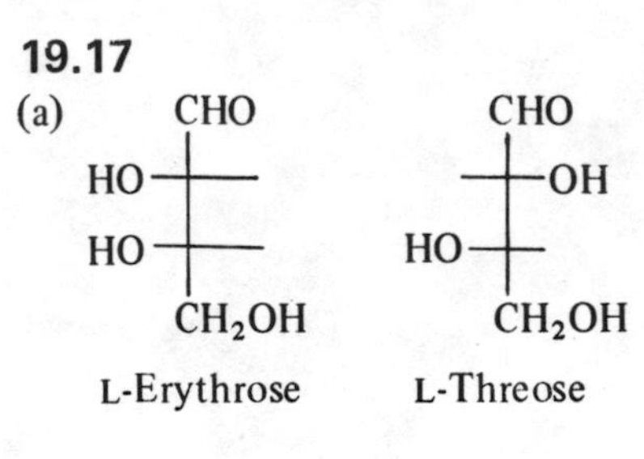

L-Erythrose L-Threose

(b) L-Glyceraldehyde, CHO / HO—H / CH_2OH

19.18

(a)

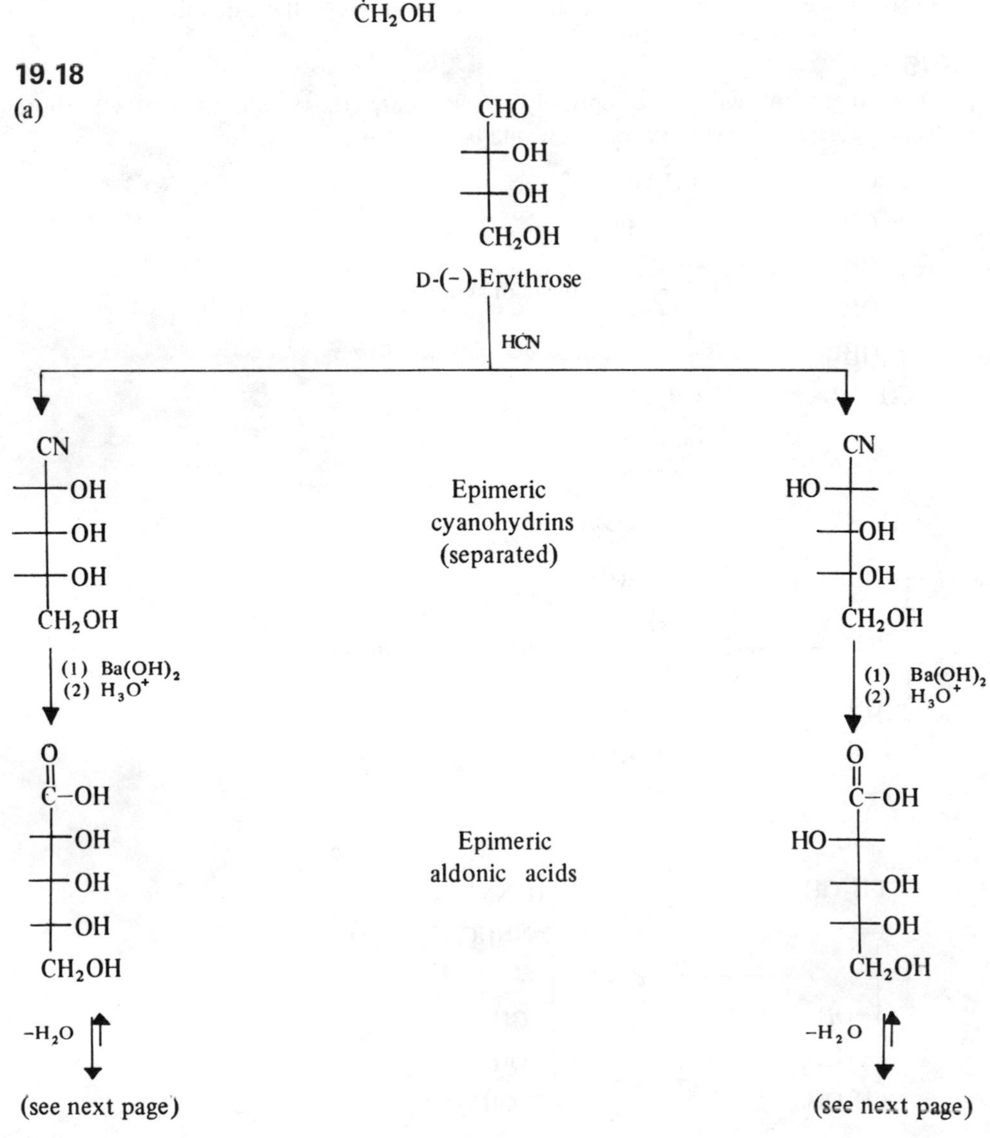

(see next page) (see next page)

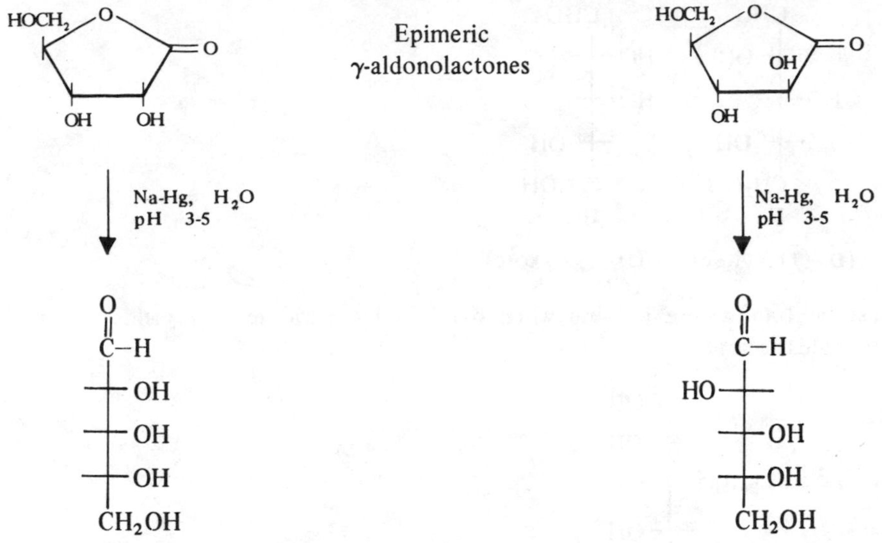

HOCH$_2$ (γ-lactone ring, =O; OH, OH) — Epimeric γ-aldonolactones — HOCH$_2$ (γ-lactone ring, =O; OH, OH)

Na-Hg, H_2O
pH 3-5

$$\begin{array}{c} O \\ \| \\ C{-}H \\ {-}OH \\ {-}OH \\ {-}OH \\ CH_2OH \end{array} \qquad \begin{array}{c} O \\ \| \\ C{-}H \\ HO{-} \\ {-}OH \\ {-}OH \\ CH_2OH \end{array}$$

(b)

$$\begin{array}{c} O \\ \| \\ C{-}H \\ {-}OH \\ {-}OH \\ {-}OH \\ CH_2OH \end{array} \xrightarrow{HNO_3} \begin{array}{c} O \\ \| \\ C{-}OH \\ {-}OH \\ {-}OH \\ {-}OH \\ C{-}OH \\ \| \\ O \end{array}$$

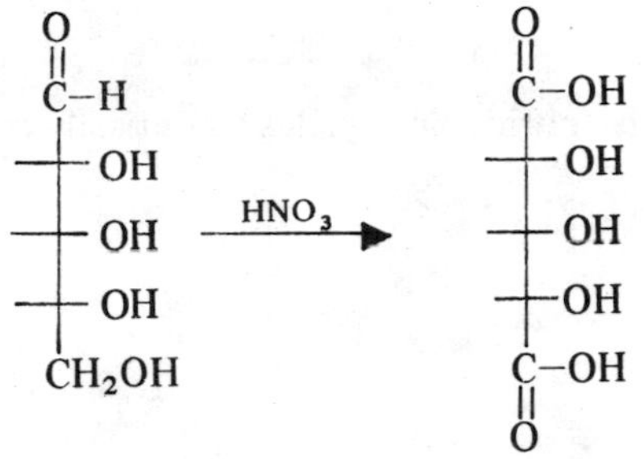

D-(−)-Ribose — optically inactive

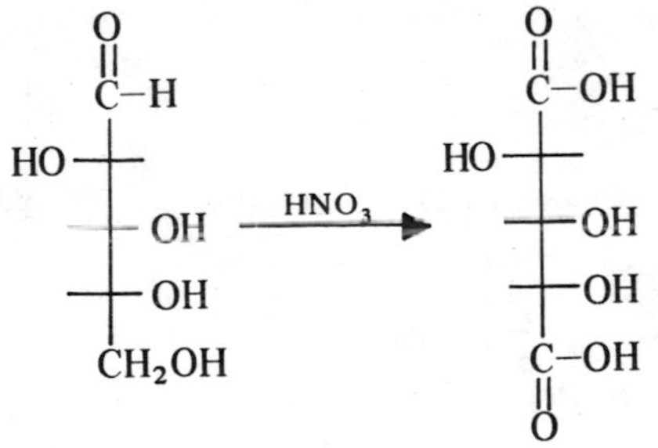

$$\begin{array}{c} O \\ \| \\ C{-}H \\ HO{-} \\ {-}OH \\ {-}OH \\ CH_2OH \end{array} \xrightarrow{HNO_3} \begin{array}{c} O \\ \| \\ C{-}OH \\ HO{-} \\ {-}OH \\ {-}OH \\ C{-}OH \\ \| \\ O \end{array}$$

D-(−)-Arabinose — optically active

19.19

A Kiliani-Fischer synthesis starting with D-(−)-threose would yield **I** and **II**.

$$
\begin{array}{c}
\text{CHO} \\
-\!\!\!+\!\!\!-\text{OH} \\
\text{HO}-\!\!\!+\!\!\!- \\
-\!\!\!+\!\!\!-\text{OH} \\
\text{CH}_2\text{OH} \\
\textbf{I} \\
(\text{D})\text{-}(+)\text{-Xylose})
\end{array}
\qquad
\begin{array}{c}
\text{CHO} \\
\text{HO}-\!\!\!+\!\!\!- \\
\text{HO}-\!\!\!+\!\!\!- \\
-\!\!\!+\!\!\!-\text{OH} \\
\text{CH}_2\text{OH} \\
\textbf{II} \\
(\text{D-}(-)\text{-Lyxose})
\end{array}
$$

I must be D-(+)-xylose because when oxidized by nitric acid, it yields an optically inactive aldaric acid:

$$
\textbf{I} \xrightarrow{\text{HNO}_3}
\begin{array}{c}
\text{COOH} \\
-\!\!\!+\!\!\!-\text{OH} \\
\text{HO}-\!\!\!+\!\!\!- \\
-\!\!\!+\!\!\!-\text{OH} \\
\text{COOH} \\
\text{optically} \\
\text{inactive}
\end{array}
$$

II must be D-(−)-lyxose because when oxidized by nitric acid it yields an optically active aldaric acid:

$$
\textbf{II} \xrightarrow{\text{HNO}_3}
\begin{array}{c}
\text{COOH} \\
\text{HO}-\!\!\!+\!\!\!- \\
\text{HO}-\!\!\!+\!\!\!- \\
-\!\!\!+\!\!\!-\text{OH} \\
\text{COOH} \\
\text{optically} \\
\text{active}
\end{array}
$$

19.20

$$
\begin{array}{c}
\text{CHO} \\
\text{HO}-\!\!\!+\!\!\!- \\
\text{HO}-\!\!\!+\!\!\!- \\
\text{HO}-\!\!\!+\!\!\!- \\
\text{CH}_2\text{OH} \\
\text{L-}(+)\text{-Ribose}
\end{array}
\quad
\begin{array}{c}
\text{CHO} \\
-\!\!\!+\!\!\!-\text{OH} \\
\text{HO}-\!\!\!+\!\!\!- \\
\text{HO}-\!\!\!+\!\!\!- \\
\text{CH}_2\text{OH} \\
\text{L-}(+)\text{-Arabinose}
\end{array}
\quad
\begin{array}{c}
\text{CHO} \\
\text{HO}-\!\!\!+\!\!\!- \\
-\!\!\!+\!\!\!-\text{OH} \\
\text{HO}-\!\!\!+\!\!\!- \\
\text{CH}_2\text{OH} \\
\text{L-}(-)\text{-Xylose}
\end{array}
\quad
\begin{array}{c}
\text{CHO} \\
-\!\!\!+\!\!\!-\text{OH} \\
-\!\!\!+\!\!\!-\text{OH} \\
\text{HO}-\!\!\!+\!\!\!- \\
\text{CH}_2\text{OH} \\
\text{L-}(+)\text{-Lyxose}
\end{array}
$$

19.21

Since D-(+)-galactose yields an optically inactive aldaric acid it must have either structure

III or structure **IV** below.

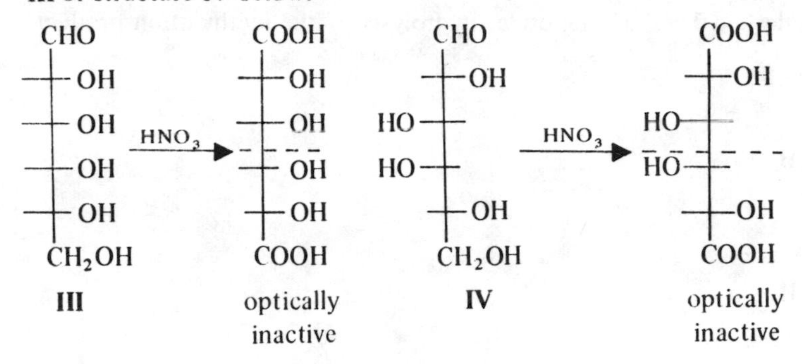

A Ruff degradation beginning with **III** would yield D-(−)-ribose

$$\text{III} \xrightarrow[H_2O]{Br_2} \xrightarrow[Fe_2(SO_4)_3]{H_2O_2}$$

CHO, OH, OH, OH, CH_2OH

D-(−)-Ribose

A Ruff degradation beginning with IV would yield D-(−)-lyxose; thus D-(+)-galactose must have structure **IV**.

$$\text{IV} \xrightarrow[H_2O]{Br_2} \xrightarrow[Fe_2(SO_4)_3]{H_2O_2}$$

CHO, HO, HO, OH, CH_2OH

D-(−)-Lyxose

19.22

D-(+)-glucose, as shown below.

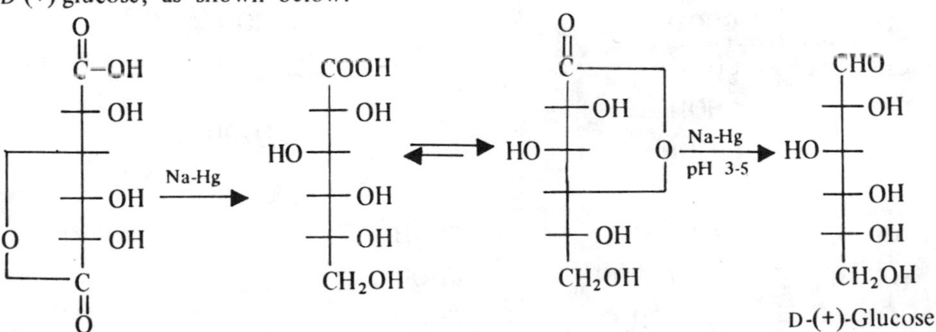

The other γ-lactone of D-glucaric acid

19.23

If the methyl glucoside had been a furanoside, hydrolysis of the methylation product would have given:

```
        CHO
         |
        -+- OCH3
CH3O -+-
        -+- OH
        -+- OCH3
        CH2CH3
```

And, oxidation would have given:

```
       CHO                       COOH             COOH              COOH
        |                         |                |                 |
       -+- OCH3                  -+- OCH3         -+- OCH3   +      -+- OCH3
CH3O -+-        --HNO3--> CH3O -+-          +     CH2OCH3            COOH
 - - - - -                        |
       -+- OH                    COOH
 - - - - -
       -+- OCH3
       CH2OCH3
```

A dimethoxysuccinic acid; A dimethoxypropanoic acid; Methoxymalonic acid

Methoxymalonic acid $\xrightarrow{-CO_2}$

```
COOH
 |
CH2OCH3
```

Methoxyacetic acid

19.24

(a)
```
CHO
 |
CHOH
 |
CHOH
 |
CHOH
 |
CH2OH
```

(b)
```
CH2OH
 |
C=O
 |
CHOH
 |
CHOH
 |
CHOH
 |
CH2OH
```

(c)
```
   CHO
    |
  (CHOH)n
    |
HO -+- H
    |
   CH2OH
```

or

```
   CH2OH
    |
   C=O
    |
  (CHOH)n
    |
HO -+- H
    |
   CH2OH
```

(d)
```
 ┌──────────┐
CHOR        |
 |          |
(CHOH)n     O
 |          |
CH ─────────┘
 |
CH2OH
```

(e)
```
COOH
 |
(CHOH)n
 |
CH2OH
```

(f)
```
COOH
 |
(CHOH)n
 |
COOH
```

(g) O ‖ C — $(CHOH)_n$ — CH — CH_2OH, ring closed through O between C and CH

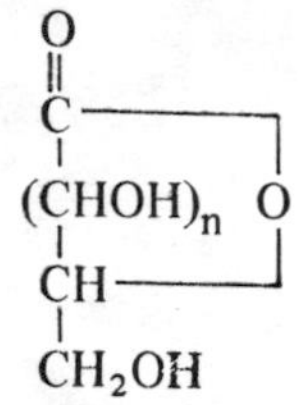

(h) OH | CH — CHOH — CHOH — CHOH — CH — CH_2OH, ring closed through O between the first and last CH

or

CH_2OH | CH—O, CHOH, CHOH, CHOH–CHOH (six-membered ring)

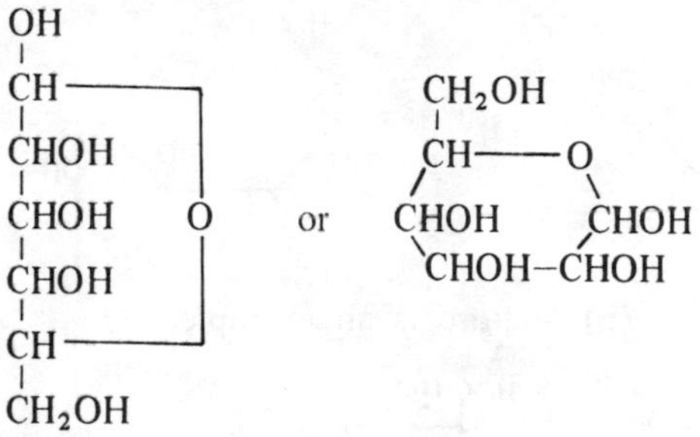

(i) OH | CH — CHOH — CHOH — CH — CHOH — CH_2OH, ring closed through O between the first and second CH

or

CH_2OH | CHOH | CH, O, CHOH, CHOH–CHOH (five-membered ring)

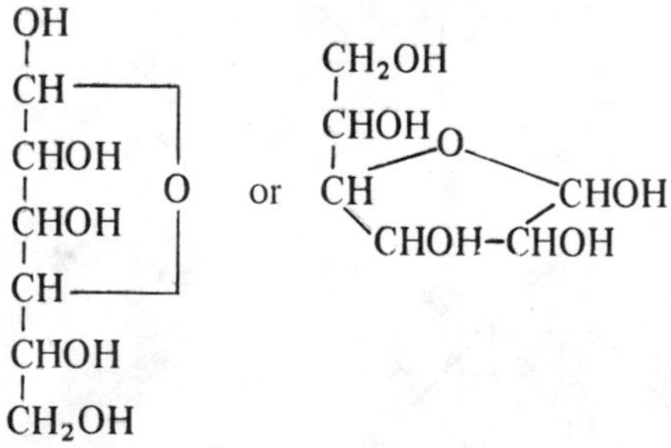

(j) Any sugar that has a free aldehyde or ketone group or one that exists as a cyclic hemiacetal or hemiketal. Examples are:

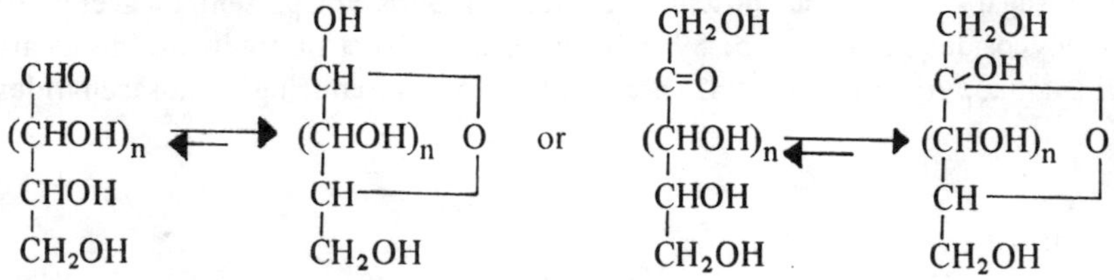

CHO | $(CHOH)_n$ | CHOH | CH_2OH ⇌ OH | CH | $(CHOH)_n$ | CH | CH_2OH (ring through O)

or

CH_2OH | C=O | $(CHOH)_n$ | CHOH | CH_2OH ⇌ CH_2OH | C–OH | $(CHOH)_n$ | CH | CH_2OH (ring through O)

(k) CH_2OH | CH—O, CHOH, CHOR, CHOH–CHOH (six-membered ring)

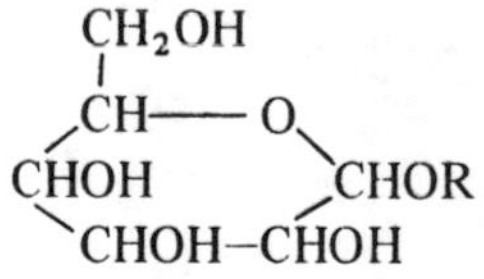

(l) CH_2OH | CHOH | CH, O, CHOR, CHOH–CHOH (five-membered ring)

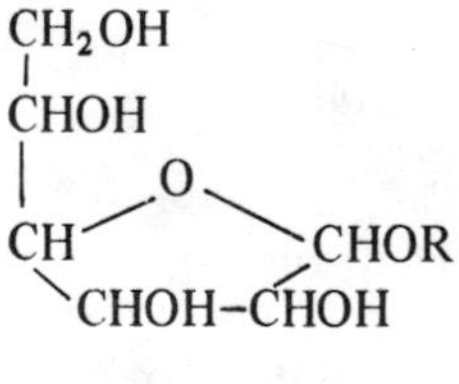

(m) Any two aldoses that differ only in configuration at C-2. (See also page 891 for a broader definition.) D-Erythrose and D-threose are examples.

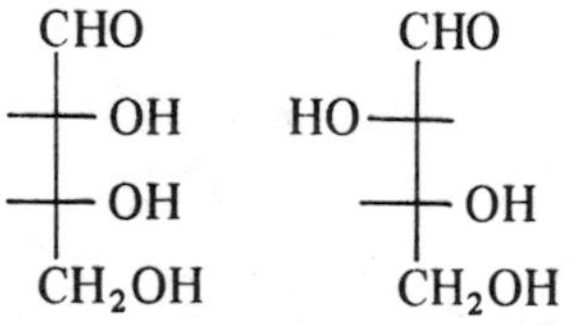

D-Erythrose		D-Threose	
CHO		CHO	
	OH	HO	
	OH		OH
CH_2OH		CH_2OH	

D-Erythrose D-Threose

(n) Cyclic sugars that differ only in the configuration of C-1. Examples are:

HOCH$_2$ O OH HO OH OH

and

HOCH$_2$ O HO HO OH OH

(o) 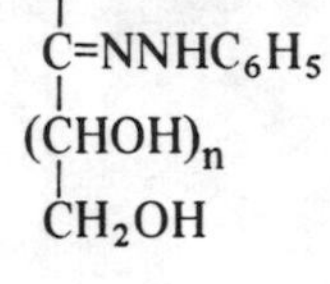

$CH{=}NNHC_6H_5$
$C{=}NNHC_6H_5$
$(CHOH)_n$
CH_2OH

(p) Maltose is an example:

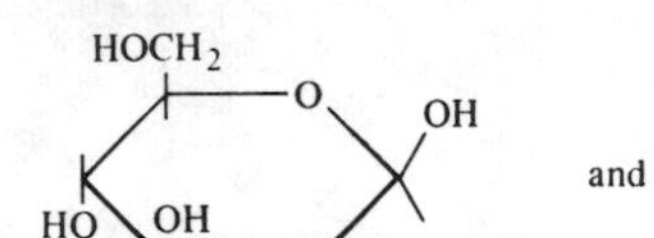

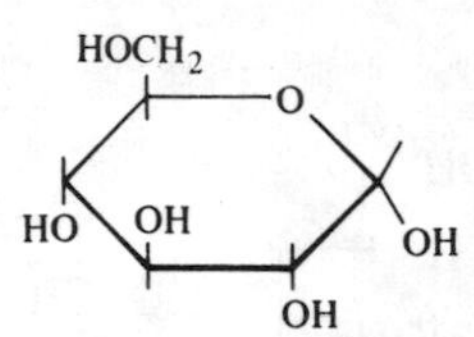

(q) Amylose is an example:

HOCH$_2$ O HO OH OH O [HOCH$_2$ O OH OH]$_n$ O HOCH$_2$ O OH OH OH

(r) Any sugar in which all potential carbonyl groups are present as acetals or ketals (i.e., as glycosides). Sucrose (p. 899) is an example of a nonreducing disacharide; the methyl-D-glucopyranosides (p. 882) are examples of nonreducing monosaccharides.

19.25

(a) HOCH$_2$ O HO OH HO OH

OH CH$_2$ O HO HO OH OH

(b) HOCH$_2$ O OCH$_3$ HO OH OH

OH CH$_2$ O HO OCH$_3$ OH OH

(c) CH$_3$OCH$_2$ O OCH$_3$ CH$_3$O CH$_3$O OCH$_3$

OCH$_3$ CH$_2$ O CH$_3$—O OCH$_3$ O CH$_3$ O CH$_3$

19.26

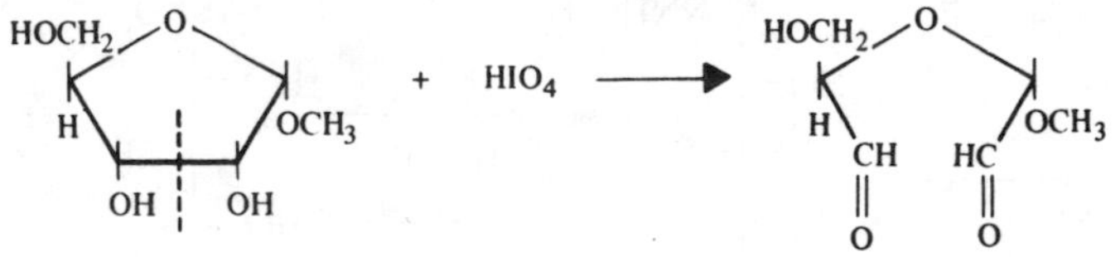

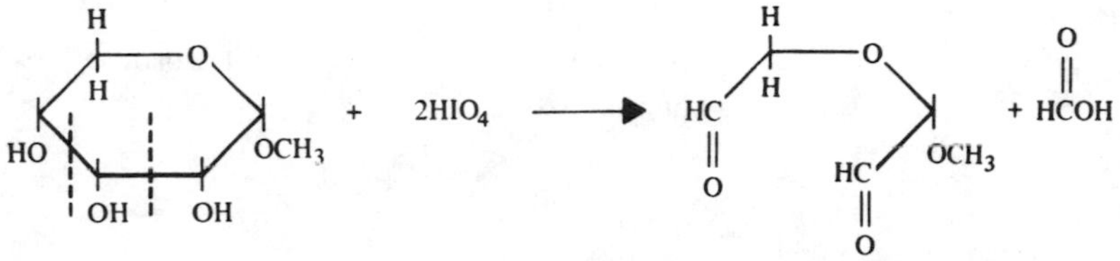

$$\text{methyl ribofuranoside} + HIO_4 \longrightarrow \text{dialdehyde}$$

$$\text{methyl ribopyranoside} + 2HIO_4 \longrightarrow \text{dialdehyde} + HCOH \text{ (formic acid, } HC(=O)OH)$$

A methyl ribofuranoside would consume only one mole of HIO_4; a methyl ribopyranoside would consume two moles of HIO_4 and would also produce one mole of formic acid.

19.27

One anomer of D-mannose is dextrorotatory ($[\alpha]_D = +29.3°$), the other is levorotatory ($[\alpha]_D = -17.0°$).

19.28

The microorganism selectively oxidizes the $-CHOH$ group of D-glucitol that corresponds to C-5 of D-glucose.

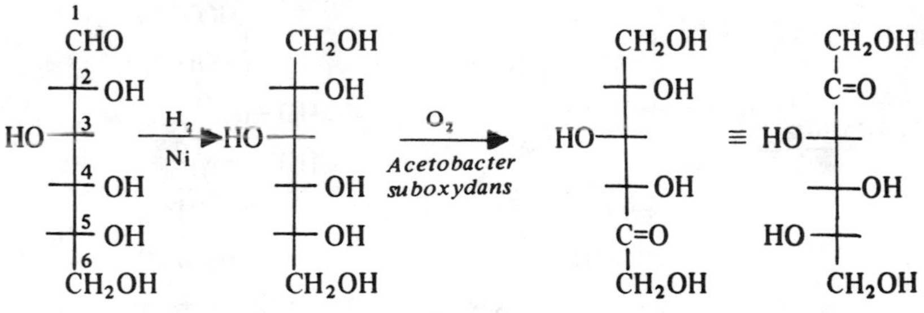

$$\text{D-Glucose} \xrightarrow[\text{Ni}]{H_2} \text{D-Glucitol} \xrightarrow[\textit{Acetobacter suboxydans}]{O_2} \text{L-Sorbose}$$

D-Glucose D-Glucitol L-Sorbose

19.29

L-Gulose and L-idose would yield the same phenylosazone as L-sorbose.

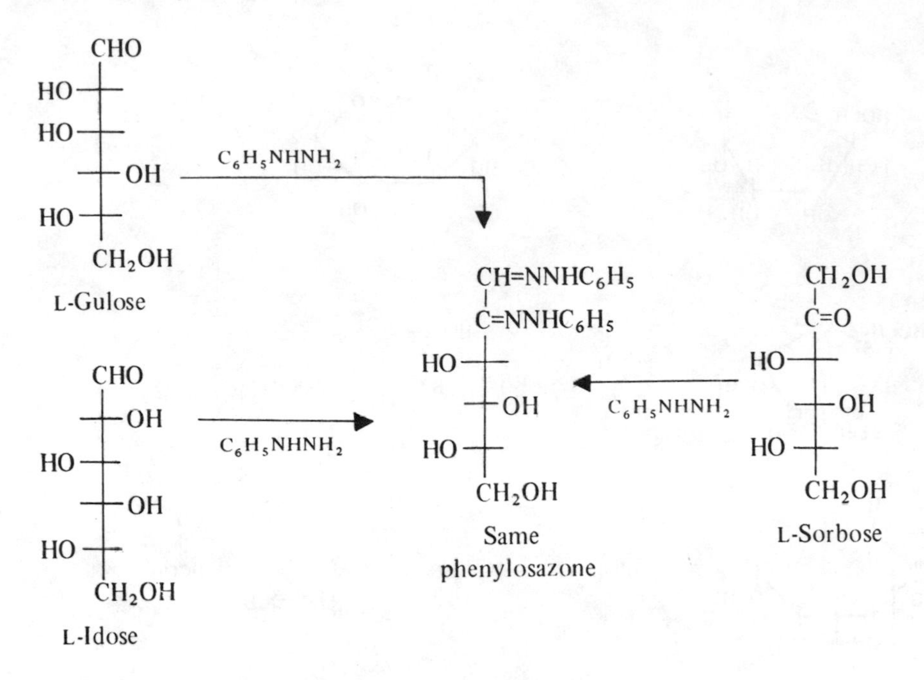

19.30

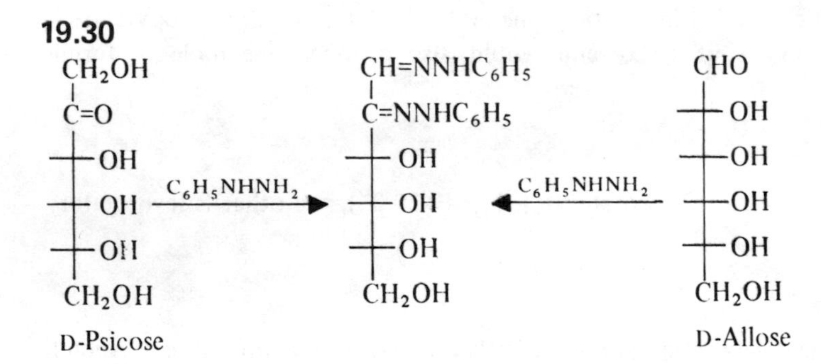

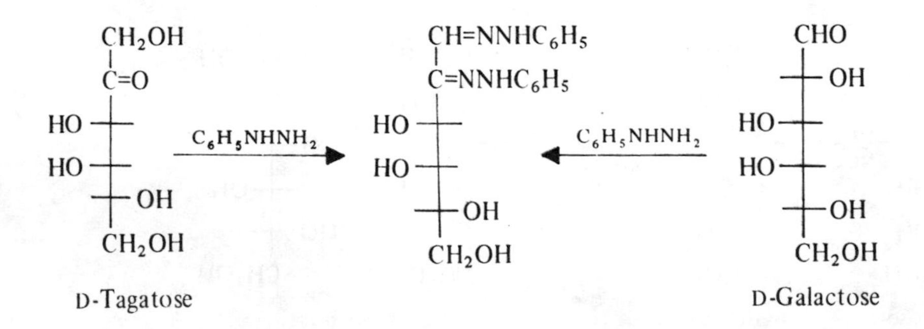

19.31

A is D-altrose, **B** is D-talose, and **C** is D-galactose:

CHO
HO—
—OH
—OH
—OH
CH_2OH
D-Altrose
A

$\xrightarrow{H_2,\ Ni}$

CH_2OH
HO—
—OH
—OH
—OH
CH_2OH

≡

CH_2OH
HO—
HO—
HO—
—OH
CH_2OH

Same alditol

$\xleftarrow{H_2,\ Ni}$

CHO
HO—
HO—
HO—
—OH
CH_2OH
D-Talose
B

A $\xrightarrow{C_6H_5NHNH_2}$

$CH{=}NNHC_6H_5$
$C{=}NNHC_6H_5$
—OH
—OH
—OH
CH_2OH

⟸ different phenylosazones ⟹

B $\xrightarrow{C_6H_5NHNH_2}$

$CH{=}NNHC_6H_5$
$C{=}NNHC_6H_5$
HO—
HO—
—OH
CH_2OH

CHO
—OH
HO—
HO—
—OH
CH_2OH
D-galactose
C

$\xrightarrow{C_6H_5NHNH_2}$

$CH{=}NNHC_6H_5$
$C{=}NNHC_6H_5$
HO—
HO—
—OH
CH_2OH

Same phenylosazone

$\xleftarrow{C_6H_5NHNH_2}$

CHO
HO—
HO—
HO—
—OH
CH_2OH
D-Talose
B

C $\xrightarrow{H_2,Ni}$

CH_2OH
—OH
HO—
HO—
—OH
CH_2OH

⟸ different alditols ⟹

B $\xrightarrow{H_2,Ni}$

CH_2OH
HO—
HO—
HO—
—OH
CH_2OH

(Note: If we had designated D-talose as **A**, and D-altrose as **B**, then **C** is D-allose)

19.32

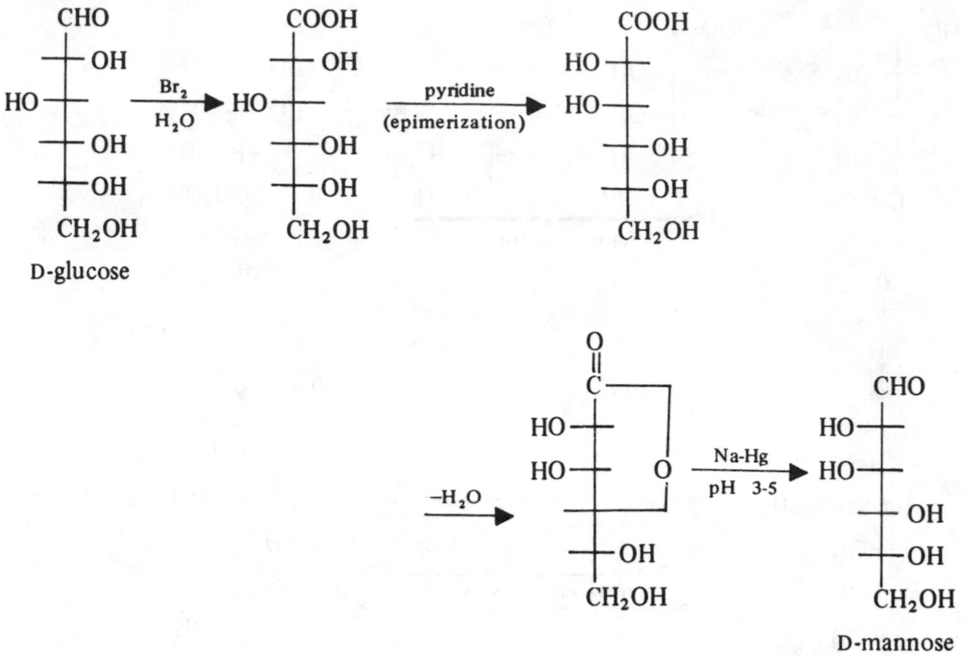

D-glucose $\xrightarrow[H_2O]{Br_2}$ $\xrightarrow[\text{(epimerization)}]{\text{pyridine}}$ $\xrightarrow{-H_2O}$ $\xrightarrow[\text{pH 3-5}]{\text{Na-Hg}}$ D-mannose

19.33

The conformation of D-idopyranose with four equatorial –OH groups and an axial $-CH_2OH$ group is more stable than the one with four axial –OH groups and an equatorial $-CH_2OH$ group.

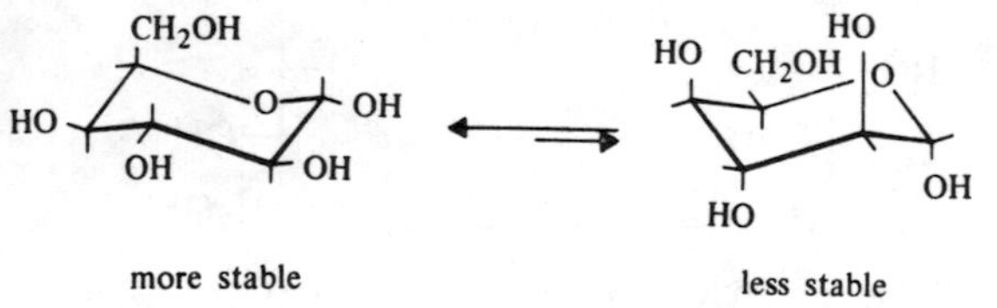

more stable

less stable

4 equatorial –OH groups
1 axial $-CH_2OH$

4 axial –OH groups
1 equatorial $-CH_2OH$

19.34

(a) The anhydro sugar is formed when the axial $-CH_2OH$ group reacts with C-1 to form a cyclic acetal.

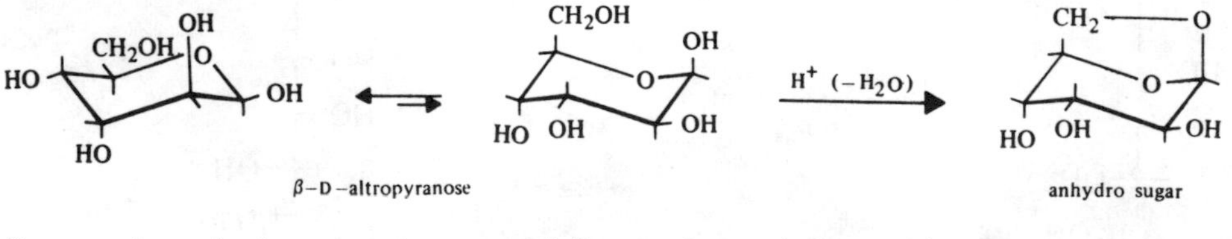

β-D-altropyranose $\xrightarrow{H^+ (-H_2O)}$ anhydro sugar

Because the anhydro sugar is an acetal (i.e., an internal glycoside), it is a non-reducing sugar.

Methylation followed by acid hydrolysis converts the anhydro sugar to 2,3,4-tri-O-methyl-D-altrose:

anhydro-β-D-altropyranose

$(CH_3)_2SO_4$, OH^-; H^+, H_2O

2, 3, 4-tri-O-methyl-D-altrose

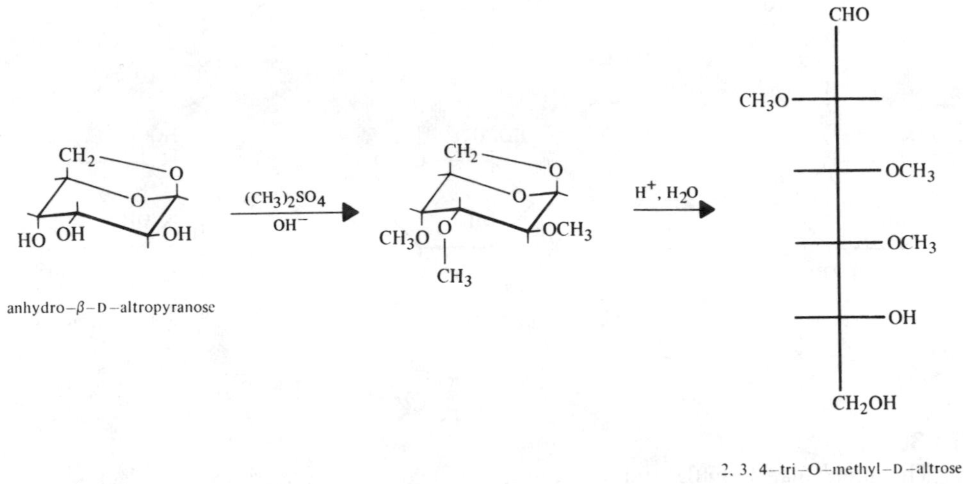

(b) Formation of an anhydro sugar requires that the monosaccharide adopt a chair conformation with the $-CH_2OH$ group axial. With β-D-altropyranose this requires that two $-OH$ groups be axial as well. With β-D-glucopyranose, however, it requires that all four $-OH$ groups become axial, and thus that the molecule adopt a very unstable conformation:

highly unstable conformation

H^+, $(-H_2O)$

β-D-glucopyranose

anhydro-β-D-glucopyranose

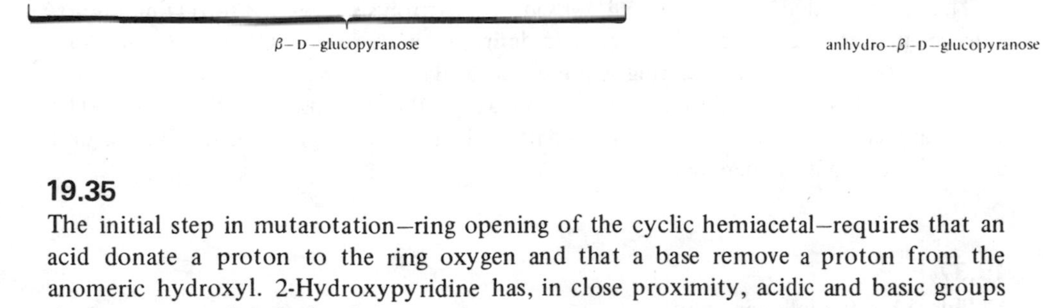

19.35

The initial step in mutarotation–ring opening of the cyclic hemiacetal–requires that an acid donate a proton to the ring oxygen and that a base remove a proton from the anomeric hydroxyl. 2-Hydroxypyridine has, in close proximity, acidic and basic groups that can accomplish both of these tasks.

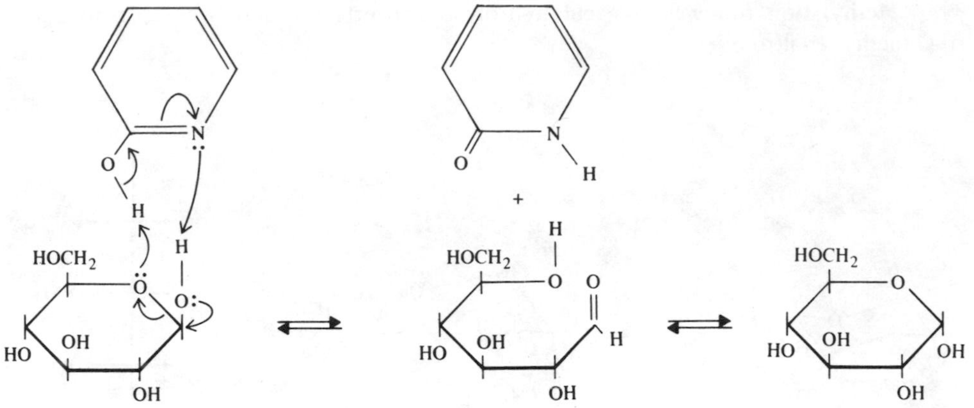

β-D-glucopyranose aldehyde form α-D-glucopyranose

19.36

1. The molecular formula and the results of acid hydrolysis show that lactose is a disaccharide composed of D-glucose and D-galactose. The fact that lactose is hydrolyzed by a *β-galactosidase* indicates that galactose is present as a glycoside and that the glycosidic linkage is *beta* to the galactose ring.

2. That lactose is a reducing sugar, forms a phenylosazone, and undergoes mutarotation indicates that one ring (presumably that of D-glucose) is present as a hemiacetal and thus is capable of existing to a limited extent as an aldehyde.

3. This experiment confirms that the D-glucose unit is present as a cyclic hemiacetal and that the D-galactose unit is present as a cyclic glycoside.

4. That 2,3,4,6-tetra-O-methyl-D-galactose is obtained in this experiment indicates (by virture of the free –OH at C-5) that the galactose ring of lactose is present as a pyranoside. That the methylated gluconolactone obtained from this experiment has a free –OH group at C-4 indicates that the C-4 oxygen of the glucose unit is connected in a glycosidic linkage to the galactose unit.

Now only the size of the glucose ring remains in question and the answer to this is provided by experiment 5.

5. That methylation of lactose and subsequent hydrolysis gives 2,3,6-tri-O-methyl-D-glucose–that it gives a methylated glucose derivative with a free –OH at C-4 and C-5–demonstrates that the glucose ring is present as a pyranose. (We know already that the oxygen at C-4 is connected in a glycosidic linkage to the galactose unit; thus a free –OH at C-5 indicates that the C-5 oxygen is a part of the hemiacetal group of the glucose unit and that the ring is six-membered.)

19.37

Melibiose has the following structure:

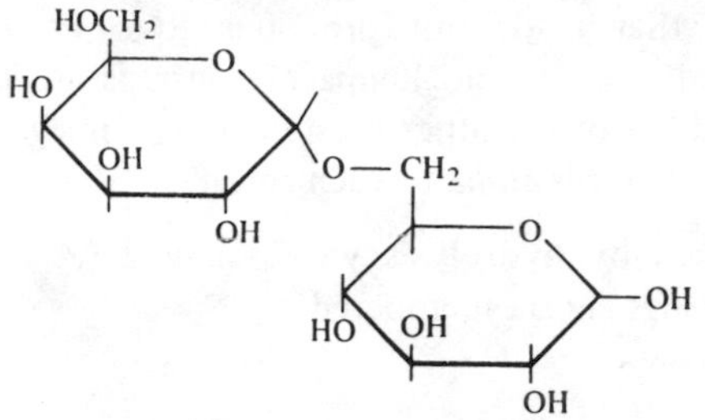

HOCH$_2$ ... (structure of melibiose)

6-O-(α-D-galactopyranosyl)-D-glucopyranose

We arrive at this conclusion from the data given:

1. That melibiose is a reducing sugar, that it undergoes mutarotation and forms a phenylosazone indicates that one monosaccharide is present as a cyclic hemiacetal.

2. That acid hydrolysis gives D-galactose and D-glucose indicates that melibiose is a disaccharide composed of one D-galactose unit and one D-glucose unit. That melibiose is hydrolyzed by an α-galactosidase suggests that melibiose is an α-D-galactosyl-D-glucose.

3. Oxidation of melibiose to melibionic acid and subsequent hydrolysis to give D-galactose and D-gluconic acid confirms that the glucose unit is present as a cyclic hemiacetal and that the galactose unit is present as a glycoside. (Had the reverse been true, this experiment would have yielded D-glucose and D-galactonic acid.)

Methylation and hydrolysis of melibionic acid produces 2,3,4,6-tetra-O-methyl-D-galactose and 2,3,4,5-tetra-O-methyl-D-gluconic acid. Formation of the first product—a galactose derivative with a free —OH at C-5—demonstrates that the galactose ring is six-membered; formation of the second product—a gluconic acid derivative with a free —OH at C-6—demonstrates that the oxygen at C-6 of the glucose unit is joined in a glycosidic linkage to the galactose unit.

4. That methylation and hydrolysis of melibiose gives a glucose derivative (2,3,4-tri-O-methyl-D-glucose) with free —OH groups at C-5 and C-6 shows that the glucose ring is also six-membered. Melibiose is, therefore, 6-O-(α-D-galactopyranosyl)-D-glucopyranose.

19.38

Trehalose has the following structure:

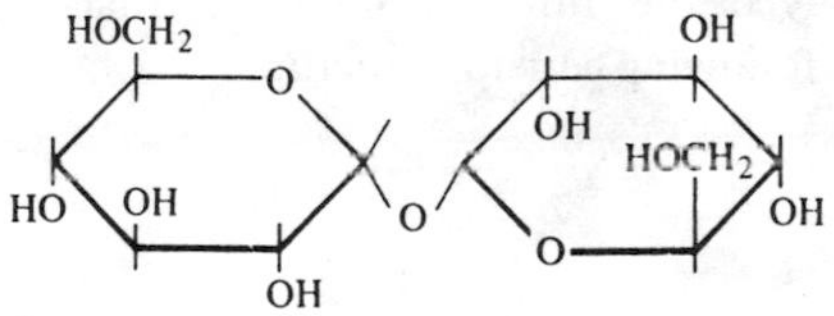

α-D-glucopyranosyl-D-glucopyranoside

We arrive at this structure in the following way:

1. Acid hydrolysis shows that trehalose is a disaccharide consisting only of D-glucose units.

2. Hydrolysis by α-glucosidases and not by β-glucosidases shows that the glycosidic linkages are *alpha*.

3. That trehalose is a non-reducing sugar, that it does not form phenylosazone, and that it does not react with bromine water indicate that no hemiacetal groups are present. This means that C-1 of one glucose unit and C-1 of the other must be joined in a glycosidic linkage. Fact 2 (above) indicates that this linkage is *alpha* to each ring.

4. That methylation of trehalose followed by hydrolysis yields only 2,3,4,6-tetra-O-methyl-D-glucose demonstrates that both rings are six-membered.

19.39

(a) Tollens' reagent or Benedict's reagent will give a positive test with D-glucose but will give no reaction with D-glucitol.

(b) D-Glucaric acid will give an acidic aqueous solution that can be detected with blue litmus paper. D-Glucitol will give a neutral aqueous solution.

(c) D-Glucose will be oxidized by bromine water and the red brown color of bromine will disappear. D-Fructose will not be oxidized by bromine water since it does not contain an aldehyde group.

(d) Nitric acid oxidation will produce an *optically active* aldaric acid from D-glucose but an *optically inactive* aldaric acid will result from D-galactose.

(e) Maltose is a reducing sugar and will give a positive test with Tollens' or Benedict's solution. Sucrose is a nonreducing sugar and will not react.

(f) Maltose will give a positive Tollens' or Benedict's test; maltonic acid will not.

(g) 2,3,4,6-Tetra-*O*-methyl-β-D-glucopyranose will give a positive test with Tollens' or Benedict's solution; methyl β-D-glucopyranoside will not.

(h) Periodic acid will react with methyl α-D-ribofuranoside because it has hydroxyl groups on adjacent carbons. Methyl 2-deoxy-α-D-ribofuranoside will not react.

19.40

That the Schardinger dextrins are nonreducing shows that they have no free aldehyde or hemiacetal groups. This strongly suggests the presence of a *cyclic* structure. That methylation and subsequent hydrolysis yields only 2,3,6-tri-*O*-methyl-D-glucose indicates that the glycosidic linkages all involve C-1 of one glucose unit and C-4 of the next. That α-glucosidases cause hydrolysis of the glycosidic linkages indicates that they are α-glycosidic linkages. Thus we are led to the following general structure.

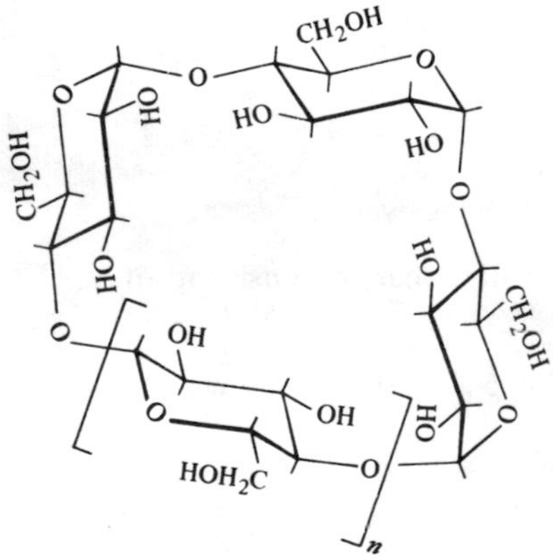

n = 3, 4, or 5

Note: Schardinger dextrins are extremely interesting compounds. They are able to form complexes with a wide variety of compounds by incorporating these compounds in the cavity in the middle of the cyclic dextrin structure. Complex formation takes place, however, only when the cyclic dextrin and the guest molecule are the right size. Anthracene molecules, for example, will fit into the cavity of a cyclic dextrin with eight glucose units but will not fit into one with seven. For more information about these fascinating compounds, see R. J. Bergeron, "Cycloamyloses," *J. Chem. Educ.*, **54**, 204 (1977).

19.41

Isomaltose has the following structure:

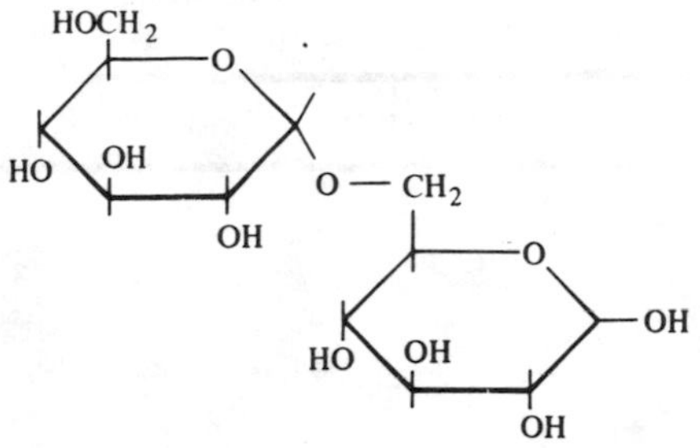

6-*O*-(α-D-glucopyranosyl)-D-glucopyranose

(1) The acid and enzymic hydrolysis experiments tell us that isomaltose has two glucose units linked by an α-linkage.

(2) That isomaltose is a reducing sugar indicates that one glucose unit is present as a cyclic hemiacetal.

(3) Methylation of isomaltonic acid followed by hydrolysis gives us information about the size of the nonreducing pyranoside ring and about its point of attachment to the reducing ring. The formation of the first product (2,3,4,6-tetra-*O*-methyl-D-glucose)–a compound with an —OH at C-5–tells us that the nonreducing ring is present as a pyranoside. The formation of 2,3,4,5-tetra-*O*-methyl-D-gluconic acid–a compound with an —OH at C-6–shows that the nonreducing ring is linked to C-6 of the reducing ring.

(4) Methylation of maltose itself tells the size of the reducing ring. That 2,3,4-tri-*O*-methyl-D-glucose is formed shows that the reducing ring is also six membered; we know this because of the free —OH at C-5.

19.42

Stachyose has the following structure:

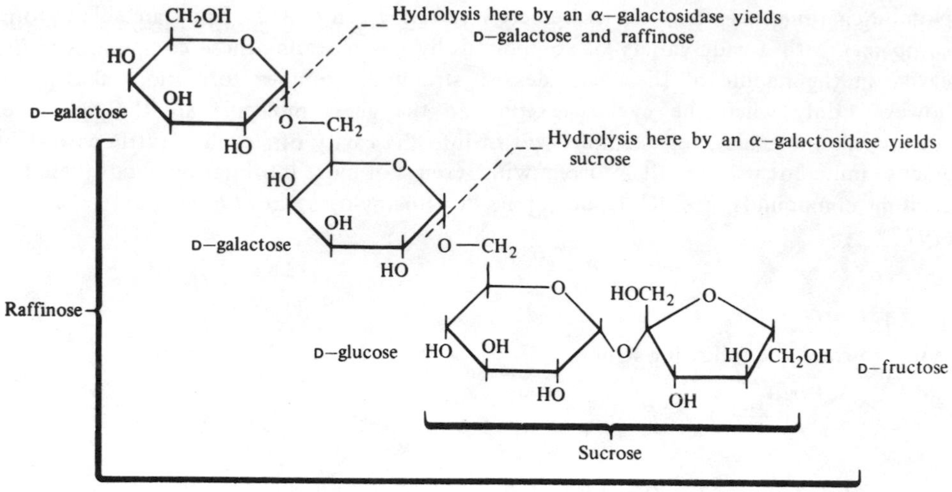

Hydrolysis here by an α–galactosidase yields D–galactose and raffinose

Hydrolysis here by an α–galactosidase yields sucrose

Raffinose — D-galactose, D-galactose, D-glucose, D-fructose; Sucrose

Raffinose has the following structure:

Hydrolysis here by an α–galactosidase yields D–galactose and sucrose

Hydrolysis here by an invertase yields melibiose and fructose

Melibiose — D-galactose, D-glucose; D-fructose; Sucrose

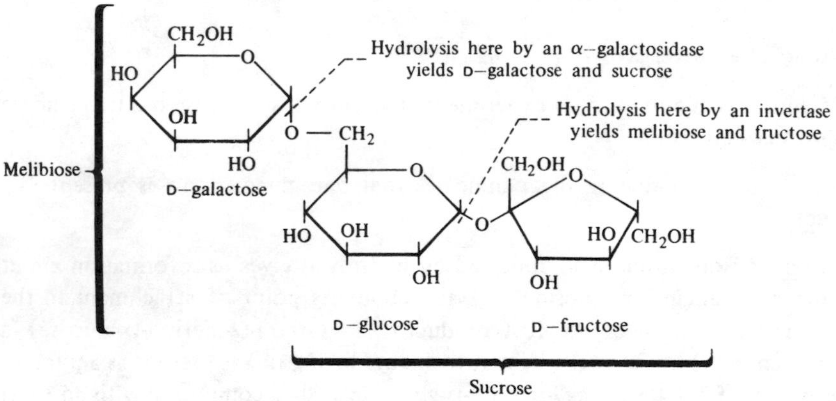

The enzymic hydrolyses (as indicated above) give the basic structure of stachyose and raffinose. The only remaining question is the ring size of the first galactose unit of stachyose. That methylation of stachyose and subsequent hydrolysis yields 2,3,4,6-tetra-*O*-methyl-D-galactose establishes that this ring is a pyranoside.

19.43

Arbutin has the following structure.

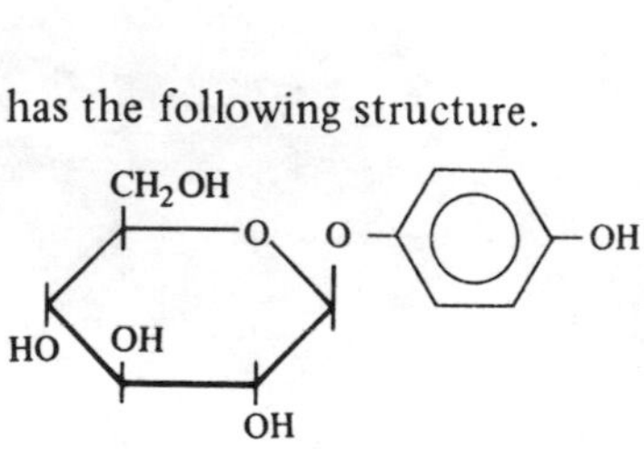

p–hydroxyphenyl–β–D–glucopyranoside

Compounds **X**, **Y**, and **Z** are hydroquinone, *p*-methoxyphenol, and *p*-dimethoxybenzene respectively.

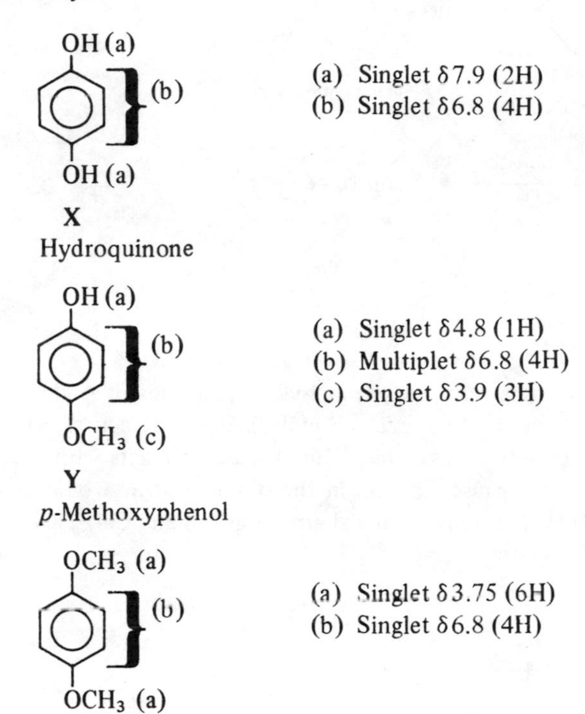

Z

p-Dimethoxybenzene

The reactions that take place are the following:

CH_2OH ... O — C₆H₄ — OH $\xrightarrow[\beta\text{-glucosidase}]{H^+, H_2O \text{ or}}$ D-glucose + HO — C₆H₄ — OH (**X**, Hydroquinone)

Arbutin $\xrightarrow[OH^-]{(CH_3)_2SO_4 \text{ (excess)}}$ (CH_2OCH_3, CH_3O, OCH_3, OCH_3) glucoside — O — C₆H₄ — OCH_3

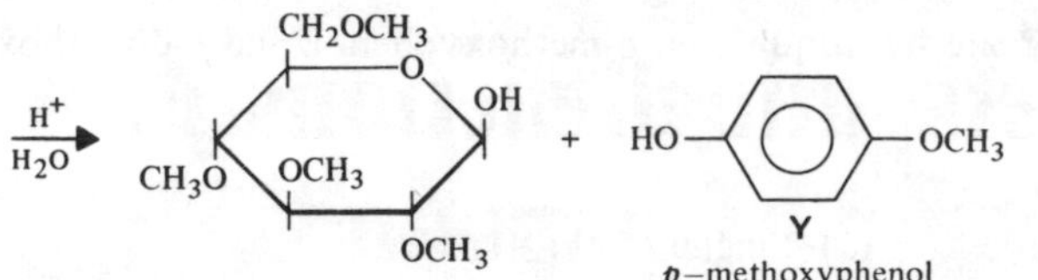

$\xrightarrow[H_2O]{H^+}$ 2,3,4,6-tetra-*O*-methyl D-glucose + HO–C₆H₄–OCH₃ (**Y**, *p*-methoxyphenol)

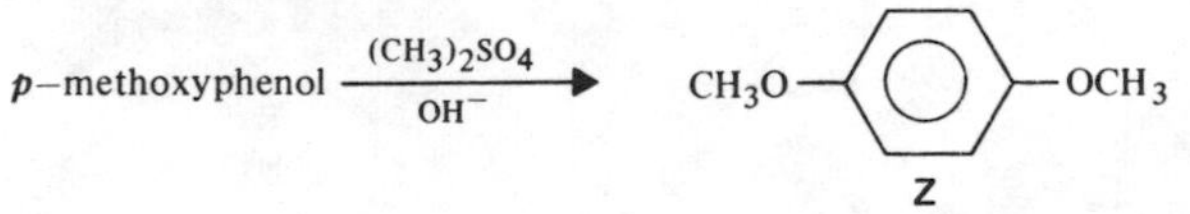

p-methoxyphenol $\xrightarrow[OH^-]{(CH_3)_2SO_4}$ CH_3O–C₆H₄–OCH_3 (**Z**, *p*-dimethoxybenzene)

19.44

(a) and (b) Two molecules of acetone react with four hydroxyl groups of D-glucose to yield a compound (below) containing two cyclic ketal linkages. Reaction with *cis* hydroxyl groups is preferred in reactions like this. Thus D-glucose reacts with acetone preferentially in the furanose form because reaction in the pyranose form would require the formation of a cyclic ketal from the *trans* hydroxyl groups at C-3 and C-4. This would introduce greater strain into the product.

W-50

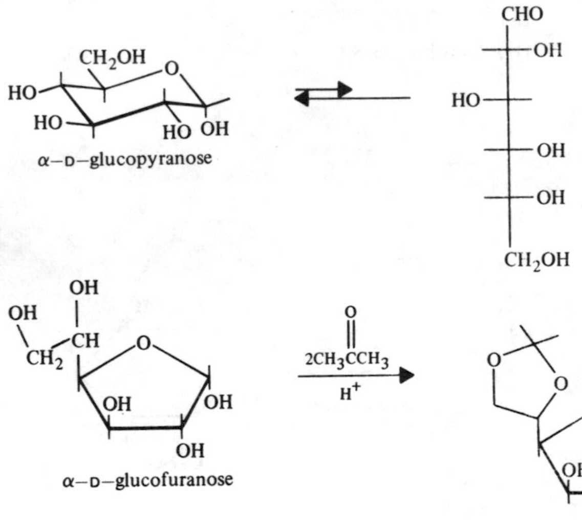

α-D-glucopyranose ⇄ open-chain D-glucose (CHO, OH, HO, OH, OH, CH_2OH)

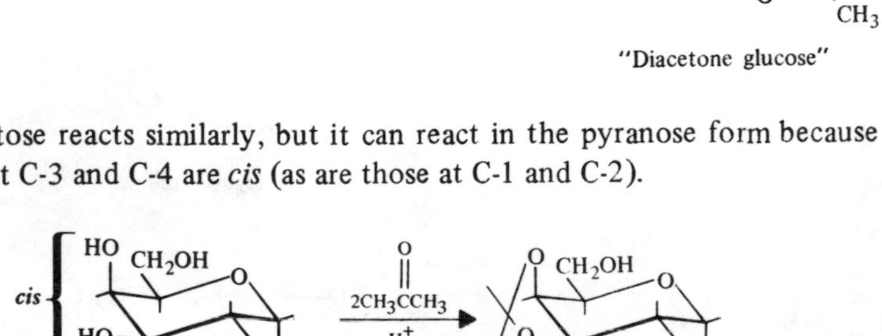

α-D-glucofuranose $\xrightarrow[H^+]{2CH_3\overset{O}{\overset{\|}{C}}CH_3}$ "Diacetone glucose"

D-Galactose reacts similarly, but it can react in the pyranose form because the hydroxyl groups at C-3 and C-4 are *cis* (as are those at C-1 and C-2).

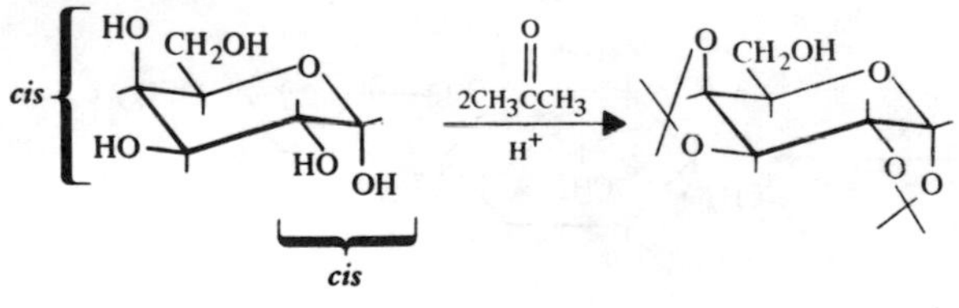

α-D-galactopyranose (*cis*, *cis*) $\xrightarrow[H^+]{2CH_3\overset{O}{\overset{\|}{C}}CH_3}$ diacetone galactose

CHAPTER TWENTY

Synthesis and Reactions of β-Dicarbonyl Compounds: More Chemistry of Enolate Ions

20.1

(a) Step 1

$$CH_3CH(H)\text{-}\overset{O}{\overset{\|}{C}}OC_2H_5 + {}^{-}OC_2H_5 \rightleftharpoons CH_3\overset{..}{\underset{-}{C}}H\text{-}\overset{O}{\overset{\|}{C}}OC_2H_5 + C_2H_5OH$$

$$\updownarrow$$

$$CH_3CH{=}\overset{O^-}{\overset{|}{C}}OC_2H_5$$

Step 2

$$CH_3CH_2\overset{O}{\overset{\|}{C}}OC_2H_5 + {}^{-}{:}\underset{CH_3}{\underset{|}{C}H}\overset{O}{\overset{\|}{C}}OC_2H_5 \rightleftharpoons CH_3CH_2\overset{O^-}{\overset{|}{\underset{C_2H_5O}{\underset{|}{C}}}}\text{-}\underset{CH_3}{\underset{|}{C}H}\text{-}\overset{O}{\overset{\|}{C}}OC_2H_5$$

$$\updownarrow$$

$$C_2H_5O^- + CH_3CH_2\overset{O}{\overset{\|}{C}}\text{-}\underset{CH_3}{\underset{|}{C}H}\text{-}\overset{O}{\overset{\|}{C}}OC_2H_5$$

Step 3

$$CH_3CH_2\overset{O}{\overset{\|}{C}}\text{-}\overset{H}{\overset{|}{\underset{CH_3}{\underset{|}{C}}}}\text{-}\overset{O}{\overset{\|}{C}}OC_2H_5 + {}^{-}OC_2H_5 \rightleftharpoons CH_3CH_2\overset{O}{\overset{\|}{C}}{=}\underset{CH_3}{\underset{|}{C}}{=}\overset{O}{\overset{\|}{C}}OC_2H_5 \;(-) + C_2H_5OH$$

(b)

$$CH_3CH_2\overset{O}{\overset{\|}{C}}\underset{CH_3}{\underset{|}{C}H}\overset{O}{\overset{\|}{C}}OC_2H_5 + CH_3CH_2\overset{OH}{\overset{|}{C}}{=}\underset{CH_3}{\underset{|}{C}}\overset{O}{\overset{\|}{C}}OC_2H_5$$

20.2

(a)

$$C_2H_5O\overset{O}{\overset{\|}{C}}CH_2CH_2CH_2CH_2\overset{O}{\overset{\|}{C}}OC_2H_5 \underset{+H^+}{\overset{-H^+}{\rightleftharpoons}} \begin{matrix} & C_2H_5O\text{-}\overset{O}{\overset{\|}{C}} & \\ CH_2 & & {:}\overset{-}{C}HCO_2C_2H_5 \\ | & & | \\ CH_2 & \text{—} & CH_2 \end{matrix}$$

$$\rightleftharpoons \text{(cyclopentane with } C_2H_5O, O^-, H, \overset{O}{\overset{\|}{C}}OC_2H_5) \rightleftharpoons \text{(cyclopentanone with } H, \overset{O}{\overset{\|}{C}}OC_2H_5) + C_2H_5O^-$$

$$\rightleftharpoons \text{(resonance-stabilized enolate of 2-carbethoxycyclopentanone)} + C_2H_5OH \xrightarrow{H^+} \text{(2-carbethoxycyclopentanone, } \overset{O}{\overset{\|}{C}}OC_2H_5) + \text{enol form}$$

(b) 2-carbethoxycyclohexanone ($\overset{O}{\overset{\|}{C}}OC_2H_5$ on cyclohexanone)

(c) To undergo a Dieckmann condensation, diethyl glutarate would have to form a highly strained four-membered ring.

20.3

$$CH_3\overset{O}{\overset{\|}{C}}OC_2H_5 + C_2H_5O^- \rightleftharpoons \bar{:}CH_2\overset{O}{\overset{\|}{C}}OC_2H_5 + C_2H_5OH$$

$$C_6H_5\overset{O}{\overset{\|}{C}}OC_2H_5 + \bar{:}CH_2\overset{O}{\overset{\|}{C}}OC_2H_5 \rightleftharpoons C_6H_5\underset{OC_2H_5}{\overset{O^-}{\overset{|}{\underset{|}{C}}}}-CH_2\overset{O}{\overset{\|}{C}}OC_2H_5$$

$$\rightleftharpoons C_6H_5\overset{O}{\overset{\|}{C}}CH_2\overset{O}{\overset{\|}{C}}OC_2H_5 + C_2H_5O^- \rightleftharpoons C_6H_5\overset{O}{\overset{\|}{C}}{=}CH{=}\overset{O}{\overset{\|}{C}}OC_2H_5 \text{ (resonance-stabilized anion)} + C_2H_5OH$$

$$\xrightarrow{H^+} C_6H_5\overset{O}{\overset{\|}{C}}CH_2\overset{O}{\overset{\|}{C}}OC_2H_5$$

$$C_6H_5CH_2\overset{O}{\overset{\|}{C}}OC_2H_5 + C_2H_5O^- \rightleftharpoons C_6H_5\ddot{C}H\overset{O}{\overset{\|}{C}}OC_2H_5 + C_2H_5OH$$

$$C_6H_5\ddot{C}H\overset{O}{\overset{\|}{C}}OC_2H_5 + C_2H_5O\overset{O}{\overset{\|}{C}}OC_2H_5 \rightleftharpoons C_2H_5O-\overset{O^-}{\overset{|}{C}}-OC_2H_5 \text{ bonded to } C_6H_5CH\text{, which bears } \overset{O}{\overset{\|}{C}}OC_2H_5$$

$$\rightleftharpoons C_6H_5CH(\overset{O}{\overset{\|}{C}}OC_2H_5)_2 + C_2H_5O^- \rightleftharpoons C_6H_5\bar{C}:(\overset{O}{\overset{\|}{C}}OC_2H_5)_2 + C_2H_5OH$$

resonance stabilized

$$\xrightarrow{H^+} C_6H_5CH(COC_2H_5)_2 \quad \text{(both C=O)}$$

20.4

(a) $CH_3CH_2\overset{O}{\overset{\|}{C}}OC_2H_5 + C_2H_5O\overset{O}{\overset{\|}{C}}-\overset{O}{\overset{\|}{C}}OC_2H_5 \xrightarrow[(2)\ H^+]{(1)\ NaOC_2H_5} CH_3CH(\overset{O}{\overset{\|}{C}}OC_2H_5)-\overset{O}{\overset{\|}{C}}-\overset{O}{\overset{\|}{C}}OC_2H_5$

(b) $CH_3\overset{O}{\overset{\|}{C}}OC_2H_5 + H\overset{O}{\overset{\|}{C}}OC_2H_5 \xrightarrow[(2)\ H^+]{(1)\ NaOC_2H_5} H\overset{O}{\overset{\|}{C}}CH_2\overset{O}{\overset{\|}{C}}OC_2H_5$

20.5

(a) cyclohexanone $+ H\overset{O}{\overset{\|}{C}}OC_2H_5 \xrightarrow[(2)\ H^+]{(1)\ NaOC_2H_5}$ 2-formylcyclohexanone (ring C=O, $\overset{O}{\overset{\|}{C}}H$)

(b) $CH_3CH_2\overset{O}{\overset{\|}{C}}CH_2CH_2CH_2\overset{O}{\overset{\|}{C}}OC_2H_5 \xrightarrow[(2)\ H^+]{(1)\ NaOC_2H_5}$ 2-methylcyclohexane-1,3-dione (CH_3)

(c) $C_2H_5O_2CCH_2C(CH_3)_2CH_2CO_2C_2H_5 + C_2H_5O\overset{O}{\overset{\|}{C}}-\overset{O}{\overset{\|}{C}}OC_2H_5 \xrightarrow[(2)\ H^+]{(1)\ NaOC_2H_5}$

$C_2H_5O-\overset{O}{\overset{\|}{C}}-\overset{O}{\overset{\|}{C}}-CH(CO_2C_2H_5)-C(CH_3)_2-CH_2CO_2C_2H_5 \xrightarrow[(2)\ H^+]{(1)\ NaOC_2H_5}$ cyclopentane-1,2-dione ring with $CO_2C_2H_5$, CH_3, CH_3, $CO_2C_2H_5$

20.6

(a) $CH_3\overset{O}{\overset{\|}{C}}CH_2CH_2CH_2CH_2\overset{O}{\overset{\|}{C}}OC_2H_5 + {}^-OC_2H_5 \underset{(-C_2H_5OH)}{\rightleftharpoons}$ $CH_2(\overset{O}{\overset{\|}{C}}OC_2H_5)-CH_2-CH_2-\ddot{C}H^- -\overset{O}{\overset{\|}{C}}-CH_3$

$\rightleftharpoons$ cyclopentane with C_2H_5O, O^-, H, $\overset{O}{\overset{\|}{C}}-CH_3$ $\rightleftharpoons$ 2-acetylcyclopentanone (O, H, $\overset{O}{\overset{\|}{C}}-CH_3$) $+ C_2H_5O^-$ (cont. on p. 348)

$$\rightleftharpoons \text{[enolate ring]} + C_2H_5OH \xrightarrow{H^+} \text{2-acetylcyclopentanone}$$

(b) $CH_3\overset{O}{\overset{\|}{C}}CH_2CH_2CH_2\overset{O}{\overset{\|}{C}}OC_2H_5 + {}^-OC_2H_5 \underset{(-C_2H_5OH)}{\rightleftharpoons}$ [$\overset{O}{\overset{\|}{C}}OC_2H_5$, CH_2, CH_2, CH_2, $:\overset{-}{C}H_2$, $C{=}O$ chain]

$\rightleftharpoons$ [^-O, OC_2H_5 cyclohexanone adduct] $\rightleftharpoons$ [1,3-cyclohexanedione] $+ {}^-OC_2H_5$

$\rightleftharpoons$ [enolate resonance structures] $+ C_2H_5OH$

$\downarrow H^+$

[1,3-cyclohexanedione]

Ionization of a hydrogen alpha to the ketone group might, in the first step, also yield $CH_3\overset{O}{\overset{\|}{C}}\overset{..}{\overset{-}{C}}HCH_2CH_2\overset{O}{\overset{\|}{C}}OC_2H_5$. Cyclization of this enolate anion does not occur to any appreciable extent, however, because to do so it would yield a (highly-strained) four-membered ring.

20.7

The partially negative oxygen of sodioacetoacetic ester acts as the nucleophile.

$$CH_3\overset{O}{\overset{\|}{C}}-\overset{..}{\overset{-}{C}}H-\overset{O}{\overset{\|}{C}}-OC_2H_5 \longleftrightarrow CH_3\overset{O^-}{\overset{|}{C}}=CH-\overset{O}{\overset{\|}{C}}-OC_2H_5$$

20.8

Again, working backward,

(a) $CH_3\overset{O}{\overset{\|}{C}}CH_2CH_2CH_3 \xleftarrow[-CO_2]{heat} CH_3\overset{O}{\overset{\|}{C}}\underset{\underset{CH_3}{|}}{\underset{CH_2}{\underset{|}{CH}}}-\overset{O}{\overset{\|}{C}}OH \xleftarrow[(2)\ H_3O^+]{(1)\ dil.\ NaOH,\ heat}$

$CH_3\overset{O}{\overset{\|}{C}}\underset{\underset{CH_3}{|}}{\underset{CH_2}{\underset{|}{C}}}H\overset{O}{\overset{\|}{C}}OC_2H_5 \xleftarrow[(2)\ CH_3CH_2Br]{(1)\ NaOC_2H_5} CH_3\overset{O}{\overset{\|}{C}}CH_2\overset{O}{\overset{\|}{C}}OC_2H_5$

(b) $CH_3\overset{O}{\overset{\|}{C}}\underset{CH_2CH_2CH_3}{\underset{|}{C}}HCH_2CH_2CH_3 \xleftarrow[-CO_2]{heat} CH_3\overset{O}{\overset{\|}{C}}-\underset{CH_2CH_2CH_3}{\overset{CH_3CH_2CH_2}{C}}-COOH \xleftarrow[(2)\ H_3O^+]{(1)\ dil.\ NaOH,\ heat}$

$CH_3\overset{O}{\overset{\|}{C}}-\underset{CH_2CH_2CH_3}{\overset{CH_3CH_2CH_2}{C}}-COOC_2H_5 \xleftarrow[(2)\ CH_3CH_2CH_2Br]{(1)\ (CH_3)_3COK} CH_3\overset{O}{\overset{\|}{C}}-\underset{CH_2CH_2CH_3}{\underset{|}{C}}H\overset{O}{\overset{\|}{C}}OC_2H_5$

$\xleftarrow[(2)\ CH_3CH_2CH_2Br]{(1)\ NaOC_2H_5} CH_3\overset{O}{\overset{\|}{C}}CH_2\overset{O}{\overset{\|}{C}}OC_2H_5$

(c) $CH_3\overset{O}{\overset{\|}{C}}CH_2CH_2C_6H_5 \xleftarrow[-CO_2]{heat} CH_3\overset{O}{\overset{\|}{C}}\underset{\underset{C_6H_5}{|}}{\underset{CH_2}{\underset{|}{C}}}H\overset{O}{\overset{\|}{C}}OH \xleftarrow[(2)\ H_3O^+]{(1)\ NaOH,\ heat} CH_3\overset{O}{\overset{\|}{C}}\underset{\underset{C_6H_5}{|}}{\underset{CH_2}{\underset{|}{C}}}H\overset{O}{\overset{\|}{C}}OC_2H_5$

$\xleftarrow[(2)\ C_6H_5CH_2Br]{(1)\ NaOC_2H_5} CH_3\overset{O}{\overset{\|}{C}}CH_2\overset{O}{\overset{\|}{C}}OC_2H_5$

20.9

(a) The first alkylation produces ethyl methylacetoacetate; it then reacts with sodium ethoxide to produce an anion that can undergo a second alkylation.

$$\underset{\text{Ethyl methylacetoacetate}}{CH_3\overset{O}{\overset{\|}{C}}-\underset{CH_3}{\underset{|}{\overset{H}{\overset{|}{C}}}}-\overset{O}{\overset{\|}{C}}OC_2H_5} \xrightarrow{NaOC_2H_5} CH_3\overset{O}{\overset{\|}{C}}-\underset{CH_3}{\underset{|}{\overset{..}{C}}}-\overset{O}{\overset{\|}{C}}OC_2H_5 \xrightarrow{CH_3I} \underset{\text{Ethyl dimethylacetoacetate}}{CH_3\overset{O}{\overset{\|}{C}}-\underset{CH_3}{\underset{|}{\overset{CH_3}{\overset{|}{C}}}}-COOC_2H_5}$$

(b) It favors monoalkylation.

(c) Methyl and ethyl halides present less steric hindrance to the attacking anion.

(d) Carry out the first alkylation with the larger alkyl halide, that is, with propyl bromide. This will minimize dialkylation in the first step.

20.10

(a) Reactivity is the same as with any S_N2 reaction. With primary halides substitution is highly favored, with secondary halides elimination competes with substitution, and with tertiary halides elimination is the exclusive course of reaction.

(b) Acetoacetic ester and 2-methylpropene.

(c) Bromobenzene is unreactive toward nucleophile substitution (Cf. Section 14.3 of the text).

20.11

$$CH_3CH_2CH_2\overset{O}{\overset{\|}{C}}OC_2H_5 \xrightarrow[(2)\ H^+]{(1)\ NaOC_2H_5} CH_3CH_2CH_2\overset{O}{\overset{\|}{C}}\underset{\underset{CH_3}{|}}{\underset{CH_2}{\underset{|}{C}}}H\overset{O}{\overset{\|}{C}}OC_2H_5$$

$$\xrightarrow[(2)\ H_3O^+]{(1)\ NaOH,\ H_2O,\ heat} CH_3CH_2CH_2\overset{O}{\overset{\|}{C}}\underset{\underset{CH_3}{|}}{\underset{CH_2}{\underset{|}{C}}}H\overset{O}{\overset{\|}{C}}OH \xrightarrow[-CO_2]{heat} CH_3CH_2CH_2\overset{O}{\overset{\|}{C}}CH_2CH_2CH_3$$

20.12

The carboxyl group that is lost most readily is the one that is *beta* to the keto group (cf. page 783 of the text).

20.13

$$CH_3\overset{O}{\overset{\|}{C}}CH_2CH_2\overset{O}{\overset{\|}{C}}C_6H_5 \xleftarrow[-CO_2]{heat} CH_3\overset{O}{\overset{\|}{C}}\underset{\underset{C_6H_5}{\underset{|}{C=O}}}{\underset{|}{\underset{CH_2}{\underset{|}{C}}}}H\overset{O}{\overset{\|}{C}}OH \xleftarrow[(2)\ H_3O^+]{(1)\ OH^-,\ H_2O,\ heat}$$

$$CH_3\overset{O}{\overset{\|}{C}}\underset{\underset{\underset{C_6H_5}{|}}{\underset{C=O}{|}}}{\underset{CH_2}{|}}{CH}\overset{O}{\overset{\|}{C}}OC_2H_5 \xleftarrow[(2)\ C_6H_5COCH_2Br]{(1)\ NaOC_2H_5} CH_3\overset{O}{\overset{\|}{C}}CH_2\overset{O}{\overset{\|}{C}}OC_2H_5$$

20.14

$$CH_3\overset{O}{\overset{\|}{C}}CH_2\overset{O}{\overset{\|}{C}}C_6H_5 \xleftarrow[-CO_2]{heat} CH_3\overset{O}{\overset{\|}{C}}\underset{\underset{C_6H_5}{|}}{\underset{C=O}{|}}{CH}\overset{O}{\overset{\|}{C}}OH \xleftarrow[(2)\ H_3O^+]{(1)\ OH^-,\ H_2O,\ heat}$$

$$CH_3\overset{O}{\overset{\|}{C}}\underset{\underset{C_6H_5}{|}}{\underset{C=O}{|}}{CH}\overset{O}{\overset{\|}{C}}OC_2H_5 \xleftarrow[(2)\ C_6H_5COCl]{(1)\ NaH} CH_3\overset{O}{\overset{\|}{C}}CH_2\overset{O}{\overset{\|}{C}}OC_2H_5$$

20.15

Here we alkylate the dianion,

$$CH_3-\overset{O}{\overset{\|}{C}}-CH_2-\overset{O}{\overset{\|}{C}}OC_2H_5 \xrightarrow[liq\ NH_3]{2KNH_2} \bar{:}CH_2-\overset{O}{\overset{\|}{C}}-\overset{..}{\bar{C}}H-\overset{O}{\overset{\|}{C}}OC_2H_5$$

$$\xrightarrow[(2)\ NH_4Cl]{(1)\ C_6H_5CH_2Cl} C_6H_5CH_2CH_2\overset{O}{\overset{\|}{C}}CH_2\overset{O}{\overset{\|}{C}}OC_2H_5$$

20.16

Working backward,

(a) $$CH_3CH_2CH_2CH_2COOH \xleftarrow[-CO_2]{heat} CH_3CH_2CH_2CH(COOH)_2 \xleftarrow[(2)\ H_3O^+]{(1)\ OH^-,\ H_2O,\ heat}$$

$$CH_3CH_2CH_2CH(COOC_2H_5)_2 \xleftarrow[CH_3CH_2CH_2Br]{NaOC_2H_5} CH_2(COOC_2H_5)_2$$

(b) $$CH_3CH_2CH_2\underset{\underset{CH_3}{|}}{C}HCOOH \xleftarrow[-CO_2]{heat} (CH_3CH_2CH_2)(CH_3)C(COOH)_2 \xleftarrow[(2)\ H_3O^+]{(1)\ OH^-,\ H_2O,\ heat}$$

$$(CH_3CH_2CH_2)(CH_3)C(COOC_2H_5)_2 \xleftarrow[NaOC_2H_5]{CH_3I} CH_3CH_2CH_2CH(COOC_2H_5)_2$$

$$CH_2(COOC_2H_5)_2 \xrightarrow[NaOC_2H_5]{CH_3CH_2CH_2Br}$$

(c) $$CH_3CH(CH_3)CH_2CH_2COOH \xleftarrow[-CO_2]{heat} CH_3CH(CH_3)CH_2CH(COOH)_2 \xleftarrow[(2)\ H_3O^+]{(1)\ OH^-,\ H_2O,\ heat}$$

$$CH_3CH(CH_3)CH_2CH(COOC_2H_5)_2 \xleftarrow[CH_3CH(CH_3)CH_2Br]{NaOC_2H_5} CH_2(COOC_2H_5)_2$$

20.17

(a) Formaldehyde, $H{-}\overset{O}{\overset{\|}{C}}{-}H$

(b) 1,3-dithiane (S, S; H, H) $\xrightarrow[(-C_4H_{10})]{C_4H_9Li}$ 2-lithio-1,3-dithiane (H, Li^+) $\xrightarrow[(-LiBr)]{C_6H_5CH_2Br}$ 2-benzyl-1,3-dithiane (H, $CH_2C_6H_5$)

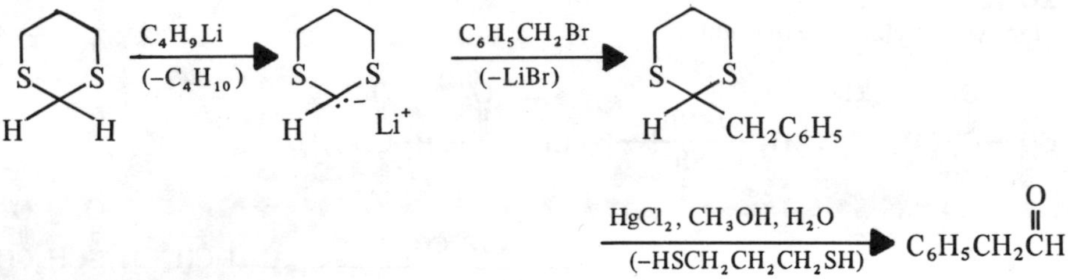

$$\xrightarrow[(-HSCH_2CH_2CH_2SH)]{HgCl_2,\ CH_3OH,\ H_2O} C_6H_5CH_2\overset{O}{\overset{\|}{C}}H$$

(c) $C_6H_5\overset{O}{\overset{\|}{C}}H + HSCH_2CH_2CH_2SH \xrightarrow{H^+}$ 1,3-dithiane (C_6H_5, H) $\xrightarrow[(2)\ CH_3I]{(1)\ C_4H_9Li}$ 1,3-dithiane (C_6H_5, CH_3)

$$\xrightarrow{HgCl_2,\ CH_3OH,\ H_2O} C_6H_5\overset{O}{\overset{\|}{C}}CH_3 + HSCH_2CH_2CH_2SH$$

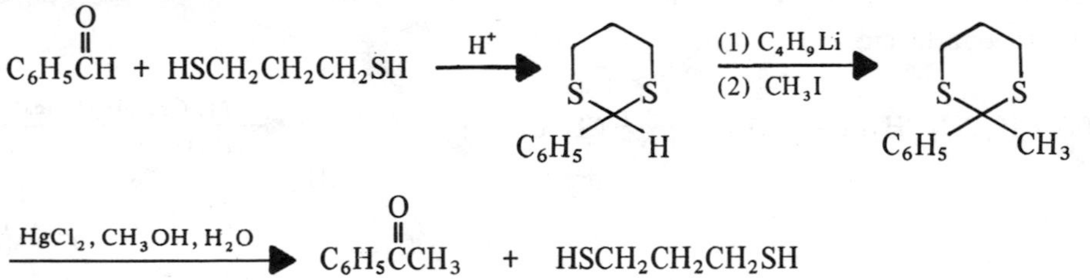

20.18

By treating the thioketal with Raney-nickel.

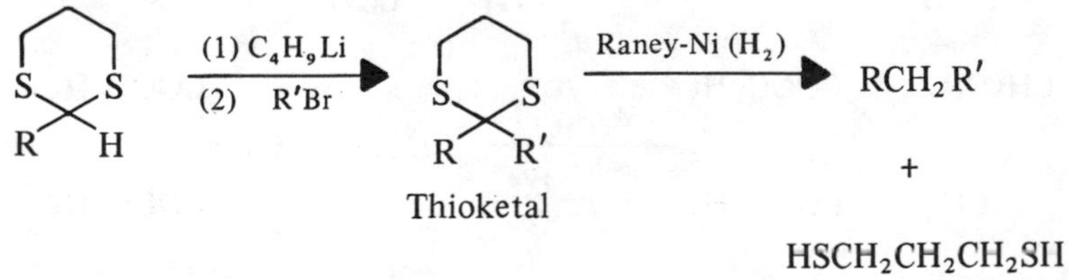

1,3-dithiane (R, H) $\xrightarrow[(2)\ R'Br]{(1)\ C_4H_9Li}$ 1,3-dithiane (R, R') Thioketal $\xrightarrow{Raney\text{-}Ni\ (H_2)}$ RCH_2R' + $HSCH_2CH_2CH_2SH$

20.19

(a) HC(=O)–C₆H₄–CH(=O) $\xrightarrow[H^+]{2\ HSCH_2CH_2CH_2SH}$ A

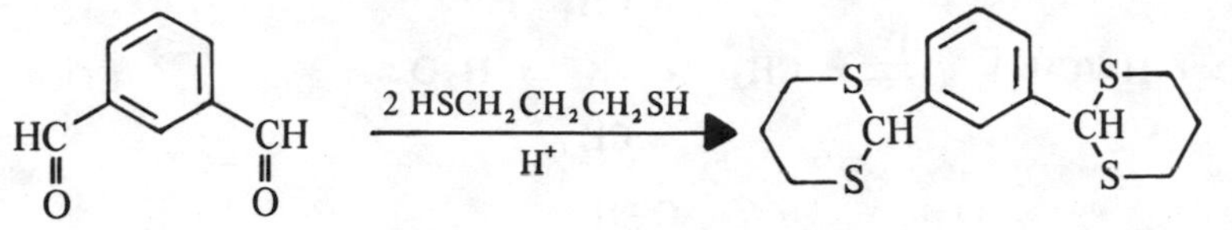

A

$\xrightarrow[(2)\ BrCH_2-C_6H_4-CH_2Br]{(1)\ 2\ C_4H_9Li}$ B

CH₂ CH₂

B

$\xrightarrow{Hydrolysis}$ C $\xrightarrow{Na\ BH_4}$

C

HO OH

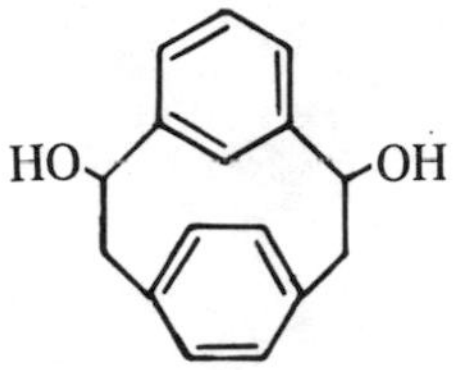

D

(b)

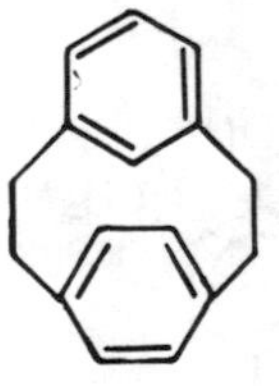

20.20

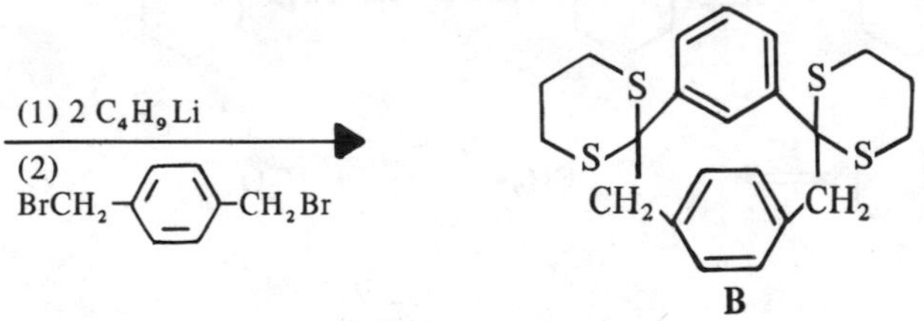

$$CH_3-\underset{\underset{O=C(OC_2H_5)\ \ C(=O)OC_2H_5}{CH}}{\overset{CH_3}{C}}CH_2\overset{O}{\overset{\|}{C}}OC_2H_5 \xrightarrow[(2)\ H_3O^+]{(1)\ OH^-,\ H_2O,\ heat} CH_3-\underset{\underset{O=C(OH)\ \ C(=O)OH}{CH}}{\overset{CH_3}{C}}CH_2\overset{O}{\overset{\|}{C}}OH$$

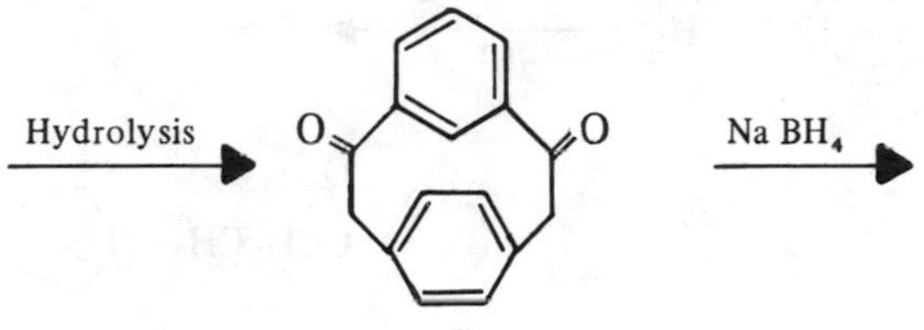

A malonic acid

$$\xrightarrow[-CO_2]{heat} HO\overset{O}{\overset{\|}{C}}CH_2\underset{CH_3}{\overset{CH_3}{C}}CH_2\overset{O}{\overset{\|}{C}}OH$$

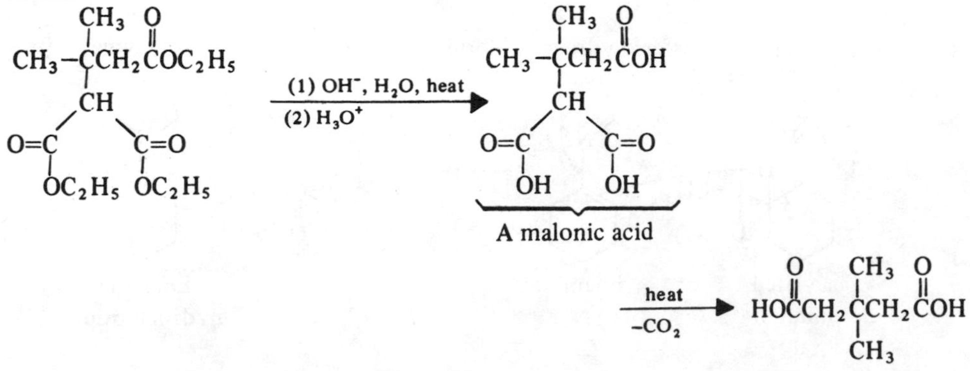

20.21

(a) $\underset{H}{\overset{H}{>}}C{=}O + HN(CH_3)_2 \overset{H^+}{\rightleftharpoons} CH_2{=}\overset{+}{N}\begin{smallmatrix}CH_3\\CH_3\end{smallmatrix} + H_2O$

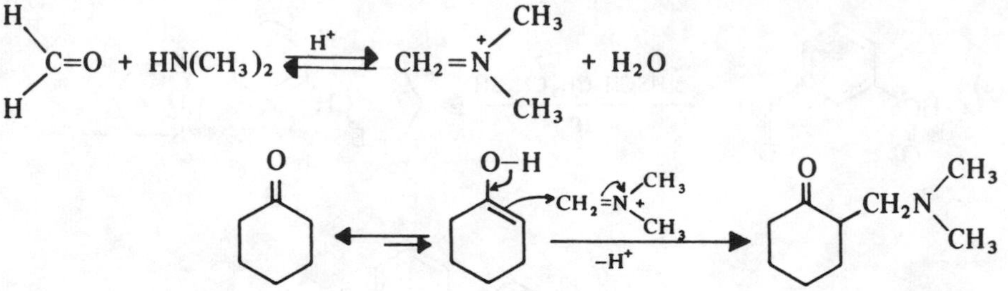

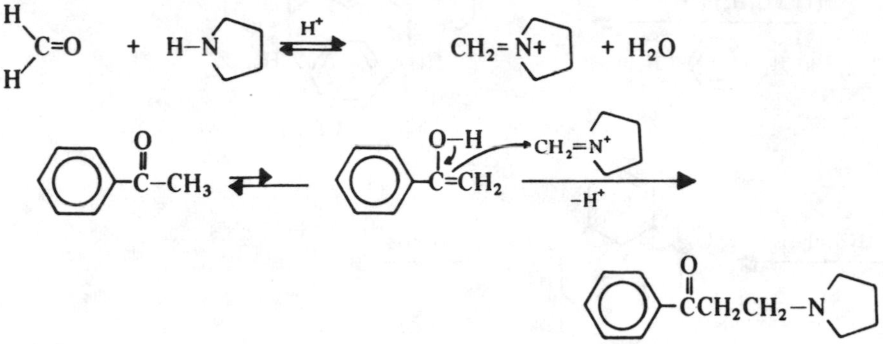

$$\text{cyclohexanone (C=O)} \rightleftharpoons \text{enol (O–H)} \; + \; CH_2{=}\overset{+}{N}(CH_3)(CH_3) \xrightarrow{-H^+} \text{2-}(CH_2N(CH_3)(CH_3))\text{cyclohexanone}$$

(b) $\underset{H}{\overset{H}{>}}C{=}O$ + H–N(pyrrolidine) $\overset{H^+}{\rightleftharpoons}$ $CH_2{=}N^+$(pyrrolidine) + H_2O

$$C_6H_5{-}\overset{O}{\overset{\|}{C}}{-}CH_3 \rightleftharpoons C_6H_5{-}\overset{O-H}{\overset{|}{C}}{=}CH_2 \; + \; CH_2{=}N^+\text{(pyrrolidine)} \xrightarrow{-H^+}$$

$$C_6H_5{-}\overset{O}{\overset{\|}{C}}CH_2CH_2{-}N\text{(pyrrolidine)}$$

(c) $\underset{H}{\overset{H}{>}}C{=}O + HN\begin{smallmatrix}CH_3\\CH_3\end{smallmatrix} \overset{H^+}{\rightleftharpoons} CH_2{=}\overset{+}{N}\begin{smallmatrix}CH_3\\CH_3\end{smallmatrix} + H_2O$

$$\text{4-}CH_3C_6H_4\ddot{O}{-}H \; + \; CH_2{=}\overset{+}{N}\begin{smallmatrix}CH_3\\CH_3\end{smallmatrix} \longrightarrow \overset{..}{O}^{+}{-}H,\; H,\; CH_2N\begin{smallmatrix}CH_3\\CH_3\end{smallmatrix},\; CH_3 \xrightarrow{-H^+}$$

$$\text{OH},\; CH_2N(CH_3)_2,\; CH_3 \xrightarrow[\text{of similar steps}]{\text{repetition}} (CH_3)_2NCH_2,\; \text{OH},\; CH_2N(CH_3)_2,\; CH_3$$

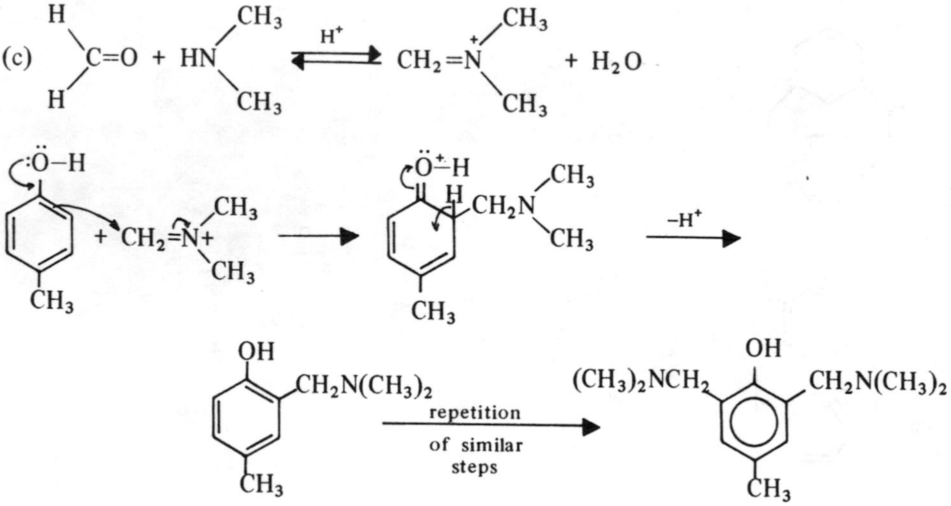

20.22

(a) The iminium salt transfers hydrogen chloride to the enamine in the following way.

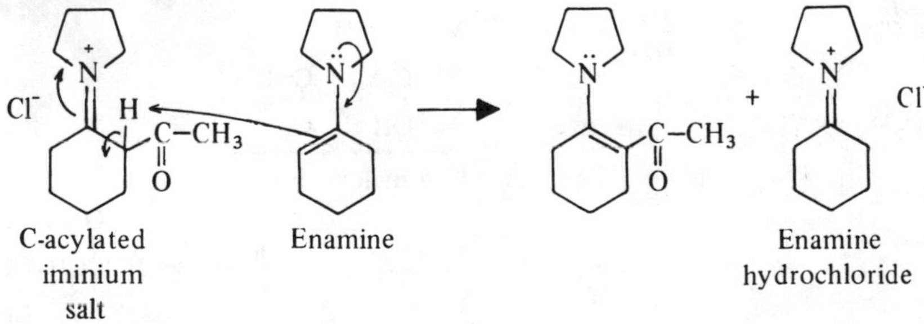

$$Cl^- \; [\overset{+}{N}\text{(pyrrolidinium)}{=}\text{cyclohexane},\; H,\; \underset{O}{\underset{\|}{C}}{-}CH_3] \; + \; [\ddot{N}\text{(pyrrolidine)}{-}\text{cyclohexene}] \longrightarrow [\ddot{N}\text{(pyrrolidine)}{-}\text{cyclohexene}{-}\underset{O}{\underset{\|}{C}}{-}CH_3] \; + \; [\overset{+}{N}\text{(pyrrolidinium)}{=}\text{cyclohexane}] \; Cl^-$$

C-acylated iminium salt

Enamine

Enamine hydrochloride

(b) Since the enamine hydrochloride formed in this acid-base reaction is not susceptible to acylation, the overall yield of the enamine synthesis will be decreased.

20.23

These syntheses are easy to see if we work backward.

(a) 2-hexanoylcyclohexanone $\xleftarrow{H_2O}$ iminium salt of 2-$\overset{O}{\overset{\|}{C}}(CH_2)_4CH_3$-cyclohexylidene pyrrolidinium Cl^- $\xleftarrow[(C_2H_5)_3N]{Cl\overset{O}{\overset{\|}{C}}(CH_2)_4CH_3}$

1-pyrrolidinylcyclohexene $\xleftarrow[H^+,\ -H_2O]{\text{pyrrolidine (N–H)}}$ cyclohexanone

(b) 2-($CH_2CH{=}CHCH_3$)cyclopentanone $\xleftarrow{H_2O}$ iminium salt with $CH_2CH{=}CHCH_3$, Br^- $\xleftarrow{BrCH_2CH{=}CHCH_3}$

1-pyrrolidinylcyclopentene $\xleftarrow[H^+,\ -H_2O]{\text{pyrrolidine (N–H)}}$ cyclopentanone

(c) 2-($CH_2\overset{O}{\overset{\|}{C}}CH_3$)cyclopentanone $\xleftarrow[(2)\ H_2O]{(1)\ BrCH_2\overset{O}{\overset{\|}{C}}CH_3}$ 1-pyrrolidinylcyclopentene $\xleftarrow[H^+,\ -H_2O]{\text{pyrrolidine (N–H)}}$ cyclopentanone

(d) 1-($CH_2\overset{O}{\overset{\|}{C}}OC_2H_5$)-2-tetralone $\xleftarrow[(2)\ H_2O]{(1)\ BrCH_2\overset{O}{\overset{\|}{C}}OC_2H_5}$ pyrrolidine enamine of 2-tetralone $\xleftarrow[H^+,\ -H_2O]{\text{pyrrolidine (N–H)}}$ 2-tetralone

20.24

(a) $CH_3CH_2CH_2\overset{O}{\overset{\|}{C}}\underset{\underset{CH_3}{|}}{\underset{CH_2}{|}}{CH}\overset{O}{\overset{\|}{C}}OC_2H_5 \xleftarrow[(2)\ H^+]{(1)\ NaOC_2H_5} CH_3CH_2CH_2\overset{O}{\overset{\|}{C}}OC_2H_5$

(b)
$$CH_3CH_2CH_2\overset{O}{\overset{\|}{C}}CH_2CH_2CH_3 \xleftarrow[-CO_2]{heat} CH_3CH_2CH_2\overset{O}{\overset{\|}{C}}\underset{CH_2CH_3}{\underset{|}{C}H}\overset{O}{\overset{\|}{C}}OH$$

$$\xleftarrow[(2)\ H_3O^+]{(1)\ OH^-,\ H_2O,\ heat} \text{Product of (a)}$$

(c)
$$C_6H_5\overset{CH_3}{\overset{|}{C}H}COOH \xleftarrow[-CO_2]{heat} (CH_3)(C_6H_5)C(COOH)_2 \xleftarrow[(2)\ H_3O^+]{(1)\ OH^-,\ H_2O,\ heat}$$

$$(CH_3)(C_6H_5)C(COOC_2H_5)_2 \xleftarrow[CH_3I]{NaOC_2H_5} C_6H_5{-}CH(COOC_2H_5)_2$$

$$\xleftarrow[(2)\ H^+]{(1)\ C_2H_5O\overset{O}{\overset{\|}{C}}OC_2H_5,\ NaOC_2H_5} C_6H_5CH_2\overset{O}{\overset{\|}{C}}OC_2H_5$$

(d)
$$CH_3CH_2\underset{\underset{O\ \ O}{\overset{\|\ \ \|}{C-COC_2H_5}}}{\underset{|}{C}H}\overset{O}{\overset{\|}{C}}OC_2H_5 \xleftarrow[(2)\ H^+]{(1)\ C_2H_5O\overset{O}{\overset{\|}{C}}-\overset{O}{\overset{\|}{C}}OC_2H_5,\ NaOC_2H_5} CH_3CH_2CH_2\overset{O}{\overset{\|}{C}}OC_2H_5$$

(e)
$$CH_3CH_2CH_2\overset{O}{\overset{\|}{C}}-\overset{O}{\overset{\|}{C}}OC_2H_5 \xleftarrow[C_2H_5OH]{H^+} CH_3CH_2CH_2\overset{O}{\overset{\|}{C}}-\overset{O}{\overset{\|}{C}}OH$$

$$\xleftarrow[-CO_2]{heat} CH_3CH_2\underset{\underset{O\ \ O}{\overset{\|\ \ \|}{C-COH}}}{\underset{|}{C}H}COOH \xleftarrow[(2)\ H_3O^+]{(1)\ OH^-,\ H_2O,\ heat} \text{Product of (d)}$$

(f)
$$C_6H_5\underset{\underset{O}{\overset{\|}{CH}}}{\underset{|}{C}H}\overset{O}{\overset{\|}{C}}OC_2H_5 \xleftarrow[(2)\ H^+]{(1)\ H\overset{O}{\overset{\|}{C}}OC_2H_5,\ NaOC_2H_5} C_6H_5CH_2\overset{O}{\overset{\|}{C}}OC_2H_5$$

(g)
$$\text{2-acetylcyclopentanone} \xleftarrow{H_2O} \text{(iminium salt of pyrrolidine, with } \overset{O}{\overset{\|}{C}}CH_3\text{)} \xleftarrow[(R_3N)]{CH_3\overset{O}{\overset{\|}{C}}Cl} \text{(1-pyrrolidinylcyclopentene)} \xleftarrow[H^+,\ -H_2O]{\text{pyrrolidine (N–H)}} \text{cyclopentanone}$$

(h)
$$\text{2-methyl-2-acetylcyclopentanone } (\overset{CH_3}{C}-\overset{O}{\overset{\|}{C}}CH_3) \xleftarrow[NaOC_2H_5]{CH_3I} \text{Product of (g)}$$

(i) 2-Ethylcyclohexanone $\xleftarrow[-CO_2]{\text{heat}}$ 2-ethyl-2-carboxycyclohexanone (CH$_2$CH$_3$, COOH) $\xleftarrow[(2)\ H_3O^+]{(1)\ OH^-, H_2O,\ heat}$

2-ethyl-2-(COOC$_2$H$_5$)cyclohexanone $\xleftarrow[NaOC_2H_5]{CH_3CH_2Br}$ 2-(COOC$_2$H$_5$)cyclohexanone

20.25

(a) $CH_3\overset{O}{\overset{\|}{C}}-C(CH_3)_2-CH_3 \xleftarrow{Zn,\ H^+} CH_3\overset{O}{\overset{\|}{C}}-C(CH_3)_2-CH_2Br \xleftarrow{PBr_3} CH_3\overset{O}{\overset{\|}{C}}-C(CH_3)_2-CH_2OH$

$\xleftarrow[(2)\ H_3O^+]{(1)\ LiAlH_4}$ $CH_3C(OCH_2CH_2O)-C(CH_3)_2COOC_2H_5$ $\xleftarrow[H^+]{HOCH_2CH_2OH}$ $CH_3\overset{O}{\overset{\|}{C}}-C(CH_3)_2-COOC_2H_5 \xleftarrow[NaOC_2H_5]{CH_3I}$

$CH_3\overset{O}{\overset{\|}{C}}-CH(CH_3)-COOC_2H_5 \xleftarrow[NaOC_2H_5]{CH_3I} CH_3\overset{O}{\overset{\|}{C}}CH_2\overset{O}{\overset{\|}{C}}OC_2H_5$

(b) $CH_3\overset{O}{\overset{\|}{C}}CH_2CH_2CH_2CH_2CH_3 \xleftarrow[-CO_2]{heat} CH_3\overset{O}{\overset{\|}{C}}CH(CH_2CH_2CH_3)\overset{O}{\overset{\|}{C}}OH \xleftarrow[(2)H_3O^+]{(1)\ OH^-,\ H_2O,\ heat}$

$CH_3\overset{O}{\overset{\|}{C}}CH(CH_2CH_2CH_3)\overset{O}{\overset{\|}{C}}OC_2H_5 \xleftarrow[CH_3CH_2CH_2Br]{NaOC_2H_5} CH_3\overset{O}{\overset{\|}{C}}CH_2\overset{O}{\overset{\|}{C}}OC_2H_5$

(c) $CH_3\overset{O}{\overset{\|}{C}}CH_2CH_2\overset{O}{\overset{\|}{C}}CH_3 \xleftarrow[-CO_2]{heat} CH_3\overset{O}{\overset{\|}{C}}CH(CH_2\overset{O}{\overset{\|}{C}}CH_3)\overset{O}{\overset{\|}{C}}OH \xleftarrow[(2)\ H_3O^+]{(1)\ OH^-,\ H_2O,\ heat}$

$CH_3\overset{O}{\overset{\|}{C}}CH(CH_2\overset{O}{\overset{\|}{C}}CH_3)\overset{O}{\overset{\|}{C}}OC_2H_5 \xleftarrow[CH_3COCH_2Br]{NaOC_2H_5} CH_3\overset{O}{\overset{\|}{C}}CH_2\overset{O}{\overset{\|}{C}}OC_2H_5$

(d) $$CH_3\overset{OH}{\overset{|}{C}}HCH_2CH_2COOH \xleftarrow{NaBH_4} CH_3\overset{O}{\overset{\|}{C}}CH_2CH_2\overset{O}{\overset{\|}{C}}OH \xleftarrow[-CO_2]{heat}$$

$$CH_3\overset{O}{\overset{\|}{C}}\underset{\underset{COOH}{|}}{\underset{CH_2}{\underset{|}{C}}}H\overset{O}{\overset{\|}{C}}OH \xleftarrow[(2)\ H_3O^+]{(1)\ OH^-,\ H_2O,\ heat} CH_3\overset{O}{\overset{\|}{C}}\underset{\underset{\underset{O}{\|}}{COC_2H_5}}{\underset{CH_2}{\underset{|}{C}}}H\overset{O}{\overset{\|}{C}}OC_2H_5$$

$$\xleftarrow[BrCH_2COOC_2H_5]{NaOC_2H_5} CH_3\overset{O}{\overset{\|}{C}}CH_2\overset{O}{\overset{\|}{C}}OC_2H_5$$

(e) $$CH_3\overset{OH}{\overset{|}{C}}H\underset{C_2H_5}{\underset{|}{C}}HCH_2OH \xleftarrow[(2)\ H^+]{(1)\ LiAlH_4} CH_3\overset{O}{\overset{\|}{C}}\underset{C_2H_5}{\underset{|}{C}}H\overset{O}{\overset{\|}{C}}OC_2H_5$$

$$\xleftarrow[C_2H_5Br]{NaOC_2H_5} CH_3\overset{O}{\overset{\|}{C}}CH_2\overset{O}{\overset{\|}{C}}OC_2H_5$$

(f) $$CH_3\overset{OH}{\overset{|}{C}}HCH_2\overset{OH}{\overset{|}{C}}HC_6H_5 \xleftarrow{NaBH_4} CH_3\overset{O}{\overset{\|}{C}}CH_2\overset{O}{\overset{\|}{C}}C_6H_5 \longleftarrow$$ cf. Problem 20.14

(g) $$C_6H_5\overset{OH}{\overset{|}{C}}HCH_2\overset{OH}{\overset{|}{C}}HCH_2CH_2OH \xleftarrow[(2)\ H^+]{(1)\ LiAlH_4} C_6H_5\overset{O}{\overset{\|}{C}}CH_2\overset{O}{\overset{\|}{C}}CH_2\overset{O}{\overset{\|}{C}}OC_2H_5$$

↑ cf. Problem 20.15, using C_6H_5COCl in place of $C_6H_5CH_2Cl$ in the second step.

20.26

(a) $$CH_3CH_2\underset{CH_3}{\underset{|}{C}}HCOOH \xleftarrow{-CO_2} (CH_3CH_2)(CH_3)C(COOH)_2 \xleftarrow[(2)\ H_3O^+]{(1)\ OH^-,\ H_2O,\ heat}$$

$$(CH_3CH_2)(CH_3)C(COOC_2H_5)_2 \xleftarrow[NaOC_2H_5]{CH_3I} CH_3CH_2CH(COOC_2H_5)_2$$

$$\xleftarrow[NaOC_2H_5]{CH_3CH_2Br} \underset{\underset{COOC_2H_5}{|}}{\overset{\overset{COOC_2H_5}{|}}{CH_2}}$$

(b) $$CH_3\underset{CH_3}{\underset{|}{C}}HCH_2CH_2CH_2OH \xleftarrow[(2)\ H^+]{(1)\ LiAlH_4} CH_3\underset{CH_3}{\underset{|}{C}}HCH_2CH_2COOH$$

(from Problem 20.16 c)

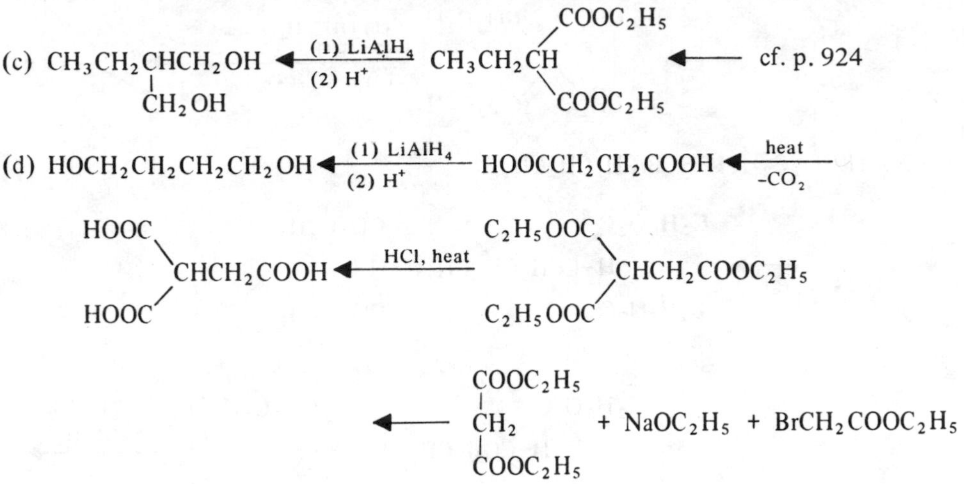

(c) $CH_3CH_2CH(CH_2OH)CH_2OH \xleftarrow[(2)\ H^+]{(1)\ LiAlH_4} CH_3CH_2CH(COOC_2H_5)_2 \longleftarrow$ cf. p. 924

(d) $HOCH_2CH_2CH_2CH_2OH \xleftarrow[(2)\ H^+]{(1)\ LiAlH_4} HOOCCH_2CH_2COOH \xleftarrow[-CO_2]{heat}$

$(HOOC)_2CHCH_2COOH \xleftarrow{HCl,\ heat} (C_2H_5OOC)_2CHCH_2COOC_2H_5$

$\longleftarrow CH_2(COOC_2H_5)_2 + NaOC_2H_5 + BrCH_2COOC_2H_5$

20.27

The following reaction took place,

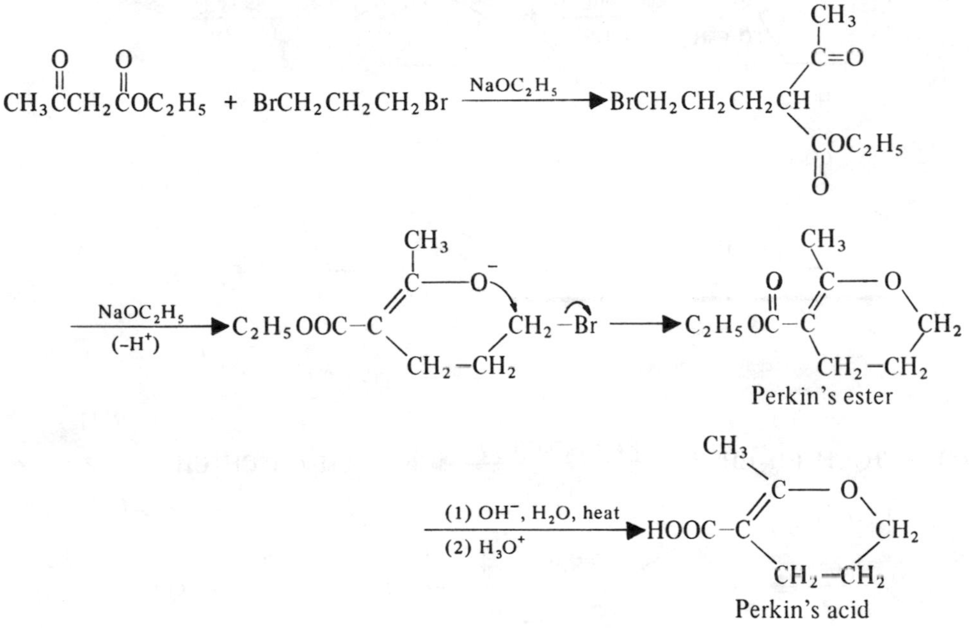

$$CH_3\overset{O}{\overset{\|}{C}}CH_2\overset{O}{\overset{\|}{C}}OC_2H_5 + BrCH_2CH_2CH_2Br \xrightarrow{NaOC_2H_5} BrCH_2CH_2CH_2CH(COCH_3)(COOC_2H_5)$$

$$\xrightarrow[(-H^+)]{NaOC_2H_5} C_2H_5OOC{-}C(=C(CH_3){-}O^-)(CH_2{-}CH_2{-}CH_2{-}Br) \longrightarrow C_2H_5O\overset{O}{\overset{\|}{C}}{-}C{=}C(CH_3){-}O{-}CH_2{-}CH_2{-}CH_2 \text{ (ring)}$$

Perkin's ester

$$\xrightarrow[(2)\ H_3O^+]{(1)\ OH^-,\ H_2O,\ heat} HOOC{-}C{=}C(CH_3){-}O{-}CH_2{-}CH_2{-}CH_2 \text{ (ring)}$$

Perkin's acid

20.28

(a) $BrCH_2CH_2Br + CH_2(COOC_2H_5)_2 + NaOC_2H_5 \longrightarrow$

$$[BrCH_2CH_2{-}CH(COOC_2H_5)_2] \xrightarrow[(-H^+)]{NaOC_2H_5} [BrCH_2CH_2{-}\ddot{C}^-(COOC_2H_5)_2]$$

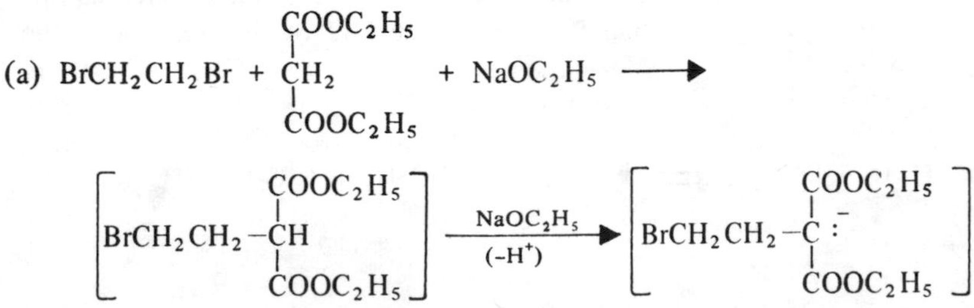

$$\longrightarrow \begin{matrix} CH_2 \\ | \\ CH_2 \end{matrix} \!\!>\! C \!<\! \begin{matrix} CO_2C_2H_5 \\ \\ CO_2C_2H_5 \end{matrix} \xrightarrow[(3)\ \text{heat},\ -CO_2]{(1)\ OH^-,\ H_2O,\ \text{heat};\ (2)\ H_3O^+} \triangleright\!-COOH$$

(b) $2NaCH(CO_2C_2H_5)_2 + BrCH_2CH_2CH_2Br$

$$\longrightarrow (C_2H_5O_2C)_2C\text{H}-CH_2CH_2CH_2-CH(CO_2C_2H_5)_2$$

A

$$\xrightarrow[Br_2]{NaOC_2H_5} \left[(C_2H_5O_2C)_2CH-CH_2CH_2CH_2-CBr(CO_2C_2H_5)_2 \right] \xrightarrow{NaOC_2H_5}$$

B (cyclopentane-1,1,2,2-tetracarboxylic acid tetraethyl ester: $CO_2C_2H_5$, $C_2H_5O_2C$, $CO_2C_2H_5$, $CO_2C_2H_5$) $\xrightarrow[(2)\ H_3O^+]{(1)\ OH^-,\ H_2O}$ (tetra-acid: CO_2H, CO_2H, CO_2H, CO_2H) $\xrightarrow[(-2CO_2)]{\text{heat}}$

B

(cyclopentane with CO_2H and CO_2H) + (CO_2H and HO_2C) — **D**, racemic modification

(cyclopentane with HOOC and COOH, cis) — **E**, *meso*-compound

(c) $BrCH_2CH_2CH_2CH_2Br \xrightarrow{NaCH(CO_2C_2H_5)_2} BrCH_2CH_2CH_2CH_2CH(CO_2C_2H_5)_2$

$$\xrightarrow{NaOC_2H_5} \text{cyclopentane-1,1-}(CO_2C_2H_5)_2 \xrightarrow[(3)\ \text{heat}]{(1)\ OH^-,\ H_2O;\ (2)\ H_3O^+} \text{cyclopentane}-COOH$$

20.29

The α-hydrogens of a thiol ester are more acidic than those of an ordinary ester, and the RS^- group is a better leaving group than RO^-. (See page 787 of the text for further explanation.)

20.30

(a) $CH_2(COOC_2H_5)_2 + {}^-OC_2H_5 \rightleftharpoons {}^-\!:CH(COOC_2H_5)_2 + C_2H_5OH$

$$C_6H_5CH{=}CH{-}\overset{O}{\overset{\|}{C}}OC_2H_5 + {}^{-}{:}CH(COOC_2H_5)_2 \rightleftharpoons C_6H_5\underset{CH(COOC_2H_5)_2}{\underset{|}{C}}HCH{=}\overset{O^-}{\overset{|}{C}}OC_2H_5$$

$$\xrightleftharpoons{+H^+} C_6H_5\underset{CH(COOC_2H_5)_2}{\underset{|}{C}}HCH_2\overset{O}{\overset{\|}{C}}OC_2H_5$$

(b) $$CH_3\ddot{N}H_2 + CH_2{=}CH{-}\overset{O}{\overset{\|}{C}}OCH_3 \rightleftharpoons CH_3{-}\overset{H}{\overset{|+}{\underset{H}{\underset{|}{N}}}}{-}CH_2{-}CH{=}\overset{O^-}{\overset{|}{C}}OCH_3$$

$$\rightleftharpoons CH_3\underset{H}{\underset{|}{N}}{-}CH_2{-}CH_2{-}\overset{O}{\overset{\|}{C}}OCH_3 \xrightleftharpoons{CH_2=CH-\overset{O}{\overset{\|}{C}}OCH_3}$$

$$CH_3N(CH_2CH_2COOCH_3)_2 \xrightleftharpoons{\text{base}} CH_3{-}N\begin{cases}CH_2{-}\ddot{C}H^{-}{-}COOCH_3 \\ CH_2{-}CH_2{-}C(=O)OCH_3\end{cases}$$

$$\xrightarrow[\text{(several steps)}]{\text{Dieckmann condensation}} CH_3{-}N\text{(piperidine ring bearing } COOCH_3 \text{ and } {=}O)$$

(c) $$CH_3{-}\overset{CH_3}{\overset{|}{\underset{CH(CO_2C_2H_5)_2}{\underset{|}{C}}}}{-}CH_2{-}\overset{O}{\overset{\|}{C}}OC_2H_5 + C_2H_5O^- \rightleftharpoons CH_3{-}\overset{CH_3}{\overset{|}{\underset{CH(CO_2C_2H_5)_2}{\underset{|}{C}}}}{-}CH{=}\overset{O^-}{\overset{|}{C}}OC_2H_5$$

$$+ C_2H_5OH$$

$$CH_3{-}\overset{CH_3}{\overset{|}{\underset{CH(CO_2C_2H_5)_2}{\underset{|}{C}}}}{-}CH{=}\overset{O^-}{\overset{|}{C}}OC_2H_5 \rightleftharpoons CH_3{-}\overset{CH_3}{\overset{|}{C}}{=}CH{-}\overset{O}{\overset{\|}{C}}OC_2H_5 + {}^{-}{:}CH(CO_2C_2H_5)_2$$

The Michael reaction is reversible and the reaction just given is an example of a reverse Michael reaction.

(d)

$$\text{(1-pyrrolidinylcyclohexene)} + CH_2{=}CH{-}\overset{O}{\overset{\|}{C}}OC_2H_5 \rightleftharpoons \text{(cyclohexylidene-pyrrolidinium, } \overset{+}{N}{=}\text{, with H, H)}{-}CH_2{-}CH{=}\overset{O^-}{\overset{|}{C}}OC_2H_5$$

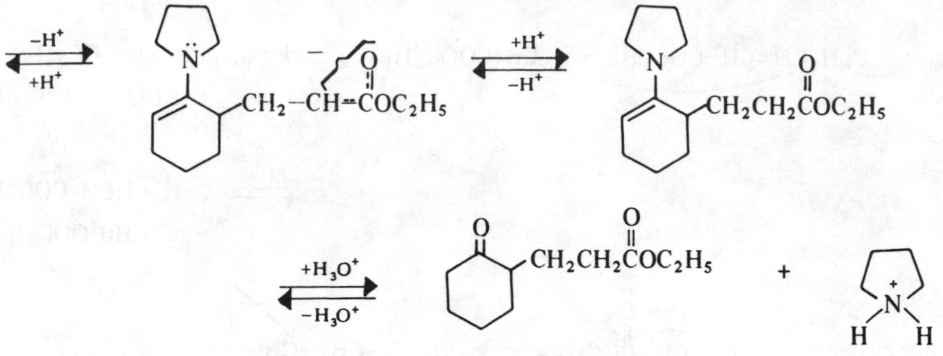

$\underset{+H^+}{\overset{-H^+}{\rightleftarrows}}$ [enamine with $CH_2-\overset{-}{C}H-\overset{O}{\overset{\|}{C}}OC_2H_5$] $\underset{-H^+}{\overset{+H^+}{\rightleftarrows}}$ [enamine with $CH_2CH_2\overset{O}{\overset{\|}{C}}OC_2H_5$]

$\underset{-H_3O^+}{\overset{+H_3O^+}{\rightleftarrows}}$ [2-($CH_2CH_2\overset{O}{\overset{\|}{C}}OC_2H_5$)cyclohexanone] + [pyrrolidinium, $\overset{+}{N}H_2$]

(e) This one is a real challenge.

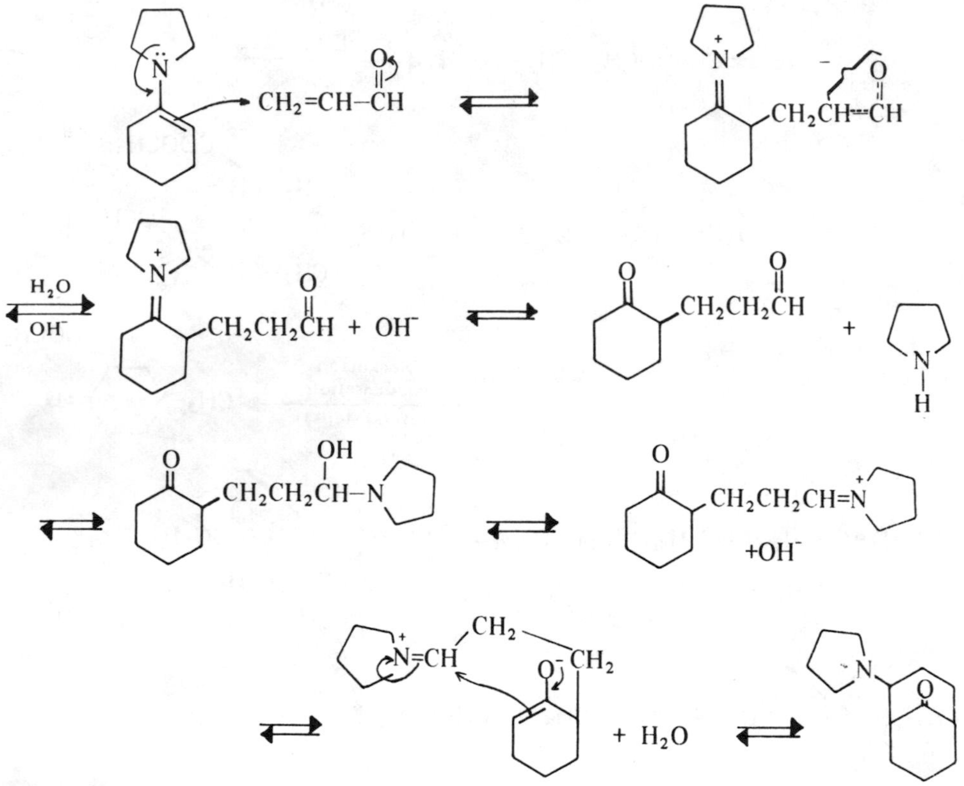

[pyrrolidine enamine of cyclohexanone] + $CH_2{=}CH{-}\overset{O}{\overset{\|}{C}}H$ $\rightleftarrows$ [iminium with $CH_2\overset{-}{C}H{-}\overset{O}{\overset{\|}{C}}H$]

$\underset{OH^-}{\overset{H_2O}{\rightleftarrows}}$ [iminium with $CH_2CH_2\overset{O}{\overset{\|}{C}}H$] + OH^- $\rightleftarrows$ [2-($CH_2CH_2\overset{O}{\overset{\|}{C}}H$)cyclohexanone] + [pyrrolidine, N–H]

$\rightleftarrows$ [2-($CH_2CH_2\overset{OH}{\overset{|}{C}}H{-}N$-pyrrolidinyl)cyclohexanone] $\rightleftarrows$ [2-($CH_2CH_2CH{=}\overset{+}{N}$-pyrrolidinium)cyclohexanone] +OH^-

$\rightleftarrows$ [$\overset{+}{N}{=}CH{-}CH_2{-}CH_2$ tethered enolate, O^-] + H_2O $\rightleftarrows$ [bicyclic ketone with pyrrolidinyl substituent]

20.31

Two reactions take place. The first is a normal Knoevenagel condensation,

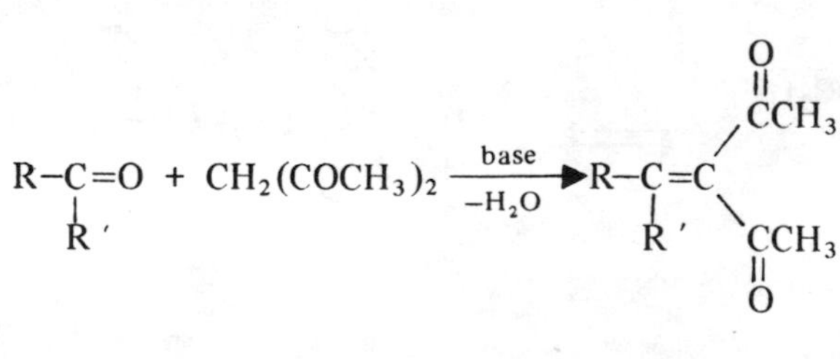

$$\underset{\substack{|\\R'}}{R{-}C{=}O} + CH_2(COCH_3)_2 \xrightarrow[-H_2O]{\text{base}} \underset{\substack{|\\R'}}{R{-}C}{=}C\begin{matrix}\overset{O}{\overset{\|}{C}}CH_3\\ \underset{\overset{\|}{O}}{C}CH_3\end{matrix}$$

Then the α,β-unsaturated diketone reacts with a second mole of the active methylene compound in a Michael addition.

$$\mathrm{R{-}C(R'){=}C(CCH_3{=}O)_2} + \mathrm{CH_2(COCH_3)_2} \xrightarrow{\text{base}} \mathrm{R{-}C(R')H{-}CH(COCH_3)_2 \ldots CH(COCH_3)_2}$$

20.32

$$\mathrm{CH_3CH_2\overset{O}{\overset{\|}{C}}OC_2H_5} \xrightarrow[\mathrm{C_2H_5ONa}]{\mathrm{H\overset{O}{\overset{\|}{C}}OC_2H_5}} \mathrm{CH_3CH(CHO)\overset{O}{\overset{\|}{C}}OC_2H_5}$$

$$\xrightarrow[\mathrm{C_2H_5ONa}]{\mathrm{H_2N{-}\overset{O}{\overset{\|}{C}}{-}NH_2}} \left[\text{cyclic intermediate: } \mathrm{C_2H_5O{-}C(O^-)},\ \mathrm{HN},\ \mathrm{C{-}CH_3},\ \mathrm{CH},\ \mathrm{C{=}O},\ \mathrm{N{-}H}\right] \longrightarrow \text{thymine}$$

20.33

$$\text{cycloheptanone} \xrightarrow[(2)\ \mathrm{H_3O^+}]{(1)\ \mathrm{CH_3MgI}} \underset{\mathbf{A}}{\text{1-methylcycloheptanol (CH}_3\text{, OH)}} \xrightarrow{\mathrm{H^+},\ \text{heat}} \underset{\mathbf{B}}{\text{1-methylcycloheptene (CH}_3\text{)}}$$

$$\xrightarrow[(2)\ \mathrm{Zn,\ H_2O}]{(1)\ \mathrm{O_3}} \underset{\mathbf{C}}{\mathrm{CH_3\overset{O}{\overset{\|}{C}}(CH_2)_5CHO}} \xrightarrow[\text{pyridine}]{\mathrm{CH_2(COOH)_2}}$$

$$\underset{\text{Queen substance}}{\mathrm{CH_3\overset{O}{\overset{\|}{C}}(CH_2)_5CH{=}CHCOOH}} \xrightarrow[\mathrm{Pd}]{\mathrm{H_2}} \underset{\mathbf{D}}{\mathrm{CH_3\overset{O}{\overset{\|}{C}}(CH_2)_7COOH}}$$

$$\xrightarrow[(2)\ \mathrm{H_3O^+}]{(1)\ \mathrm{I_2/NaOH}} \underset{\mathbf{E}}{\mathrm{HOOC(CH_2)_7COOH}}$$

20.34

$$\mathrm{CH_2{=}\underset{CH_3}{\underset{|}{C}}{-}CH{=}CH_2} + \mathrm{HBr} \longrightarrow \underset{\mathbf{F}}{\mathrm{CH_3\underset{CH_3}{\underset{|}{C}}{=}CHCH_2Br}}$$

$$\xrightarrow{Na^{+}\ ^{-}:CH(COCH_3)(CO_2C_2H_5)} CH_3C(CH_3){=}CHCH_2CH(CO_2C_2H_5)CCH_3(=O) \xrightarrow[(2)\ H_3O^{+},\ (3)\ heat]{(1)\ dil.\ NaOH}$$

G

$$CH_3C(CH_3){=}CHCH_2CH_2CCH_3(=O) \xrightarrow[(2)\ H_3O^{+}]{(1)\ LiC{\equiv}CH}$$

H

$$CH_3C(CH_3){=}CHCH_2CH_2C(OH)(CH_3)C{\equiv}CH \xrightarrow[\text{Lindlar's catalyst}]{H_2} \text{Linalool}$$

I

20.35

$$(C_2H_5OC(=O))_2HC:^{-}\ Na^{+} + BrCH_2CH_2CH_2Br \longrightarrow \left[(C_2H_5OC(=O))_2CH{-}CH_2CH_2CH_2Br\right]$$

$$(C_{10}H_{17}BrO_4)$$

$$\xrightarrow{NaOC_2H_5} (C_2H_5OC(=O))_2C\langle(CH_2)_3\rangle \xrightarrow[(2)\ H_2O]{(1)\ LiAlH_4}$$

$$(C_{10}H_{16}O_4)$$

$$\text{cyclobutane-1,1-}(CH_2OH)_2 \xrightarrow{HBr} \text{cyclobutane-1,1-}(CH_2Br)_2$$

$$(C_6H_{12}O_2) \qquad (C_6H_{10}Br_2)$$

$$\xrightarrow[2NaOC_2H_5]{CH_2(CO_2C_2H_5)_2} \text{spiro[3.3]heptane-2,2-}(CO_2C_2H_5)_2$$

$$(C_{13}H_{20}O_4)$$

$$\xrightarrow[(2)\ H^{+}]{(1)\ OH^{-},\ H_2O} \text{spiro[3.3]heptane-2,2-}(COOH)_2 \xrightarrow{heat}$$

$$(C_9H_{12}O_4)$$

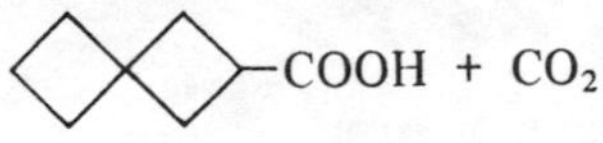

—COOH + CO_2

F

($C_8H_{12}O_2$)

20.36

(a) $ClCH_2COOC_2H_5 + C_2H_5O^- \rightleftharpoons Cl-\ddot{C}^-HCOOC_2H_5 + C_2H_5OH$

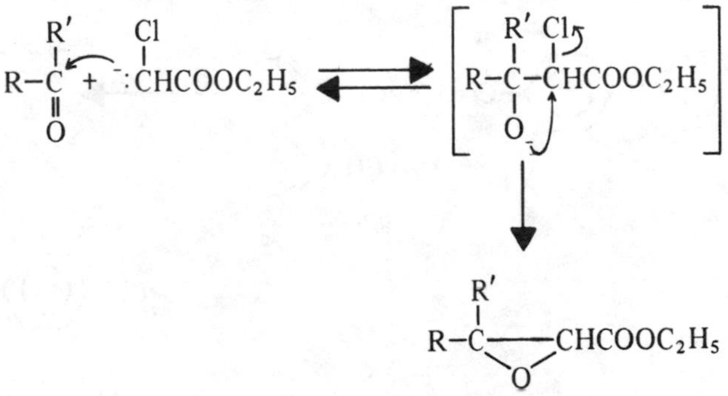

R'–C(=O)–R + $^-$:CH(Cl)COOC$_2$H$_5$ ⇌ [R–C(R')(O$^-$)–CH(Cl)COOC$_2$H$_5$] → R–C(R')–CHCOOC$_2$H$_5$ (epoxide, –O–)

(b) Decarboxylation of the epoxy acid gives an enol anion which, on protonation, gives an aldehyde.

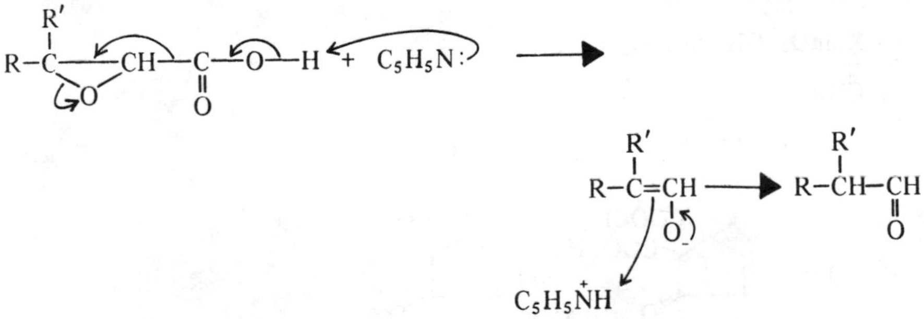

R–C(R')–CH–C(=O)–O–H (epoxide, –O–) + C_5H_5N: → R–C(R')=CH–O$^-$ + $C_5H_5\overset{+}{N}H$ → R–CH(R')–CH=O

(c)

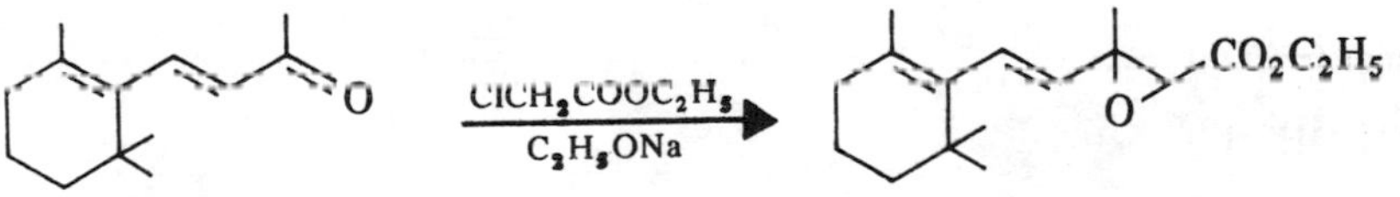

β-Ionone $\xrightarrow[C_2H_5ONa]{ClCH_2COOC_2H_5}$ epoxy ester (–$CO_2C_2H_5$)

$\xrightarrow[(2)\ H_3O^+]{(1)\ OH^-,\ H_2O}$ epoxy acid (–COOH) $\xrightarrow[\text{heat},\ -CO_2]{C_5H_5N}$

[enolate (=CH–O$^-$) + $C_5H_5\overset{+}{N}H$] → aldehyde (–CH=O)

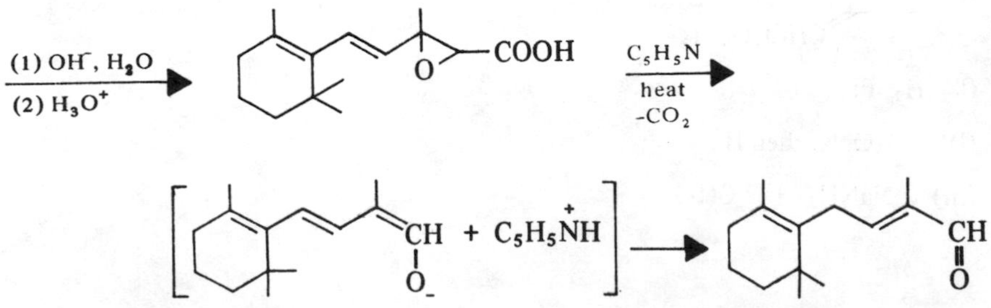

20.37

(a) 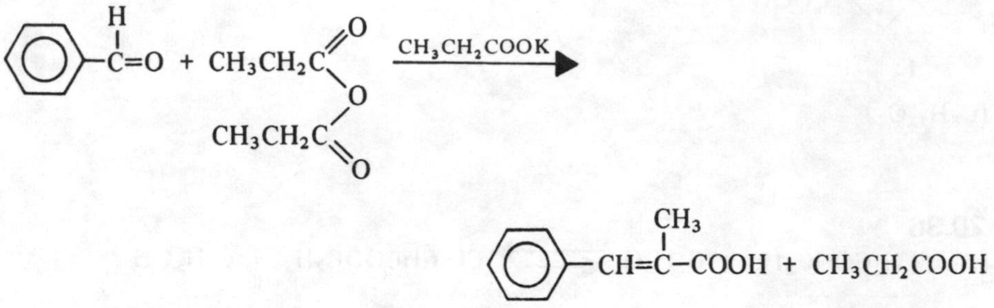 $C_6H_5CH{=}O + (CH_3CH_2CO)_2O \xrightarrow{CH_3CH_2COOK} C_6H_5CH{=}C(CH_3){-}COOH + CH_3CH_2COOH$

(b) 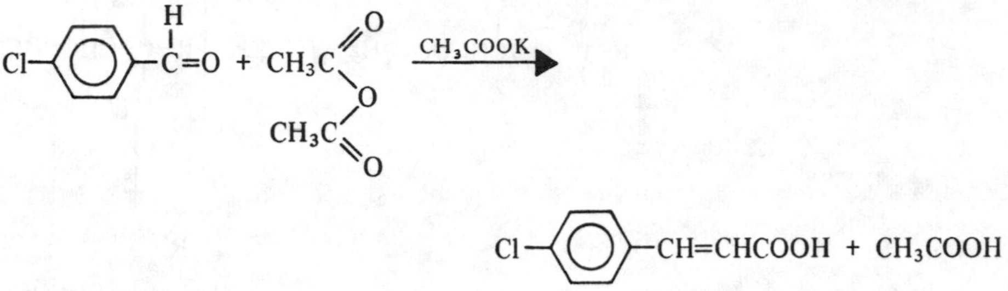 $Cl{-}C_6H_4{-}CH{=}O + (CH_3CO)_2O \xrightarrow{CH_3COOK} Cl{-}C_6H_4{-}CH{=}CHCOOH + CH_3COOH$

20.38

(a) $CH_2{=}C(CH_3){-}COOCH_3$

(b) $KMnO_4$, OH^-, then H_3O^+

(c) CH_3OH, H^+

(d) CH_3ONa, then H^+

(e) and (f) 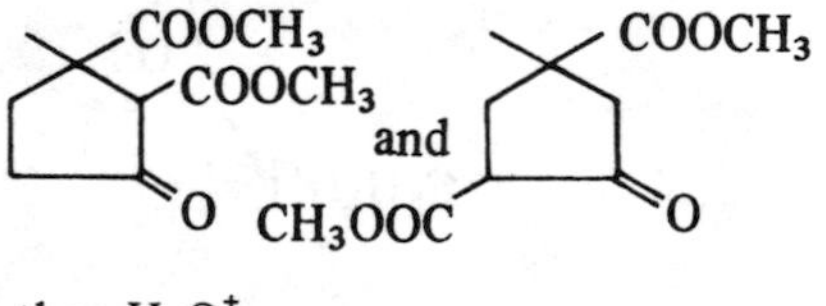 1-methyl-1,2-bis($COOCH_3$)cyclopentan-3-one [ring: C(CH₃)(COOCH₃)–CH(COOCH₃)–C(=O)] and 1-methyl-1,3-bis($COOCH_3$)cyclopentan-4-one [ring: C(CH₃)(COOCH₃)–CH₂–C(=O)–CH(COOCH₃)]

(g) OH^-, H_2O, then H_3O^+

(h) heat ($-CO_2$)

(i) CH_3OH, H^+

(j) 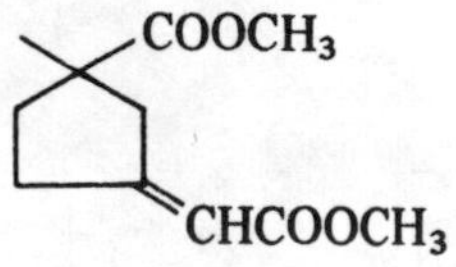 1-methyl-1-($COOCH_3$)cyclopentane with $=CHCOOCH_3$ at C-3

(k) H_2, Pt

(l) CH_3ONa, then H^+

(m) 2 $NaNH_2$ + 2 CH_3I

20.39

(a) A Grignard synthesis.

(b) An organozinc compound, CH_3OOCCH_2ZnBr.

(c) The organozinc compound adds to the carbonyl group of a ketone (or aldehyde) but it does not add to the carbonyl group of an ester. Hence the Reformatsky reagent does not react with itself (or with other ester groups).

(d) Aldehydes and ketones are more reactive toward nucleophilic attack than esters are because esters are stabilized by the type of resonance shown below.

$$R-\overset{\overset{\displaystyle O}{\|}}{C}-OR \longleftrightarrow R-\overset{\overset{\displaystyle O^-}{|}}{C}=O^+R$$

Organozinc compounds are less reactive than organomagnesium compounds because zinc is less electropositive. Hence organozinc compounds (Reformatsky reagents) add to the more reactive carbonyl groups of aldehydes and ketones, but unlike organomagnesium compounds (Grignard reagents) they do not react at the less reactive carbonyl groups of esters.

(e) $$C_6H_5CHO + CH_3\underset{\underset{\displaystyle Br}{|}}{C}HCOOCH_3 \xrightarrow[(2)\ H_3O^+]{(1)\ Zn,\ ether}$$

$$C_6H_5\underset{\underset{\displaystyle OH}{|}}{C}H\overset{\overset{\displaystyle CH_3}{|}}{C}HCOOCH_3 \xrightarrow[heat]{H^+} C_6H_5CH=\overset{\overset{\displaystyle CH_3}{|}}{C}COOCH_3 \xrightarrow[Pt]{H_2}$$

$$C_6H_5CH_2\overset{\overset{\displaystyle CH_3}{|}}{C}HCOOCH_3$$

20.40

If we look at the molecule as consisting of the following pieces,

[Structure: cyclohexanone ring with O, two C_6H_5 groups and COOH, pieces circled]

we can begin to see how it might be constructed from acetone, malonic ester, and two moles of benzaldehyde.

We can begin by carrying out a double Claisen-Schmidt condensation using acetone, and two moles of benzaldehyde.

$$2\ C_6H_5\overset{\overset{\displaystyle H}{|}}{C}=O + CH_3-\overset{\overset{\displaystyle O}{\|}}{C}-CH_3 \xrightarrow{OH^-} C_6H_5CH=CH-\overset{\overset{\displaystyle O}{\|}}{C}-CH=CHC_6H_5$$

Then we carry out a double Michael addition, followed by hydrolysis and decarboxylation.

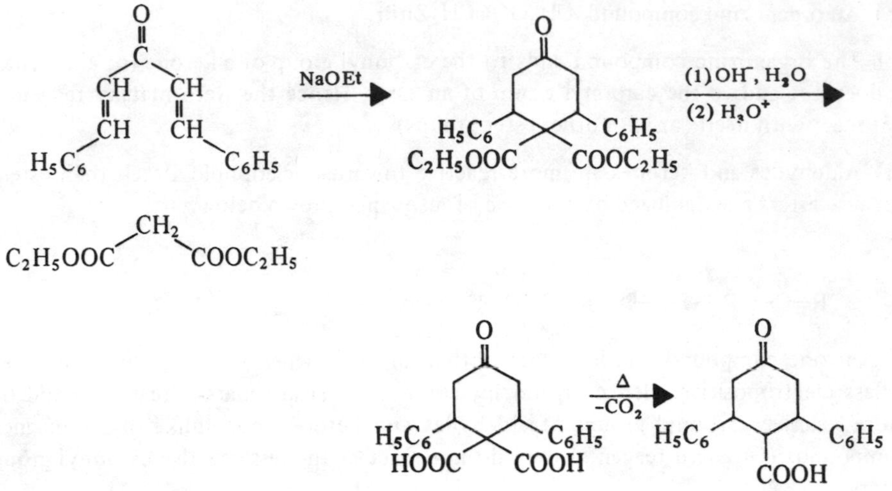

O, C, CH, CH, CH, CH, H_5C_6, C_6H_5 ; CH_2, C_2H_5OOC, $COOC_2H_5$ — NaOEt → O, H_5C_6, C_6H_5, C_2H_5OOC, $COOC_2H_5$ — (1) OH^-, H_2O (2) H_3O^+ →

O, H_5C_6, C_6H_5, HOOC, COOH — Δ, $-CO_2$ → O, H_5C_6, C_6H_5, COOH

20.41

(a) CH_3, C=O, CH_2, CH_2, C=O, CH_3 + $(NH_4)_2CO_3$ —100°→ CH_3, N–H, CH_3 + $2H_2O$ + NH_4HCO_3

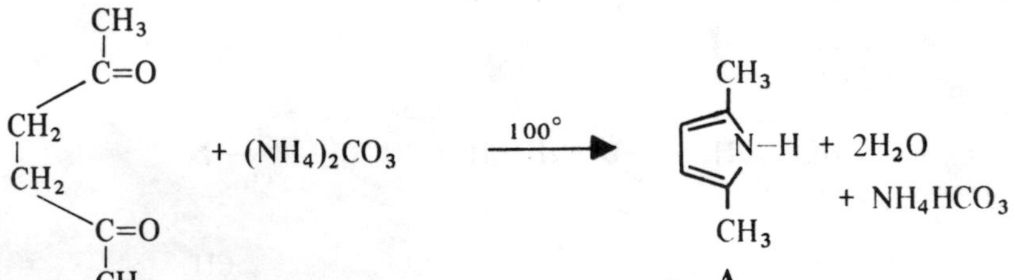

A

(b) CH_3, C, O, CH_2, NH_2 ; CH_3, C=O, CH_3 —base→ CH_3, N–H, CH_3 + $2H_2O$

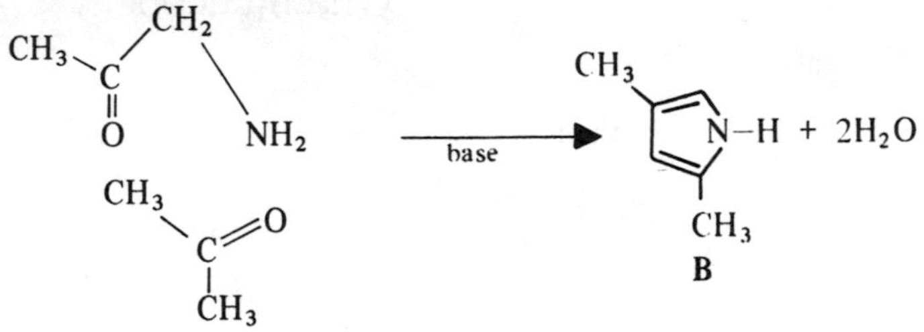

B

(c) $CH(OCH_3)_2$, CH_2, $CH(OCH_3)_2$ + NH_2–NH–CH_3 —H^+, H_2O→ N, N, CH_3 + $4CH_3OH$

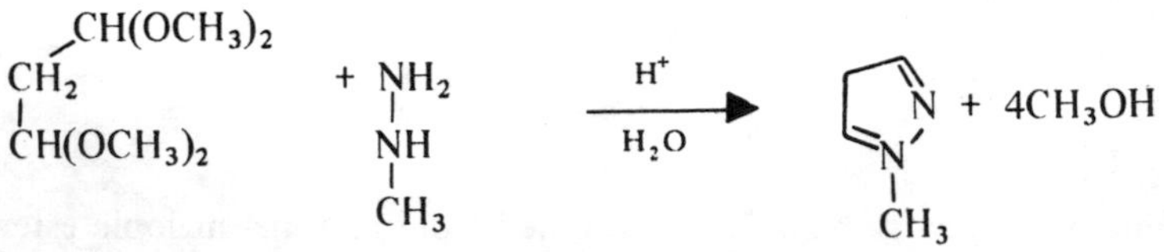

C

(d) CH_3, C, O, CH_2, CH_2, O, C, CH_3 + H_2N–H_2N → CH_3, C, N, CH_2, CH_2, N, C, CH_3 —O_2→ CH_3, N, N, CH_3

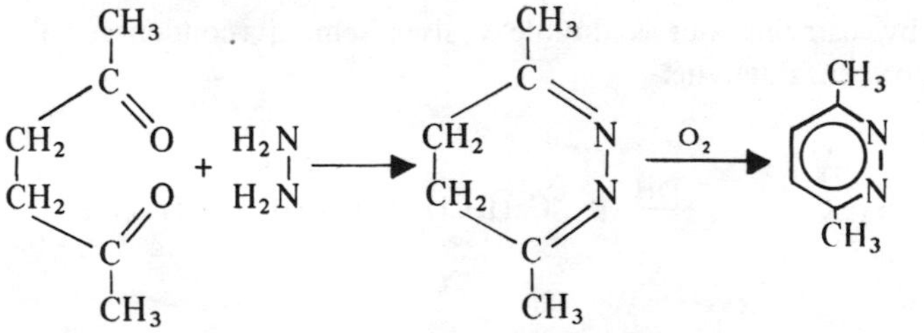

D **E**

(e) Aniline + $CH_3-\overset{O}{\overset{\|}{C}}-CH=CH_2$ $\xrightarrow[FeCl_3]{ZnCl_2}$ [$C_6H_5NH-CH_2-CH_2-\overset{O}{\overset{\|}{C}}-CH_3$] ⟶ (4-methylquinoline) + H_2O

F

(f) (3-pyridyl)$\underset{I}{CH}-CH_2-CH_2-CH_2-\underset{CH_3}{NH}$ $\xrightarrow{heat}$ (N-methyl-N-H pyrrolidinium iodide) I^- $\xrightarrow{OH^-}$

G Nicotine $\xrightarrow[(2)\ H^+]{(1)\ KMnO_4,\ OH^-}$ **H** Nicotinic acid (3-pyridinecarboxylic acid, COOH)

20.42

$CH_3\overset{O}{\overset{\|}{C}}CH_2\overset{O}{\overset{\|}{C}}OC_2H_5 + C_6H_5NHNH_2 \xrightarrow{heat} CH_3\underset{N-NH-C_6H_5}{\underset{\|}{C}}-CH_2\overset{O}{\overset{\|}{C}}OC_2H_5$

$\xrightarrow{\text{Further heating}}$ (pyrazolone ring: CH_3C-CH_2, N, $C=O$, $N-C_6H_5$) $\xrightarrow{CH_3OSO_2OCH_3}$ (N-methylated pyrazolonium, CH_3, C_6H_5) $^-OSO_2OCH_3$ $\xrightarrow{NaOH}$ Antipyrine (CH_3, CH_3-N, $N-C_6H_5$, $C=O$)

TWENTY ONE
Lipids

21.1

(a) There are two sets of enantiomers, giving a total of four stereoisomers

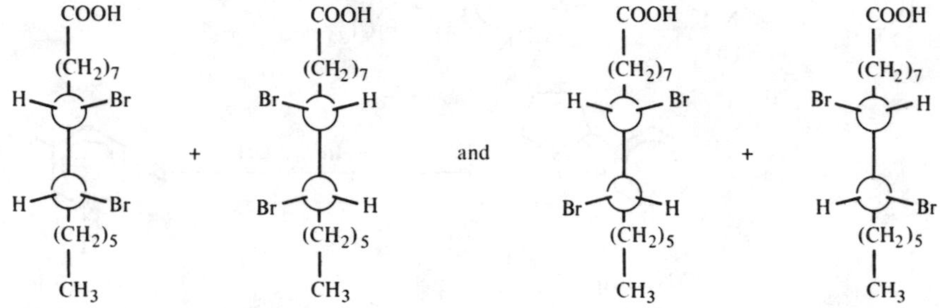

COOH
(CH2)7
H Br
H Br
(CH2)5
CH3

+

COOH
(CH2)7
Br H
Br H
(CH2)5
CH3

and

COOH
(CH2)7
H Br
Br H
(CH2)5
CH3

+

COOH
(CH2)7
Br H
H Br
(CH2)5
CH3

(b) $CH_3(CH_2)_5(H)C{=}C(H)(CH_2)_7COOH + Br_2 \longrightarrow$

$CH_3(CH_2)_5$ H (CH2)7COOH H Br+ Br− (a) (b)

(a) (b)

Br (CH2)7COOH H CH3(CH2)5 H Br

|||

COOH
(CH2)7
H Br
Br H
(CH2)5
CH3

CH3(CH2)5 H Br Br (CH2)7COOH H

|||

COOH
(CH2)7
Br H
H Br
(CH2)5
CH3

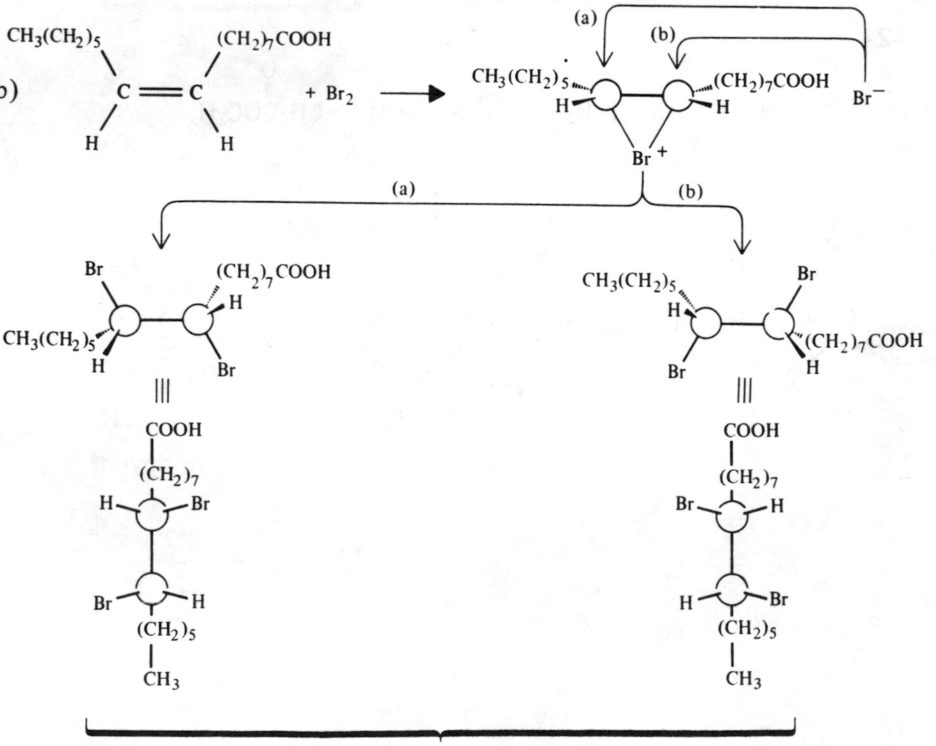

(±)-*threo*-9, 10-dibromohexadecanoic acids

Formation of a bromonium ion at the other face of palmitoleic acid gives a result such that the *threo*-enantiomers are obtained as a racemic modification.

21.2

$HOOC(CH_2)_7$(H)C=C(H)$(CH_2)_7CH_3$ —[$KMnO_4$ or OsO_4; syn hydroxyation]→ COOH, $(CH_2)_7$, H–C–OH, H–C–OH, $(CH_2)_7$, CH_3 + COOH, $(CH_2)_7$, HO–C–H, HO–C–H, $(CH_2)_7$, CH_3

(±)–*erythro*–9, 10–dihydroxyocta–decanoic acids

$HOOC(CH_2)_7$(H)C=C(H)$(CH_2)_7CH_3$ —[(1) HCOOH (O above C) (2) H_3O^+; anti hydroxylation]→ COOH, $(CH_2)_7$, H–C–OH, HO–C–H, $(CH_2)_7$, CH_3 + COOH, $(CH_2)_7$, HO–C–H, H–C–OH, $(CH_2)_7$, CH_3

(±)–*threo*–9, 10–dihydroxyocta–decanoic acids

The designations *erythro*- and *threo*- come from the names of the sugars called *erythrose* and *threose*.

21.3

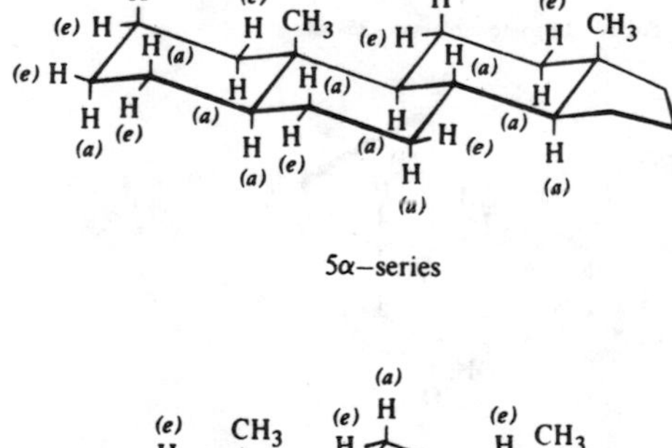

5α–series

5β–series

21.4

(a)

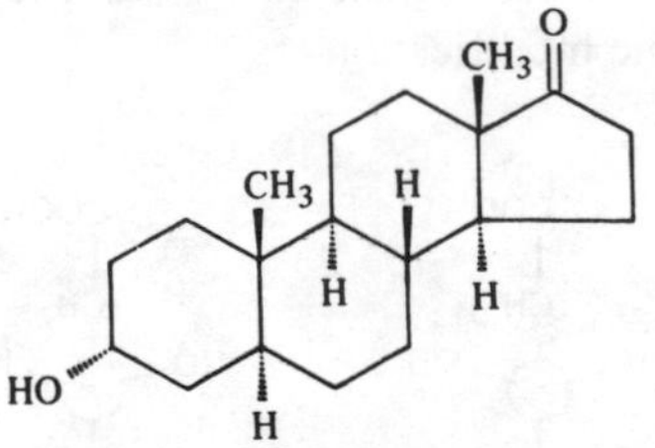

androstan–3α– ol–17–one
(androsterone)

(b)

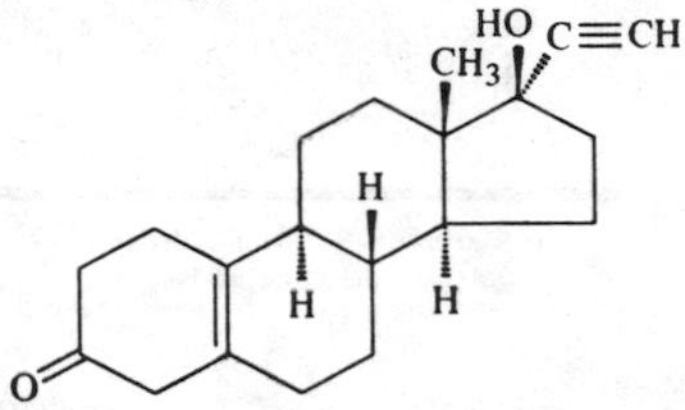

17α–ethynyl–17β–hydroxy–5(10)–estren–3–one
(norethynodrel)

21.5

Estrone and estradiol are ***phenols*** and thus are soluble in aqueous sodium hydroxide. Extraction with aqueous sodium hydroxide separates the estrogens from the androgens.

21.6

(a)

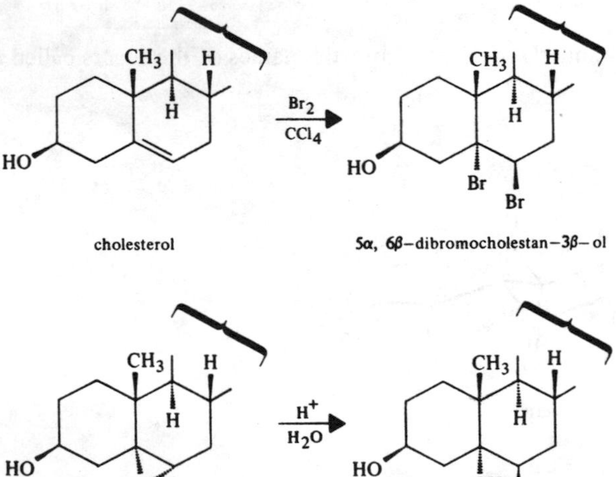

$\xrightarrow[CCl_4]{Br_2}$

cholesterol

5α, 6β–dibromocholestan–3β– ol

(b)

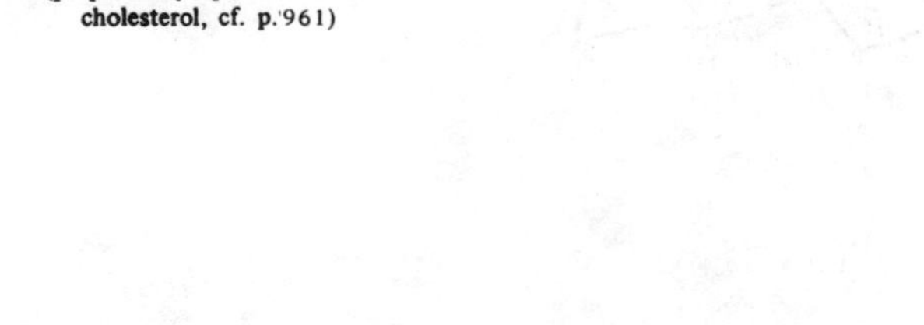

$\xrightarrow[H_2O]{H^+}$

5α, 6α–oxidocholestan–3β–ol
(prepared by epoxidation of
cholesterol, cf. p. 961)

cholestan–3β, 5α, 6β– triol

(c)

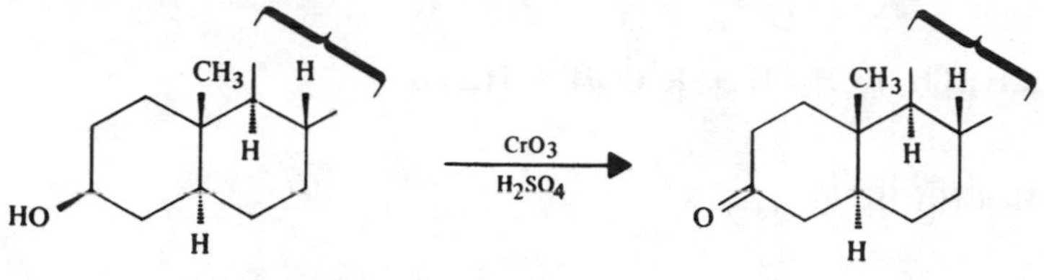

$\xrightarrow[H_2SO_4]{CrO_3}$

5α–cholestan–3β–ol
(prepared by hydrogenation of cholesterol, cf. p. 961)

5α–cholestan–3–one

(d)

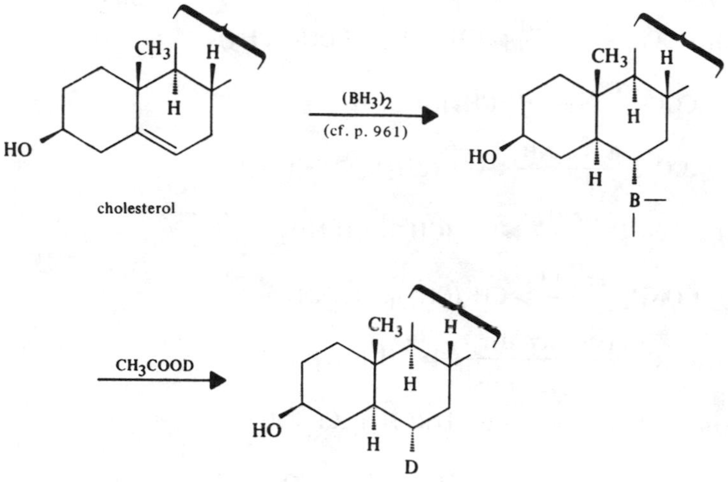

$\xrightarrow[\text{(cf. p. 961)}]{(BH_3)_2}$

cholesterol

$\xrightarrow{CH_3COOD}$

6α–deuterio–5α–cholestan–3β–ol

(e)

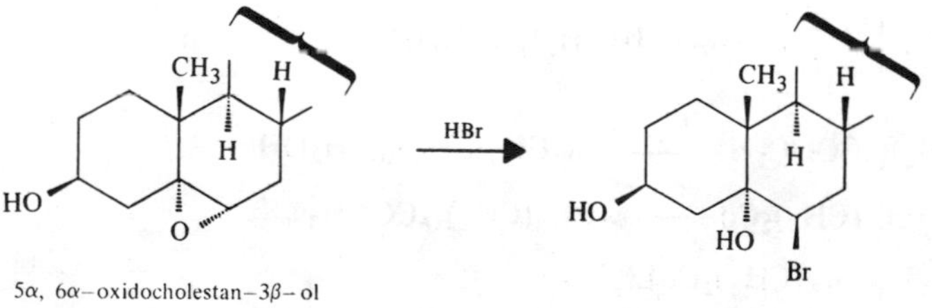

$\xrightarrow{HBr}$

5α, 6α–oxidocholestan–3β–ol

6β–bromocholestan–3β, 5α–diol

21.7

(a) $\begin{matrix} CH_2OH \\ | \\ CHOH \\ | \\ CH_2OH \end{matrix} + R\overset{O}{\overset{||}{C}}OH + R'\overset{O}{\overset{||}{C}}OH + H_3PO_4 + HOCH_2CH_2\overset{+}{N}(CH_3)_3\ X^-$

(b) $\begin{matrix} CH_2OH \\ | \\ CHOH \\ | \\ CH_2OH \end{matrix} + R\overset{O}{\overset{||}{C}}OH + R'\overset{O}{\overset{||}{C}}OH + H_3PO_4 + HOCH_2CH_2NH_2$

$$\text{(c)}\ \begin{array}{l}CH_2OH\\ |\\ CHOH\\ |\\ CH_2OH\end{array} + CH_3(CH_2)_nCH_2\overset{O}{\overset{\|}{C}}H + R^1\overset{O}{\overset{\|}{C}}OH + H_3PO_4$$

$$+\ HOCH_2CH_2\overset{+}{N}(CH_3)_3\ X^-$$

21.8

$$\text{(a)}\ CH_3(CH_2)_{16}COOH + C_2H_5OH \overset{H^+}{\rightleftharpoons} CH_3(CH_2)_{16}COOC_2H_5 + H_2O$$

$$CH_3(CH_2)_{16}COOH \xrightarrow{SOCl_2} CH_3(CH_2)_{16}COCl \xrightarrow{C_2H_5OH} CH_3(CH_2)_{16}COOC_2H_5$$

$$\text{(b)}\ CH_3(CH_2)_{16}COCl \xrightarrow{(CH_3)_3COH} CH_3(CH_2)_{16}COOC(CH_3)_3$$

$$\text{(c)}\ CH_3(CH_2)_{16}COCl \xrightarrow{NH_3} CH_3(CH_2)_{16}CONH_2$$

$$\text{(d)}\ CH_3(CH_2)_{16}COCl \xrightarrow{(CH_3)_2NH} CH_3(CH_2)_{16}CON(CH_3)_2$$

$$\text{(e)}\ CH_3(CH_2)_{16}CONH_2 \xrightarrow{LiAlH_4} CH_3(CH_2)_{16}CH_2NH_2$$

$$\text{(f)}\ CH_3(CH_2)_{16}CONH_2 \xrightarrow{Br_2,OH^-} CH_3(CH_2)_{15}CH_2NH_2$$

$$\text{(g)}\ CH_3(CH_2)_{16}COCl \xrightarrow{LiAlH[OC(CH_3)_3]_3} CH_3(CH_2)_{16}CHO$$

$$\text{(h)}\ CH_3(CH_2)_{16}COOC_2H_5 \xrightarrow{H_2,\ Ni} CH_3(CH_2)_{16}CH_2OH$$

$$CH_3(CH_2)_{16}CH_2OH + CH_3(CH_2)_{16}COCl \longrightarrow CH_3(CH_2)_{16}COOCH_2(CH_2)_{16}CH_3$$

$$\text{(i)}\ CH_3(CH_2)_{16}COOH \xrightarrow[(2)\ H_2O]{(1)\ LiAlH_4} CH_3(CH_2)_{16}CH_2OH$$

$$CH_3(CH_2)_{16}COOC_2H_5 \xrightarrow{H_2,\ Ni} CH_3(CH_2)_{16}CH_2OH$$

$$\text{(j)}\ CH_3(CH_2)_{16}COCl + (CH_3)_2Cd \longrightarrow CH_3(CH_2)_{16}COCH_3$$

$$\text{or}\ (CH_3)_2CuLi$$

$$\text{(k)}\ CH_3(CH_2)_{16}CH_2OH \xrightarrow{PBr_3} CH_3(CH_2)_{16}CH_2Br$$

$$\text{(l)}\ CH_3(CH_2)_{16}CH_2Br \xrightarrow[(2)\ H^+,\ H_2O,\ heat]{(1)\ NaCN} CH_3(CH_2)_{16}CH_2COOH$$

21.9

$$\text{(a)}\ CH_3(CH_2)_{11}CH_2COOH \xrightarrow{Br_2,\ P} CH_3(CH_2)_{11}\underset{Br}{\underset{|}{C}}HCOOH$$

$$\text{(b)}\ CH_3(CH_2)_{11}\underset{Br}{\underset{|}{C}}HCOOH \xrightarrow[(2)\ H^+]{(1)\ OH^-,\ heat} CH_3(CH_2)_{11}\underset{OH}{\underset{|}{C}}HCOOH$$

$$\text{(c)}\ CH_3(CH_2)_{11}\underset{Br}{\underset{|}{C}}HCOOH \xrightarrow[(2)\ H^+]{NaCN} CH_3(CH_2)_{11}\underset{CN}{\underset{|}{C}}HCOOH$$

(d) $CH_3(CH_2)_{11}CH(Br)COOH \xrightarrow[(2)\ H^+]{NH_3\,(excess)} CH_3(CH_2)_{11}CH(NH_2)COOH$ or $CH_3(CH_2)_{11}CH(\overset{+}{N}H_3)COO^-$

(cf. p. 952)

21.10

(a) $CH_3(CH_2)_5CH{=}CH(CH_2)_7COOH \xrightarrow{I_2} CH_3(CH_2)_5CHICHI(CH_2)_7COOH$

(b) $CH_3(CH_2)_5CH{=}CH(CH_2)_7COOH \xrightarrow{H_2,\ Ni} CH_3(CH_2)_{14}COOH$

(c) $CH_3(CH_2)_5CH{=}CH(CH_2)_7COOH \xrightarrow{KMnO_4} CH_3(CH_2)_5CHOHCHOH(CH_2)_7COOH$

(d) $CH_3(CH_2)_5CH{=}CH(CH_2)_7COOH \xrightarrow{HCl} CH_3(CH_2)_7CH_2CHCl(CH_2)_7COOH$
$+$
$CH_3(CH_2)_7CHClCH_2(CH_2)_7COOH$

21.11

Elaidic acid is *trans*-9-octadecenoic acid:

$$\begin{array}{c} CH_3(CH_2)_7 \quad\quad H \\ C{=}C \\ H \quad\quad (CH_2)_7COOH \end{array}$$

It is formed by the isomerization of oleic acid.

21.12

(a)
$$\begin{array}{c} CH_3(CH_2)_9 \quad\quad (CH_2)_7COOH \\ C{=}C \\ H \quad\quad H \end{array} \quad \text{and} \quad \begin{array}{c} CH_3(CH_2)_9 \quad\quad H \\ C{=}C \\ H \quad\quad (CH_2)_7COOH \end{array}$$

(b) Infrared spectroscopy

(c) A peak in the 675-730 cm^{-1} region would indicate that the double bond is *cis*; a peak in the 960-975 cm^{-1} region would indicate that it is *trans*.

21.13

$CH_3(CH_2)_5C{\equiv}CH + NaNH_2 \xrightarrow[NH_3]{} CH_3(CH_2)_5C{\equiv}CNa$ (**A**)

$\xrightarrow{ICH_2(CH_2)_7CH_2Cl} CH_3(CH_2)_5C{\equiv}CCH_2(CH_2)_7CH_2Cl$ (**B**) $\xrightarrow{NaCN}$

$CH_3(CH_2)_5C{\equiv}CCH_2(CH_2)_7CH_2CN$ (**C**) $\xrightarrow{KOH,\ H_2O} CH_3(CH_2)_5C{\equiv}CCH_2(CH_2)_7CH_2COOK$ (**D**)

$\xrightarrow{H_3O^+} CH_3(CH_2)_5C{\equiv}CCH_2(CH_2)_7CH_2COOH$ (**E**) $\xrightarrow{H_2,\ Pd\text{-}BaSO_4}$

$$\begin{array}{c} CH_3(CH_2)_5 \quad\quad CH_2(CH_2)_7CH_2COOH \\ C{=}C \\ H \quad\quad H \end{array}$$

Vaccenic acid

21.14

$$FCH_2(CH_2)_6CH_2Br + HC{\equiv}CNa \longrightarrow \underset{\textbf{F}}{FCH_2(CH_2)_6CH_2C{\equiv}CH}$$

$$\xrightarrow[(2)\ I(CH_2)_7Cl]{(1)\ NaNH_2} \underset{\textbf{G}}{FCH_2(CH_2)_6CH_2C{\equiv}C(CH_2)_7Cl} \xrightarrow[DMSO]{NaCN}$$

$$\underset{\textbf{H}}{FCH_2(CH_2)_6CH_2C{\equiv}C(CH_2)_7CN} \xrightarrow[(2)\ H^+]{(1)\ KOH} \underset{\textbf{I}}{FCH_2(CH_2)_6CH_2C{\equiv}C(CH_2)_7COOH}$$

$$\xrightarrow[Pd\text{-}BaSO_4]{H_2} FCH_2(CH_2)_6CH_2(H)C{=}C(H)(CH_2)_7COOH$$

21.15

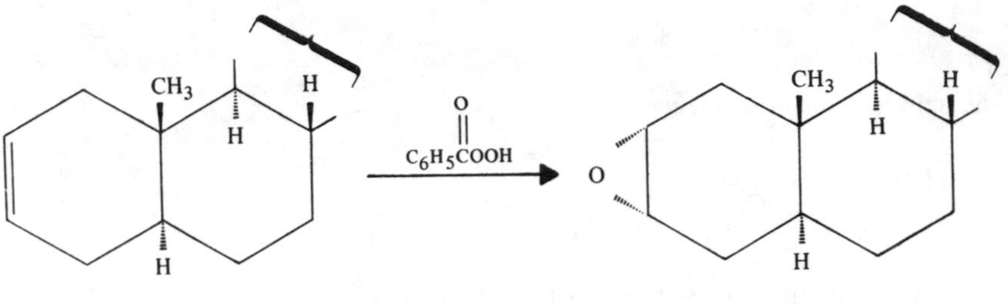

5α–cholest–2–ene $\xrightarrow{C_6H_5COOH\ (\text{peroxy, } C_6H_5CO_3H)}$ **A** $\xrightarrow{HBr}$ **B**

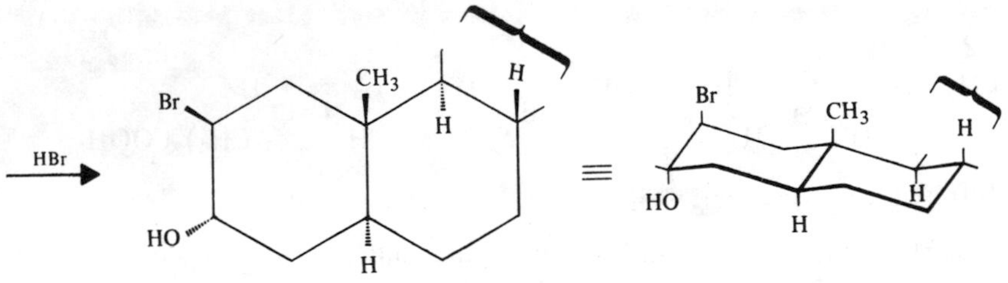

Here we find that epoxidation takes place at the less hindered α face (cf. p. 961). Ring opening by HBr takes place in an *anti* fashion to give a product with diaxial substituents.

21.16

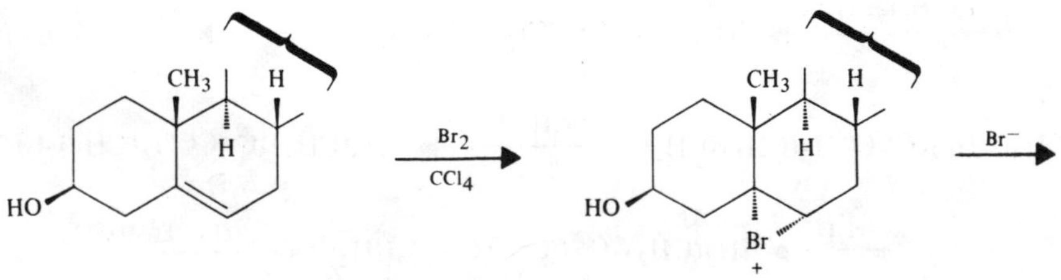

cholesterol $\xrightarrow[CCl_4]{Br_2}$ 5α, 6α–bromonium ion $\xrightarrow{Br^-}$

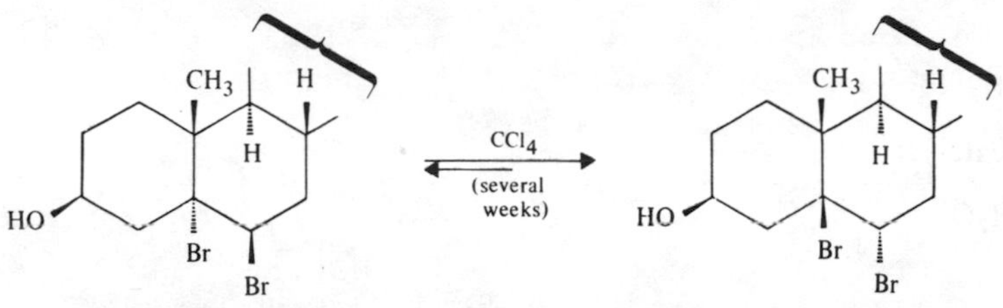

5α, 6β–dibromocholestan–3β–ol ⇌ (CCl_4, several weeks) 5β, 6α–dibromocholestan–3β–ol

Here formation of the bromonium ion takes place preferentially at the α face. Ring opening takes place in an *anti* manner (with a bromide ion attacking the 6-position from above) to give, initially, the 5α,6β-dibromo compound. The 5α,6β-dibromocholestan-3β-ol, however is a *diaxial* dibromide and is, therefore, unstable. It isomerizes to the 5β, 6α-dibromocholestan-3β-ol (below)–a compound in which the bromine substituents are both equatorial.

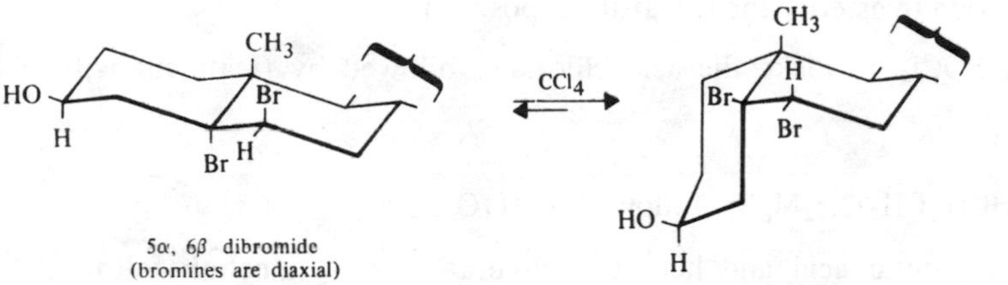

5α, 6β dibromide (bromines are diaxial) ⇌ (CCl_4) 5β, 6α–dibromide (bromines are diequatorial)

This isomerization does not result from a simple "flipping" of the cylohexane rings; it requires an inversion of configuration at carbons 5 and 6. One mechanism that has been proposed for the isomerization involves the formation of a "bromonium-bromide" ion pair:

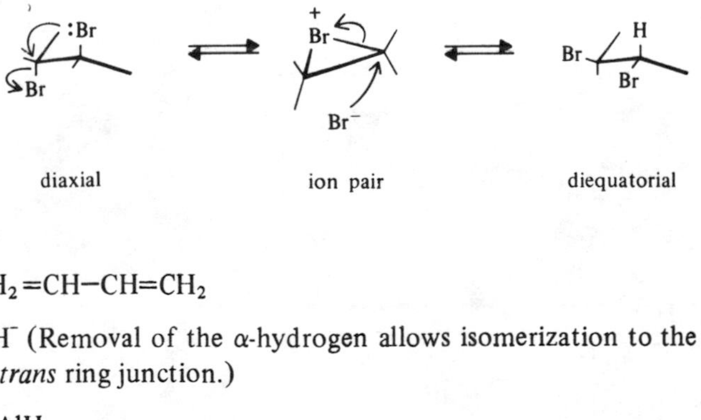

diaxial ⇌ ion pair ⇌ diequatorial

21.17

(a) $CH_2{=}CH{-}CH{=}CH_2$

(b) OH^- (Removal of the α-hydrogen allows isomerization to the more stable compound with a *trans* ring junction.)

(c) $LiAlH_4$

(d) H_3O^+ and heat. (Hydrolysis of the enol ether is followed by dehydration of one alcohol group.)

(e) $HCOOC_2H_5$, C_2H_5ONa

(f) OsO_4

(g) $CH_3\overset{\overset{O}{\|}}{C}CH_3$, H^+

(h) H_2, Pd catalyst

(i) H_3O^+, H_2O

(j) HIO_4

(k) Base and heat (This is an aldol condensation.)

(l) and (m) Na_2CrO_4, CH_3COOH to oxidize the aldehyde to an acid, followed by esterification.

(n) H_2 and Pt (Hydrogen addition takes place from the less hindered α-face of the molecule.)

(o), (p), (q) $NaBH_4$ to reduce the keto group; OH^-, H_2O to hydrolyze the ester; and acetic anhydride to esterify the OH at the 3-position.

(r) and (s) $SOCl_2$ to make the acid chloride, followed by treatment with $(CH_3)_2Cd$.

(t) $CH_3\overset{\overset{CH_3}{|}}{C}HCH_2CH_2CH_2MgBr$, followed by H_3O^+.

(u), (v), (w) acetic acid and heat to dehydrate the 3° alcohol; followed by acetic anhydride to acetylate the 2° alcohol; followed by H_2, Pt to hydrogenate the double bond.

21.18

(a) $CH_3(CH_2)_4\overset{\overset{O}{\|}}{C}H$

(b) C_4H_9Li

(c) 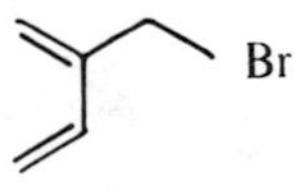 [structure: 2-methylene-3-butenyl bromide] Br

(d) $NC(CH_2)_6$ [structure: trans alkene with H, H, NO_2]

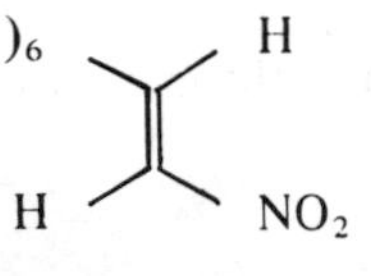

(e) Michael addition using a basic catalyst.

TWENTY TWO

Amino Acids and Proteins

22.1

(a) $HOOCCH_2CH_2\underset{|}{C}HCOOH$ with $^+NH_3$ on the α-carbon

$$HOOCCH_2CH_2\underset{^+NH_3}{\underset{|}{C}}HCOOH$$

(b)

$$^-OOCCH_2CH_2\underset{NH_2}{\underset{|}{C}}HCOO^-$$

(c) $HOOCCH_2CH_2\underset{^+NH_3}{\underset{|}{C}}HCOO^-$ predominates at the isoelectric point rather than $^-OOCCH_2-CH_2\underset{^+NH_3}{\underset{|}{C}}HCOOH$ because of the acid-strengthening inductive effect of the α-amino group.

(d) Since glutamic acid is a dicarboxylic acid, acid must be added (i.e., the pH must be made lower) to suppress the ionization of the second carboxyl group and thus achieve the isoelectric point. Glutamine, with only one carboxyl group, is similar to glycine or phenylalanine and has its isoelectric point at a higher pH.

22.2

The conjugate acid is highly stabilized by resonance.

$$R-\ddot{N}H-\overset{\overset{\ddot{N}H}{\|}}{C}-\ddot{N}H_2 \xrightarrow{H^+} R-\ddot{N}H-\overset{\overset{NH_2^+}{\|}}{C}-\ddot{N}H_2 \longleftrightarrow R-\ddot{N}H-\overset{\overset{\ddot{N}H_2}{|}}{C}=NH_2^+$$

$$\longleftrightarrow R-\overset{+}{N}H=\overset{\overset{NH_2}{|}}{C}-NH_2$$

22.3

(a)

$$C_6H_5CONHCH(CO_2C_2H_5)_2 \xrightarrow[C_6H_5CH_2Br]{NaOCH_2CH_3}$$

$$C_6H_5CONH\underset{\underset{C_6H_5}{|}}{\underset{CH_2}{\underset{|}{C}}}(CO_2C_2H_5)_2 \xrightarrow[heat]{HBr} \left[H_3\overset{+}{N}-\overset{\overset{COOH}{|}}{\underset{\underset{\underset{C_6H_5}{|}}{CH_2}}{\underset{|}{C}}}-COO^- \right] \xrightarrow{-CO_2} C_6H_5CH_2\underset{^+NH_3}{\underset{|}{C}}HCOO^-$$

Phenylalanine

(b) $C_6H_5CONHCH(CO_2C_2H_5)_2 \xrightarrow[BrCH_2CO_2C_2H_5]{NaOCH_2CH_3}$

$C_6H_5CONHC(CO_2C_2H_5)_2(CH_2CO_2C_2H_5) \xrightarrow{HBr} [H_3\overset{+}{N}-C(COOH)(COO^-)-CH_2COOH]$

$\xrightarrow{-CO_2} HOOCCH_2CH(\overset{+}{N}H_3)COO^-$

Aspartic acid

(c) $C_6H_5CONHCH(CO_2C_2H_5)_2 \xrightarrow[(CH_3)_2CHBr]{NaOCH_2CH_3}$

$C_6H_5CONHC(CO_2C_2H_5)_2(CHCH_3CH_3) \xrightarrow[heat]{HBr} [H_3\overset{+}{N}-C(COOH)(COO^-)-CHCH_3CH_3]$

$\xrightarrow{-CO_2} CH_3CH(CH_3)-CH(\overset{+}{N}H_3)COO^-$

Valine

22.4

(a) Potassium phthalimide (NK) $+ BrCH(CO_2C_2H_5)_2 \longrightarrow$

Phthalimido-N$-CH(CO_2C_2H_5)_2 \xrightarrow[(CH_3)_2CHCH_2Br]{NaOCH_2CH_3}$

Phthalimido-N$-C(CO_2C_2H_5)_2-CH_2CH(CH_3)_2 \xrightarrow[heat]{NaOH}$ $C_6H_4(COO^-)-C(=O)-NHC(COO^-)_2CH_2CH(CH_3)_2$

$\xrightarrow[Heat]{HCl} (CH_3)_2CHCH_2CH(\overset{+}{N}H_3)COO^- + CO_2 + C_6H_4(COOH)_2$

Leucine

(b) Phthalimide $N{-}CH(CO_2C_2H_5)_2$ $\xrightarrow[CH_3I]{NaOCH_2CH_3}$

Phthalimide $N{-}C(CH_3)(CO_2C_2H_5)_2$ $\xrightarrow[heat]{NaOH}$ $C_6H_4(COO^-){-}C(=O){-}NHC(CH_3)(COO^-)_2$

$\xrightarrow[Heat]{HCl}$ $CH_3CH(^+NH_3)COO^-$ + CO_2 + $C_6H_4(COOH)_2$

Alanine

(c) Phthalimide $NCH(CO_2C_2H_5)_2$ $\xrightarrow[C_6H_5CH_2Br]{NaOCH_2CH_3}$

Phthalimide $N{-}C(CH_2C_6H_5)(CO_2C_2H_5)_2$ $\xrightarrow[heat]{NaOH}$ $C_6H_4(COO^-){-}C(=O){-}NHC(CH_2C_6H_5)(COO^-)_2$

$\xrightarrow[heat]{HCl}$ $C_6H_5CH_2CH(^+NH_3)COO^-$ + CO_2 + $C_6H_4(COOH)_2$

Phenylalanine

22.5

Because of the presence of an electron-withdrawing 2,4-dinitrophenyl group, the labeled amino acid is relatively non-basic and is, therefore, insoluble in dilute aqueous acid. The other amino acids (those that are not labeled) dissolve in dilute aqueous acid.

22.6

(a) $H_3\overset{+}{N}CH(CH(CH_3)_2)CONHCH(CH_3)CONHCH_2COO^-$ $\xrightarrow[HCO_3^-]{O_2N{-}C_6H_3(NO_2){-}F}$

Val·Ala·Gly

$O_2N{-}C_6H_3(NO_2){-}NHCH(CH(CH_3)_2)CONHCH(CH_3)CONHCH_2COO^-$ $\xrightarrow[heat]{H_3O^+}$

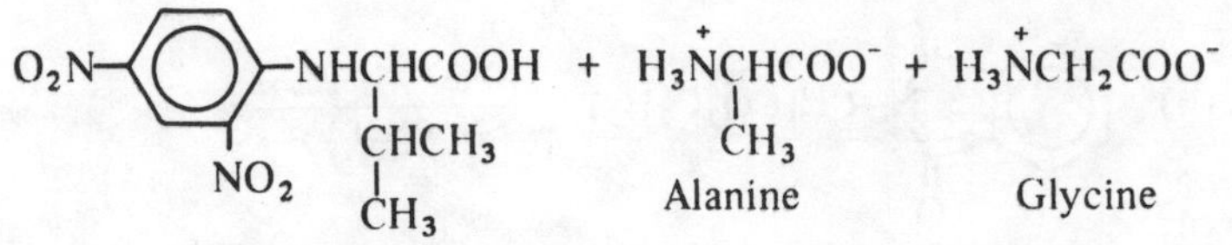

$$O_2N\text{-}C_6H_3(NO_2)\text{-}NHCH(CH(CH_3)CH_3)COOH + H_3\overset{+}{N}CH(CH_3)COO^- + H_3\overset{+}{N}CH_2COO^-$$

Labeled valine (separate and identify) — Alanine — Glycine

(b) $$O_2N\text{-}C_6H_3(NO_2)\text{-}NHCH(CH(CH_3)CH_3)COOH + O_2N\text{-}C_6H_3(NO_2)\text{-}NHCH_2CH_2CH_2CH_2CH(\overset{+}{N}H_3)COO^-$$

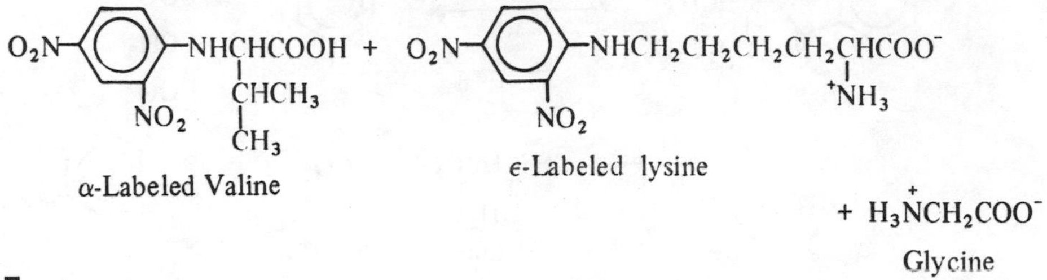

α-Labeled Valine — ε-Labeled lysine

$$+ H_3\overset{+}{N}CH_2COO^-$$

Glycine

22.7

$$C_6H_5\text{-}N{=}C{=}S + H_2\ddot{N}\text{-}CH(CH_2CH_2SCH_3)CO\text{-}NHCH(CH(CH_3)CH_2CH_3)CO\text{-}NHCH(CH_2CH_2CH_2NHC(=NH)NH_2)COO^- \xrightarrow{OH^-}$$

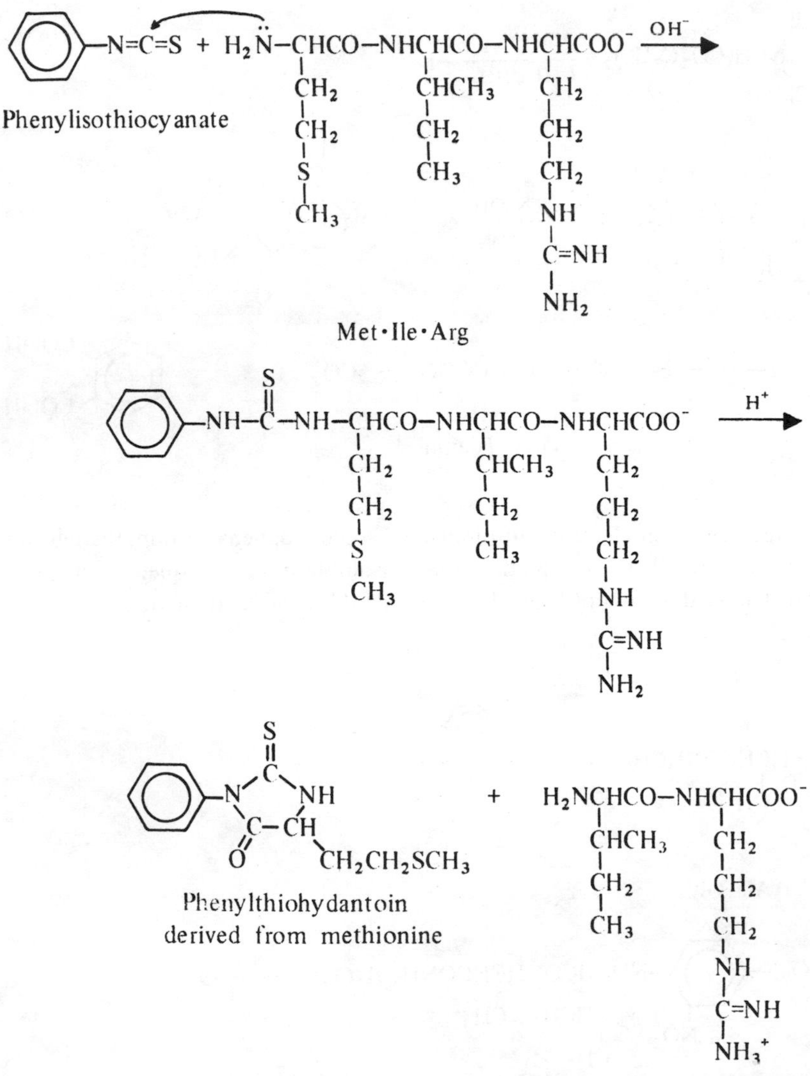

Phenylisothiocyanate — Met·Ile·Arg

$$C_6H_5\text{-}NH\text{-}C(=S)\text{-}NH\text{-}CH(CH_2CH_2SCH_3)CO\text{-}NHCH(CH(CH_3)CH_2CH_3)CO\text{-}NHCH(CH_2CH_2CH_2NHC(=NH)NH_2)COO^- \xrightarrow{H^+}$$

$$\text{(phenylthiohydantoin: } C_6H_5\text{-}N\text{-}C(=S)\text{-}NH\text{-}CH(CH_2CH_2SCH_3)\text{-}C(=O)\text{, ring)} + H_2NCH(CH(CH_3)CH_2CH_3)CO\text{-}NHCH(CH_2CH_2CH_2NHC(=NH)NH_3^+)COO^-$$

Phenylthiohydantoin derived from methionine

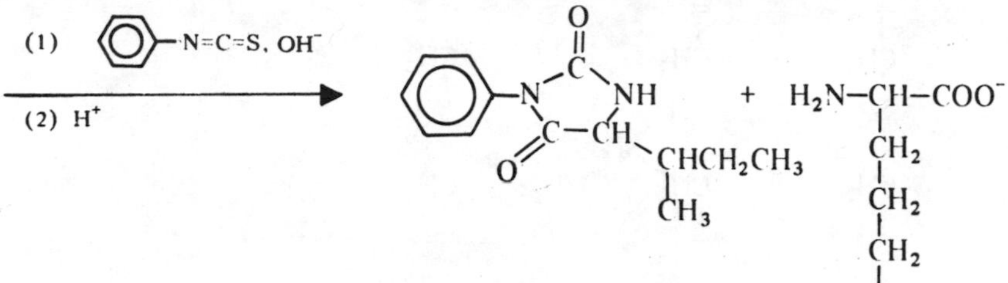

$$\xrightarrow[\text{(2) } H^+]{\text{(1) } C_6H_5{-}N{=}C{=}S,\ OH^-}$$ Phenylthiohydantoin (C_6H_5–N ring with C=O, NH, C=O, CH–CH(CH_3)CH_2CH_3) + $H_2N{-}CH{-}COO^-$ with side chain $CH_2CH_2CH_2NH{-}C({=}NH){-}NH_3^+$

Phenylthiohydantoin
derived from isoleucine

22.8

(a) Two structures are possible with the sequence Glu•Cys•Gly. Glutamic acid may be linked to cysteine through its α-carboxyl group,

$$HOOCCH_2CH_2\underset{{}^+NH_3}{CH}CO{-}NH\underset{CH_2SH}{CH}CO{-}NHCH_2COO^-$$

or through its γ-carboxyl group,

$$H_3\overset{+}{N}\underset{COO^-}{CH}CH_2CH_2CO{-}NH\underset{CH_2SH}{CH}CO{-}NHCH_2COO^-$$

(b) This shows that the second structure above is correct, that in glutathione the γ-carboxyl group is linked to cysteine.

22.9

We look for points of overlap to determine the amino acid sequence in each case.

(a)

	Ser ·	Thr	
		Thr ·	Hyp
Pro ·	Ser		
Pro ·	Ser ·	Thr ·	Hyp

(b)

	Ala ·	Cys		
		Cys ·	Arg	
			Arg ·	Val
Leu ·	Ala			
Leu ·	Ala ·	Cys ·	Arg ·	Val

22.10

Sodium in liquid ammonia brings about reductive cleavage of the disulfide linkage of oxytocin to two thiol groups, then air oxidizes the two thiol groups back to a disulfide linkage:

$$\begin{array}{c} | \\ CH_2 \\ | \\ S \\ \diagdown \\ S \\ | \\ CH_2 \\ | \end{array} \xrightarrow[NH_3]{Na} \begin{array}{c} | \\ CH_2 \\ | \\ SH \\ \\ SH \\ | \\ CH_2 \\ | \end{array} \xrightarrow{O_2} \begin{array}{c} | \\ CH_2 \\ | \\ S \\ \diagdown \\ S \\ | \\ CH_2 \\ | \end{array}$$

See also pages 693-694 in the text.

22.11

$$\underset{\text{Glycine}}{H_3\overset{+}{N}CH_2COO^-} + \underset{\substack{tert\text{-Butoxy-}\\ \text{carbonyl azide}}}{(CH_3)_3CO\overset{O}{\overset{\|}{C}}N_3} \xrightarrow[25^\circ]{OH^-}$$

$$\underset{\text{Boc-Gly}}{(CH_3)_3C{-}O\overset{O}{\overset{\|}{C}}NHCH_2COOH} \xrightarrow[(2)\ ClCO_2C_2H_5]{(1)\ (C_2H_5)_3N}$$

$$\underset{\text{Mixed anhydride}}{(CH_3)_3CO\overset{O}{\overset{\|}{C}}NHCH_2\overset{O}{\overset{\|}{C}}O\overset{O}{\overset{\|}{C}}OC_2H_5} \xrightarrow[(-CO_2,\ -C_2H_5OH)]{H_3\overset{+}{N}CHCOO^-(CHCH_3)(CH_3)\ \text{Valine}}$$

$$\underset{\text{Boc-Gly·Val}}{(CH_3)_3CO\overset{O}{\overset{\|}{C}}NHCH_2\overset{O}{\overset{\|}{C}}NH\underset{\substack{|\\ CHCH_3\\ |\\ CH_3}}{C}HCOOH} \xrightarrow[(2)\ ClCO_2C_2H_5]{(1)\ (C_2H_5)_3N}$$

$$\underset{\text{Mixed anhydride}}{(CH_3)_3CO\overset{O}{\overset{\|}{C}}NHCH_2\overset{O}{\overset{\|}{C}}NH\underset{\substack{|\\ CHCH_3\\ |\\ CH_3}}{C}H\overset{O}{\overset{\|}{C}}O\overset{O}{\overset{\|}{C}}OC_2H_5} \xrightarrow{H_3\overset{+}{N}CHCOO^-(CH_3)\ \text{Alanine}}$$

$$\underset{\text{Boc-Gly·Val·Ala}}{(CH_3)_3CO\overset{O}{\overset{\|}{C}}NHCH_2\overset{O}{\overset{\|}{C}}NH\underset{\substack{|\\ CHCH_3\\ |\\ CH_3}}{C}H\overset{O}{\overset{\|}{C}}NH\underset{\substack{|\\ CH_3}}{C}HCOOH} \xrightarrow[CH_3COOH,\ 25^\circ]{CF_3COOH}$$

(cont. on next page)

$$(CH_3)_2C{=}CH_2 + CO_2 + H_3\overset{+}{N}CH_2\overset{O}{\overset{\|}{C}}NH\underset{\underset{CH_3}{|}}{\underset{CHCH_3}{|}}{CH}\overset{O}{\overset{\|}{C}}NH\underset{CH_3}{\underset{|}{CH}}COO^-$$

Gly·Val·Ala

22.12

(a)
$$2C_6H_5CH_2O\overset{O}{\overset{\|}{C}}Cl + H_2NCH_2CH_2CH_2CH_2\underset{NH_2}{\underset{|}{C}}HCOO^- \xrightarrow[25^\circ]{OH^-}$$

Benzyl chloro-carbonate — Lysine

$$C_6H_5CH_2O\overset{O}{\overset{\|}{C}}NHCH_2CH_2CH_2CH_2\underset{\underset{C_6H_5CH_2OC{=}O}{|}}{\underset{NH}{\underset{|}{C}}}HCOOH \xrightarrow[(2)\ ClCOOC_2H_5]{(1)\ (C_2H_5)_3N}$$

$$C_6H_5CH_2O\overset{O}{\overset{\|}{C}}NHCH_2CH_2CH_2CH_2\underset{\underset{C_6H_5CH_2OC{=}O}{|}}{\underset{NH}{\underset{|}{C}}}H\overset{O}{\overset{\|}{C}}O\overset{O}{\overset{\|}{C}}OC_2H_5 \xrightarrow[(-CO_2,\ -C_2H_5OH)]{CH_3CH_2\underset{CH_3}{\underset{|}{C}}H-\underset{NH_3^+}{\underset{|}{C}}HCOO^-}$$

$$C_6H_5CH_2O\overset{O}{\overset{\|}{C}}NHCH_2CH_2CH_2CH_2\underset{\underset{C_6H_5CH_2OC{=}O}{|}}{\underset{NH}{\underset{|}{C}}}H\overset{O}{\overset{\|}{C}}NH\underset{\underset{\underset{CH_3}{|}}{\underset{CH_2}{|}}}{\underset{CHCH_3}{\underset{|}{C}}}HCOO^- \xrightarrow[\substack{CH_3COOH \\ cold}]{HBr}$$

$$2C_6H_5CH_2Br + 2CO_2 + H_3\overset{+}{N}CH_2CH_2CH_2CH_2\underset{NH_2}{\underset{|}{C}}H\overset{O}{\overset{\|}{C}}NH\underset{\underset{\underset{CH_3}{|}}{\underset{CH_2}{|}}}{\underset{CHCH_3}{\underset{|}{C}}}HCOO^-$$

Lys·Ile

(b)
$$3C_6H_5CH_2O\overset{O}{\overset{\|}{C}}Cl + H_2N\overset{NH}{\overset{\|}{C}}NHCH_2CH_2CH_2\underset{NH_2}{\underset{|}{C}}HCOO^- \xrightarrow[25^\circ]{OH^-}$$

$$C_6H_5CH_2O\overset{O}{\overset{\|}{C}}NH\overset{NH}{\overset{\|}{C}}\underset{\underset{C_6H_5CH_2O}{|}}{\underset{C{=}O}{\underset{|}{N}}}CH_2CH_2CH_2\underset{\underset{\underset{C_6H_5CH_2O}{|}}{\underset{C{=}O}{|}}}{\underset{NH}{\underset{|}{C}}}HCOOH \xrightarrow[(2)\ ClCOOC_2H_5]{(1)\ (C_2H_5)_3N}$$

$$\underset{\substack{| \\ C=O \\ | \\ C_6H_5CH_2O}}{C_6H_5CH_2O\overset{O}{\overset{\|}{C}}NH\overset{NH}{\overset{\|}{C}}N}CH_2CH_2CH_2\underset{\substack{| \\ NH \\ | \\ C=O \\ | \\ C_6H_5CH_2O}}{CH}\overset{O}{\overset{\|}{C}}\overset{O}{\overset{\|}{C}}OC_2H_5 \xrightarrow[(-CO_2,\ -C_2H_5OH)]{CH_3CH(^+NH_3)COO^-}$$

$$\underset{\substack{| \\ C=O \\ | \\ C_6H_5CH_2O}}{C_6H_5CH_2O\overset{O}{\overset{\|}{C}}NH\overset{NH}{\overset{\|}{C}}N}CH_2CH_2CH_2\underset{\substack{| \\ NH \\ | \\ C=O \\ | \\ C_6H_5CH_2O}}{CH}\overset{O}{\overset{\|}{C}}NH\underset{\substack{| \\ CH_3}}{CH}COOH \xrightarrow[\substack{CH_3COOH \\ cold}]{HBr}$$

$$3C_6H_5CH_2Br + 3CO_2 + {}^+H_3N\overset{NH}{\overset{\|}{C}}NHCH_2CH_2CH_2\underset{\substack{| \\ NH_2}}{CH}CONH\underset{\substack{| \\ CH_3}}{CH}COO^-$$

Arg·Ala

22.13

The weakness of the benzyl-oxygen bond (p. 999) allows these groups to be removed by catalytic hydrogenolysis.

22.14

(a) An electrophilic aromatic substitution reaction:

$$\underset{\substack{| \\ C_6H_5}}{+CH_2CH+_n} + CH_3OCH_2Cl \xrightarrow{BF_3} \underset{\substack{| \\ C_6H_4 \\ | \\ CH_2Cl}}{+CH_2CH+_n} + CH_3OH$$

(b) The linkage between the resin and the polypeptide is a benzylic ester. It is cleaved by HBr in CF_3COOH at room temperature because the carbocation that is formed initially is the relatively stable, benzylic cation.

22.15

$$\text{(resin)}-CH_2Cl + HO\overset{O}{\overset{\|}{C}}\underset{\substack{| \\ CH_3}}{CH}NH\overset{O}{\overset{\|}{C}}OC(CH_3)_3$$

$\downarrow$ base — 1 Add Boc·Ala

$$\text{(resin)}-CH_2O\overset{O}{\overset{\|}{C}}\underset{\substack{| \\ CH_3}}{CH}NH\overset{O}{\overset{\|}{C}}OC(CH_3)_3$$

2 Purify by washing

$\downarrow$ CF_3COOH, CH_2Cl_2 — 3 Remove protecting group

(cont. on next page)

$\text{(resin)}-CH_2OC(=O)CH(CH_3)NH_2$

4 Purify by washing

↓ $HOC(=O)CH(CH_2C_6H_5)NHC(=O)OC(CH_3)_3$ and dicyclohexylcarbodiimide

5 Add Boc·Phe

$\text{(resin)}-CH_2OC(=O)CH(CH_3)NHC(=O)CH(CH_2C_6H_5)NHC(=O)OC(CH_3)_3$

6 Purify by washing

↓ CF_3COOH, CH_2Cl_2

7 Remove protecting group

$\text{(resin)}-CH_2OC(=O)CH(CH_3)NHC(=O)CH(CH_2C_6H_5)NH_2$

8 Purify by washing

↓ $HOC(=O)CH(NHC(=O)OC(CH_3)_3)CH_2CH_2CH_2CH_2NHC(=O)OC(CH_3)_3$ and dicyclohexylcarbodiimide

9 Add Protected Lys

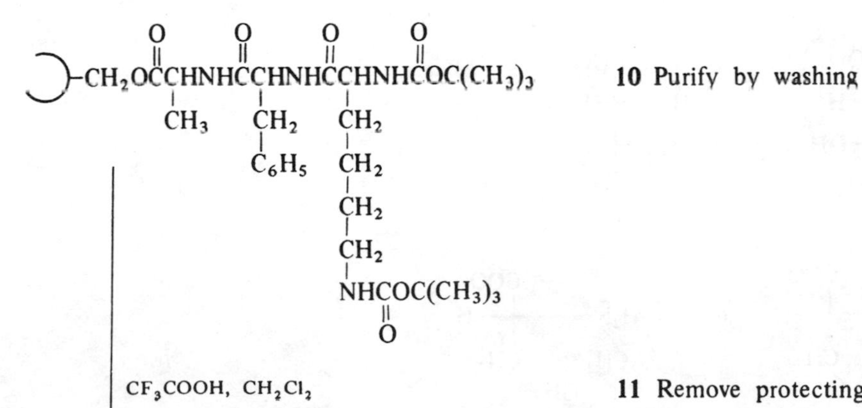

10 Purify by washing

11 Remove protecting groups

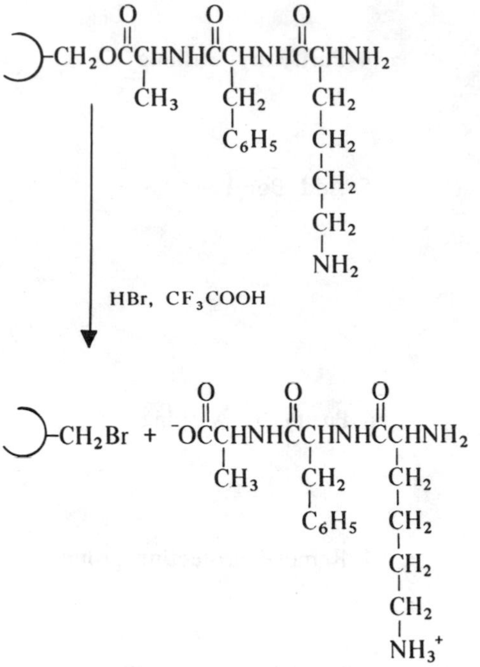

$$\text{(resin)}-CH_2O\overset{O}{\overset{\|}{C}}\underset{CH_3}{\underset{|}{C}}HNH\overset{O}{\overset{\|}{C}}\underset{\underset{C_6H_5}{|}}{\underset{CH_2}{\underset{|}{C}}}HNH\overset{O}{\overset{\|}{C}}\underset{CH_2CH_2CH_2CH_2NH_2}{\underset{|}{C}}HNH_2$$

12 Purify by washing

$\xrightarrow{HBr,\ CF_3COOH}$

13 Detach tripeptide

$$\text{(resin)}-CH_2Br + {}^{-}O\overset{O}{\overset{\|}{C}}\underset{CH_3}{\underset{|}{C}}HNH\overset{O}{\overset{\|}{C}}\underset{\underset{C_6H_5}{|}}{\underset{CH_2}{\underset{|}{C}}}HNH\overset{O}{\overset{\|}{C}}\underset{CH_2CH_2CH_2CH_2NH_3^+}{\underset{|}{C}}HNH_2$$

Lys·Phe·Ala

14 Isolate product

22.16

(a) Isoleucine, threonine, hydroxyproline, and cystine.

(b) Fischer projections (top to bottom):

COO^-; $H_3\overset{+}{N}$—C—H; CH_3—C—H; CH_2; CH_3 and COO^-; $H_3\overset{+}{N}$—C—H; H—C—CH_3; CH_2; CH_3

COO^-; $H_3\overset{+}{N}$—C—H; H—C—OH; CH_3 and COO^-; $H_3\overset{+}{N}$—C—H; HO—C—H; CH_3

COO^-; $H_2\overset{+}{N}$—C—H; ring CH_2, CH_2, C(H)(OH) with H up, OH down and COO^-; $H_2\overset{+}{N}$—C—H; ring CH_2, CH_2, C(OH)(H) with OH up, H down

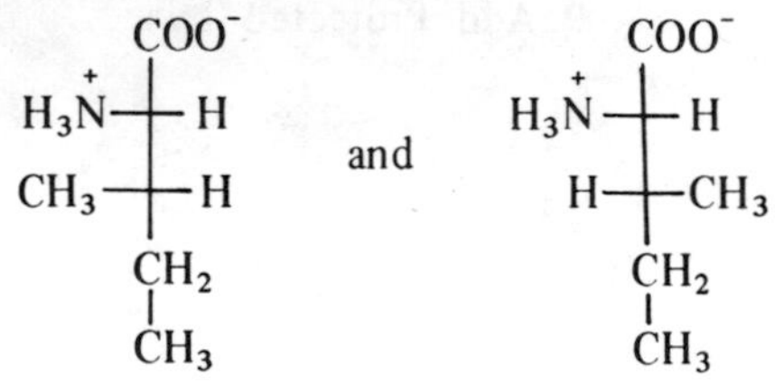

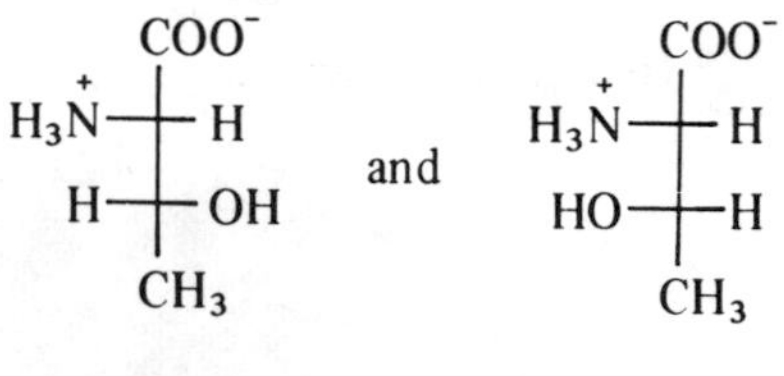

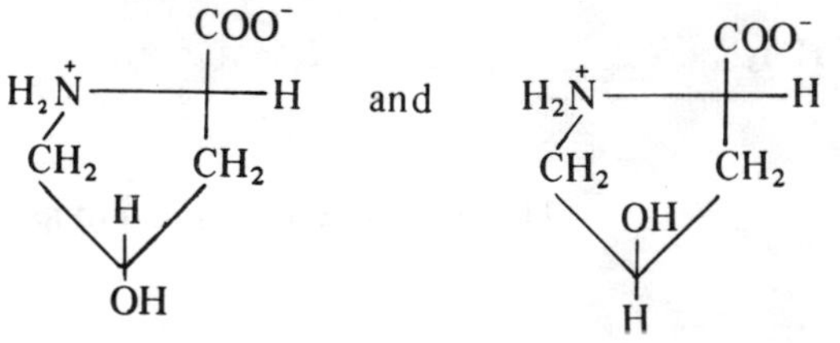

(With cystine, both chiral carbons are α-carbons, thus according to the problem, both must have the L-configuration, and no isomers of this type can be written.)

(c) Diastereomers

22.17

(a) Alanine

$$CH_3\underset{\underset{{}^{+}NH_3}{|}}{C}HCOO^- + HONO \longrightarrow CH_3\underset{\underset{OH}{|}}{C}HCOOH + N_2$$

(b) Proline and hydroxyproline. All of the other amino acids have at least one primary amino group.

(c) 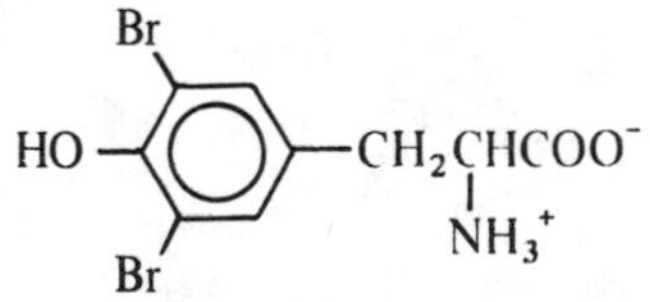

(d) 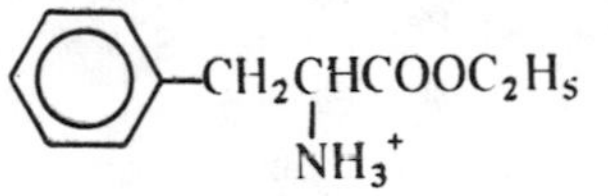

(e) 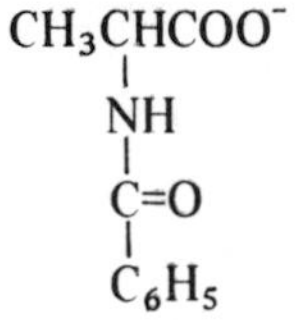

22.18

(a)

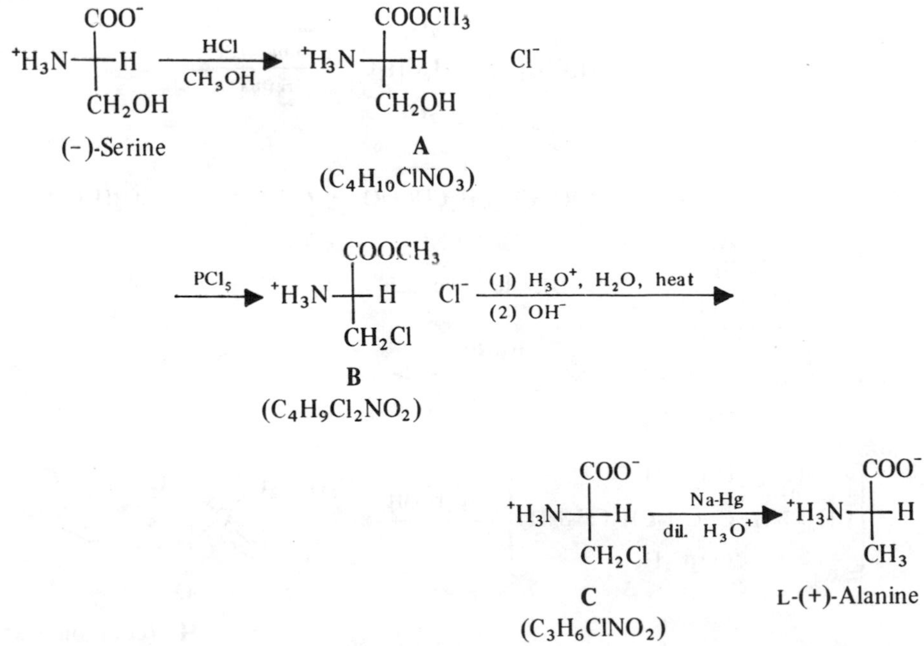

(b)

$$B \xrightarrow{OH^-} H_2N\text{—}\underset{CH_2Cl}{\overset{COOCH_3}{C}}\text{—}H \xrightarrow{NaSH} H_2N\text{—}\underset{CH_2SH}{\overset{COOCH_3}{C}}\text{—}H$$

D ($C_4H_8ClNO_2$) E ($C_4H_9NO_2S$)

$$\xrightarrow[(2)\ OH^-]{(1)\ H_3O^+,\ H_2O,\ heat} {}^+H_3N\text{—}\underset{CH_2SH}{\overset{COO^-}{C}}\text{—}H$$

L-(+)-Cysteine

(c)

$$\overset{+}{H_3N}\text{—}\underset{CH_2\overset{O}{\overset{\|}{C}}NH_2}{\overset{COO^-}{C}}\text{—}H \xrightarrow{NaOBr,\ OH^-} H_2N\text{—}\underset{CH_2NH_2}{\overset{COO^-}{C}}\text{—}H$$

L-Asparagine F ($C_3H_7N_2O_2$)

$$ {}^+H_3N\text{—}\underset{CH_2Cl}{\overset{COO^-}{C}}\text{—}H \xrightarrow{NH_3} F$$

C (from part A)

22.19

(a)

$$CH_3\overset{O}{\overset{\|}{C}}NHCH(CO_2C_2H_5)_2 + CH_2{=}CH{-}C{\equiv}N \xrightarrow[C_2H_5OH]{NaOC_2H_5}$$

$$CH_3\overset{O}{\overset{\|}{C}}NH\text{—}\underset{CO_2C_2H_5}{\overset{CO_2C_2H_5}{C}}\text{—}CH_2CH_2C{\equiv}N \xrightarrow[reflux]{conc\ HCl}$$

G

$$HOOCCH_2CH_2\underset{NH_3^+}{CH}COO^- + CH_3COOH + 2C_2H_5OH + NH_4^+ + CO_2$$

DL-Glutamic acid

(b)

$$CH_3\overset{O}{\overset{\|}{C}}NH\text{—}\underset{CO_2C_2H_5}{\overset{CO_2C_2H_5}{C}}\text{—}CH_2CH_2C{\equiv}N \xrightarrow[68^\circ,\ 1000\ psi]{H_2(Ni)}$$

$$\left[CH_3\overset{O}{\overset{\|}{C}}NH\text{—}\underset{CO_2C_2H_5}{\overset{CO_2C_2H_5}{C}}\text{—}CH_2CH_2CH_2NH_2\right] \xrightarrow{-C_2H_5OH}$$

(cyclic product H: ring C(CO₂C₂H₅)(NHCOCH₃)—CH₂—CH₂—CH₂—NH—C(=O)—, i.e. $C_2H_5O_2C$ and $CH_3\overset{O}{\overset{\|}{C}}NH$ on C; C—CH_2—CH_2—CH_2—NH—$\overset{O}{\overset{\|}{C}}$)

H (cont. on next page)

$$\xrightarrow[\text{reflux}]{\text{conc. HCl}} \underset{\text{Cl}^-}{\text{H}_3\overset{+}{\text{N}}}\text{CH}_2\text{CH}_2\text{CH}_2\underset{\text{NH}_3^+}{\underset{|}{\text{C}}}\text{HCOO}^- + \text{CH}_3\text{COOH} + \text{CO}_2 + \text{C}_2\text{H}_5\text{OH}$$

DL-Ornithine hydrochloride

22.20

(a)
$$\text{C}_6\text{H}_5\text{CH}_2\overset{\text{O}}{\overset{\|}{\text{C}}}\text{H} \xrightarrow[\text{HCN}]{\text{NH}_3} \text{C}_6\text{H}_5\text{CH}_2\underset{\text{NH}_2}{\underset{|}{\text{C}}}\text{HC}{\equiv}\text{N} \xrightarrow{\text{H}_3\text{O}^+} \text{C}_6\text{H}_5\text{CH}_2\underset{\text{NH}_3^+}{\underset{|}{\text{C}}}\text{HCOO}^-$$

Phenyl acetaldehyde — DL-Phenylalanine

(b)
$$\text{CH}_3\text{SH} + \text{CH}_2{=}\text{CH}{-}\overset{\text{O}}{\overset{\|}{\text{C}}}\text{H} \xrightarrow{\text{base}} \text{CH}_3\text{SCH}_2\text{CH}_2\overset{\text{O}}{\overset{\|}{\text{C}}}\text{H}$$

$$\xrightarrow[\text{HCN}]{\text{NH}_3} \text{CH}_3\text{SCH}_2\text{CH}_2\underset{\text{NH}_2}{\underset{|}{\text{C}}}\text{HC}{\equiv}\text{N} \xrightarrow{\text{H}_3\text{O}^+} \text{CH}_3\text{SCH}_2\text{CH}_2\underset{\text{NH}_3^+}{\underset{|}{\text{C}}}\text{HCOO}^-$$

DL-Methionine

22.21

$$\text{C}_6\text{H}_5\text{CH}_2\underset{\text{NH}_3^+}{\underset{|}{\text{C}}}\text{HCOO}^- + \text{HOOCCH}_2\text{CH}_2\underset{\text{O}}{\underset{\|}{\text{C}}}\text{COOH} \underset{}{\overset{\text{transaminase}}{\rightleftharpoons}}$$

Phenylalanine — α-Ketoglutaric acid

$$\text{C}_6\text{H}_5\text{CH}_2\underset{\text{O}}{\underset{\|}{\text{C}}}\text{COOH} + \text{HOOCCH}_2\text{CH}_2\underset{\text{NH}_3^+}{\underset{|}{\text{C}}}\text{HCOO}^-$$

Phenylpyruvic acid — Glutamic acid

Then:

$$\text{HOOCCH}_2\text{CH}_2\underset{\text{NH}_3^+}{\underset{|}{\text{C}}}\text{HCOO}^- + \text{HOOCCH}_2\underset{\text{O}}{\underset{\|}{\text{C}}}\text{COOH} \overset{\text{transaminase}}{\rightleftharpoons}$$

Glutamic acid — Oxaloacetic acid

$$\text{HOOCCH}_2\text{CH}_2\underset{\text{O}}{\underset{\|}{\text{C}}}\text{COOH} + \text{HOOCCH}_2\underset{\text{NH}_3^+}{\underset{|}{\text{C}}}\text{HCOO}^-$$

α-Ketoglutaric acid — Aspartic acid

This amounts to:

Phenylalanine + α-Ketoglutaric acid $\rightleftharpoons$ Phenylpyruvic acid + Glutamic acid
Glutamic acid + Oxaloacetic acid $\rightleftharpoons$ α-Ketoglutaric acid + Aspartic acid

---

Net: Phenylalanine + Oxaloacetic acid $\rightleftharpoons$ Phenylpyruvic acid + Aspartic acid

22.22

We look for points of overlap:

Phe · Ser
Pro · Gly · Phe
Pro · Pro Ser · Pro · Phe
Arg · Pro Phe · Arg

Arg · Pro · Pro · Gly · Phe · Ser · Pro · Phe · Arg

Bradykinin

22.23

1. This shows that valine is the N-terminal amino acid and that valine is attached to leucine. (Lysine labeled at the ϵ-amino group is to be expected if lysine is not the N-terminal amino acid and if it is linked in the polypeptide through its α-amino group.)

2. This shows that alanine is the C-terminal amino acid and that it is linked to glutamic acid.

At this point, then, we have the following information about the structure of the heptapeptide.

Val · Leu (Ala, Lys, Phe) Glu · Ala

the sequence here is unknown

3. (a) This shows that the dipeptide, **A**, is

Leu · Lys

(b) The carboxypeptidase reaction shows that the C-terminal amino acid of the tripeptide, **B**, is glutamic acid; the DNP labeling experiment shows that the N-terminal amino acid is phenylalanine. Thus the tripeptide **B** is:

Phe · Ala · Glu

Putting these pieces together in the only way possible, we arrive at the following amino acid sequence for the heptapeptide.

Val · Leu
Leu · Lys
Phe · Ala · Glu
Glu · Ala

Val · Leu · Lys · Phe · Ala · Glu · Ala

22.24

At pH 2-3 the γ-carboxyl groups of polyglutamic acid are uncharged (they are present as $-COOH$ groups). At pH 5 the γ-carboxyl groups ionize and become negatively charged (they become γ-COO^- groups). The repulsive forces between these negatively charged groups cause an unwinding of the α-helix and the formation of a random coil.

22.25

The observation that the pmr spectrum taken at room temperature shows two different signals for the methyl groups suggests that they are in different environments. This would be true if rotation about the carbon-nitrogen bond was not taking place.

$$\underset{\delta 8.05}{H}\diagdown\!\!\underset{O\diagup\!\!\diagup}{C}-N\begin{matrix}\diagup CH_3 \;\; \delta 2.95 \\ \diagdown CH_3 \;\; \delta 2.80\end{matrix}$$

We assign the $\delta 2.80$ signal to the methyl group that is on the same side as the electronegative oxygen.

The fact that the methyl signals appear as doublets (and that the formyl signal is a multiplet) indicates that long-range coupling is taking place between the methyl protons and the formyl proton.

That the two doublets are not simply the result of spin-spin coupling is indicated by the observation that the distance that separates one doublet from the other changes when the applied magnetic field strength is lowered. [Remember the magnitude of a chemical shift is proportional to the strength of the applied magnetic field while the magnitude of a coupling constant is not.]

That raising the temperature (to 111°) causes the doublets to coalesce into a single signal indicates that at higher temperatures the molecules have enough energy to surmount the energy barrier of the carbon-nitrogen bond. Above 111°, rotation is taking place so rapidly that the spectrometer is unable to discriminate between the two methyl groups.

0

Special Topic

0.1

(a) The first step is similar to a crossed-Claisen condensation:

$$\text{3-pyridyl-}\overset{O}{\overset{\|}{C}}OC_2H_5 + \text{N-methyl-2-pyrrolidinone} \xrightarrow{C_2H_5ONa} \text{3-pyridyl-}\overset{O}{\overset{\|}{C}}\text{-(3-position of N-methyl-2-pyrrolidinone, } N\text{-}CH_3)$$

(b) This step involves hydrolysis of an amide (lactam) and can be carried out with either acid or base. Here we use acid.

$$\xrightarrow[H_2O]{H_3O^+} \text{3-pyridyl-}\overset{O}{\overset{\|}{C}}-\underset{COOH}{\underset{|}{C}}HCH_2CH_2NHCH_3$$

(c) This step is the decarboxylation of a substituted malonic acid; it requires only the application of heat and takes place during the acid hydrolysis of step (b).

(d) This is the reduction of a ketone to a secondary alcohol. A variety of reducing agents can be used, sodium borohydride, for example.

$$\xrightarrow{NaBH_4} \text{3-pyridyl-}\overset{OH}{\overset{|}{C}}HCH_2CH_2CH_2NHCH_3$$

(e) Here we convert the secondary alcohol to an alkyl bromide with hydrogen bromide; this also gives a hydrobromide salt of the aliphatic amine.

$$\xrightarrow[heat]{HBr} \text{3-pyridyl-}\overset{Br}{\overset{|}{C}}HCH_2CH_2CH_2\overset{H}{\underset{H}{N^+}}-CH_3 \quad Br^-$$

(f) Treating the salt with base produces the secondary amine; it then acts as a nucleophile and attacks the carbon bearing the bromine. This leads to the formation of a five-membered ring and (±) nicotine.

$\xrightarrow[(-HBr)]{base}$ 3-pyridyl–$\overset{Br}{CH}CH_2CH_2CH_2\overset{H}{\ddot{N}}-CH_3$ $\xrightarrow[(-HBr)]{base}$

(±) nicotine

O.2

(a) The chiral carbon adjacent to the ester carbonyl group is racemized by base (probably through the formation of an anion that can undergo inversion of configuration, cf. Sect. 16.12A).

(b)

N–CH_3, O, O=C, C_6H_5, C, H, CH_2OH, H

O.3

(a)

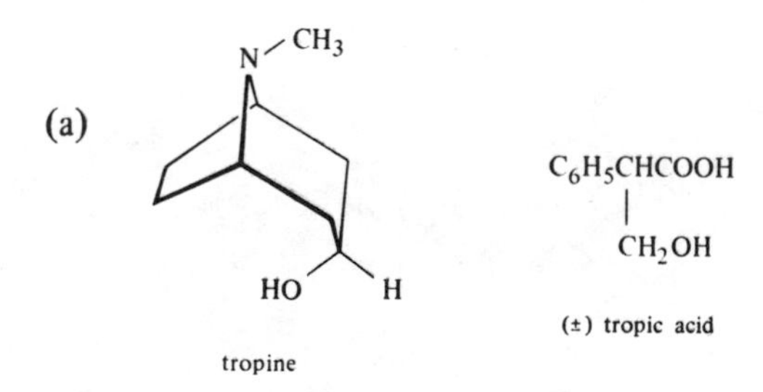

tropine

$C_6H_5CHCOOH$
|
CH_2OH

(±) tropic acid

(b) Tropine is a meso compound; it has a plane of symmetry that passes through the $\gt CHOH$ group, the $\gt NCH_3$ group, and between the two $-CH_2-$ groups of the five-membered ring.

CH———CH_2

CH_2
– – – NCH_3– – – $CHOH$– – plane of symmetry
CH_2

CH———CH_2

(c)

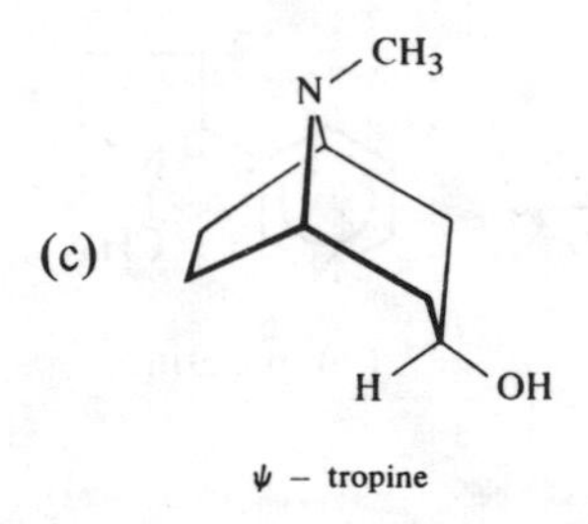

O.4

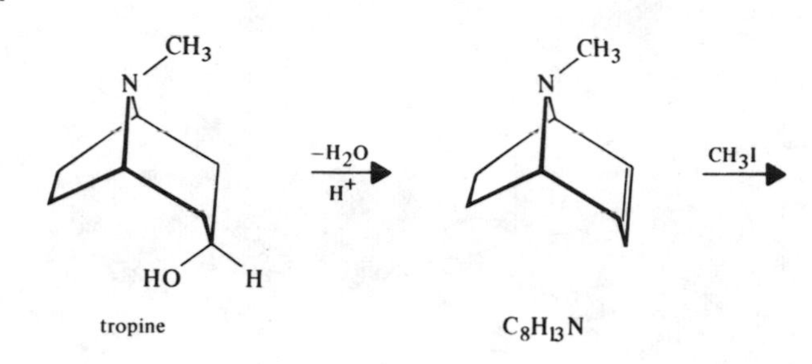

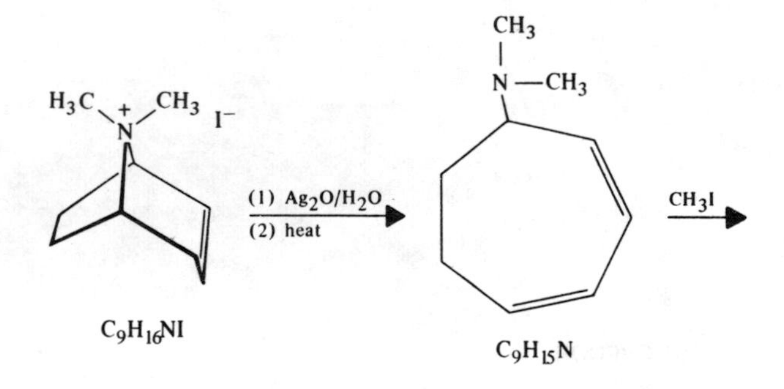

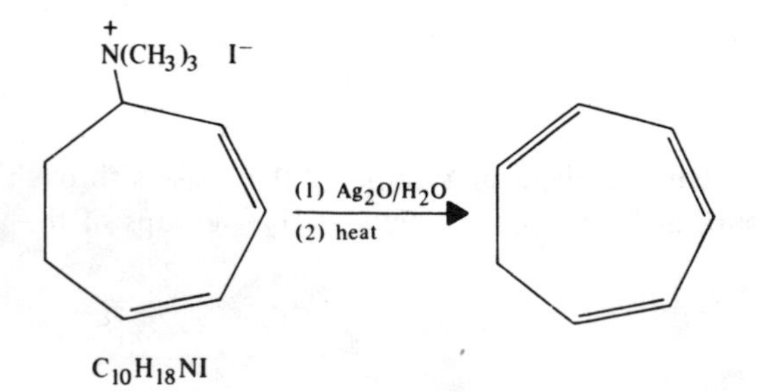

O.5

One possible sequence of steps is the following:

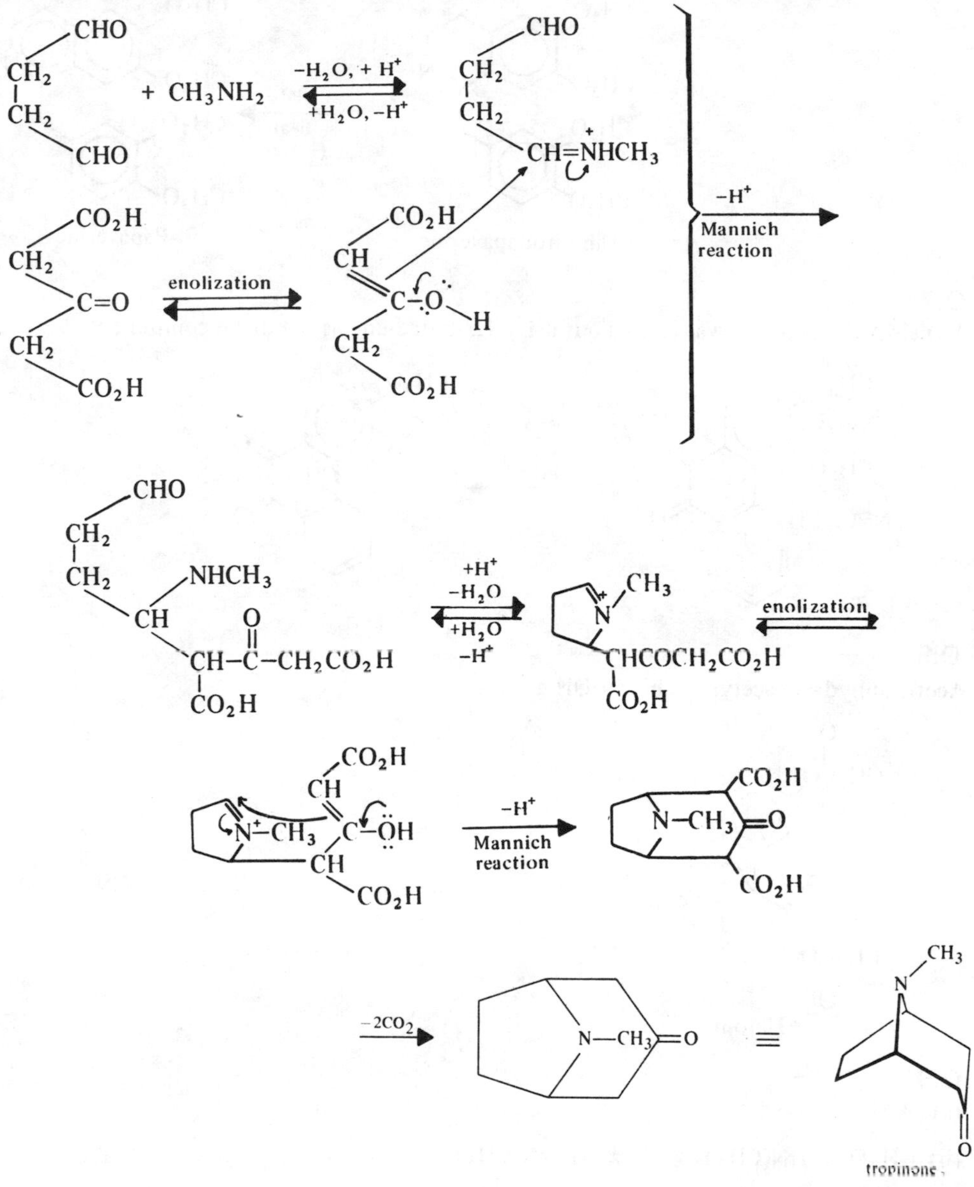

O.6

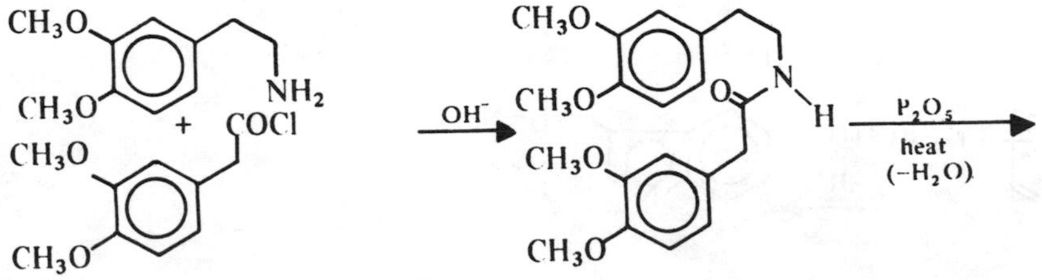

$C_{20}H_{25}NO_5$

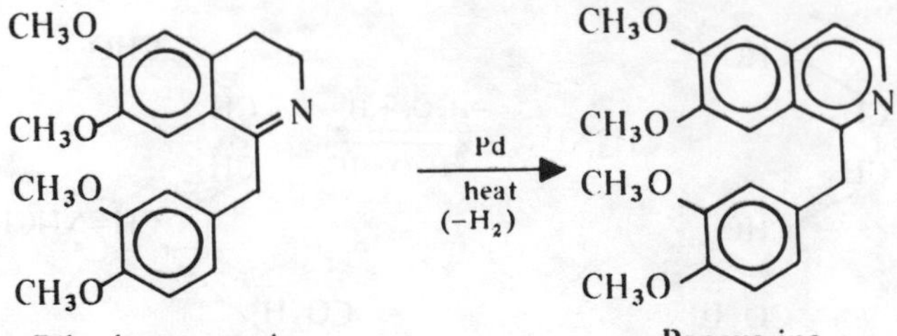

CH_3O, CH_3O, CH_3O, CH_3O — Dihydropapaverine $\xrightarrow[\text{heat }(-H_2)]{\text{Pd}}$ Papaverine

O.7

A Diels-Alder reaction was carried out using 1,3-butadiene as the diene component.

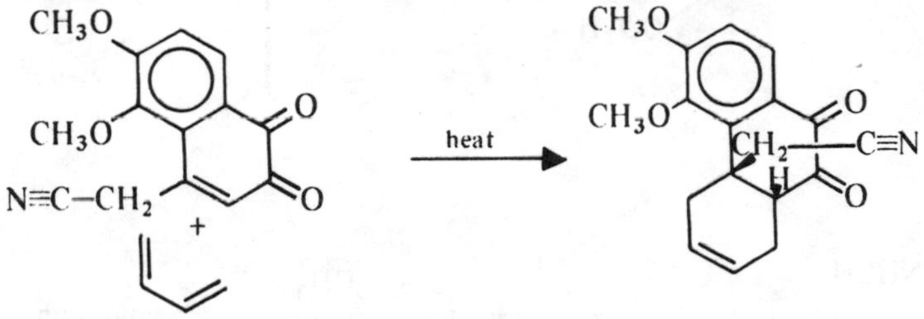

CH_3O, CH_3O, $N{\equiv}C{-}CH_2$ + (butadiene) $\xrightarrow{\text{heat}}$ CH_3O, CH_3O, $CH_2{-}C{\equiv}N$

O.8

Acetic anhydride acetylates both —OH groups.

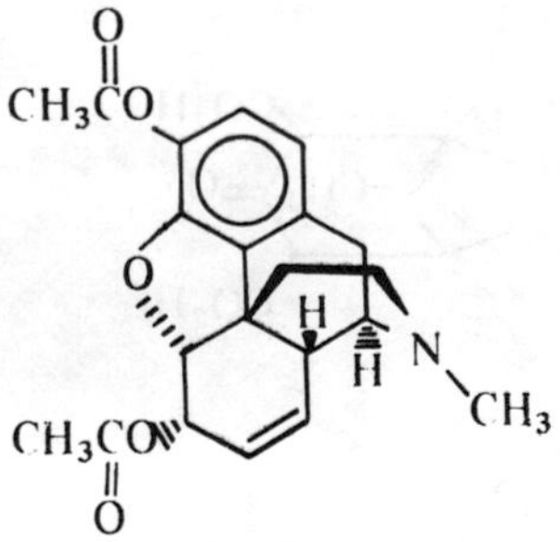

Heroin

O.9

(a) A Mannich reaction.

(b) $CH_2O + HN(CH_3)_2 \underset{}{\overset{+H^+,\ -H_2O}{\rightleftharpoons}} CH_2{=}\overset{+}{N}(CH_3)_2$

$\xrightarrow{-H^+}$ $CH_2{-}N(CH_3)_2$

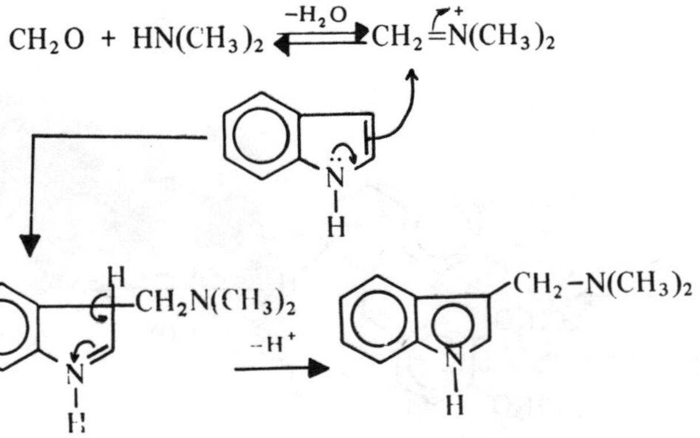

Gramine

O.10

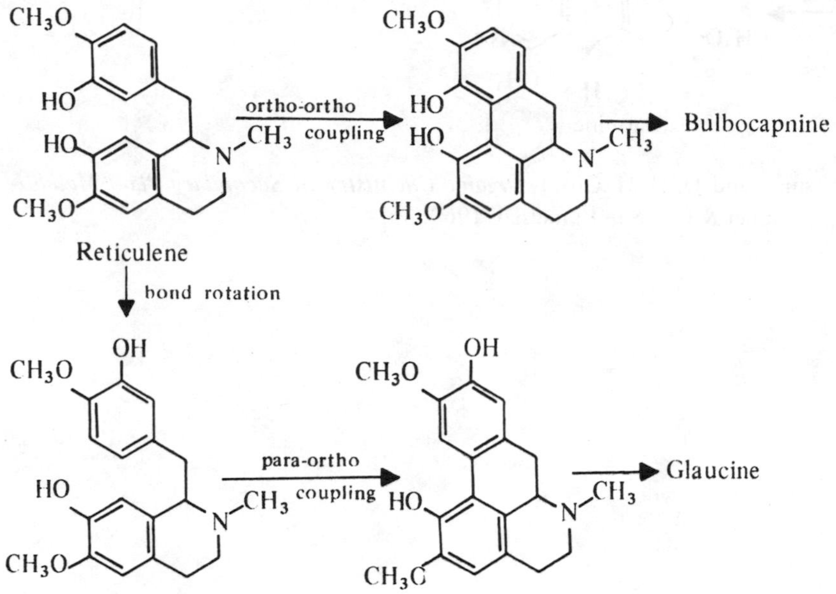

Reticulene $\xrightarrow{\text{ortho-ortho coupling}}$ → Bulbocapnine

Reticulene ↓ bond rotation

$\xrightarrow{\text{para-ortho coupling}}$ → Glaucine

O.11

Yes, because according to the pathway given on pages 1020-1021, carbons 1 and 3 of papaverine arise from the α-carbons of two molecules of tyrosine.

(b) Yes.

(c) Methylation of the four phenolic hydroxyl groups and dehydrogenation of the nitrogen-containing ring.

O.12

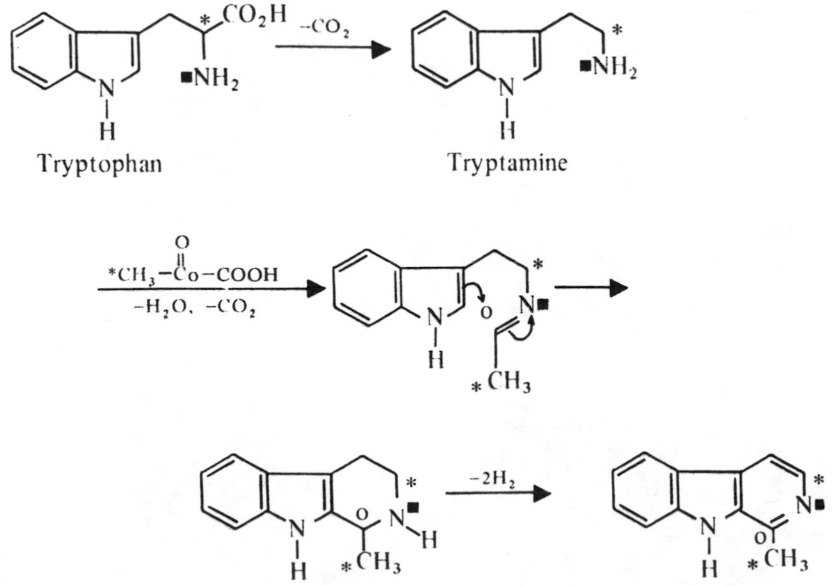

Tryptophan $\xrightarrow{-CO_2}$ Tryptamine

$\xrightarrow[-H_2O,\ -CO_2]{*CH_3-\overset{O}{\overset{\|}{C}}-COOH}$ →

$\xrightarrow{-2H_2}$

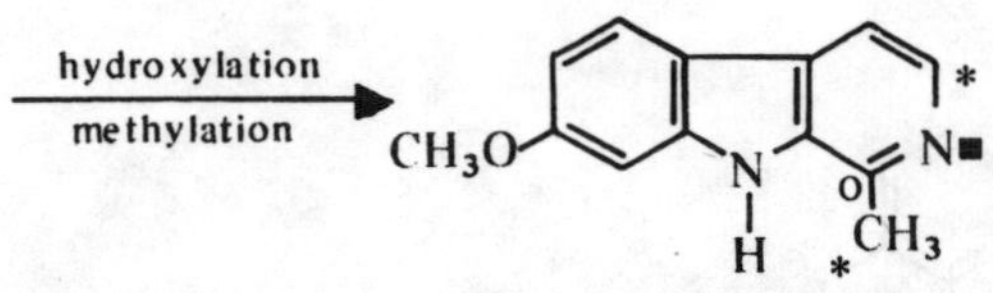

hydroxylation
methylation

Harmine

Cf. T. A. Geissman and D. H. G. Crout, *Organic Chemistry of Secondary Plant Metabolism*, Freeman, Cooper & Co., San Francisco, 1969, pp. 473-474.

P

Special Topic

P.1

Adenine:

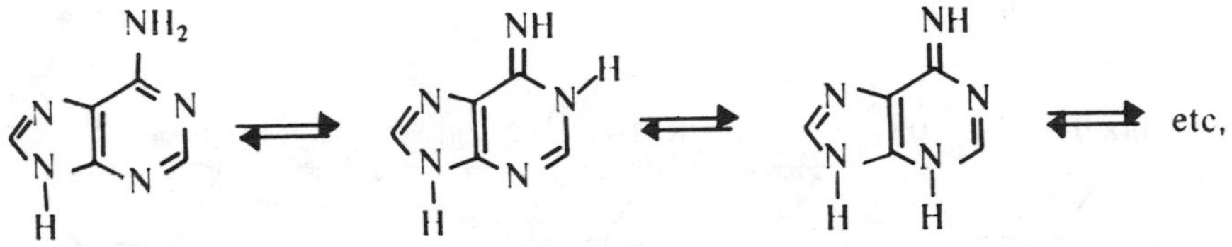

NH$_2$ ⇄ NH ⇄ NH ⇄ etc,

Guanine:

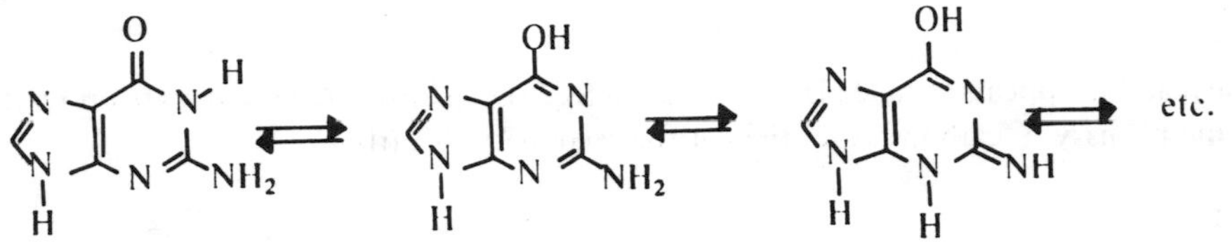

⇄ ⇄ ⇄ etc.

Cytosine:

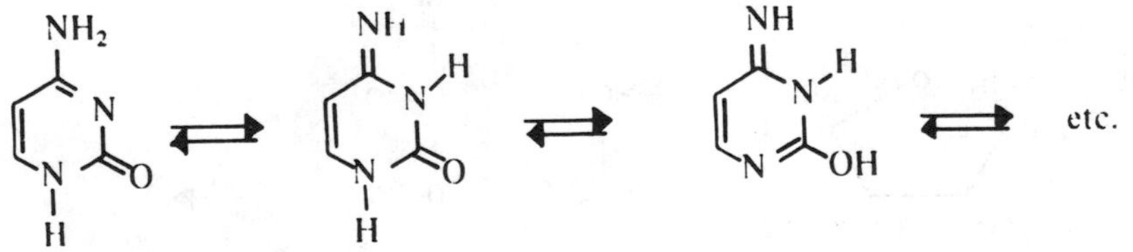

⇄ ⇄ ⇄ etc.

Thymine (R = CH_3) or Uracil (R = H):

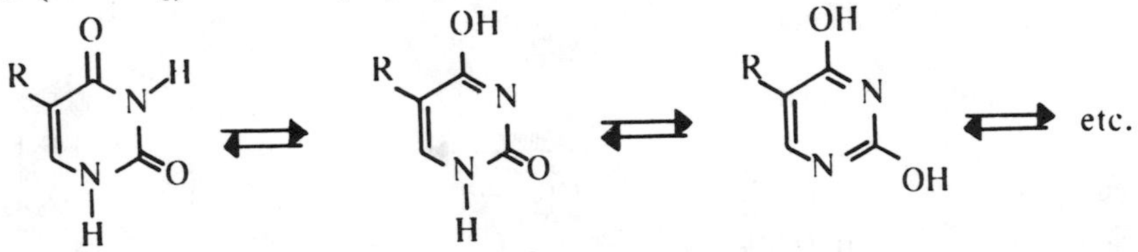

⇄ ⇄ ⇄ etc.

P.2

(a) The nucleosides have an N-glycosidic linkage that (like an O-glycosidic linkage) is rapidly hydrolyzed by aqueous acid but one that is stable in aqueous base.

(b)

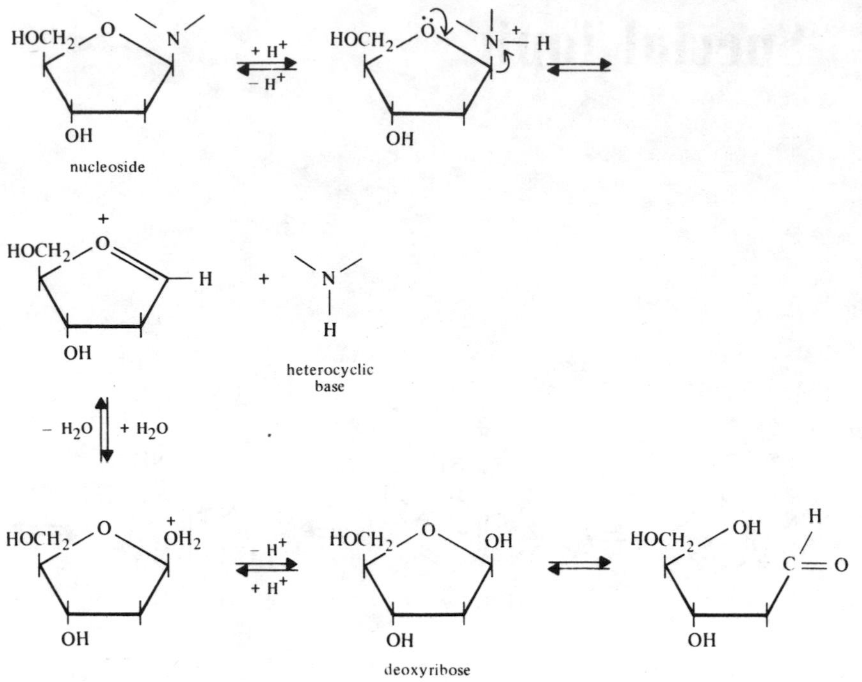

nucleoside

$+H^+$ / $-H^+$

heterocyclic base

$-H_2O$ / $+H_2O$

$-H^+$ / $+H^+$

deoxyribose

P.3

The reaction appears to take place through an S_N2 mechanism. Attack occurs preferentially at the primary 5′-carbon rather than at the secondary 3′-carbon.

P.4

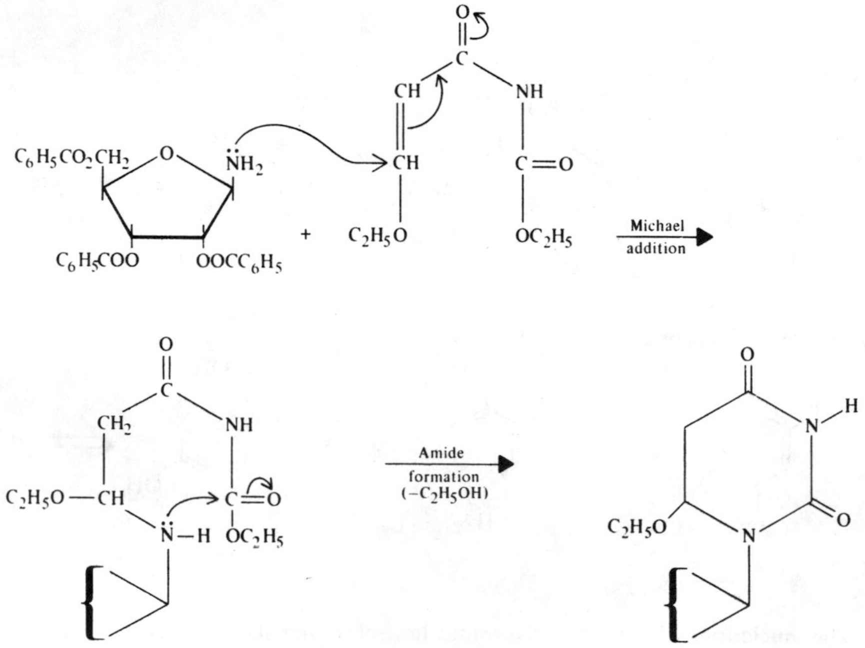

Michael addition

Amide formation ($-C_2H_5OH$)

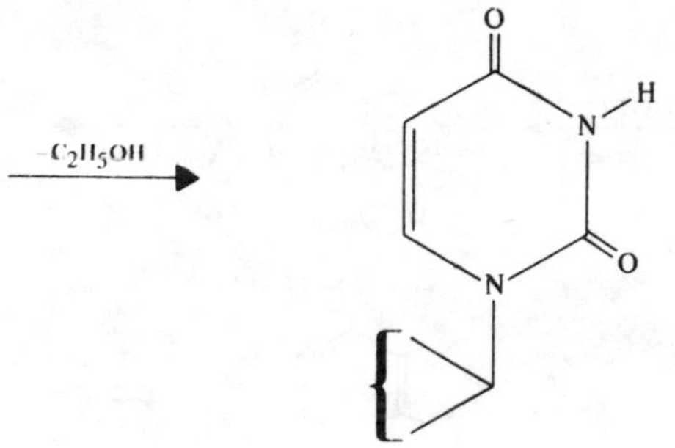

$\xrightarrow{-C_2H_5OH}$

P.5

(a) The isopropylidene group is a cyclic ketal.

(b) It can be installed by treating the nucleoside with acetone and a trace of acid and by simultaneously removing the water that is produced.

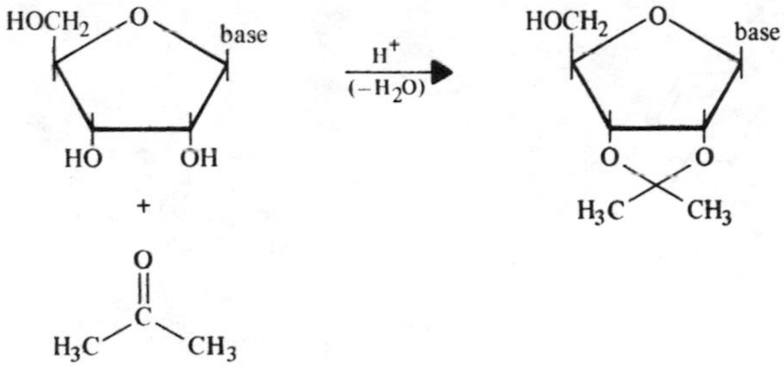

HOCH₂ … base / HO OH + $H_3C-C(=O)-CH_3$ $\xrightarrow[(-H_2O)]{H^+}$ HOCH₂ … base / O O, H_3C CH_3

P.6

(a) 6×10^9 base pairs $\times \dfrac{34\text{Å}}{10 \text{ base pairs}} \times \dfrac{10^{-10} \text{meters}}{\text{Å}} \cong 2$ meters

(b) $6 \times 10^{-12} \dfrac{\text{g}}{\text{ovum}} \times 3 \times 10^9 \text{ ova} = 1.8 \times 10^{-2}\text{g}$

P.7

(a)

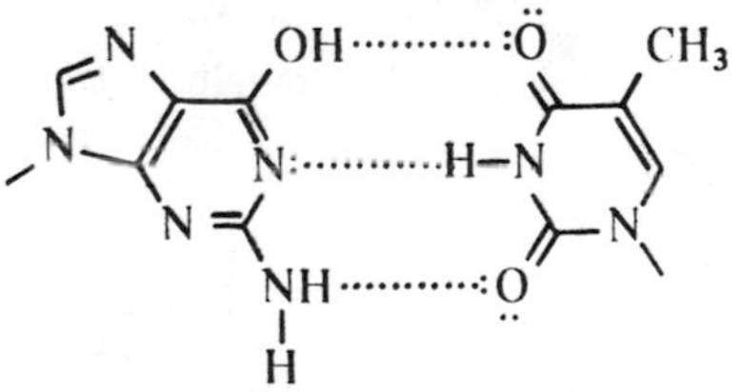

Lactim form of guanine   Thymine

(b) Thymine would pair with adenine and thus adenine would be introduced into the complementary strand where guanine should occur.

P.8

(a) A diazonium salt and a heterocyclic analog of a phenol.

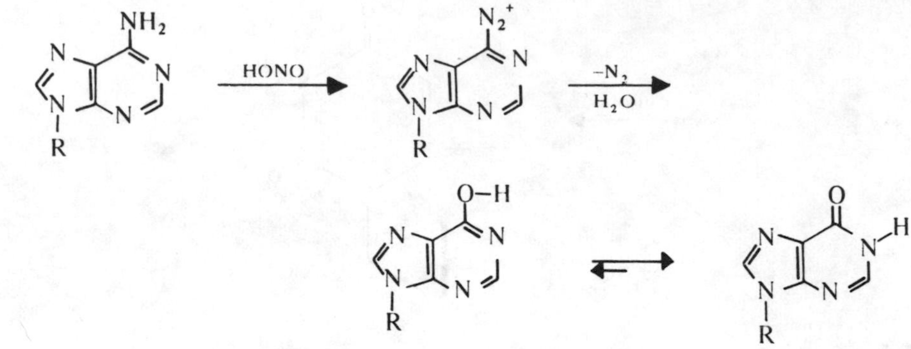

NH_2 HONO N_2^+ $-N_2$ H_2O

O–H

O

Hypoxanthine nucleotide

(b)

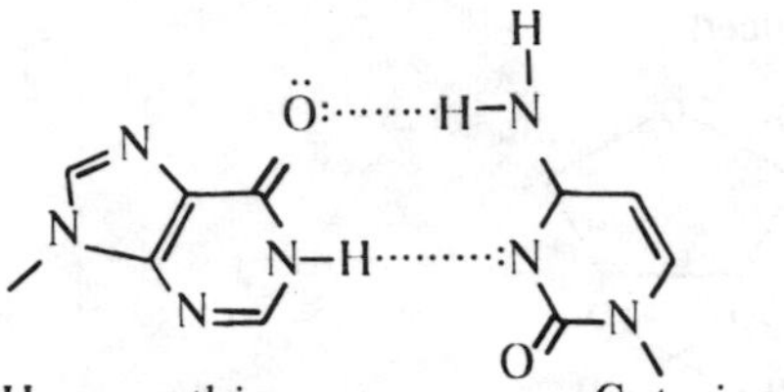

Hypoxanthine

Cytosine

(c) Original double strand

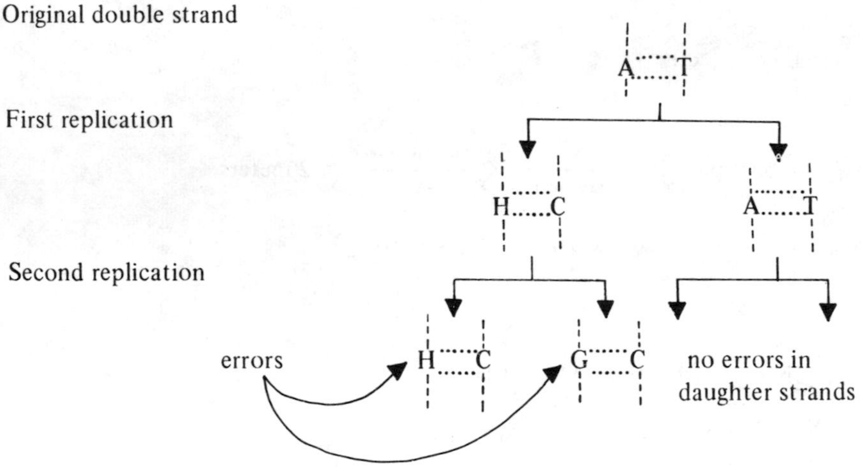

A.....T

First replication

H.....C

A.....T

Second replication

errors

H.....C

G.....C

no errors in daughter strands

P.9

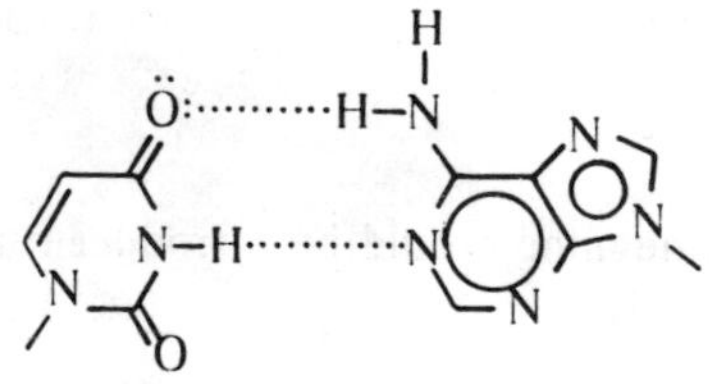

Uracil (in mRNA)

Adenine (in DNA)

P.10

(a)	UGG	GGG	UUU	UAC	AGC	mRNA
(b)	Tyr	Gly	Phe	Tyr	Ser	Amino acids
(c)	ACC	CCC	AAA	AUG	UCG	Anticodons

P.11

	Arg ·	Ile ·	Cys ·	Tyr ·	Val	Amino acids
(a)	AGA	AUA	UGC	UGG	GUA	mRNA
(b)	TCT	TAT	ACG	ACC	CAT	DNA
(c)	UCU	UAU	ACG	ACC	CAU	anticodons

P.12

A change from C–T–T to C–A–T or a change from C–T–C to C–A–C.

Q

Special Topic

Q.1

Conrotatory motion of the type shown would lead to increasingly unfavorable interaction of the methyl groups as the transition state is approached. Thus this path is not followed to any appreciable extent.

Q.2

According to the Woodward-Hoffmann rule for electrocyclic reactions of $4n$ π electron systems (p. 1048), the photochemical cyclization of *cis,trans*-2,4-hexadiene should proceed with *disrotatory motion*. Thus it should yield *trans*-3,4-dimethylcyclobutene:

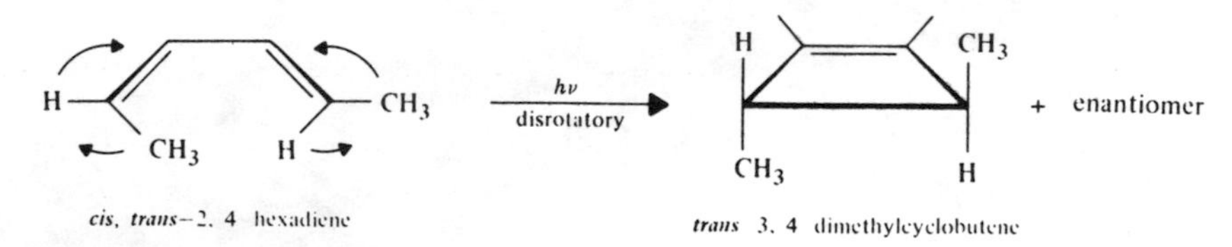

Q.3

(a)

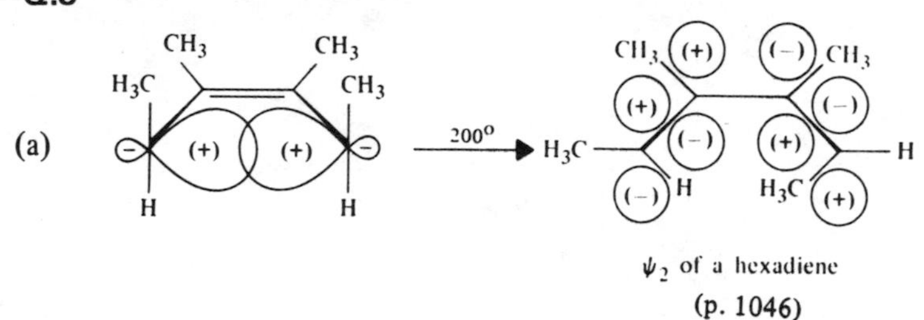

(b) This is a thermal electrocyclic reaction of a $4n$ π electron system; it should, *and does*, proceed with conrotatory motion.

Q.4

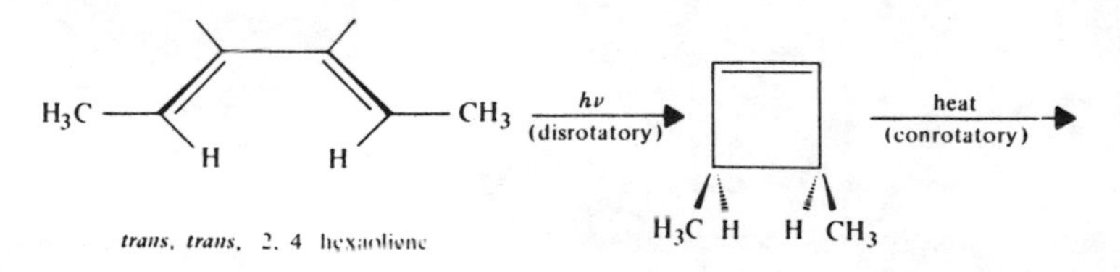

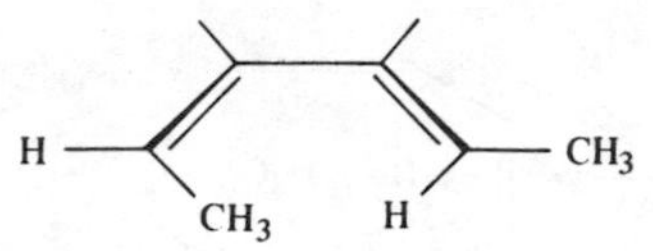

cis, trans-2, 4-hexadiene

Here we find that two consecutive electrocyclic reactions (the first photochemical, the second thermal), provide a stereospecific synthesis of *cis,trans*-2,4-hexadiene from *trans,trans*-2,4-hexadiene.

Q.5

(a) This is a photochemical electrocyclic reaction of an eight π electron system—a $4n$ π system where $n = 2$. It should, therefore, proceed with disrotatory motion.

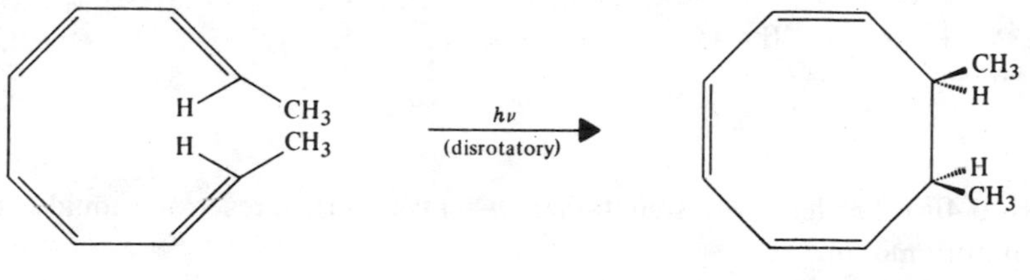

cis-7, 8-dimethyl-1, 3, 5-cyclooctatriene

(b) This is a thermal electrocyclic reaction of the eight π electron system. It should proceed with conrotatory motion.

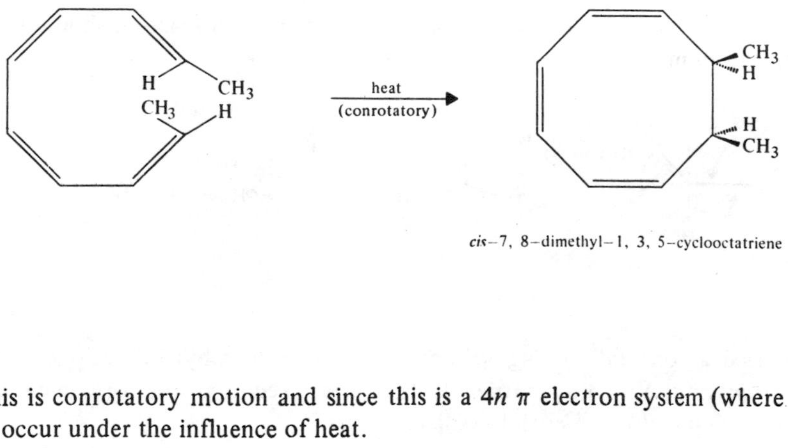

cis-7, 8-dimethyl-1, 3, 5-cyclooctatriene

Q.6

(a) This is conrotatory motion and since this is a $4n$ π electron system (where $n = 1$) it should occur under the influence of heat.

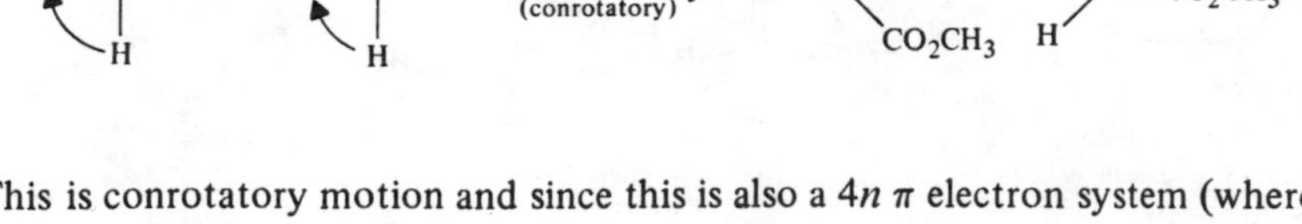

(b) This is conrotatory motion and since this is also a $4n$ π electron system (where $n = 2$) it should occur under the influence of heat.

(c) This is disrotatory motion. This, too is a $4n$ π electron system (where $n = 1$), thus it should occur under the influence of light.

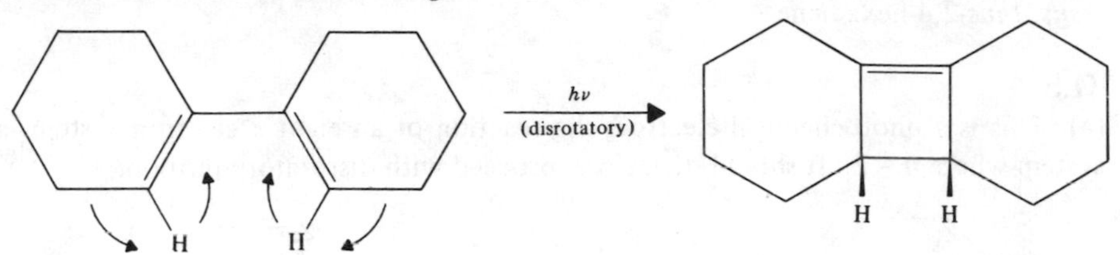

Q.7

(a) This is a $4n + 2$ π electron system (where $n = 1$); a thermal reaction should take place with disrotatory motion:

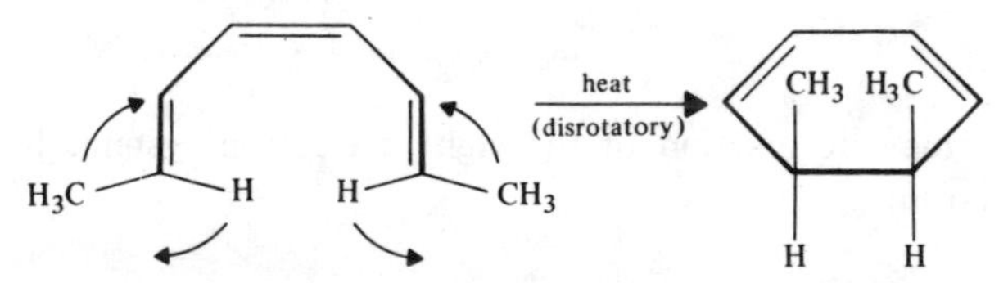

(b) This is also a $4n + 2$ π electron system; a photochemical reaction should take place with conrotatory motion.

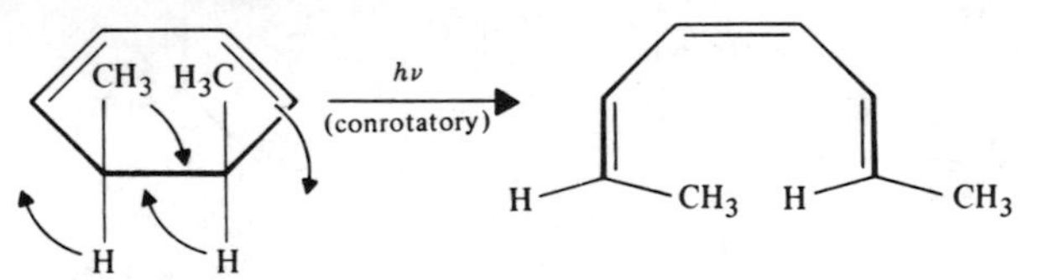

Q.8

Here we need a conrotatory ring-opening of *trans*-5,6-dimethyl-1,3-cyclohexadiene (to produce *trans,cis,trans*-2,4,6-octatriene), then we need a disrotatory cyclization to produce *cis*-5,6-dimethyl-1,3-cyclohexadiene.

trans-5, 6-dimethyl-1, 3-cyclohexadiene

trans, cis, trans-2, 4, 6-octatriene

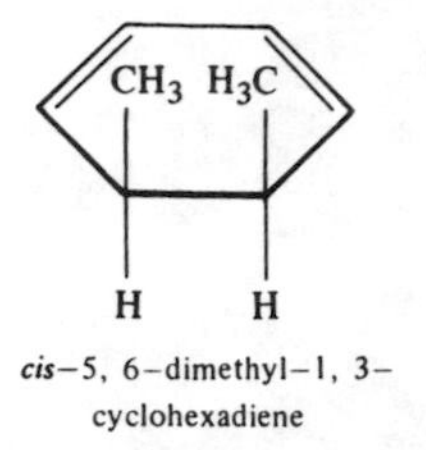

Since both reactions involve $4n + 2$ π electron systems we apply light to accomplish the first step and heat to accomplish the second. It would also be possible to use heat to produce *trans,cis,cis*-2,4,6-octatriene then use light to produce the desired product.

Q.9

The first electrocyclic reaction is a thermal, conrotatory ring opening of a $4n$ π electron system. The second electrocyclic reaction is a thermal, disrotatory ring closure of a $4n + 2$ π electron system.

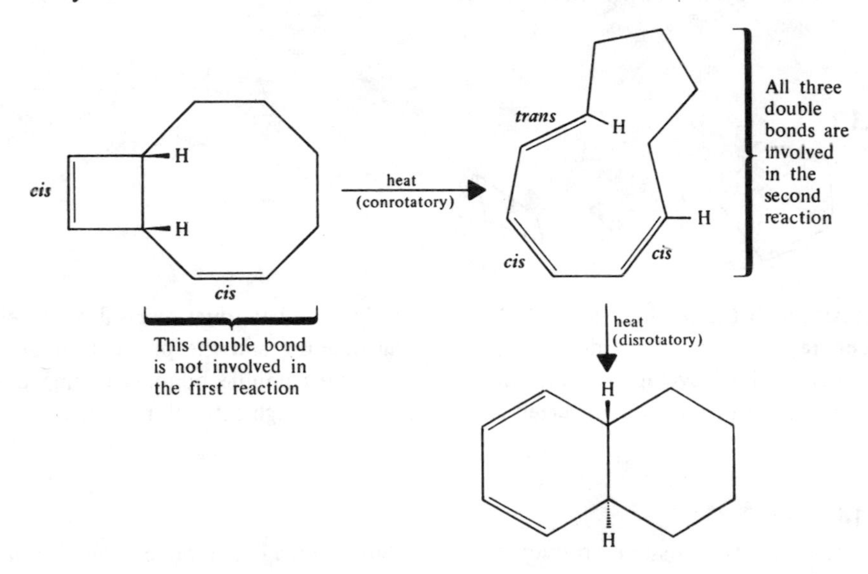

Q.10

(a) This reaction involves two π electrons, thus it is a $4n + 2$ π system where $n = 0$.

(b) An allylic cation is formed.

(c) The cyclopropyl anion is a $4n$ π system (where $n = 1$), thus a thermal reaction should take place with conrotatory motion.

Q.11

(a) This is a $4n + 2$ π electron system undergoing disrotatory motion. Heat is required.

(b) This is a $4n + 2$ π electron system undergoing conrotatory motion. Light is required.

(c) This is a $4n + 2$ π electron system undergoing conrotatory motion. Light is required.

(d) This is a $4n + 2$ π electron system undergoing disrotatory motion. Heat is required.

Q.12

1
Cis,trans-cyclonona-1,3-diene

2
Cis,cis-cyclonona-1,3-diene

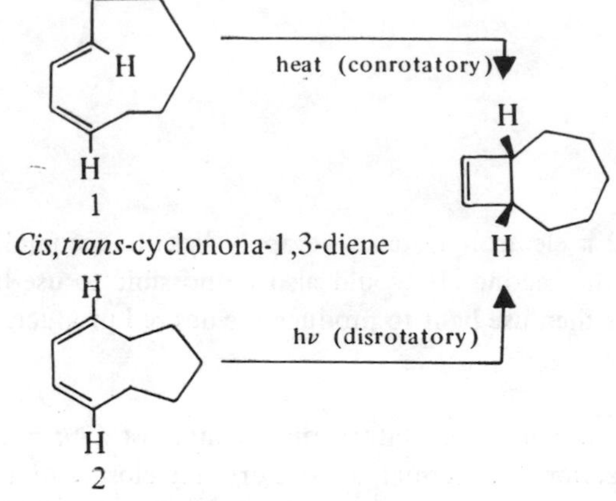

Q.13

(a)

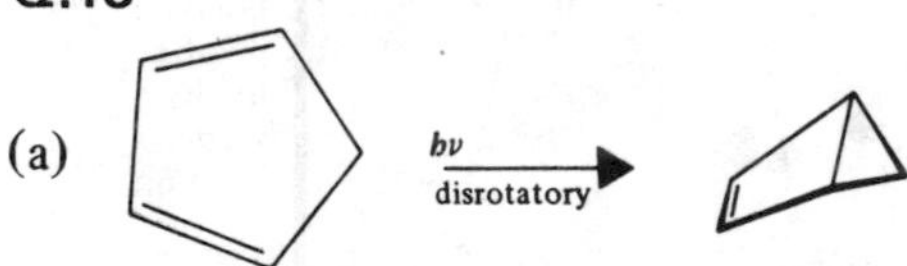

(b) Although the product is highly strained, a concerted thermal reversal of the electrocyclic reaction would require conrotatory ring opening and the production of an impossibly strained cyclopentadiene ring with one *trans* double bond. The ring opening, therefore, is probably nonconcerted and has a relatively high activation energy.

Q.14

(a) There are two possible products that can result from a concerted cycloaddition. They are formed when *cis*-2-butene molecules come together in the following ways:

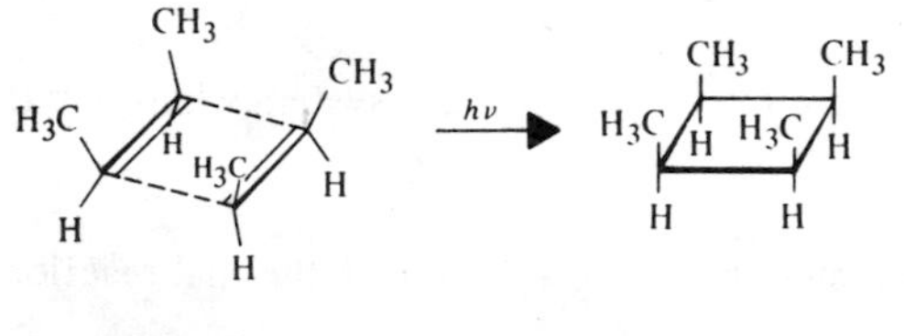

and

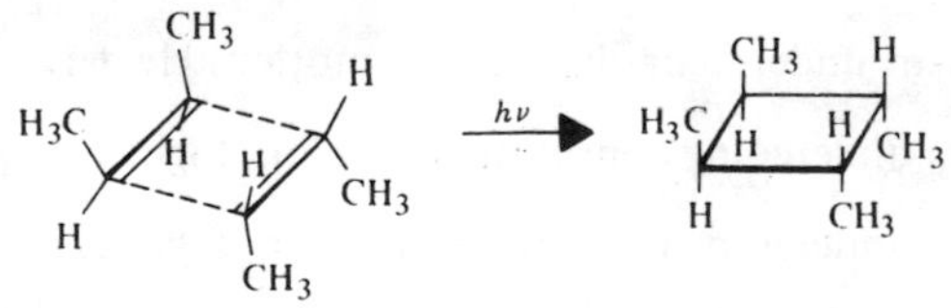

(b) There are two possible products that can be obtained from *trans*-2-butene as well.

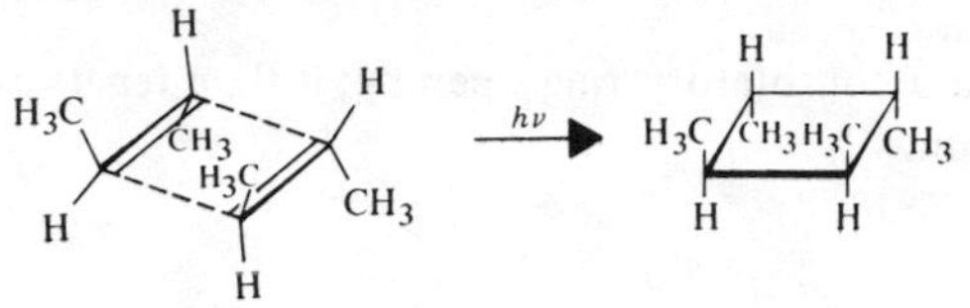

and

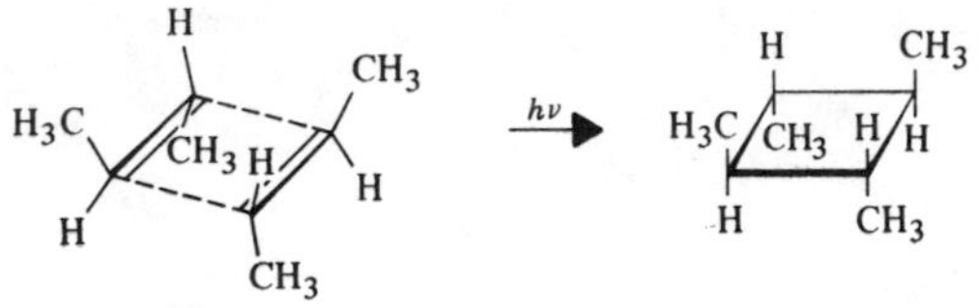

Q.15

This is an intramolecular [2 + 2] cycloaddition.

Q.16

(a)

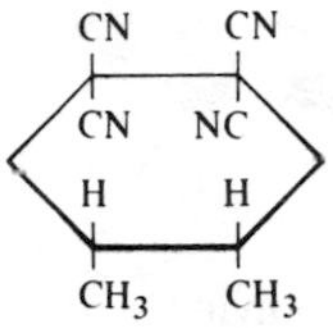

(b)

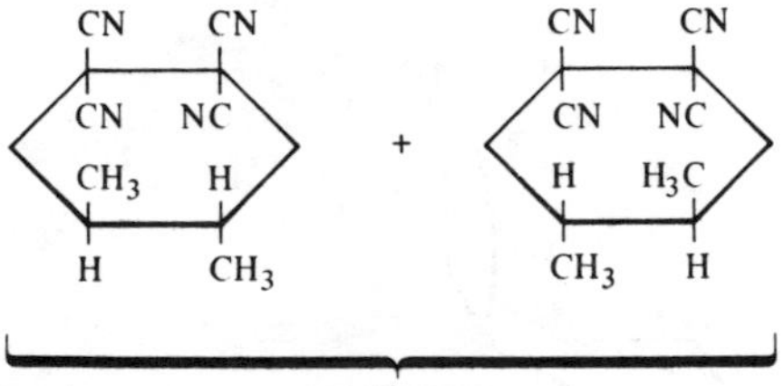

enantiomers

Q.17

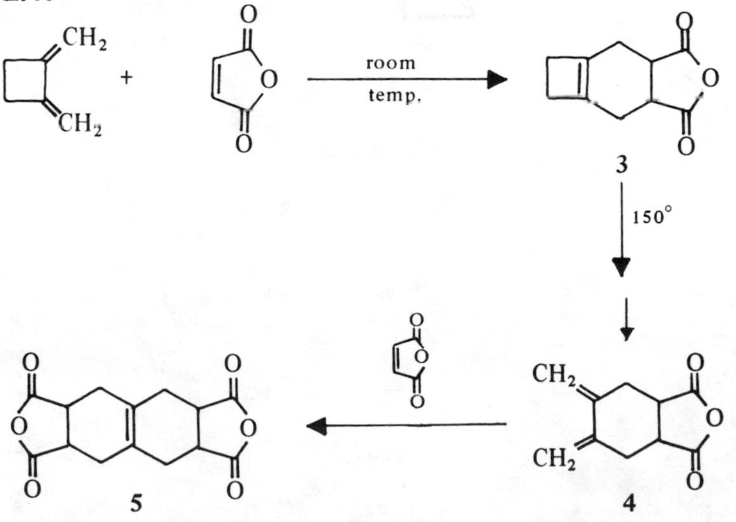

room temp.

3

150°

4

5

Q.18

Compound 7 (below) results from a conrotatory ring opening; it then reacts as the diene component of a Diels-Alder reaction.

H R H R

7

Q.19

A is

B and **C** are H H and H H

Q.20

A is the product of a disrotatory thermal electrocyclic reaction involving a 6π electron segment of cyclooctatetraene. **B** is the Diels-Alder adduct.

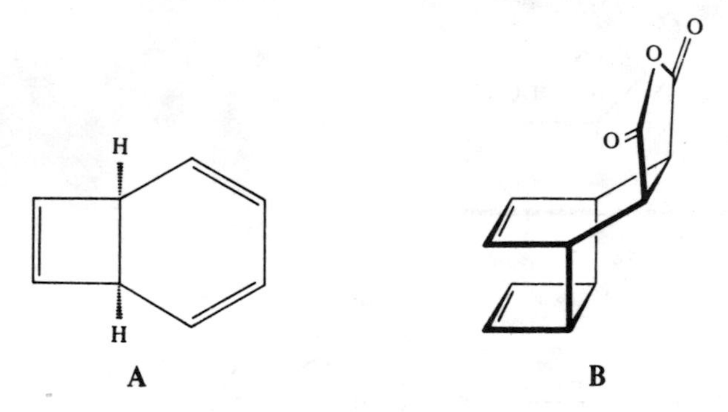